Lecture Notes in Computer Science 13235

More information about this series at https://link.springer.com/bookseries/558

Hasti Seifi · Astrid M. L. Kappers ·
Oliver Schneider · Knut Drewing ·
Claudio Pacchierotti · Alireza Abbasimoshaei ·
Gijs Huisman · Thorsten A. Kern (Eds.)

Haptics: Science, Technology, Applications

13th International Conference on Human Haptic Sensing
and Touch Enabled Computer Applications, EuroHaptics 2022
Hamburg, Germany, May 22–25, 2022
Proceedings

Springer

Editors
Hasti Seifi
University of Copenhagen
Copenhagen, Denmark

Astrid M. L. Kappers
Eindhoven University of Technology
Eindhoven, The Netherlands

Oliver Schneider
University of Waterloo
Waterloo, ON, Canada

Knut Drewing
Justus-Liebig-University
Giessen, Germany

Claudio Pacchierotti
IRISA, Inria Rennes – Bretagne Atlantique
Rennes, France

Alireza Abbasimoshaei
Hamburg University of Technology
Hamburg, Germany

Gijs Huisman
Delft University of Technology
Delft, The Netherlands

Thorsten A. Kern
Hamburg University of Technology
Hamburg, Germany

ISSN 0302-9743   ISSN 1611-3349 (electronic)
Lecture Notes in Computer Science
ISBN 978-3-031-06248-3   ISBN 978-3-031-06249-0 (eBook)
https://doi.org/10.1007/978-3-031-06249-0

This Springer imprint is published by the registered company Springer Nature Switzerland AG
The registered company address is: Gewerbestrasse 11, 6330 Cham, Switzerland

Preface

In this volume, you will find the proceedings of the EuroHaptics 2022 conference, held during May 22–24, 2022, in Hamburg, Germany. EuroHaptics is a major international conference, organized every two years, that features the latest advances in haptics science, technology, and applications. This year's proceedings are open access. We follow the example given by the EuroHaptics 2020 team in Leiden and hope that future organizers will decide to do the same.

Organizing a conference of this size and international impact is always special and unique, but our conference in particular faced many challenges due to the circumstances. When Hamburg applied to host the event in 2020, we all assumed that the COVID-19 pandemic would be over by 2022. So, we originally planned optimistically for an on-site conference with global participation. However, the virus proved to be more persistent than expected, so we had to plan three versions of the conference in parallel: an on-site conference, a hybrid conference, and a virtual conference. Fortunately, the situation stabilized such that, with some reassurance from the scientific community, we were able to decide on an on-site conference. However, just as decisions were being made, geopolitical challenges arose from Russia's war in Ukraine. Since this scientific community is traditionally more interested in interaction than in exclusion and separation, we again decided to keep open the opportunity for scientific cross-cultural exchanges. At the time of submitting this conference volume to production, more than 180 participants have confirmed their attendance, and a conference program has been designed with more than 140 papers ranging from oral presentations and posters to numerous hands-on demonstrations.

More than 350 (co-)authors submitted their contributions to the conference. In these uncertain times and after almost two years of the COVID-19 pandemic, the long paper category was comparatively small, with 65 papers, 36 of which were accepted for publication, yielding an acceptance rate of 55%. Impressively, 62 Work-In-Progress (WIP) short papers were submitted, of which 51 were accepted for inclusion in the conference, yielding an acceptance rate of 82%. In addition, we received 39 submissions for hands-on demonstrations and several industry exhibitors at the conference. We are excited to see how our scientific communication will evolve in the coming years, what will remain and what will revert to earlier times.

The entire evaluation and peer review process could not have been accomplished without the enthusiastic help of our haptics research community. In addition to the editorial board, 39 associate editors and 140 reviewers invested their time to provide valuable input to the organizers and authors. We are grateful for everyone's help, especially with so many things happening around us.

Haptics research is a growing community. There is no conference or journal on robotics, actuators, psychology, neurology, or human-computer interaction without at least a few papers related to our research area. This proliferation can be seen as a threat to general haptics conferences. However, we see it as an opportunity, since there are no conferences like EuroHaptics where technology, psychology, neuroscience, and applications coexist on an equal footing in a single-track event that promotes true interdisciplinary exchange. We are grateful to have been able to contribute to the progress of this community with a great team of editors and reviewers.

April 2022

Thorsten A. Kern
Hasti Seifi
Astrid M. L. Kappers
Oliver Schneider
Knut Drewing
Claudio Pacchierotti
Alireza Abbasimoshaei
Gijs Huisman

Organization

General Chair

Thorsten A. Kern   Hamburg University of Technology, Germany

Program Chairs

Hasti Seifi   University of Copenhagen, Denmark
Astrid M. L. Kappers   Eindhoven University of Technology, The Netherlands

Conference Editorial Board

Oliver Schneider   University of Waterloo, Canada
Knut Drewing   Justus Liebig University Giessen, Germany
Claudio Pacchierotti   CNRS, France
Alireza Abbasimoshaei   Hamburg University of Technology, Germany

Publication Chair

Fady Youssef   Hamburg University of Technology, Germany

Work-In-Progress Chair

Gijs Huisman   Delft University of Technology, The Netherlands

Workshop Chair

Ilana Nisky   Ben Gurion University of the Negev, Israel

Demo Chair

William Frier   Ultraleap, UK

Awards Committee Chair

Jan van Erp   University of Twente, The Netherlands

Local Chairs

Frank Steinicke   University of Hamburg, Germany
Dennis Kähler   Hamburg University of Technology, Germany

Student-Volunteer Chairs

Femke van Beek   Eindhoven University of Technology, The Netherlands
Sabrina Panëels   University of Kent, UK

Communication Chairs

Jana Ihrens   Hamburg University of Technology, Germany
Timon Hartwich   Hamburg University of Technology, Germany
Sravan Shelam   Hamburg University of Technology, Germany

Industry and Sponsorship Chairs

Julius Harms   Hamburg University of Technology, Germany
Elisa Santella   Grewus, Germany

Registration and Finance

Janick Lokocz   Hamburg University of Technology, Germany
Petra Schlegel   Hamburg University of Technology, Germany

Associate Editors

Yuki Ban   University of Tokyo, Japan
Cagatay Basdogan   Koc University, Turkey
Matteo Bianchi   University of Pisa, Italy
Mike Chen   National Taiwan University, Taiwan
Ildar Farkhatdinov   Queen Mary University of London, UK
Francesco Ferrise   Politecnico di Milano, Italy
William Frier   University of Sussex, UK
Greg Gerling   University of Virginia, USA
David Gueorguiev   Sorbonne University, France
Matthias Harders   University of Innsbruck, Austria
Keyvan Hastrudi-Zaad   University of British Columbia, Canada
Ali Israr   Purdue University, USA
Seokhee Jeon   Kyung Hee University, South Korea
Hiroyuki Kajimoto   University of Electro-Communications, Japan
Ayse Kucukyilmaz   University of Nottingham, UK
Yoshihiro Kuroda   University of Tsukuba, Japan
Scinob Kuroki   NTT Communication Science Laboratories, Japan
Vincent Levesque   McGill University, Canada
Pedro Lopes   University of Chicago, USA
Monica Malvezzi   University of Siena, Italy
Anna Metzger   Justus-Liebig-Universität Gießen, Germany
Kouta Minamizawa   Keio University, Japan
Alessandro Moscatelli   University of Rome Tor Vergata, Italy

Shogo Okamoto	Nagoya Institute of Technology, Japan
Melisa Orta Martinez	Carnegie Mellon University, USA
Hannes P. Saal	University of Sheffield, UK
Sabrina Panëels	University of Kent, UK
Gunhyuk Park	Gwangju Institute of Science and Technology, South Korea
Evren Samur	Boğaziçi University, Turkey
Jeroen Smeets	Vrije University Amsterdam, The Netherlands
Massimiliano Solazzi	Institute of Communication Information and Perception Technologies, Italy
Eckehard Steinbach	Technical University of Munich, Germany
Yoshihiro Tanaka	Nagoya Institute of Technology, Japan
Jean-Louis Thonnard	University of Louvain, Belgium
Yasemin Vardar	Delft University of Technology, The Netherlands
Dangxiao Wang	Beihang University, China
Mounia Ziat	Northern Michigan University, USA
Femke van Beek	Eindhoven University of Technology, The Netherlands
Jan van Erp	University of Twente, The Netherlands

Reviewers

Yusuf Aydin	Massimiliano Di Luca
Arsen Abdulali	Catherine Dowell
Rochelle Ackerley	Mihai Dragusanu
Mehmet Ayyildiz	Knut Drewing
Hector Barreiro	Basil Duvernoy
Edoardo Battaglia	Jonathan Eden
Alexis Block	Mohamad Eid
Rebecca Boehme	Sonia Elizondo
Jonathan Browder	Sara Falcone
Rachael Burns	Simone Fani
Antonio Cataldo	Shane Forbrigger
Nathalia Cespedes	Euan Freeman
Francesco Chinello	Rebecca Friesen
Seungmoon Choi	Bruno Fruchard
Giorgos Christopoulos	Gregory Gerling
Simone Ciotti	Kaj Gijsbertse
Ed Colgate	Frederic Giraud
Giulia Corniani	Lili Golmohammadi
Luke Cox	David Gueorguiev
Heather Culbertson	Taku Hachisu
Nicole D'Aurizio	Reza Haghighi Osgouei
Thomas Daunizeau	Amy Kyungwon Han
Yuri De Pra	Jess Hartcher-O'Brien
Jonathan Delafield-Butt	Waseem Hassan
Benoit Delhaye	Alice Haynes

Hsin-Ni Ho
Mehdi Hojatmadani
Thomas Howard
Thomas Hulin
Nicolas Huloux
Inwook Hwang
Seokhee Jeon
Anis Kaci
Hiroyuki Kajimoto
Takaaki Kamigaki
HyeongYeop Kang
Mehrdad Kermani
Thorsten A. Kern
Humayun Khan
Konstantina Kilteni
Yeongmi Kim
Roberta Klatzky
Krishna Dheeraj Kommuri
Ayse Kucukyilmaz
Irene Kuling
Yuichi Kurita
Ki-Uk Kyung
Silvia Logozzo
Pablo Lopez-Custodio
Hojin Lee
Yanan Li
Tommaso Lisini Baldi
Pedro Lopes
Ottavia Maddaluno
Monica Malvezzi
Paul Marasco
Bogdan Maris
Stephen Mascaro
Anna Metzger
Mayumi Mohan
Rafael Morales
Tania Morimoto
Manivannan Muniyandi
Evelyn Muschter
Cara Nunez

Seungjae Oh
Stefano Papetti
Gunhyuk Park
Dario Pittera
Myrthe Plaisier
Mithun Poozhiyil
Narjes Pourjafarian
Pornthep Preechayasomboon
Ismo Rakkolainen
Teresa Ramundo
Ben Richardson
Colleen Ryan
Semin Ryu
Camille Sallaberry
Massimo Satler
Félicien Schiltz
Immo Schütz
Cecile Scotto
Yitian Shao
Craig Shultz
Benjamin Stephens Fripp
Minghui Sun
Hong Tan
Marc Teyssier
Alexander Toet
Matteo Toscani
Jan Van Erp
Jean Vroomen
Laurence Willemet
Judith Weda
Yannick Weiss
Alan Wing
Xiao Xu
Takumi Yokosaka
Yongjae Yoo
Kyle Yoshida
Shunsuke Yoshimoto
Hong Zeng
Tao Zeng
Loes van Dam

Contents

Haptic Science

Haptic Technology

Appendix

Haptic Science

Haptic Discrimination of Different Types of Soft Materials

Müge Cavdan[1]([⊠]), Katja Doerschner[1,2], and Knut Drewing[1]

[1] Giessen University, 35390 Giessen, Germany
muege.cavdan@psychol.uni-giessen.de
[2] National Magnetic Resonance Research Center, Bilkent University, 06800 Ankara, Turkey

Abstract. We interact with different types of soft materials on a daily basis such as salt, hand cream, etc. Recently we have shown that soft materials can be described using four perceptual dimensions which are deformability, granularity, viscosity, and surface softness [1]. Here, we investigated whether humans can actually perceive systematic differences in materials that selectively vary along one of these four dimensions as well as how judgments on the different dimensions are correlated to softness judgments. We selected at least two material classes per dimension (e.g., hair gel and hand cream for viscosity) and varied the corresponding feature (e.g., the viscosity of hair gel). Participants ordered four to ten materials from each material class according to their corresponding main feature, and in addition, according to their softness. Rank orders of materials according to the main feature were consistent across participants and repetitions. Rank orders according to softness were correlated either positively or negatively with the judgments along the associated four perceptual dimensions. These findings support our notion of multiple softness dimensions and demonstrate that people can reliably discriminate materials which are artificially varied along each of these softness dimensions.

Keywords: Haptics · Softness perception · Material perception · Granularity

1 Introduction

Studies focusing on haptic softness traditionally equated softness with the compliance of materials [2–4]. However, we have recently shown that everyday materials that appear more or less soft do not only differ in their perceived compliance (i.e., deformability in response to normal pressure) but also in their surface softness, granularity, and viscosity [1]. Thus, material features in quite different perceptual dimensions seem to underlie the perception of softness. While there are a number of studies that focused on deformability (i.e., compliance), often using custom-made rubber silicone stimuli [3, 4], the other dimensions have been rarely studied, if at all. One exception is a study [5] on the haptic

Research was supported by the EU Marie Curie Initial Training Network "DyVito" (H2020-ITN, Grant Agreement: 765121), Deutsche Forschungsgemeinschaft (DFG, German Research Foundation) – project number 222641018 – SFB/TRR 135, A5, and EU FET-OPEN Project "ChronoPilot" (H2020 – Grant Agreement: 964464.

© The Author(s) 2022
H. Seifi et al. (Eds.): EuroHaptics 2022, LNCS 13235, pp. 3–11, 2022.
https://doi.org/10.1007/978-3-031-06249-0_1

discrimination of viscous materials by stirring every day-like liquids with spatula vs. with index finger. This study showed a much better viscosity perception when exploring directly with the finger as compared to the indirect exploration with a tool (see [6] for another viscosity example).

Here, we investigate whether people can actually perceive systematic and consistent differences in materials that vary along each one of the four different perceptual dimensions that we previously found to be associated with soft, everyday materials. More importantly we tested whether participants judgments along these single dimensions would correlate with their softness judgements as we suppose that these dimensions underly perceived softness. As in [5] we aim to use classes of everyday materials that can be selectively varied along a single perceptual dimension such as grain sizes for granularity. This allowed us to measure discrimination within a single dimension. We selected at least two material classes per dimension and varied the levels of the corresponding dimension with each material class (e.g., therapy dough varied in deformability). Participants were blindfolded and asked to order each material once according to the main dimension feature (the most distinctive softness feature of a dimension, e.g., order based on decreasing viscosity) and once according to softness. Using Spearman's rho correlations, we tested the consistency between ordering as well as the relationship between softness judgements and the four perceptual dimensions.

2 Methods

2.1 Participants

Twenty students (10 males) with mean age of 22.5 years (age range: 19–28, *SE*: .63) from Giessen University participated in the study. They were naïve to the aim of the experiment and none of them reported any sensory or motor impairments. All participants had two-point touch discrimination better than 4 mm at the index finger. They gave written informed consent prior to the experiment in accordance with the Declaration of Helsinki except preregistration (2013).

2.2 Setup and Materials

Materials were presented on plastic plates (diameter: 21.5 cm). Participants sat at a table on which the materials were presented (see Fig. 1). They were blindfolded with sleep masks. Earplugs and active noise cancelling headphones (Sennheiser HD 4.50 BTNC) were used to block any noises which could be generated from exploring the materials. The experimenter stood next to the participants, exchanged the stimuli, and collected the participant's responses. A standard computer with MATLAB 2019b (MathWorks Inc. 2007) was used to guide the experimenter and collect the responses. All materials (except silicone which has sufficient lifetime) were replaced after each participant warranting that the materials felt similar for each participant.

Material classes were selected based on findings on the four softness-associated dimensions in [1]. We chose basis materials that had high values in one of the four dimensions (i.e., high dimension scores in the PCA that had revealed the 4 dimensions) and built a material class by varying the main feature of the dimension for each material. High-scoring materials for each dimension were as follows; granularity: salt and sand; viscosity: hand cream and hair gel; surface softness: velvet, fur, and cotton; deformability: therapy dough and custom-made silicone rubber stimuli, (see [2, 3] for some examples). For each material we selected several exemplars with different levels of main dimension feature (e.g., varying granularity of salt, by varying grain size). We intended to collect as many feature levels as possible within each material class which resulted in unequal number of levels across materials. We either created or purchased (if commercially available, see Appendix A Table 1.) the materials in different levels. Our aim was to create materials which lie along the four different dimensions and are discriminable from each other. To this end, we formed an initial material set based on our own perception, and piloted the materials to check whether the intermediate levels could also be distinguished by others. If some steps were indistinguishable we increased the differences between stimuli (e.g., adding more water to diluted hand cream). The proper experiment here, then served to corroborate the piloting results for a representative sample and to test the correlation with softness judgments.

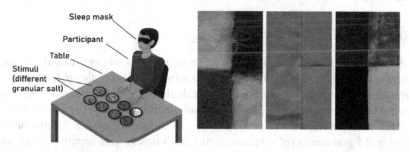

Fig. 1. Illustration of the setup and an example material set (i.e., granular material: salt) in the experiment. A blindfolded participant explores salt in different levels of granularity. Samples of fur, cotton, and velvet used in the experiment (right).

In order to vary the granularity levels of sand (six levels) and salt (seven levels), we mixed materials of different grain sizes (e.g., material with grain size of 0.1–0.3 mm was mixed with material of grain size 0.4–0.8 mm). None of the participants noticed that the granular materials are mixture. We diluted hand cream and hair gel with water as well as mixing different types hair gel (five levels) or hand cream (five levels) with each other to manipulate the viscosity. Silicone rubber stimuli were created in ten different compliance levels by mixing two components of silicone rubber solution (AlpaSil EH 10:1) in addition to different amounts of silicone oil (polydimethylsiloxane). All details on these mixtures can be found in the appendix. The therapy doughs (Theraputty, ABH Webshop) were purchased in eight strength levels from *very soft* to *very hard*. For material classes belonging to surface softness, namely for cotton, fur, velvet, we arbitrarily selected the levels based on the hair length.

2.3 Design and Procedure

We divided participants and the materials into two different groups in order to keep the duration of the experiment within reasonable limits (~3 h per participant including 2 breaks up to 10 min each). Participants in the first group only explored hair gel, sand, velvet, and silicone while the others explored hand cream, salt, fur, cotton, and therapy dough. Upon arrival, they signed the informed consent which was followed by initial instructions. During the experiment each material set was presented eight times, four times for softness judgements and four times for judgments on its corresponding main dimension. In each trial participants were presented with the full set of materials from a single material class. The participant's task was to order the set according to the main dimension that was named and further explained by 1–2 adjectives (see Fig. 2, depiction for details) or according to softness. Softness was not further defined, because we were interested to find out how people use the term "soft" for our stimuli. There were no restrictions within trials, participants were able to go back to a previous material and re-explore it as often as they wanted. With this, we aimed to rule out the role of memory in the rating task. While the participants in the first group performed 32 trials (4 material sets × 4 repetition × 2 judgment types) participants in the second group performed 40 trials (5 material sets × 4 repetition × 2 judgment types). The presentation order of material sets and the position of each material within a set were randomized.

2.4 Data Analysis

First, the average rankings of each participant for the material sets were calculated across four trials separately for judgments according to softness and according to the dimension corresponding to the material class. Then we calculated the average ranking values of each material within the material set across participants. We correlated these average values with each participant's responses. In 28 (10 surface softness, 5 deformability, 8 viscosity, and 5 granularity) of 90 cases, participant's orders were negatively correlated with the average order, meaning that these participants had ordered materials in the opposite direction as compared to other participants (e.g., coarser to finer instead of finer to coarser for granular materials). We hence inverted the individual data that were negatively correlated with the average order. We looked at the switches in overall data of a person. We assumed that participants were consistent in their orders, and did not consider inverting data trial-wise. In single-trial data reversed orders can hardly be distinguished from confusions anyway. After inverting, we again calculated average responses across participants in order to determine the perceptual levels within each material set. Next, we correlated each individual's data with the grand average in order to check the consistency across participants. Finally, using Spearman's rho analyses we correlated judgments according to softness and according to the dimension corresponding to each material class.

3 Results

Figure 2 shows the confusion matrices of rank orders for the softness-associated dimensions. By averaging rank orders across participants and repetitions, we determined average rank orders of the different material levels (x-axis), i.e., each column corresponds to one material. Along the y axes we depicted the actual (relative) frequency that a material level was assigned a certain rank in each individual trial (e.g., sum of all responses). As can be seen, participants highly agreed in how they order the levels of salt, sand, velvet, and silicone material, while there was some confusion for the other material sets. For instance, people frequently confused the current level of cotton or hair gel (i.e., level two) with the previous (i.e., level one) or the next level (i.e., level three), which indicates that here successive levels were perceptually close to each other. Individual rankings mostly differ in hand cream and hair gel. In order to statistically test the consistency between participants, we calculated the Cohen's weighted kappa using a leave-one-out approach in which each person's responses are compared to the average excluding that person for each material class. Overall, the average of the weighted kappa (10 per class) indicated at least substantial interindividual consistency; fur = .67, cotton = .71, velvet = .79, hair gel = .64, hand cream = .57, therapy dough = .80, silicone = .89, sand = .92, and salt = .87.

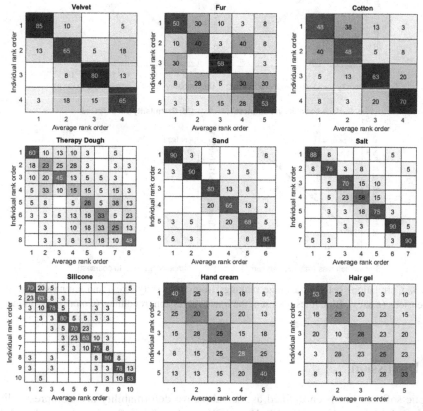

Fig. 2. The disagreement matrices show the percentages of individual trial-wise rankings. Hues correspond to dimensions (blue: surface softness [most velvety/furry to least], purple: deformability [most elastic/flexible to least], pink: viscosity [most viscous/sticky to least], red: granularity [finest to coarsest]). The shades get darker with higher agreement between individual and average ranks while they get lighter with less agreement (due to rounding rows do not necessarily add up to exactly 100%) (Color figure online).

In order to look at the relationship between the average softness rankings and associated dimensions (i.e., granularity, viscosity, surface softness, and deformability) we correlated the average rankings across feature levels and material classes for each dimension (Fig. 3). Softness and granularity were correlated positively $r_s = .99, p < .01$. Similarly, the correlation between softness and surface softness was significantly positive $r_s = .94$, $p < .01$ as well as the relationship between deformability and softness, $r_s = .99, p < .01$. Finally, viscosity and softness rankings were correlated negatively, $r_s = -.91, p < .01$.

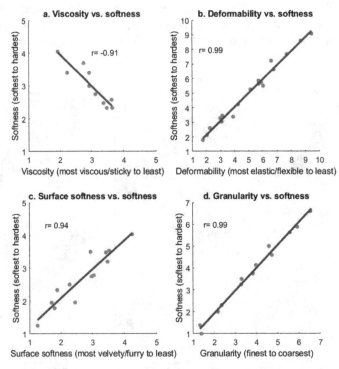

Fig. 3. Correlations between average rankings from softness (y axes) and softness-associated dimensions (x axes). Blue lines represent the linear regression trendlines. (Color figure online)

4 Discussion

Haptic softness has been defined as the subjective deformability and compressibility degree of objects and materials [7]. How different materials deform varies widely: while elastic rubber stimuli quickly returning its original form, therapy dough keeps its shape for a while even after the force is removed. Other materials such as velvet or salt deform yet in their own ways – all these materials can be characterized as *soft* from this definition. For many everyday materials, perceived softness seems to be associated with four perceptual dimensions which are surface softness, granularity, viscosity, and deformability [1]. Here, we tested whether humans perceive consistent and systematic differences in materials that we created to vary *within* each given dimension, and how judgments about the dominant dimension feature correlate with softness judgments. In general, participants were able to order different levels of stimuli within each dimension in substantially coherent manners, and their judgments were correlated with the softness. The former finding can be considered a general proof that the claimed dimensions indeed represent perceptual continua along that humans can discriminate materials. The latter finding fits with the idea that softness has multiple dimensions. Previous practice equate haptic softness mainly to compliance of materials [2–4]. In line with those studies we found a positive correlation between softness and (perceived) deformability in response to normal forces. However, we also found correlations with other dimensions:

1. Results show a significant positive correlation between granularity and softness judgments in granular materials (i.e., salt and sand). Specifically, granular materials felt more granular and softer when the grain sizes were smaller. When the grain size of the stimuli decreases, our receptors might not be able to sense the edges of the salt or sand grains. This, in turn, might lead us to associate the finer particles with a softer feeling and coarser particles with a harder feeling.
2. Decreased levels of viscosity was correlated with increased softness for viscous materials: the more diluted a viscous material is the softer it felt. The dilution of the viscous materials resulted in overall reduction in both density and viscosity. It seems like both are negatively related to softness. However, the current results cannot discern whether the relationship depends on density or viscosity. By carefully manipulating one of the two variables, future studies can investigate both possibilities in detail. A general limitation of the current study was the lack of physical measurements of the materials. In future, we will focus on the physical characteristics of the stimuli.
3. We also found a positive relationship between surface softness and softness. This relationship might be affected by the length of the hairs or the thread size of the fabric. While a fabric might feel softer with the increased hair length, it might feel harder with the increased thread size.

Here, we used a ranking task which is proven to save time, yet tends to be complex and might yield confusion as observed in inversed sorting directions. Future studies might consider using a less complex task such as magnitude estimation. Also, demand characteristics might play a role in ranking tasks and future studies should investigate this point as well. Further, a multidimensional nature of softness, implies an important caveat for softness experiments. When asking participants to judge softness of different materials, the meaning of the softness should carefully be defined. If an experimenter only asks participants to rate softness, what participants think of as softness might inter-individually differ.

In conclusion, we showed that different material sets that vary along each single softness dimension and along that people perceive differences can be created, and that these dimensions covary with softness judgments.

Acknowledgement. The authors thank Jonas Wellmann for data collection. The authors express their gratitude to Hatice Dokumaci and Öznur Bastürk for carefully preparing the stimuli and collecting data and Oleksandra Nychyporchuk for the illustration of the setup.

Appendix A

Table 1. Recipes of sand, salt, hair gel, hand cream, and silicone rubber stimuli. Numbers in the parentheses correspond to different materials in a recipe. Sand: 1 = 0.3–0.5 mm, 2 = 0.4–0.8 mm, 3 = 0.63–1.25 mm, 4 = 0.8–1.6 mm, 5 = 1–2 mm grain size. Salt: 1 = 0.3–1 mm, 2 = 0.5–1 mm, 3 = 1–2 mm, 4 = 2–5 mm grain size. Hair gel: 1 = Balea Men invisible look (Drogerie markt – dm), 2 = Balea Men wet look (dm), 3 = water. Hand cream: 1 = Balea hand cream urea (dm), 2 = Balea pH 5.5 (dm), 3 = water. Silicone rubber stimuli: 1 = component A (AlpaSil EH 10:1), 2 = component B (AlpaSil EH 10:1), 3 = silicone oil.

Levels	Sand	Salt	Hair gel
1	84.8 g(1) + 24.6 (2) + 2.6 g(3)	38.4 g (1) + 52.4 g(2)	10 g (1) + 3.8 g (2) + 1 g (3)
2	40.5 g(1) + 64.4 g(2) + 7.1g (3)	14.1g (1) + 59.7g (2) + 17 g (3)	10g(1) + 3.8g(2) + 2.4 g(3)
3	45.4 g (2) + 64.5g(3) + 2.1 g (4)	10.5g(1) + 37.8g(2) + 42.5g (3)	10 g (1) + 3.8 g(2) + 3.8 g (3)
4	6.5 g (2) + 90 g (3) + 15.5g (4)	34.4 g (2) + 56.4 g (3)	10 g (1) + 3.8 g (2) + 5.2g (3)
5	21.9g (2) + 46.4 g (3) + 43.7g (4)	20.8 g (2) + 50 g(3) + 20 g (4)	10 g (1) + 3.8 g(2) + 6.6g (3)
6	45.7 g (3) + 53.2g(4) + 12.2 g (5)	17.9 g (2) + 30g(3) + 42.9 g (4)	-
7	-	14.4g(2) + 22.2 g (3) + 54.2g (4)	-
	Hand cream	Silicone rubber stimuli	
1	10 g (1) + 3.8 g (2) + 1 g (3)	56.96 g (1) + 5.70 g (2) + 106.68 g (3)	
2	10 g (1) + 3.8 g (2) + 2.4 g (3)	61.06 g (1) + 6.11 g (2) + 102.17 g (3)	
3	10 g (1) + 3.8 g (2) + 3.8 g (3)	65.25 g (1) + 6.52 g (2) + 97.56 g (3)	
4	10 g (1) + 3.8 g (2) + 5.2 g (3)	69.52 g (1) + 6.95 g (2) + 92.86 (3)	
5	10 g (1) + 3.8 g (2) + 6.6 g (3)	73.90 g (1) + 7.39 g (2) + 88.05 g (3)	
6	-	78.37 g (1) + 7.84 (2) + 83.13 (3)	

(*continued*)

Table 1. (*continued*)

Levels	Sand	Salt	Hair gel
7	-	82.94 g (1) + 8.29 g (2) + 78.10 g (3)	
8	-	87.62 g (1) + 8.76 g (2) + 72.95 g (3)	
9	-	92.42 g (1) + 9.24 g (2) + 67.67 g (3)	
10	-	97.33 g (1) + 9.73 (2) + 62.27 (3)	

References

1. Cavdan, M., Doerschner, K., Drewing, K.: Task and material properties interactively affect softness explorations along different dimensions. IEEE Trans. Haptics **14**, 603–614 (2021). https://doi.org/10.1109/TOH.2021.3069626
2. Kaim, L., Drewing, K.: Exploratory strategies in haptic softness discrimination are tuned to achieve high levels of task performance. IEEE Trans. Haptics **4**, 242–252 (2011)
3. Srinivasan, M., LaMotte, R.: Tactual discrimination of softness. J. Neurophysiol. **73**, 88–101 (1995). https://doi.org/10.1152/jn.1995.73.1.88
4. Xu, C., Wang, Y., Gerling, G.: An elasticity-curvature illusion decouples cutaneous and proprioceptive cues in active exploration of soft objects. PLoS Comput. Biol. **17**, e1008848 (2021). https://doi.org/10.1371/JOURNAL.PCBI.1008848
5. Bergmann Tiest, W., Vrijling, A., Kappers, A.: Haptic discrimination and matching of viscosity. IEEE Trans. Haptics **6**, 24–34 (2013)
6. Caldiran, O., Tan, H., Basdogan, C.: Visuo-Haptic discrimination of viscoelastic materials. IEEE Trans. Haptics **12**, 438–450 (2019)
7. Di Luca, M.: Multisensory Softness. Springer London (2014). https://doi.org/10.1007/978-1-4471-6533-0

Moving Hands Feel Stimuli Before Stationary Hands

Knut Drewing[1]([⊠]) [iD] and Jean Vroomen[2] [iD]

[1] Department of Experimental Psychology—HapLab, JLU Giessen, Giessen, Germany
Knut.Drewing@psychol.uni-giessen.de
[2] Department Cognitive Neurosychology, Tilburg University, Tilburg, The Netherlands

Abstract. In the flash lag effect (FLE), a moving object is seen to be ahead of a brief flash that is presented at the same spatial location; a haptic analogue of the FLE has also been observed [1, 2]. Some accounts of the FLE relate the effect to temporal delays in the processing of the stationary stimulus as compared to that of the moving stimulus [3–5]. We tested for movement-related processing effects in haptics. People judged the temporal order of two vibrotactile stimuli at the two hands: One hand was stationary, the other hand was executing a fast, medium, or slow hand movement. Stimuli at the moving hand had to be presented around 36 ms later, to be perceived to be simultaneous with stimuli at the stationary hand. In a control condition, where both hands were stationary, perceived simultaneity corresponded to physical simultaneity. We conclude that the processing of haptic stimuli at moving hands is accelerated as compared to stationary ones–in line with assumptions derived from the FLE.

Keywords: Time perception · Movement · Vibrotactile stimuli

1 Introduction

Perception is subject to a manifold of spatial and temporal distortions, which can be highly informative for our understanding of basic perceptual processes. One of these distortions is the flash-lag effect (FLE). In the original visual FLE a moving stimulus is shown; at some point in time a brief stationary visual stimulus (the flash) is presented next to its trajectory. If the flash is presented spatially aligned with the moving stimulus, the moving stimulus is perceived ahead in direction of movement [1–6]. Similar effects were reported for auditory, visual-auditory and haptic stimuli [2, 7–9]: In the haptic FLE, participants swayed one of their hands from right to left and vice versa, while they kept the other hand stationary below that trajectory. They judged whether the moving hand was left or right to the stationary one when a brief tactile stimulus ("flash") was applied to the moving finger. If the two fingers were spatially aligned the moving finger was perceived to be ahead. Potentially linked to this haptic effect are mislocalizations of tactile stimuli during unseen hand movement in direction of movement [10, 11].

Some explanations for the FLE, and for haptic mislocalization relate the positional effects to temporal delays in the processing of the flash relative to that of the moving

© The Author(s) 2022
H. Seifi et al. (Eds.): EuroHaptics 2022, LNCS 13235, pp. 12–20, 2022.
https://doi.org/10.1007/978-3-031-06249-0_2

stimulus. Differential latency theory assumes that processing time for stationary stimuli (i.e., the flash) is longer than for moving stimuli ([3], cf. [10, 11]): Hence, when the stationary flash becomes aware, the moving stimulus is perceived at a time point after the onset of the flash, and thus it is perceived shifted in direction of movement. In turn, a moving stimulus should be perceived to be temporally simultaneous with a stationary stimulus, when the moving stimulus is physically delayed. Such a hypothetical delay has been computed (from the positional FLE and stimulus velocity) to be around 50 ms for the visual FLE, and around 56 ms for the haptic analogue [2, 12]. The attention-shift account of the FLE [4] makes a similar prediction: It assumes that attention is initially focused on the moving stimulus, and that it takes time to shift attention to the stationary flash, when it appears. During this time, the moving stimulus continues to move, and so the flash is perceived as lagging the moving stimulus. Again, a stationary stimulus should be perceived temporally delayed relative to an attended moving stimulus, because the attention shift takes time. Finally, the temporal-sampling account also assumes that the magnitude of FLE depends on differential processing times—even if these differences are not the main reason for the FLE according to this theory [5].

In the present study, we studied movement-related processing time differences in haptics. We investigate delays in the perception of the time point of a stimulus that is applied to a stationary hand as compared to a moving hand. FLE studies computed hypothetical delays between stationary and movement-related stimuli from positional judgments. Here, we directly measure the delay avoiding potential problems with localizing unseen hands. Participants performed a temporal order judgments task for vibrotactile stimuli presented at the two hands (TOJ task, "Which hand was stimulated first?"): one hand was always stationary, for the other hand we varied movement speed (fast, medium, slow, stationary control). The hands never crossed in order to avoid known confounds [13]. The task included a wide range of stimulus onset asynchronies (SOAs). We analyzed the point of subjective simultaneity (PSS) and the just noticeable differences (JNDs). We predicted that stimuli at the moving hand, must be given later than at the stationary hand to be perceived to be simultaneous—but of course not in the stationary control condition. We also checked for an effect of movement speed, but note that accounts for the FLE do not predict speed effects on temporal delay [2, 3].

2 Methods

2.1 Participants

We collected data from 10 students from Giessen (20–24 years, average 22, 6 females, right-handed). 8 participated for course credit, 2 for pay (8€/hour). All were naïve to the purpose of the study, reported no tenosynovitis in the past and showed no motor or cutaneous impairments. Data from one participant were excluded from analysis due to imprecise judgments (2 JNDs > mean + 2 SD, average JND 251 ms). Participants provided written informed consent, the experiment was approved by the local ethics committee LEK FB06 and conducted in accordance with 2013 Declaration of Helsinki.

2.2 Stimuli and Setup

Participants sat at a visuo-haptic workbench (Fig. 1) including a PHANToM 1.5A force feedback device (resolution 0.003 cm, 1000 Hz; $38 \times 27 \times 20$ cm^3 workspace), a 22″ LCD screen (Samsung, 120 Hz), wireless stereo glasses (Nvidia 3D Vision kit), two tactile actuators (Haptuator Mark II, Tactile Labs) and headphones (Sony MDR-XD100). Tactile actuators were embedded in thimble-like holders, in which participants inserted the distal phalanx of their left and right index fingers. One finger holder was fixed to the PHANToM device, the other was on the table, 38 cm left of the center of the PHANToM's workspace. The position of the left holder warranted that the right finger never crossed the left one when moving along the x-axis (=sideways, see Fig. 1). The screen was viewed through a mirror via stereo glasses (40-cm viewing distance; head stabilized by chinrest). The mirror occluded vision from the hands. The visual display served to guide participants through the experiment, but turned blacked during the movements. All devices were connected to a PC which controlled the experiment, collected responses and recorded finger positions from the PHANToM.

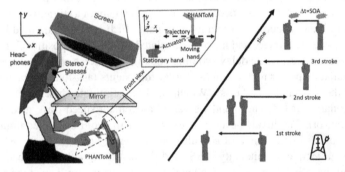

Fig. 1. Setup (left), and single trial (right); in half of the trials the first stroke started at the right side as depicted, in the other half it started at the left side.

During the experiment, the left index finger was held stationary, the right index finger moved periodically back and forth along the x-axis (Fig. 1). In line with previous literature [13] we call each unidirectional movement segment a "stroke". Participants produced 3 strokes, and in the last/3rd stroke one vibrotactile stimulus (sine-wave 50 Hz, 10 ms duration) was presented to each index finger at various SOAs. Auditory metronome signals (sine-wave 698 Hz, 20 ms duration) set the different paces for the moving index finger by defining time points for reversing movement direction. Additional auditory feedback (sine-wave 500 Hz, 20 ms duration) was used to inform participants about reversing points (14 cm left & right to center). As a result, during stimulation in the 3rd stroke participants perform active movements that are temporally and spatially well defined [2]. White noise masked the sounds of actuators and PHANToM. The PHANToM confined movements by a virtual corridor of 0.5 cm depth (z) and 35 cm length (x). The corridor was 4 cm above (y) the stationary finger.

2.3 Design and Procedure

In an experimental trial, participants performed 3 strokes with the right index finger, and during the 3rd stroke two vibrotactile stimuli were presented, to the moving right and the stationary left index finger, respectively. The task was to judge which of the two stimuli was presented first (temporal order judgment task, TOJ). We varied the movement speed of the right hand (within participants): The metronome was set to 1.80 Hz (fast), 1.16 Hz (medium), or 0.75 Hz (slow), corresponding to stroke durations of 556 ms, 862 ms and 1333 ms, respectively. In half of the trials, movement started on the left side (1st stroke to right), in the other half on the right side (1st stroke to left). In addition, we presented a stationary control condition, where the "moving finger" was fixed by PHANToM forces in one of two positions (for sake of consistency the finger connected to the PHANToM is still referred to as "moving"). In all conditions including the stationary one, the standard stimulus was delivered to the right finger either when it was 2 cm left to the center of the movement path or 2 cm right to the center. The comparison was delivered to the stationary left index finger with stimulus onset asynchronies (SOAs) of ±10 ms, ±20 ms, ±40 ms, ±80 ms, ±140 ms or ±180 ms with respect to the standard (negative sign indicating that stationary comparison was pre-sented first). Using the method of constant stimuli we determined relative frequencies of the response that the moving finger was stimulated first as a function of SOA.

Each block of the experiment comprised trials from only one of the four movement speed conditions (fast, medium, slow, control)–namely two trials per each combination of SOA, starting position and position of the standard stimulus in random order (2 × 12 × 2 × 2 = 96 trials); each session included one block per movement speed condition, the order of which was balanced across participants according to a Latin square design. There were four different sessions on different days (96 × 4 × 4 = 1536 trials over-all, 384 per condition), lasting overall 8 h. Session 1 started with movement training. Initially, participants trained only the movement in synchrony with the slow, medium and fast metronome–without the TOJ task. Further, before each experimental block in this session, participants performed a corresponding movement training including the TOJ task. Each training ended when the movement was performed with maximally two movement errors in 20 successive trials. There was no training in sessions 2 to 4.

Each experimental trial started with a voice saying "left" or "right" and a visual landmark, which both indicate the starting position of the right moving finger (14 cm left/ right to center). Once the participants reached the starting position, in the stationary control condition another landmark appeared (−2, 2 cm from center), participants moved the finger to that position, and after about 1000 ms the vibrotactile stimuli were presented. In the other conditions, metronome and white noise started, participants were instructed to wait for two metronome signals and afterwards to move forth and back along the x-axis in synchrony with the metronome; auditory feedback signals were given when the moving finger reached reversing points. The screen went black during movement. In the 3rd stroke the vibrotactile stimuli were presented. In all conditions, participants responded by moving the right finger to the extreme right and then up to respond that the first stimulus was presented at the right index finger, and to the extreme left and up to respond that the first stimulus was at the left index finger. In case of a "movement error", participants obtained feedback about their error without responding, and the trial was repeated later. Movement errors were defined by a root-mean-squared error >45% in stroke duration or length (targets 1000 ms and 28 cm). If the comparison stimulus at the stationary finger had to be given prior to the standard stimulus at the moving finger (negative SOA), we had to predict when the moving finger would reach the target position for the vibrotactile stimulus (−2, 2 cm). We modeled finger movement using the following function (data from [2, 13]).

$$x_t = -\frac{1}{2}A_M \cos(\frac{2\pi t}{T}) \tag{1}$$

x_t: position of moving finger; T: period of 2 strokes; t: time from stroke onset; A_M: movement amplitude (31 cm). We calculated at which position the moving finger would be, when it required the time span specified in the SOA to move to its target position. When the moving finger was at that position, we delivered the stimulus at the stationary finger. The stimulus at the moving finger was given after the exact SOA, leading to some jitter of presentation around the proper target position. We mimicked the positional jitter when the stimulus at the moving finger was delivered first. Jitter also reduced spatial predictability of the standard stimulus.

2.4 Data Analysis

We determined condition-wise individual psychometric functions as percentages of trials in which the moving stimulus was perceived first. We fitted cumulative Gaussians using Bayesian methods in psignifit 4 [14]; μ assessed the point of subjective simultaneity (PSS, SOA at which the standard is equally often perceived first and second); σ assessed just noticeable differences (JND = 84% discrimination threshold). PSSs and JNDs were analysed by paired-sample t-tests and repeated-measures ANOVAs.

3 Results

Figure 2a depicts psychometric functions; Fig. 2b and c depict average PSSs and JNDs, respectively. Because we had clear hypotheses on the PSSs, we directly tested these using specific planned t-tests, and no unspecific preceding omnibus analysis was required. As expected, PSSs in the actual movement conditions were significantly below 0, slow: $t(8)$ = 2.2, p = .030, PSS = −31 ms; medium: $t(8)$ = 2.4, p = .021, PSS = −29 ms; fast: $t(8)$ = 2.0, p = .041, PSS = −49 ms (one-sided), indicating that the stimulus at the stationary finger needs to be delivered before the stimulus at the moving finger to be perceived to be simultaneous. Also as expected, the PSS in the control condition (both fingers stationary), did not deviate from 0, $t(8)$ = 0.3, p = .790 (two-sided), PSS = − 2 ms. Further planned t-tests confirmed that PSSs were larger in the stationary condition as compared to the actual movement conditions, slow: $t(8)$ = 2.4, p = .023; medium $t(8)$ = 2.7, p = .013; fast $t(8)$ = 2.3, p = .024 (one-sided; Fig. 2b). PSSs did not differ between actual movement conditions; slow-medium: $t(8)$ = 0.4, p = .688; medium-fast: $t(8)$ = 1.1, p = .291; slow-fast: $t(8)$ = 1.1, p = .293 (two-sided). Results suggest that movement per se affects processing time, whereas the movement's speed did not have a significant effect—as expected [2].

We analysed JNDs, for which we did not have hypotheses, first in an ANOVA with the variable Movement Speed (control, slow, medium, fast). JNDs differed significantly, $F(3, 24) = 20.0$, $p < .001$. Bonferroni-corrected t-tests between each two conditions (overall-$\alpha < 5\%$) revealed significant differences between the stationary and each actual movement condition, and between the fast and the slow conditions, suggesting that JNDs were lower in the control condition, and higher in the fast movement condition.

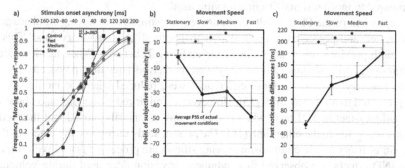

Fig. 2. Psychometric functions from data collapsed across all participants (a), average PSSs (b) and JNDs (c) with standard errors. Asterisks indicate significant differences between conditions. Negative SOAs mean that the stationary comparison was presented first.

4 Discussion

We tested whether tactile stimuli at a moving hand are perceived to occur earlier than those presented at a stationary hand. The results confirm this assumption: Stimuli at a moving hand had to be given around 36 ms later, on average, to be perceived to be simultaneous with stimuli at a stationary hand. Movement speed did not have a significant influence on the PSSs, but it had a large and clear effect on JNDs. When both hands were stationary physical and perceived simultaneity coincided. We conclude that movement per se speeds up processing of associated haptic signals, as expected.

We had derived our hypotheses from theories accounting for the flash-lag effect: Differential latency theory claims that processing time is shorter for moving as compared to stationary stimuli [3]. In our case, processing time was shorter for a stimulus applied to a moving hand as compared to when applied to a stationary hand. The tactile stimulus at the moving hand can be considered a moving stimulus in that it moves in reference to the environment (but not relative to the observer). Thus, our results are in line with this view. In [4] it has been assumed that time is required for an attentional shift from the moving towards the stationary stimulus, which also fits with our finding that stimuli at the stationary hand are perceived later than stimuli at the moving hand (assuming attention is at the moving hand). Thus, findings are in line with differential-latency, attention-shift and temporal-sampling accounts [3–5]. They though cannot unequivocally be applied to the situations under which FLEs actually have occurred in the previous studies due to a number of differences. Most studies used visual stimuli. In most visual studies, the moving stimulus was continuously visible, moved in reference to the observer and the stationary stimulus suddenly appeared. Here, both stimuli suddenly appeared, and the moving stimulus did not move in reference to the observer. There are also differences to the situation for that the haptic FLE was observed [2, 9]: There, participants were asked for the position of the moving hand at the moment of a tactile signal at that hand. Here, we found slower processing of tactile signals from a stationary as compared to a moving hand, but it is not clear whether this can be extended to delays between sudden tactile as compared to motion signals from a moving hand.

Independent of the FLE, results demonstrate that stimuli at a moving hand are processed faster, and perceived to occur earlier than stimuli at a stationary hand. This is in line with previous observations in tactile-auditory TOJ tasks: When tactile stimuli were presented to a stationary limb, they were perceived later than physically simultaneous auditory stimuli, but these differences were reduced when the tactile stimuli were presented to a moving limb [15]. However, a similar reduction was also observed in an auditory-tactile TOJ task, when one limb was voluntarily moved, but the tactile stimulus was given to the other limb [16]. The present study unequivocally confirms that in the intrasensory haptic comparison moving stimuli are processed faster. With respect to the JND effects, we observed that judgment precision decreased during voluntary movements, the more so the faster the movement was. Previous studies typically did not report a corresponding effect in JNDs [15, 16]. However, in these previous studies typically only a single movement was performed. Here, we asked participants to perform three strokes in synchrony with a metronome, in order to present stimuli during active movements that are temporally and spatially well defined [2]. Probably, it is these additional timing demands that costed precision in the TOJ task.

Why did we observe a processing difference between stimuli on the moving as compared to the stationary hand? Note first that there was no processing difference between left and right hand in the stationary condition, so effects cannot be explained by the hands being used. Secondly, PSSs effects of movement are independent of JND effects of movement, in that JNDs significantly increase with movement speed, but PSSs do not. Also, in the previous study on audio-tactile TOJs similar effects of movement on PSSs were found, but not on JNDs [15]. Another candidate for speeding up could be temporal anticipation of signals on the moving limb, but this also appears unlikely: Signal presentation was spatially defined, and the time point of signals was obscured by noise in motor system and signal position. However, efference copies from voluntary movement could play a role for PSS effects, because they modulate sensory processing of movement-related skin areas and might have enhanced stimulus processing on the moving hand (sensory attenuation can also occur, but would have slowed down processing [15, 17]). Alternatively, information on the moving hand may be processed faster, because the moving hand receives more attention, e.g., to allow for precise movement control. Future experiments will distinguish between these options.

Our finding of a temporal bias during hand movement fits into a number of findings showing that time perception is subject to perturbations during the observation of moving/changing stimuli [18]. Technologically, such findings can be used to modulate time perception when required, for instance to expand time in stressful situations [19]. Knowing that certain stimuli are processed faster than others can, e.g., be used to subjectively lengthen intervals between events marked by the stimuli.

Acknowledgment. Research was supported by EU (ChronoPilot, Horizon 2020 FET Open Programme, grant agreement 964464) and by DFG (SFB/TRR135/1-3, A05, project 222641018). Thanks to Leandra Ruloff for conducting the experiment as part of her thesis work.

References

1. Nijhawan, R.: Motion extrapolation in catching. Nature **370**, 256–257 (1994)
2. Cellini, C., Scocchia, L., Drewing, K.: The buzz-lag effect. Exp. Brain Res. **234**(10), 2849–2857 (2016). https://doi.org/10.1007/s00221-016-4687-4
3. Whitney, D., Murakami, I., Cavanagh, P.: Illusory spatial offset of a flash relative to a moving stimulus is caused by differential latencies for moving and flashed stimuli. Vis. Res. **40**, 137–149 (2000)
4. Baldo, M.V., Klein, S.A.: Extrapolation or attention shift? Nature **378**, 565–566 (1995)
5. Brenner, E., van Beers, E.R., Rotman, G., Smeets, J.B.: The role of uncertainty in the systematic spatial mislocalization of moving objects. J. Exp. Psychol.: HPP **32**, 811–825 (2006)
6. Hubbard, T.L.: The flash-lag effect and related mislocalizations: findings, properties, and theories. Psychol. Bull. **140**, 308–338 (2006)
7. Alais, D., Burr, D.: The "Flash-Lag" effect occurs in audition and cross-modally. Curr. Biol. **13**, 59–63 (2003)
8. Vroomen, J., de Gelder, B.: Temporal ventriloquism: sound modulates the flash-lag effect. J. Exp. Psychol.: HPP **30**, 513–518 (2004)
9. Drewing, K., Hitzel, E., Scocchia, L.: The haptic and the visual flash-lag effect and the role of flash characteristics. PLoS ONE **13**(1), e0189291 (2018)
10. Watanabe, J., Nakatani, M., Ando, H., Tachi, S.: Haptic localizations for onset and offset of vibro-tactile stimuli are dissociated. Exp. Brain Res. **193**, 483–489 (2009)
11. Maij, F., Wing, A.M., Medendorp, W.P.: Spatiotemporal integration for tactile localization during arm movements: a probabilistic approach. J. Neurophysiol. **110**, 2661–2669 (2013)
12. Lopez-Moliner, J., Linares, D.: The flash-lag effect is reduced when the flash is perceived as a sensory consequence of our action. Vis. Res. **46**(13), 2122–2129 (2006)
13. Drewing, K., Hartmann, F., Vroomen, J.H.: The crossed-hands deficit in temporal order judgments occurs for present, future, and past hand postures. In: 2019 IEEE World Haptics Conference (WHC), pp. 145–150. IEEE (2019)
14. Schütt, H.H., Harmeling, S., Macke, J.H., Wichmann, F.A.: Painfree and accurate Bayesian estimation of psychometric functions for (potentially) overdispersed data. Vis. Res. **122**, 105–123 (2016)
15. Hao, Q., Ogata, T., Ogawa, K.I., Kwon, J., Miyake, Y.: The simultaneous perception of auditory–tactile stimuli in voluntary movement. Front. Psychol. **6**, 1429 (2015)
16. Hao, Q., Ora, H., Ogawa, K.I., Ogata, T., Miyake, Y.: Voluntary movement affects simultaneous perception of auditory and tactile stimuli presented to a non-moving body part. Sci. Rep. **6**(1), 1–8 (2016)

17. Brown, H., Adams, R.A., Parees, I., Edwards, M., Friston, K.: Active inference, sensory attenuation and illusions. Cogn. Process. **14**(4), 411–427 (2013). https://doi.org/10.1007/s10339-013-0571-3
18. Matthews, W.J.: How do changes in speed affect the perception of duration? J. Exp. Psychol.: HPP **37**, 1617–1627 (2011)
19. Botev, J., Drewing, K., Hamann, H., Khaluf, Y., Simoens, P., Vatakis, A.: ChronoPilot – Modulating time Perception. In: IEEE AIVR 2021, the 4th International Conference on Artificial Intelligence and Virtual Reality (AIVR), pp. 1–4. IEEE (2021)

Perception of Friction in Tactile Exploration of Micro-structured Rubber Samples

Maja Fehlberg[1,2] ⓘ, Kwang-Seop Kim[3,4] ⓘ, Knut Drewing[5] ⓘ, René Hensel[1] ⓘ,
and Roland Bennewitz[1,2(✉)] ⓘ

[1] INM Leibniz-Institute for New Materials, Saarbrücken, Germany
roland.bennewitz@leibniz-inm.de
[2] Physics Department, Saarland University, Saarbrücken, Germany
[3] Nanomechatronics, University of Science and Technology (UST), Daejeon, Republic of Korea
[4] Nano-Convergence Mechanical Systems Research Division, Korea Institute of Machinery and
Materials, (KIMM), Daejeon, Republic of Korea
[5] Department of Psychology, Justus Liebig University, Giessen, Germany

Abstract. Fingertip friction and the related shear of skin are key mechanical
mechanisms in tactile perception, but the perception of friction itself is rarely
explored except for the flat surfaces of tactile displays. We investigated the perception of friction for tactile exploration of a unique set of samples whose fabric-like
surfaces are equipped with regular arrays of flexible micropillars. The measured
fingertip friction increases with decreasing bending stiffness, where the latter is
controlled by radius (20–75 μm) and aspect ratio of the micropillars. In forced-
choice tasks, participants noticed relative differences in friction as small as 0.2,
and even smaller when a sample with less than 100 μm distance between pillars
is omitted from the analysis. In an affective ranking of samples upon active touch,
the perception of pleasantness is anticorrelated with the measured friction. Our
results offer insights towards a rational design of materials with well-controlled
surface microstructure which elicit a dedicated tactile appeal.

Keywords: Tactile perception · Friction · Materials

1 Introduction

Friction is the force which resists sliding of the fingertip over a sample surface in tactile
exploration. Its strength indicates shear deformation in the skin which leads to activation
of mechanoreceptors and thus contributes to the process of tactile perception [1]. Friction, often referred to by the word pair sticky/slippery, has been invoked as one of the
important dimensions in the tactile perception of surface textures [2, 3] and in the perception of similarity or distinction between materials [4–8]. The perception of fingertip
friction also plays a key role in the adjustment of prehensile forces, securing grip when
lifting objects [9–11]. Despite the frequent discussion of friction as important channel
in tactile perception, there are but few studies on the perception of friction itself. Smith
and Scott asked participants to rate their tactile perception on a scale between "most

© The Author(s) 2022
H. Seifi et al. (Eds.): EuroHaptics 2022, LNCS 13235, pp. 21–29, 2022.
https://doi.org/10.1007/978-3-031-06249-0_3

slippery" and "most sticky" and found an average correlation of 0.85 with the kinetic friction coefficient in a wide range from 0.4 to 2.8 [12]. Grierson and Carnahan reported significant correlation between perceived slipperiness and the measured friction coefficient only if a tangential motion of the fingertip over the surface was involved, in contrast to static or tapping touch [9]. Little correlation between measured and perceived friction was reported in two studies on surfaces of consumer goods [13, 14].

The perception of friction can be entangled with the perception of surface texture. The mutual influence is manifest in the structure of the perceptual space derived from tactile exploration of materials, where the tactile dimensions of slipperiness and of roughness are correlated [4]. The cross-talk between resistance against lateral motion and roughness perception has been used to render roughness in tactile displays by modulation of lateral forces [15]. The entanglement of friction and roughness perception is also reflected in the finding that subjective roughness estimates decrease upon lubrication of the contact [16].

Tactile displays allow to modulate fingertip friction by imperceptible ultrasonic excitation, where an air cushion effectively lubricates the contact [17]. This technology allows to determine just noticeable differences in friction without changing roughness or surface material. Weber fractions, i.e. just noticeable changes of friction, where found to be around 0.18 for spatial variation [18] and 0.11 for transient changes [19].

Our interest lies in the understanding of role of friction in tactile perception of materials towards a design of materials with a predictable tactile appeal. Here, we focus on the physical basis of friction and the friction perception of well-controlled fabric-like surfaces in contrast to the previously studied smooth or less controlled surfaces. We prepared polymer samples with a surface structure consisting of a regular array of flexible cylindrical pillars with flat top surfaces. These samples represent fibrillar materials like fabrics and papers, however with a well-controlled structure and the option to vary the structural parameters. We asked participants to compare friction and rate pleasantness of samples and measured the forces during their tactile exploration.

2 Experiments

Micro-structured elastomer samples were prepared by replica molding using templates which were themselves replicated from a microfabricated arrays of silicon pillars (Institute of Semiconductors and Microsystems, TU Dresden, Germany). Square samples with a side length of 50 mm carried a hexagonal array of pillars (Fig. 1) with a radius of 20–75 μm and a center-to-center distance of four times the radius, i.e. 80–300 μm (see Fig. 1). In this design, the flat top surfaces of the pillars cover a fraction of $\pi/(8\sqrt{3}) \approx 22.6\%$ of the total area for all samples, i.e. the exposed surface on top of the pillar is constant for different pillar radii. Six arrays with the following radius and height of pillars were used in this study: 20 μm/120 μm, 50 μm/100 μm, 50 μm/200 μm, 50 μm/300 μm, 75 μm/350 μm, 75 μm/450 μm.

Polydimethylsiloxane (PDMS, Elastosil M4601, Wacker Chemie AG, München, Germany) templates were replicated from micropatterned silicon wafers exhibiting a micropillar array of 5 cm × 5 cm (TU Dresden, Germany). The samples were made from the polyurethane 'Neukadur high elastic A50' (Altropol, weight ratio of components

1:1, Young's modulus of 5 MPa at 1 Hz). This elastomer was poured onto the PDMS templates, degassed in a vacuum chamber for 10 min, and baked overnight in an oven at 65 °C to cure the polyurethane.

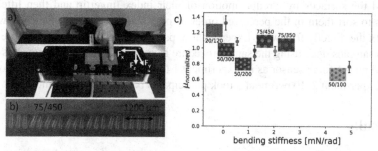

Fig. 1. a) Tactile exploration setup. Samples are mounted to a three-axis force sensor with coordinates as indicated. The visual access is blocked by an opaque screen. b) Close-up of a finger sliding over an array of pillars (75 μm radius, 300 μm distance, 450 μm height). c) Average normalized coefficient of friction as function of the calculated bending stiffness of pillars Eq. (1). Error bars indicate standard deviation across participants Labels next to the data points indicate radius and height of the pillars in μm. The top view images have a size of 550 μm × 750 μm.

For the tactile exploration experiments (Fig. 1), samples were mounted to 3-axis-force sensor (K3D120 with GSV-8 amplifier, ME-Messysteme, Germany). Forces in normal direction (F_N) and friction forces $(F_F = \sqrt{F_x^2 + F_y^2})$ were recorded at a rate of 120 Hz, the friction coefficient was determined as $\mu = F_F/F_N$ and averaged over the time of each trial. Participants were asked to explore the surfaces with the index fingertip of their dominant hand in circular movements with a straight finger. An opaque screen blocked the view of the samples, and a headphone suppressed the sound from the sliding fingertip. Participants were asked to maintain a constant normal force. As visual feedback on their actual normal force, they were shown a bar chart with a marked target range of 0.3 to 0.5 N. The fingertip moisture was recorded with a corneometer (CM 825, Courage + Khazaka electronic GmbH, Germany).

In our psychophysical study, 19 Participants (age 20 to 27, 7 males, 1 left-handed, unpaid volunteer university students of physics, engineering, psychology, and the arts with no known cutaneous or motor impairments) explored the 6 samples described above in three different experiments. The participants were naïve with respect to the goal of the study, they were instructed before the experiments in detail and gave their consent to participation. All experiments were designed to comply with the principles outlined in the Declaration of Helsinki. The study was approved by a university ethics board (proposal "Perception of micro-patterned materials (18–16)").

In Experiment 1, participants explored each of the 15 pairs of 6 samples once in random order for the time they needed, and they were allowed to switch between the two samples of one pair as often as they wanted. In a forced-choice task, they had to decide "for which of the two sample it is more difficult to move the finger over the surface, if you apply the same pressure on the sample." We did not ask directly about friction to avoid a bias in answers which could arise from a different understanding of the technical term friction. Experiment 1 took 16 to 35 min for each participant.

In Experiment 2, scheduled one week after Experiment 1, participants explored by touch all 6 samples lying next to each other on a table behind the opaque screen. They explored the surfaces by circular motion of their index fingertip and then lifted the samples to sort them in the perceived order of pleasantness in touch, typically within 5 min. In the directly following Experiment 3, participants were asked to explore each of the 6 samples once using the same procedure as in Experiment 1. The samples were mounted on the force sensor as in Experiment 1 to repeat friction measurements on the day of Experiment 2. Experiment 3 took participants between 8 and 14 min.

3 Results

In Experiments 1 and 3, mean applied normal forces were between 0.36 and 0.39 N for the participant with most constant forces, and between 0.30 and 0.63 N for the participant with the largest range of mean normal forces applied to different samples. The measured friction coefficients varied between 1.0 and 1.3 for the participant with lowest and between 2.1 and 2.9 for the one with the highest friction. We found no correlation of fingertip moisture with friction coefficients ($r = 0.06$, $p = 0.81$). For the analysis of results, we normalized the friction coefficients by division with the average friction coefficient of each participant to give equal weight to variations between samples for each participant, independent of the absolute value of the friction coefficients. The normalized average coefficient of friction is plotted as a function of the bending stiffness for pillars on each sample in Fig. 1c. While there was no clear relation of friction to either pillar height or pillar radius, we found a correlation ($r = -0.81$, $p = 0.0506$) with the bending stiffness of the pillars. In Fig. 1c, samples are ordered by the bending stiffness, which can be approximated as [20]:

$$\frac{F_{top}}{\theta} = \frac{\pi}{2} \frac{ER^4}{L^2},$$ (1)

F_{top} the lateral force acting on the top of each pillar, θ the bending angle, L the height of the pillars, R their radius, and $E = 5$ MPa the elastic modulus of the material. The normalized coefficient of friction from above 130% of each participant's mean value for most bendable pillars to below 80% for the least bendable pillars. High aspect ratio and small pillar diameter contribute to the bending flexibility of pillars. The photograph in Fig. 1b visualizes the bending. Pillar bending may increase friction by direct contact between the side walls of the pillars with the skin and by interlocking of their edges with the papillary structure of the fingertip skin [21].

The results for friction perception in Experiment 1 are summarized in Fig. 2. The psychometric curves represent the probability that participants have indicated that sample as "more difficult to move the finger", for which the higher friction coefficient was measured. This probability is plotted as function of the relative difference in friction coefficient between the two samples. The relative difference is computed for each trial, i.e. each sample pair and participant. The probability is then calculated for bins of 19 trials, where the relative difference in friction coefficient is the average for all trials in that bin. The probability increases from a value of 0.5, which indicates a choice by chance at small friction differences, to a value of 1 at large difference in friction, where all decisions on perceived higher friction agree with the measurement. In Fig. 2a, the measured forces are analyzed in form of the friction coefficient μ, which can be considered as invariant under different applied normal forces. From the level of 75% probability in the psychometric curve, we can extract a just noticeable difference of $\Delta\mu/\mu = 0.21$ for the perception of relative differences in friction (Weber fraction).

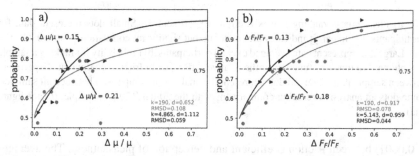

Fig. 2. Probability for indicating the sample with the higher measured friction versus a) the relative difference in friction coefficient between two samples, and b) the relative difference in the measured friction force. Red dots represent analysis of data for all samples, blue triangles analysis of data after omission of trials including the 20 μm/120 μm sample. Each data point represents the probability for a bin of 19 trials with similar relative differences. The solid lines are Weibull sigmoid functions fitted to the data points. The fit parameters and the root mean-square deviation (RMSD) of data points from the fit curves are listed. (Color figure online)

The data point representing one set of 19 trials ($\Delta\mu/\mu = 0.29$) is a peculiar outlier. We noticed that this set of trials includes a high number of samples with smallest pillar radius of 20 μm. Assuming that participants were unsuccessful in comparing friction on this sample with friction of other samples, we also present a psychometric curve for trials with all samples except the 20 μm/120 μm sample. There is no outlier and less scatter of data points with respect to the sigmoid function, reflected in a drop of the root-mean-square deviation (RMSD) from 0.108 to 0.059. The just noticeable difference in friction coefficient for the reduced set of samples decreases to $\Delta\mu/\mu = 0.15$.

We do not know if the perception of "the difficulty to move the finger over the surface" in our task reflects the friction force or the coefficient of friction, i.e. if participants directly compare friction forces or if they implicitly consider the applied normal pressure when judging the friction force. In Fig. 2b we present psychometric curves which are based on relative differences in the measured friction force. These curves follow a similar trend as the curves based on the friction coefficients with lower values for the RMSD. The just noticeable difference in the friction force is lower with 0.18 and with 0.13 after excluding comparisons with the 20 μm/120 μm sample.

The results of Experiment 2, where the six samples were ordered with respect to perceived pleasantness in touch, are analyzed in Fig. 3. The rank in pleasantness of each sample is plotted versus the normalized friction coefficient for all samples and all participants in Fig. 3a. There is a moderate but significant anticorrelation ($r = 0.444$,

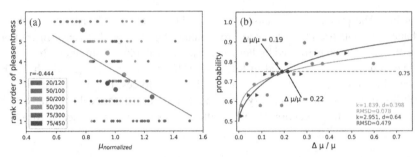

Fig. 3. a) Pleasantness ranking versus friction (Experiment 3). Small dots represent data for all participants, the linear fit indicates the moderate but significant correlation ($r = -0.44$, $p < 0.001$). Large dots represent data averaged over all participants. b) Probability for ranking a sample as less pleasant to touch as function of the relative difference in friction coefficient (Experiment 3) between sample pairs. The red dots represent the analysis of comparison between all samples, the blue dots an analysis excluding comparisons with the 20 μm/120 μm sample. (Color figure online)

$p < 0.0001$) between friction coefficient and perception of pleasantness. The averages over all participants visualize this correlation. When we correlated the rank of perceived pleasantness for each individual participant with the rank of the friction coefficient using Kendall's correlation coefficient, we find negligible correlation for 7 participants, weak or moderate positive correlation for 3 participants, and moderate to very strong negative correlation for 9 participants.

Assuming that large differences in friction cause an unequivocal decision on perceived pleasantness, the data can be analyzed in analogy to the psychometric curves (see Fig. 3b). We plot the probability to rank the sample with the higher measured friction coefficient as less pleasant to touch for all samples pairs as function of the relative difference in friction coefficient. The probability to perceive one sample as less pleasant increases with increasing difference in the friction coefficient and is larger than 75% for relative differences of more than 0.22, i.e. above the just noticeable difference in friction perception. When we exclude comparisons with the 20 μm/120 μm sample from the analysis, the 75% level is reached already at a relative difference in friction of 0.19.

Friction coefficients for all samples and participants were correlated between Experiment 1 and Experiment 3 ($r = 0.67$, $p < 0.001$). A strong correlation ($r = 0.983$, $p < 0.001$) was observed for the normalized averaged friction coefficients of the six samples between Experiment 1 and Experiment 3. We conclude that friction between fingertip and micro-structured rubber samples is consistent over time in participants and samples, with some variation between trials which is probably caused by variations of the portion of the fingertip in contact with the sample and of the angle between knuckle and surface.

4 Conclusion

We created a set of samples from one polymeric material with perceptible differences in fingertip friction by varying only the length scales of a regular array of flexible micropillars on the surface. Friction increases with decreasing bending stiffness of the pillars.

By combining a forced-choice task on friction perception with force measurements on each trial, we report for the first time just noticeable relative differences (JNDs) in the friction coefficient between samples with controlled structural variations. The JNDs are around 0.2 and thus comparable to those reported for spatial friction contrast on tactile displays [18]. They are higher than JNDs reported for transient changes in tactile displays [19], which may be explained by distraction through roughness and compliance differences or by the break when lifting the finger for a switch between samples. Our results also confirm a correlation of lower friction with pleasant touch [22] for a set of manufactured samples which differ not in material but rather in the microscopic surface structure. Psychophysical experiments with well-controlled flexible surface structures thus open new opportunities for systematic differentiation of friction from other tactile dimensions such as roughness or compliance.

We noticed that participants had difficulties judging friction differences when one sample was the one with the smallest microstructures. Similarly, this sample received widely varying rankings in the pleasantness of touch. The values of JNDs in friction dropped by 30% after omitting this sample from the analysis. It is the only one that clearly falls into the small-scale regime of the duplex theory, where different mechanisms of perception are expected [23], for example a perception of slipperiness through the subjective intensity of vibrations which are excited by small structures [24]. It would be interesting to construct a stimulus set of samples from both regimes of the duplex theory, i.e. with structures smaller and larger than 100 μm, to verify if friction differences are perceived correctly in each regime, but not between regimes.

In conclusion, the combination of materials science approaches for a full control of surface structures at the micrometer scale with the elucidation of mechanisms in fingertip friction and with the quantification of friction perception is a step towards a rational design of materials with low friction and a pleasant tactile appeal.

Acknowledgements. We thank Lisa Sold for preparing molds for replication. We acknowledge valuable discussions with Eduard Arzt and financial support by the Volkswagen Foundation and for K.-S. Kim by an internal program of the Korea Institute of Machinery and Materials (KIMM).

References

1. Willemet, L., Kanzari, K., Monnoyer, J., Birznieks, I., Wiertlewski, M.: Initial contact shapes the perception of friction. PNAS **118**, e2109109118 (2021)
2. Okamoto, S., Nagano, H., Yamada, Y.: Psychophysical dimensions of tactile perception of textures. IEEE Trans. Haptics **6**, 81–93 (2013)
3. Skedung, L., Arvidsson, M., Chung, J.Y., Stafford, C.M., Berglund, B., Rutland, M.W.: Feeling small: exploring the tactile perception limits. Sci. Rep. **3**, 2617 (2013)
4. Hollins, M., Faldowski, R., Rao, S., Young, F.: Perceptual dimensions of tactile surface texture - a multidimensional-scaling analysis. Percept. Psychophys. **54**, 697–705 (1993)
5. Bergmann Tiest, W.M., Kappers, A.M.L.: Analysis of haptic perception of materials by multidimensional scaling and physical measurements of roughness and compressibility. Acta Psychologica **121**, 1–20 (2006)

6. Gueorguiev, D., Bochereau, S., Mouraux, A., Hayward, V., Thonnard, J.L.: Touch uses frictional cues to discriminate flat materials. Sci. Rep. **6** (2016). Article No. 25553

7. Carpenter, C.W., et al.: Human ability to discriminate surface chemistry by touch. Mater. Horiz. **5**, 70–77 (2018)

8. Vardar, Y., Wallraven, C., Kuchenbecker, K.J.: Fingertip interaction metrics correlate with visual and haptic perception of real surfaces. In: 2019 IEEE World Haptics Conference (WHC), pp. 395–400 (2019)

9. Grierson, L.E.M., Carnahan, H.: Manual exploration and the perception of slipperiness. Percept. Psychophys. **68**, 1070–1081 (2006)

10. Cadoret, G., Smith, A.M.: Friction, not texture, dictates grip forces used during object manipulation. J. Neurophysiol. **75**, 1963–1969 (1996)

11. Westling, G., Johansson, R.S.: Factors influencing the force control during precision grip. Exp. Brain Res. **53**, 277–284 (1984)

12. Smith, A.M., Scott, S.H.: Subjective scaling of smooth surface friction. J. Neurophysiol. **75**, 1957–1962 (1996)

13. Schreiner, S., Rechberger, M., Bertling, J.: Haptic perception of friction—correlating friction measurements of skin against polymer surfaces with subjective evaluations of the surfaces' grip. Tribol. Int. **63**, 21–28 (2013)

14. Liu, X., Yue, Z., Cai, Z., Chetwynd, D.G., Smith, S.T.: Quantifying touch-feel perception: tribological aspects. Meas. Sci. Technol. **19** (2008). Article No. 084007

15. Klatzky, R.L., Lederman, S.J.: The perceived roughness of resistive virtual textures: I. Rendering by a force-feedback mouse. ACM Trans. Appl. Percept. **3**, 1–14 (2006)

16. Smith, A.M., Chapman, C.E., Deslandes, M., Langlais, J.S., Thibodeau, M.P.: Role of friction and tangential force variation in the subjective scaling of tactile roughness. Exp. Brain Res. **144**, 211–223 (2002)

17. Basdogan, C., Giraud, F., Levesque, V., Choi, S.: A review of surface haptics: enabling tactile effects on touch surfaces. IEEE Trans. Haptics **13**, 450–470 (2020)

18. Ben Messaoud, W., Bueno, M.A., Lemaire-Semail, B.: Relation between human perceived friction and finger friction characteristics. Tribol. Int. **98**, 261–269 (2016)

19. Gueorguiev, D., Vezzoli, E., Mouraux, A., Lemaire-Semail, B., Thonnard, J.L.: The tactile perception of transient changes in friction. J. R. Soc. Interface **14**, 10 (2017)

20. Gedsun, A., Hensel, R., Bennewitz, R.: Bending as key mechanism in the tactile perception of fibrillar surfaces. Adv. Mater. Interfaces **9**(4), 2101380 (2021)

21. Tomlinson, S.E., Carre, M.J., Lewis, R., Franklin, S.E.: Human finger contact with small, triangular ridged surfaces. Wear **271**, 2346–2353 (2011)

22. Klöcker, A., Wiertlewski, M., Theate, V., Hayward, V., Thonnard, J.L.: Physical factors influencing pleasant touch during tactile exploration. PLoS ONE **8**, e79085 (2013)

23. Hollins, M., Risner, S.R.: Evidence for the duplex theory of tactile texture perception. Percept. Psychophys. **62**, 695–705 (2000)

24. Bensmaia, S., Hollins, M.: Pacinian representations of fine surface texture. Percept. Psychophys. **67**, 842–854 (2005)

Influence of Prior Visual Information on Exploratory Movement Direction in Texture Perception

Michaela Jeschke[(✉)] [iD], Aaron C. Zöller, and Knut Drewing

Justus-Liebig University, 35390 Giessen, Germany
Michaela.Jeschke@psychol.uni-giessen.de

Abstract. When humans explore objects haptically, they seem to use prior as well as sensory information to adapt their exploratory behavior [1]. For texture discrimination, it was shown that participants adapted the direction of their exploratory movement to be orthogonal to the orientation of textures with a defined direction [2]. That is, they adapted the exploratory direction based on the sensory information gathered over the course of an exploration, and this behavior improved their perceptual precision. In the present study we examined if prior visual information that indicates a texture orientation produces a similar adjustment of exploratory movement direction. We expected an increase of orthogonal initial exploration movements with higher qualities of prior information. In each trial, participants explored two grating textures with equal amplitude, only differing in their spatial period. They had to report the stimulus with the higher spatial frequency. Grating stimuli were given in six different orientations relative to the observer. Prior visual information on grating orientation was given in five different qualities: 50% (excellent information), 35%, 25%, 15% and 0% (no information). We analyzed movement directions of the first, middle and last strokes over the textures of each trial. The results show an increase in the amount of initial orthogonal strokes and a decrease in variability of movement directions with higher qualities of prior visual information.

Keywords: Prior information · Exploration · Texture · Movement adaption

1 Introduction

Haptic perception is inherently active. To achieve high levels of perceptual performance and to behave most efficiently during haptic exploration, humans adapt their exploratory behavior to the present objects and task [3–5]. Without prior information about object properties, this motor adaptation is enabled by closed-loop processes that are based on sensory information, gathered and integrated over the sequential exploration process [2, 6, 7]. However, in many situations humans have prior information about the objects they are going to interact with. The current study investigates how they integrate prior information in a subsequent haptic exploration process.

© The Author(s) 2022
H. Seifi et al. (Eds.): EuroHaptics 2022, LNCS 13235, pp. 30–38, 2022.
https://doi.org/10.1007/978-3-031-06249-0_4

Prior information can arise e.g. from recent interactions with similar objects, from semantic [8] or visual information [9]. Most of the times humans look at objects before they interact with them; hereby they can gather information about properties as shape, size or orientation. They use this to adapt manual motor behavior already at initial contact [10]. A number of previous studies have examined the use of prior information on weight in grasping and lifting behavior [11] or on softness in exploratory behavior [1, 7]. For softness discrimination, it was shown that humans explore stimuli with higher initial peak forces when they expect harder as compared to softer stimuli. Here, we study the use of prior visual information on orientation in texture exploration.

Haptic texture exploration is a crucial subset of tools that humans employ during object discrimination and it is typically performed by moving the fingertips laterally across an object's surface several times [5]. Thus, movements produce small patterns of vibrations, i.e. temporal cues that enable humans to discriminate between fine textures by their microgeometry [12]. To investigate texture exploration, grating patterns are particularly useful: their surfaces consist of periodically repeating grooves which define a clear orientation. For patterns with a defined orientation, efficiency of the intake of temporal cues can be maximized by moving orthogonal to the grating direction. Lezkan and Drewing [2] showed that indeed exploratory direction in grating discrimination is optimized accordingly based on sensory information, gathered over the course of an exploration: In their experiment participants had to discriminate gratings by their spatial frequency. In each trial they explored two types of textures: One single grating texture (defined orientation) and a texture made of two combined gratings (no defined orientation). Participants adapted their final movement directions orthogonal to the texture orientation when exploring the single grating stimulus. In another experiment this adaption was shown to be beneficial in terms of perceptual precision as it led to lower just noticeable differences (JNDs) for spatial frequency discrimination. We used a similar method and examined whether prior visual information on grating orientation evokes the same adaption of movement direction as in [2], but already for initial movements. Moreover, according to ideal observer models, humans should weigh less noisy prior information higher than noisy prior information [8]. Hence, we hypothesized that humans show more adaptation behavior when receiving prior information with higher as compared to lower quality, and presented information with different qualities.

2 Methodology

2.1 Participants

16 right-handed students from Giessen University participated (10 female, mean age: 23 y., range: 18–27). All had normal or corrected-to-normal vision, reported no tenosynovitis in the past and no motor or cutaneous impairments. They had a 2-point discrimination threshold <4 mm on the right fingertip, were naïve to the purpose of the experiment, provided written informed consent and received financial compensation (8€/h). The experiment was approved by the local ethics committee LEK FB06 and conducted in accordance with 2013 Declaration of Helsinki, except for preregistration.

2.2 Stimuli and Setup

Participants sat at a custom-made visuo-haptic workbench (Fig. 1a), consisting of a PHANToM 1.5A haptic force feedback device (spatial resolution: 0.03 mm, temporal resolution: 1000 Hz), a force sensor to collect data of the executed finger force (682 Hz, resolution: 0.05 N), and a 24″ computer screen (120 Hz, 1600 × 900 pixel). They looked at the screen through a mirror and stereo glasses (Nvidia 3D Vision 2, viewing distance 40 cm). The visual setup displayed 3D scenes aligned with the haptic workspace and prevented that participants saw their hand. In the scenes, grating stimuli were displayed as light grey cylindrical discs on a dark green checkerboard. At the same positions two real stimuli were placed side-by-side on the force sensor in front of the participant. A small sphere (8 mm diameter) represented the participants' finger position in the scene but disappeared as soon as they touched the stimulus. A chinrest stabilized the head position. The index finger was connected to the PHANToM via a spherical magnet fixed at the fingernail, allowing to move the finger in all axes with the maximum amount of freedom in a 38 × 27 × 20 cm³ workspace. The adapter left the fingertip free for bare-finger exploration. Devices were connected to a PC where C++-based custom software controlled the experiment and processed the data. Acoustic noises were masked with passive noise-cancelling headphones and white noise. As haptic stimuli we used four 3D-printed grating stimuli (Fig. 1b; printer: Stratasys Objet 30 Pro, resolution of 600 × 600 × 1600 dpi). We printed 4 mm high (z-axis) grating discs with a texture diameter of 90.7 mm (100.7 mm including the border). A 10 × 5 mm grip helped the experimenter arranging the stimuli before each trial. All stimuli consisted of a groove pattern following a sine-wave function in height $z(x)$:

$$z = \frac{1}{2}A \sin \frac{2\pi x}{P} + \frac{1}{2}A \tag{1}$$

Textures were defined to have the same amplitude A of 0.3 mm, differing only in their period P. The stimuli had a period of 1.524 mm, 2.032 mm, 2.540 mm, and 3.048 mm. Stimuli with adjacent periods were compared (=three pairs). Every pair was presented in 6 possible orientations relative to the observer (15°, 45°, 75°, 105°, 135° and 165°; for 0° ridges would be parallel to the body, see Fig. 1b). For prior visual information, visual representations of the grating stimuli were displayed with a texture made of 100 dark grey stripes (8 × 1 mm) on their top side (Fig. 1c). The textures indicated the orientation of the following haptic gratings and were displayed before exploration. Quality of prior information was manipulated by varying the percentage of stripes following the same orientation as the gratings' ridges while the remaining ones were randomly oriented. Pilot studies (N = 16 overall) had shown that participants perceived the orientation of such stimuli excellently when 100% to 50% of the stripes were identically oriented, but variance in answers systematically increased with lower percentages. We hence presented visual stimuli with percentages of 50% or lower.

2.3 Procedure and Design

We varied the quality of prior visual information on stimulus orientation in five steps (50%, 35%, 25%, 15%, and 0% [=no information]). Stimuli were presented in 6 possible orientations (15°, 45°, 75°, 105°, 135° and 165°). We presented three different pairs of stimuli with adjacent half-periods in both possible left-right assignments. In both assignments participants started exploration equally often with the left and the right stimulus. They conducted a 2AFC discrimination task: On each trial they explored the two stimuli and had to decide which one had the higher spatial frequency.

Overall, each participant conducted 360 experimental trials (5 qualities × 6 orientations × 3 pairs × 2 stimulus location × 2 start location)in randomized order plus 8 initial practice trials. We implemented a break of 3 min every 60 trials; the experiment took about 3.5 h in a single session. During each single trial, prior information was displayed for 2500 ms before participants were allowed to explore the haptic stimuli. Exploration was initiated by a beep sound. One of the two visual stimulus representations discs was colored in a bright yellow to indicate where participants should start exploration. Participants were instructed to use the typical movement scheme for this, i.e. stroking laterally over the surface [11]. They were free to switch between stimuli as often as desired. After the exploration they indicated which stimulus they had perceived to be of higher frequency by pressing a virtual button above it. During the whole experiment white noise was presented through the headphones to mask any exploration sounds containing information about spatial frequency. After the experiment, participants filled out a questionnaire in order to check whether they had paid attention to the (not explained) visual prior information and intentionally used it for adaption purposes. They were not informed about the relation between prior information and the gratings' orientation beforehand.

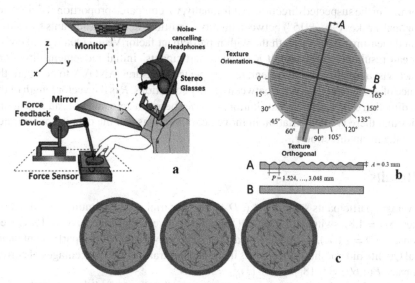

Fig. 1. a, visuo-haptic workbench **b,** haptic stimulus in orientation 165° **c,** examples of visual prior information with qualities 0%, 25% and 50% (from left to right) in orientation 135°.

2.4 Data Analysis

Raw data of individual observers is available at 10.5281/zenodo.5718623. Movement directions of the first, middle and last stroke in the exploration of each stimulus pair were analyzed. In case of an even number of total strokes, the later of the two possible ones was defined as the middle stroke. An algorithm segregated single strokes from movement data when the participants' finger was touching the stimulus area with at least 0.1 N of force for >200 ms. We detected strokes as continuous movements either from one texture border to another or between two movement turns, being extracted by zero crossings in the 1st order derivatives of the x- or y-position over time. We were particularly interested in the initial movement direction as it can be assumed to be hardly affected by sensory feedback and is thus an indicator of the use of prior information. Movement direction was determined as the orientation of the line connecting start and end point (x and z coordinates) of each stroke. Note that movement directions were processed as axial data and thus ranged only between 0°–180°. We aligned all stimulus orientations with 0° in order to collapse data over trials. We calculated circular histograms of initial movement directions (bin size: 10°) separately for each visual quality condition. Each histogram displays how many times participants moved in a specific direction. Movement directions were analyzed using circular statistics: We used V-tests on the distributions of individual average movement directions per visual quality condition and separately for the first, middle and last stroke in order to test whether the distribution is not uniform (=all directions are equally likely) but rather has a specified mean direction of 90° (=optimal adaption behavior). Significant test statistics imply a deviation from uniformity in the suspected direction. Additionally, we compared proportions of close-to-orthogonal strokes (90° ± 15°) between quality conditions across participants by using a repeated measures ANOVA with the within-participant factor Visual Quality. Likewise, the mean resultant vector length $\bar{R}$ of each participants' initial movement directions for each visual quality condition entered a repeated measures ANOVA to compare the variance of movement directions between quality conditions. $\bar{R}$ is the vector length of the mean direction $\bar{\theta}$ of a circular distribution. It varies between zero and one; an $\bar{R}$ near one implies that there is little variation in movement directions and the data is concentrated around the mean direction $\bar{\theta}$.

3 Results

On average, participants spent 5.5 s ($SD = 3.6$) per trial on the stimuli, performed 6.6 strokes ($SD = 4.8$), switched 1.7 times ($SD = 1.1$) between them and gave 91.1% correct responses ($SD = 2.8\%$). A repeated measures ANOVA with the within-participant factor Visual Quality did not show differences in the arcsine-transformed percentages of correct responses, $F(4,60) = 0.18$, $p = .951$, $\eta_p^2 = .03$.

We plotted angular distributions of movement directions of all initial strokes separately for each visual quality condition (Fig. 2a–e). V-tests were performed on the distributions of individual average movement directions per quality condition, and separately for the first, middle and last strokes (Bonferroni-corrected α-levels at 0.0033 for 15 tests). For initial strokes, movement distributions of the 25%, 35% and 50% quality condition deviated significantly from uniformity, 25%: $V = 11.71$, $p < .001$, 35%: V

$= 9.54, p < .001$, 50%: $V = 9.15, p < .001$, suggesting optimization of movement behavior. Distributions of the 0% and 15% quality condition did not deviate significantly, 0%: $V = 7.26, p = .005$, 15%: $V = 6.56, p = 0.012$. For individual average movement directions of the middle and last strokes, distributions of all quality conditions deviated significantly from uniformity (all $p < .001$, all $V > 11.37$). Proportions of movements close-to-orthogonal to the textures' surface 90° ($\pm15°$) among all initial movements across participants (Fig. 2f) entered an ANOVA with the within-participant factor Visual Quality. As expected, proportion of orthogonal movements was higher with higher visual qualities, $F(4, 60) = 4.39, p = .036, \eta_p^2 = .23$ (Greenhouse-Geisser adjusted [13]), confirmed by a linear trend, $F(1,15) = 6.11, p = .026, \eta_p^2 = .29$. For middle and last movements, the proportions of strokes close-to-orthogonal to the textures' surface did not differ between quality conditions, $F(4,60) = 2.331, p = .114, \eta_p^2 = .14$ (Greenhouse-Geisser adjusted) and $F(4,60) = 1.52, p = .208, \eta_p^2 = .092$.

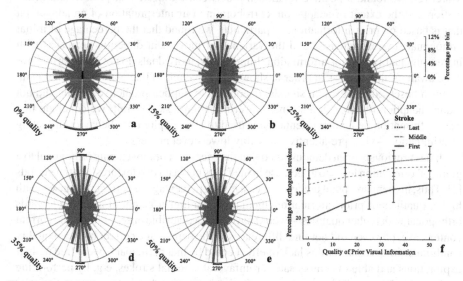

Fig. 2. a–e, Initial movement direction histograms for each quality condition including all participant data. Textures were aligned to a 0° orientation. Note, possible movement directions varied only between 0–180° and were mirrored on the lower part of each figure. Orange lines indicate mean directions, black bars in the center of each circle represent resultant vector lengths. Black circle segments mark the 95% confidence interval of the mean direction. **f,** average percentages of orthogonal movements and standard errors for each quality condition and movement type.

For the initial movements, which are at focus here, also the individual mean resulting vector lengths, being indices of variance in movement direction, entered an ANOVA with the within-participant factor Visual Quality. Vectors were longer with higher visual qualities, i.e. variability of the initial movements' directions decreased with higher visual quality, $F(4, 60) = 4.81, p = .03, \eta_p^2 = .243$ (Greenhouse-Geisser adjusted), confirmed by a linear trend, $F(1,15) = 6.16, p = .025, \eta_p^2 = .30$ (0%: .11 ± .09[M ± SD], 15%: .21 ± .18, 25%: .23 ± .27, 35%: .28 ± .27, 50%: .29 ± .28). Questionnaire data showed that only two participants were aware that prior information indicated the orientation of the upcoming haptic stimulus and intentionally used it to adapt their exploration behavior.

4 Discussion

The aim of this study was to investigate the influence of prior visual information that indicates a grating orientation on the direction of (initial) exploratory movements in a grating discrimination task. Previous findings had shown that, without prior information, participants adapt the direction of their exploratory movement to be orthogonal to the orientation of textures in later strokes of the exploration, i.e., based on the sensory information previously gathered [2]. The results here clearly show that with prior information participants adapt already the direction of their initial stroke to the stimulus orientation. Moreover, we found a positive relationship between the quality of prior visual information and the extent of adaptation, corroborating our interpretation of an adaptation that is based on prior information. In particular, we found that the percentage of initial orthogonal strokes increases, and the variance of movement direction decreases with increasing quality of prior information. Importantly, individuals do not necessarily have to be aware of this process in order to show said behavior. In line with that, Zöller et al. [1] even observed that explicit signals, in contrast to implicit predictive signals, did not induce adaption of force in a softness discrimination task. Overall, we conclude that prior information on texture orientation is used to hone exploratory behavior in grating discrimination—via a pre-attentive, possibly lower-level process.

It has previously been demonstrated that the execution of movements orthogonal to a gratings' orientation leads to a decrease in the JNDs between two textures' spatial periods [2]. Thus, higher perceptual precision, i.e., a higher percentage of correct responses, with higher qualities of prior information–mediated by a higher extent of adaption towards orthogonal with higher qualities—would have been a plausible observation in the current context. We, however, did not find differences in the percentage correct between quality conditions. One reason for this lack of effect could be that participants were free in their explorations and able to compensate for unfavorable initial strokes, e.g. by performing extra strokes later on. This is represented in the descriptive data but not in a statistically significant manner. Another reason could be task difficulty, which was relatively low in the present study as stimuli differed by approximately 2 Wb fractions from each other in their spatial period [14]. Differences in perceptual precision between quality conditions might be less pronounced due to ceiling effects, but should be present when task demands are higher. Future investigations might manipulate task difficulty by using lower differences between stimuli. High task demands might affect the use of prior information and thus the adaption of exploration behavior differently than low demands. A manipulation of task difficulty could even reveal an interaction effect: Higher task demands might increase adaption behavior compared to low demands especially for lower information qualities, as behavioral optimization is needed more under difficult circumstances and to achieve optimization, even low quality prior information might be used. Furthermore, the effect of sensory signals on movement adjustment during softness discrimination has been shown before to partly depend on reward-induced task relevance, as the signature of force tuning gets stronger with higher motivation [7]. This might be the case for prior information too, leading to maximized adaption behavior when prior information is given and motivation is highest.

In conclusion, it became evident that prior visual information is given substantial weight to in the context of exploratory movement planning; and of which the individual does not necessarily have to be aware. Whether this resembles unconscious learning or rather inherent processing in humans is an interesting question for a future study. Eventually, findings on human use of prior information can be applied in the field of e.g. neurorobotics to enhance the performance of AI-based object recognition [15]. The implementation of optimized exploration processes could facilitate a robotic system's information gathering, leading to a more efficient and more accurate decision-making.

Acknowledgments. Research was supported by the Deutsche Forschungsgemeinschaft (DFG, German Research Foundation) – project number 222641018 – SFB/TRR 135, A5.

References

1. Zoeller, A.C., Lezkan, A., Paulun, V.C., Fleming, R., Drewing, K.: Integration of prior knowledge during haptic exploration depends on information type. J. Vis. **19**(4), 20 (2019)
2. Lezkan, A., Drewing, K.: Interdependences between finger movement direction and haptic perception of oriented textures. PLoS ONE **13**(12), e0208988 (2018)
3. Kaim, L., Drewing, K.: Exploratory strategies in haptic softness discrimination are tuned to achieve high levels of task performance. Trans. Haptics **4**(4), 242–252 (2011)
4. Tanaka, Y., Bergmann Tiest, W.M., Kappers, A.M.L., Sano, A.: Contact force and scanning velocity during active roughness perception. PLoS ONE **9**(3), e93363 (2014)
5. Lederman, S.J., Klatzky, R.L.: Hand movements: a window into haptic object recognition. Cogn. Psychol. **19**(3), 342–368 (1987)
6. Saig, A., Gordon, G., Assa, E., Arieli, A., Ahissar, E.: Motor-sensory confluence in tactile perception. J. Neurosci. **32**(40), 14022–14032 (2012)
7. Lezkan, A., Metzger, A., Drewing, K.: Active haptic exploration of softness: indentation force is systematically related to prediction, sensation and motivation. Front. Integr. Neurosci. **12**, 59 (2018)
8. Ernst, M.O., Bülthoff, H.H.: Merging the senses into a robust percept. Trends Cogn. Sci. **8**(4), 162–169 (2004)
9. Newell, F.N., Ernst, M.O., Tjan, B.S., Bülthoff, H.H.: Viewpoint dependence in visual and haptic object recognition. Psychol. Sci. **12**(1), 37–42 (2001)
10. Wolpert, D.M., Flanagan, J.R.: Motor prediction. Curr. Biol.: CB **11**(18), R729–R732 (2001)
11. Johansson, R.S., Cole, K.J.: Sensory-motor coordination during grasping and manipulative actions. Curr. Opin. Neurobiol. **2**(6), 815–823 (1992)
12. Hollins, M., Risner, S.R.: Evidence for the duplex theory of tactile texture perception. Percept. Psychophys. **62**(4), 695–705 (2000)
13. Greenhouse, S.W., Geisser, S.: On methods in the analysis of profile data. Psychometrika **24**, 95–112 (1959)

14. Nefs, H.T., Kappers, A.M., Koenderink, J.J.: Amplitude and spatial-period discrimination in sinusoidal gratings by dynamic touch. Perception **30**(10), 1263–1274 (2001)
15. Fishel, J.A., Loeb, G.E.: Bayesian exploration for intelligent identification of textures. Front. Neurorobot. **6**, 4 (2012)

Guidance for the Design of Vibrotactile Patterns for Use on the Human Back

Astrid M. L. Kapppers[1]([⊠])[ID] and Myrthe A. Plaisier[1,2][ID]

[1] Eindhoven University of Technology, Eindhoven, The Netherlands
{a.m.l.kappers,m.a.plaisier}@tue.nl
[2] Delft University of Technology, Delft, The Netherlands

Abstract. In this paper, we present an overview of parameters that are of relevance for the perception of vibrotactile patterns on the back. These patterns are delivered via varying numbers of vibration motors fixed to the back rest of a chair, vests or belts. We present recent findings from the literature about vibrotactile anisotropy, timing, spacing, anchor points, resolution and intensity. From this overview, we derive recommendations that should be considered when designing a vibrotactile device for the back. The main recommendations are: 1) Use sequential stimulation for conveying spatial patterns; 2) Avoid tactors on the spine; 3) For a rectangular grid 4×4 tactors seems optimal; 4) Carefully consider relative horizontal and vertical spacing. We hope that this overview will raise awareness of several issues that play a role in perception and that our recommendations will provide guidance when designing vibrotactile communication devices.

Keywords: Vibrotactile · Illusions · Timing · Spacing · Resolution

1 Introduction

For already more than half a century, attempts have been made to convey information via haptic devices on the back [15]. Most early aims were to create aids for people with visual impairments, e.g. [1,18], but later the focus became more general on devices that could be used in circumstances where vision and/or audition were less reliable or overloaded, e.g. [6,13]. The number of vibration motors (tactors) used in these devices varies widely from only 9 to as many as 400. However, performance does not necessarily improve with this number. For example, even after several hours of training, only around 50% of block letter patterns presented via 400 tactors was recognized, whereas without any training, 87% of the letters presented via a grid of only 9 tactors were recognized [29].

For an optimal design of vibration patterns for the back, it is important to make use of existing knowledge about (mis)perceptions, anisotropies, perceptual illusions, and already published 'tricks' to improve performance. The current

This work was partially supported by the European Union's Horizon 2020 research and innovation programme under Grant 780814, Project SUITCEYES.

H. Seifi et al. (Eds.): EuroHaptics 2022, LNCS 13235, pp. 39–47, 2022.
https://doi.org/10.1007/978-3-031-06249-0_5

study is aimed at creating an inventory of all such non-veridical perceptions that could and should be considered. We will start with giving an overview of all such issues and end with a summary of design recommendations.

2   Relevant Perceptual Findings

2.1   Anisotropy

One of the first researchers to report on tactile anisotropy was Weber [26]. Measuring two-point pressure thresholds on the back (and many other body parts), he noticed that thresholds in vertical direction were larger than in horizontal direction. In his experiments, either one or two compass legs were pressed against the body and the participant had to decide whether he felt one or two points. Hoffmann et al. [10] found a similar anisotropy for vibrotactile stimulation on the back. Their stimuli consisted of two consecutive vibrations, either at the same location or slightly shifted. Participants had to decide whether the second stimulus was to the right, left or at the same location for the horizontal condition, and up, down, or same for the vertical direction. Averaged over the three inter-vibrator distances, accuracy in the horizontal condition was significantly higher than in the vertical direction. Plaisier et al. [23] asked for length estimates between sequential vibrations on the back. They found that vertical distances were perceived as larger than horizontal distances, which seems in contradiction with the results of Hoffmann et al. [10], although direction perception and length estimates do not necessarily lead to the same results. In both studies, the influence of the spine as an anchor point is given as a possible explanation of the results (see Subsect. 2.3). Interestingly, Nicula and Longo [19] obtained similar results as in [23] for pressure stimuli on the lower back; on the upper back the results were reversed, indicating that anisotropy on the back is inhomogeneous.

Kappers and colleagues [14] investigated vibrotactile direction perception on the back in 12 directions. A first vibration was always given centred on the spine and a second vibration was presented on one of 12 equally spaced locations on a circle with a radius of 11 cm. Participants had to adjust a pointer on a frontoparallel plane to indicate the perceived direction. They found that both accuracy and precision were significantly higher for the cardinal (i.e., horizontal and vertical) directions than for the oblique directions, with vertical even better than horizontal. A partial explanation comes from the results of Hoffmann et al. [10]. The differences in the perception of horizontal and vertical lengths that they found will directly influence the perception of the *direction* of oblique stimuli and thus the accuracy of the responses.

2.2   Temporal Aspects

Weber [26] observed that it is easier to discriminate sequentially presented pressure stimuli than two simultaneously presented stimuli. Eskilden et al. [7] investigated this for vibrotactile stimuli, but they did not find a significant difference in threshold between sequential and simultaneous stimulation. This non-significance was possibly due to their limited number of participants (only 5), but in any case, the difference was quite small. v. Békésy [2] showed that if the delay between two subsequent vibrations on the arm became shorter, the perceived location of the stimulations moved to halfway the actual locations. In the study of Plaisier et al. [23], participants had to estimate the distance between the locations of two vibrotactile stimulations. In the case of simultaneous stimulation, the distances were estimated to be much shorter than in the sequential condition. Moreover, there was hardly any difference between the estimates for a 4-cm and a 12-cm presented distance, indicating that simultaneous stimulated locations are hard to distinguish. These results can be understood from the findings of v. Békésy [2] as simultaneous stimulations will be perceived as halfway between the vibration locations and thus lack a clear distance. Van Erp [25] showed that a longer time between two vibrotactile stimuli (larger Stimulus Onset Asynchrony) resulted in better performance if participants had to indicate whether the second stimulus was to the right or the left of the first stimulus. Measuring two-point vibrotactile discrimination thresholds on the lower back, Stronks et al. [24] found that an SOA of 0 ms (i.e., simultaneous stimuli) resulted in significantly higher thresholds than an SOA of 200 ms.

This advantage of sequential stimulation becomes even more clear when more complicated patterns are presented. Loomis [18] tested recognition of letters presented on a 20×20 grid of tactors on the back and clearly performance was worse with letters presented statically (that is, all required tactors for a letter vibrating simultaneously) compared to conditions where a slit moved over the letter or the letter itself moved. Novich and Eagleman [20] used a 3×3 grid of tactors to compare spatial (that is, all tactors of a pattern vibrate simultaneously) with spatiotemporal stimulation. Pattern identification was significantly higher for the spatiotemporal patterns than for the spatial patterns.

An interesting effect of timing of vibrotactile stimulations was reported by Geldard and Sherrick [8]. Presenting 3 bursts of 5 brief pulses to the forearm, one burst near the wrist, one at the centre of the forearm and one near the elbow, was perceived as 15 pulses equally spaced moving from wrist to elbow. Varying the number of pulses in each burst influenced the perceived spacing of the locations. They termed this effect 'cutaneous rabbit', as it felt as if a tiny rabbit was hopping over the arm. So again, timing of vibrations has a distinct influence on perceived location.

The aim of vibrotactile stimulation is often to convey dynamic patterns or traces. Kim et al. [16] showed that a more continuous trace produced by overlap in stimulation of subsequent tactors on the foot resulted in better recognition performance. This is again an application of the findings of v. Békésy [2]. Also, Israr and Poupyrev [11] made use of this mislocalization in their sophisticated Tactile Brush algorithm. Virtual locations on their intended trace, i.e., locations on the line connecting two tactors, were simulated by an appropriate scaling of the intensity of the two tactors. In a small evaluation study, they tested 3 vibratory patterns on a device with 12 (4×3) tactors on the back generated with either the Tactile Brush algorithm or subsequent stimulation of the tactors. Participants had to decide how many strokes they felt, but in all cases, the intention was that it should feel as one continuous stroke. The more conventional stimulation resulted in a number close to 3, whereas for the Tactile Brush algorithm this number was just above 1, indicating the perception of a continuous stroke.

2.3 Anchor Points

Misperceptions of localization are often due to nearby anchor points such as wrist, elbow, and other joints. Boring [3] describes these anchor points as forming a frame of reference to which the perceptions of other points are drawn. For vibrotactile stimuli Cholewiak and Collins [5] showed that localization performance on the forearm was best for stimuli near the wrist, the elbow, or the shoulder, and worse at other locations on the arm. In a subsequent study [4], they showed that for localization around the torso both navel and spine served as anchor points, especially in conditions where the spacing between the possible vibration locations was small (i.e., 12 possible locations around the torso).

Van Erp [25] measured tactile acuity by asking participants whether a second vibration was located to the left or the right of the first vibration location. To determine thresholds, they varied the actual distance between locations. They found that thresholds were much lower (and thus performance better) near the spine and the navel. Hoffmann and colleagues [10] used a similar experimental paradigm to measure vibrotactile acuity on the back. They found that horizontal accuracy for direction perception was *lower* near or across the spine compared to more peripheral areas. As a probable explanation for this lower accuracy near the spine, they argue that there will be an increased spread of the vibrations along the spine (i.e., bone conduction), making the perception task harder. However, they did not find this effect for vertical accuracy and their vibrators were not actually placed on the spine, so it remains to be seen whether this is the real explanation. In the length estimation experiment of Plaisier et al. [23], the proximity of the spine in the vertical condition is also given as a possible explanation for their finding that vertical length estimates were larger than horizontal ones.

2.4 Resolution

The resolution of vibrotactile stimuli on the back depends on various experimental factors, such as the SOA, the exact location on the back, tactor type, the participant, and the experimental task. Eskilden et al. [7] found a median threshold of 17.8 mm in a task where participants had to say whether they felt one or two simultaneous vibrations (i.e., a two-point discrimination task). In a second experiment, participants had to estimate the distance between two vibration locations in both a simultaneous condition and a successive condition. They found thresholds of 11.36 mm and 10.15 mm, respectively, which were not significantly different. Van Erp [25] found a uniform acuity of 2 to 3 cm on the torso, except near the spine where the acuity was 1 cm. Stronks et al. [24] report two-point vibrotactile discrimination thresholds on the lower back of 51 mm for simultaneous stimulation and 28 mm for stimulation with an SOA of 200 ms.

Johannesson et al. [12] measured direction accuracy for three different inter-tactor distances: 13 mm, 20 mm and 30 mm. Accuracies for the different distances were 64%, 82% and 91%, respectively. In a subsequent study, Hoffmann et al. [10] compared several tactors and they found best performance with N ERMs (Normal rotation eccentric rotating mass motors). The accuracy in this direction experiment was 65% for 20 mm between the tactors and 50% for a 10 mm distance. Finally, in a pattern recognition task, Novich and Eagleman [20] found that an inter-tactor distance of 6 cm was necessary for a performance of 80% correct vibrotactile pattern recognition.

2.5 Intensity

The intensity of the vibrations will also play a role in how the vibrotactile stimulation is perceived. Wu and colleagues [27,28] used a 6 × 8 grid of tactors on the back to present letters and simple geometric figures. Subsequent tactors of a trace had a small overlap in activation time and tactors on the vertices of a pattern were activated with higher intensity. They found increased recognition performance if vertices were given a higher vibration intensity than the other tactors representing the pattern. In their Tactile Brush algorithm, Israr and Poupyrev [11] used the relative intensity of vibrations to vary the perceived location of the vibration in between two tactors.

An interesting new illusion was reported by Hoffmann et al. [9]. They found that a weak vibration followed by a strong vibration at the same location, was often perceived as an illusory upward movement, and vice versa. Also, if the locations of the tactors actually differed, the perceived movement could be made stronger via this illusion.

3   Design Recommendations

From the above overview, it should be clear that various parameters such as timing, spacing, and intensity will play a role in how a vibration pattern will be perceived. However, it will depend on the intended application which aspects of the stimulation are relevant. Here, we will present a list of design recommendations for a vibrotactile device on the back that should at least be considered.

1. Use sequential stimulation for conveying spatial patterns
   Several studies showed that sequential stimulation results in better performance in terms of acuity, direction perception, pattern recognition and length estimates than simultaneous stimulation [18,20,23–25]. So especially when the intention is to present a spatial pattern, sequential presentation is essential. If the strength of a stimulus but not the exact location is relevant, simultaneous stimulation would be an option. Also, a spatial pattern consisting of simultaneous symmetric stimulation at both sides of the spine will probably be recognized.

2. Avoid tactors on the spine

   Several studies showed that tactors on or very near the spine will influence acuity, length estimates and direction perception [4, 10, 23, 25]. Informal observations and introspection also indicate that stimulation on the spine feels different than stimulation at other back areas; especially persons with a hearing impairment mentioned that vibrations on their spine were uncomfortable.

3. For a rectangular grid, 4 × 4 tactors seems optimal

   The density of tactors does not have to be higher than the human resolution. Moreover, with an algorithm like the Tactile Brush [11], the density can be further reduced. Therefore, given the vibrotactile resolution on the spine [7, 10, 12, 20, 24, 25] and to avoid tactors on the spine, a 4 × 4 grid of tactors seems a good choice, although a 6 × 6 grid also lies within the resolution of the back.

4. Carefully consider relative horizontal and vertical spacing

   Tactile acuity in horizontal direction is better than in vertical direction [26], and this was also found for vibrotactile acuity [10]. It is unknown whether this holds for all areas on the back. However, it should be kept in mind that spatial patterns presented on the back might not be perceived veridically, but instead be shrunken in vertical direction.

5. Miscellaneous recommendations

   Preliminary research suggests that emphasizing corners via a stronger vibration might help recognition of patterns [27, 28]. If a pattern consists of several traces, a short break between two separate traces will improve recognition [22].

4   Conclusions

In this paper, we summarized the most relevant perceptual findings from the literature for the design of a vibrotactile device for the back. Both spatial and temporal parameters have a strong influence on how a stimulus will be perceived. Often perception of a stimulus is not veridical. Many of the results depended on the exact experimental conditions, but still, we could derive several design recommendations that seem generally valid. All recommendations are aimed at maximizing recognizability of the vibrotactile patterns and are based on published psychophysical studies.

One interesting application of vibrotactile stimulation is the possiblity to convey Social Haptic Communication (SHC) via vibration patterns on the back [22]. SHC is used for communication with persons with deafblindness, mainly to provide environmental information, such as, 'the size of the room', 'the number of people in a room', 'there is applause', etc. [17, 21]. This type of information is usually given by a second interpreter, the other interpreter translating the spoken language. Our first co-design sessions with teachers of SHC, both persons with and without deafblindness, showed that emulating SHC via vibration patterns is promising [22]. From these sessions, we also learned that efforts to create such vibrotactile devices are highly appreciated by the target population.

Of course, there are many other possible applications, such as in gaming, virtual words, navigation, etc. We hope that our design recommendations will provide some guidance to all researchers who want to create useful vibrotactile devices.

References

1. Bach-y-Rita, P., Collins, C.C., Saunders, F.A., White, B., Scadden, L.: Vision substitution by tactile image projection. Nature **221**(5184), 963–964 (1969). https://doi.org/10.1038/221963a0
2. v. Békésy, G.: Sensations on the skin similar to directional hearing, beats, and harmonics of the ear. J. Acoust. Soc. Am. **29**(4), 489–501 (1957). https://doi.org/10.1121/1.1908938
3. Boring, E.: Sensation and Perception in the History of Experimental Psychology. The Century Psychology Series, Appleton-Century-Crofts, New York (1942)
4. Cholewiak, R.W., Brill, J.C., Schwab, A.: Vibrotactile localization on the abdomen: effects of place and space. Percept. Psychophys. **66**(6), 970–987 (2004). https://doi.org/10.3758/BF03194989
5. Cholewiak, R.W., Collins, A.A.: Vibrotactile localization on the arm: effects of place, space, and age. Percept. Psychophys. **65**(7), 1058–1077 (2003). https://doi.org/10.3758/BF03194834
6. Ertan, S., Lee, C., Willets, A., Tan, H., Pentland, A.: A wearable haptic navigation guidance system. In: Digest of the Second International Symposium on Wearable Computers, pp. 164–165 (1998). https://doi.org/10.1109/ISWC.1998.729547
7. Eskildsen, P., Morris, A., Collins, C.C., Bach-y-Rita, P.: Simultaneous and successive cutaneous two-point thresholds for vibration. Psychon. Sci. **14**(4), 146–147 (1969). https://doi.org/10.3758/BF03332755
8. Geldard, F.A., Sherrick, C.E.: The cutaneous "rabbit": a perceptual illusion. Science **178**(4057), 178–179 (1972). https://doi.org/10.1126/science.178.4057.178
9. Hoffmann, R., Brinkhuis, M.A.B., Kristjánsson, Á., Unnthorsson, R.: Introducing a new haptic illusion to increase the perceived resolution of tactile displays. In: Proceedings of the 2nd International Conference on Computer-Human Interaction Research and Applications (CHIRA 2018), pp. 45–53 (2018). https://doi.org/10.5220/0006899700450053
10. Hoffmann, R., Valgeirsdóttir, V.V., Jóhannesson, Ó.I., Unnthorsson, R., Kristjánsson, Á.: Measuring relative vibrotactile spatial acuity: effects of tactor type, anchor points and tactile anisotropy. Exp. Brain Res. **236**(12), 3405–3416 (2018). https://doi.org/10.1007/s00221-018-5387-z
11. Israr, A., Poupyrev, I.: Tactile brush: Drawing on skin with a tactile grid display. In: CHI' 2011: Proceedings of the SIGCHI Conference on Human Factors in Computing Systems, pp. 2019–2028 (2011). https://doi.org/10.1145/1978942.1979235
12. Jóhannesson, Ó.I., Hoffmann, R., Valgeirsdóttir, V.V., Unnþórsson, R., Moldoveanu, A., Kristjánsson, Á.: Relative vibrotactile spatial acuity of the torso. Exp. Brain Res. **235**(11), 3505–3515 (2017). https://doi.org/10.1007/s00221-017-5073-6
13. Jones, L.A., Kunkel, J., Piateski, E.: Vibrotactile pattern recognition on the arm and back. Perception **38**(1), 52–68 (2009). https://doi.org/10.1068/p5914

14. Kappers, A.M.L., Bay, J., Plaisier, M.A.: Perception of vibratory direction on the back. In: Nisky, I., Hartcher-O'Brien, J., Wiertlewski, M., Smeets, J. (eds.) EuroHaptics 2020. LNCS, vol. 12272, pp. 113–121. Springer, Cham (2020). https://doi.org/10.1007/978-3-030-58147-3_13

15. Kappers, A.M.L., Plaisier, M.A.: Hands-Free devices for displaying speech and language in the tactile modality – Methods and approaches. IEEE Trans. Haptics 14(3), 465–478 (2021). https://doi.org/10.1109/TOH.2021.3051737

16. Kim, H., Seo, C., Lee, J., Ryu, J., bok Yu, S., Lee, S.: Vibrotactile display for driving safety information. In: IEEE Intelligent Transportation Systems Conference, pp. 573–577 (2006). https://doi.org/10.1109/ITSC.2006.1706802

17. Lahtinen, R.: Haptices and Haptemes - A case study of developmental process in social-haptic communication of acquired deafblind people. Ph.D. thesis, University of Helsinki (2008)

18. Loomis, J.M.: Tactile letter recognition under different modes of stimulus presentation. Percept. Psychophys. 16(2), 401–408 (1974). https://doi.org/10.3758/BF03203960

19. Nicula, A., Longo, M.R.: Perception of tactile distance on the back. Perception 50(8), 677–689 (2021). https://doi.org/10.1177/03010066211025384

20. Novich, S.D., Eagleman, D.M.: Using space and time to encode vibrotactile information: toward an estimate of the skin's achievable throughput. Exp. Brain Res. 233(10), 2777–2788 (2015). https://doi.org/10.1007/s00221-015-4346-1

21. Palmer, R., Lahtinen, R.M.: History of social-haptic communication. DBI Rev. 50, 68–70 (2013)

22. Plaisier, M.A., Kappers, A.M.L.: Social haptic communication mimicked with vibrotactile patterns - An evaluation by users with deafblindness. In: The 23rd International ACM SIGACCESS Conference on Computers and Accessibility, ASSETS 2021. Association for Computing Machinery, New York (2021). https://doi.org/10.1145/3441852.3476528

23. Plaisier, M.A., Sap, L.I.N., Kappers, A.M.L.: Perception of vibrotactile distance on the back. Sci. Rep. 10(1), 17876 (2020). https://doi.org/10.1038/s41598-020-74835-x

24. Stronks, H.C., Parker, D.J., Barnes, N.: Vibrotactile spatial acuity and intensity discrimination on the lower back using coin motors. IEEE Trans. Haptics 9(4), 446–454 (2016). https://doi.org/10.1109/TOH.2016.2569484

25. Van Erp, J.B.F.: Vibrotactile spatial acuity on the torso: effects of location and timing parameters. In: First Joint Eurohaptics Conference and Symposium on Haptic Interfaces for Virtual Environment and Teleoperator Systems. World Haptics Conference, pp. 80–85 (2005). https://doi.org/10.1109/WHC.2005.144

26. Weber, E.H.: De tactu. In: Ross, H.E., Murray, D.J. (eds.) E. H. Weber on the Tactile Senses. Erlbaum (UK) Taylor & Francis, Hove (1834/1996)

27. Wu, J., Song, Z., Wu, W., Song, A., Constantinescu, D.: A vibro-tactile system for image contour display. In: IEEE International Symposium on Virtual Reality Innovation 2011, 19–20 March, Singapore, pp. 145–150 (2011)

28. Wu, J., Zhang, J., Yan, J., Liu, W., Song, G.: Design of a vibrotactile vest for contour perception. Int. J. Adv. Rob. Syst. 9(166), 1–11 (2012). https://doi.org/10.5772/52373

29. Yanagida, Y., Kakita, M., Lindeman, R.W., Kume, Y., Tetsutani, N.: Vibrotactile letter reading using a low-resolution tactor array. In: 12th International Symposium on Haptic Interfaces for Virtual Environment and Teleoperator Systems, HAPTICS 2004, pp. 400–406 (2004). https://doi.org/10.1109/HAPTIC.2004.1287227

Speed Discrimination in the Apparent Haptic Motion Illusion

I. Lacôte[1]([✉]), D. Gueorguiev[2], C. Pacchierotti[3], M. Babel[1], and M. Marchal[1,4]

[1] Univ Rennes, INSA Rennes, IRISA, Inria, CNRS, Rennes, France
ines.lacote@inria.fr
[2] CNRS, Sorbonne Université, ISIR, Paris, France
[3] CNRS, Univ Rennes, Inria, IRISA, Rennes, France
[4] Institut Universitaire de France (IUF), Paris, France

Abstract. When talking about the Apparent Haptic Motion (AHM) illusion, temporal parameters are the most discussed for providing the smoothest illusion. Nonetheless, it is rare to see studies addressing the impact of changing these parameters for conveying information about the *velocity* of the elicited motion sensation. In our study, we investigate the discrimination of velocity changes in AHM and the robustness of this perception, considering two stimulating sensations and two directions of motion. Results show that participants were better at discriminating the velocity of the illusory motion when comparing stimulations with higher differences in the actuators activation delay. Results also show limitations for the integration of this approach in everyday life applications.

Keywords: Apparent haptic motion · Tactile speed · Tactile devices

1 Introduction

Haptic illusions are a major tool to enhance tactile stimulations in a large variety of domains [11,12]. They are an interesting topic of research as they enable to convey rich sensations with rather simple stimulation techniques. One major illusion is the Apparent Haptic Motion (AHM) illusion. The apparent haptic motion illusion aims at conveying a sensation of continuous movement along the skin when only discrete points are stimulated. In his original work, Burtt [1] found that two distinct vibrotactile stimuli elicited in close proximity on the skin with overlapping actuation were not perceived as localized sensations but rather as a single moving vibration.

Studies regarding AHM were conducted on different body parts [8,13] to test its robustness and understand the essential parameters driving this sensory illusion. The illusion was demonstrated to be effective in conveying directional cues and proved to be robust in both 1D and 2D patterns [14], which suggests a potential for providing directional information during navigation tasks. Besides spatial parameters, i.e., the position of the activation points and their distance to each other, temporal aspects have also been studied, so as to deepen the

© The Author(s) 2022
H. Seifi et al. (Eds.): EuroHaptics 2022, LNCS 13235, pp. 48–56, 2022.
https://doi.org/10.1007/978-3-031-06249-0_6

understanding of the illusion mechanisms [9,15]. In this respect, some studies showed that the temporal parameters, i.e., activation delays between motors, were actually not strongly constrained. Indeed, Stimuli Onset Asynchrony (SOA) and Duration of Signals (DoS) that are different from those proposed by Sherrick and Rogers [16], can also efficiently elicit this illusion [8].

1.1 Speed Perception and Impacting Parameters

Perception and discrimination of tactile speed has been studied in a large variety of conditions such as textures and vibrations [2]. These works mainly realized experiments with a surface sliding under the fingertip, creating a contact and skin stretch. Hence, the literature provides information on the influence of textures and vibrations on speed discrimination for different velocity ranges, going, e.g. from 33 to 120 mm.s^{-1} [3,6]. The results from [2,3] show that smooth surfaces are systematically felt as sliding slower than textured surfaces, even when presented with an identical sliding velocity. It was found that the Pacinian corpuscles have a crucial role in the discrimination of tactile speed [3], which explains the impact of material-induced vibrations on speed perception.

1.2 Speed Perception of the Apparent Motion Illusion

As previously mentioned, various studies confirmed the presence of the AHM illusion at different distances between the stimulation points (the position of the actuators) and with different SOA and DoS, deviating from the parameters indicated by Sherrick and Rogers [16]. Interest has been put to investigate various parameters regarding the spatial and temporal dimensions of the AHM illusion [9,13], enabling the creation of more complex and informative stimulations. Although other works have focused on determining the optimal actuation timing for conveying the most natural apparent motion, to the best of our knowledge, no study focused on the perception of speed and duration as a source of information in AHM. Understanding the parameters that make two stimulations easily distinguishable could indeed be relevant for tactile communication or navigation. For example, the speed perception of the apparent motion could help representing a moving obstacle or a safe direction to follow.

1.3 Contribution

The goal of this paper is to investigate the perception of the velocity conveyed during the AHM illusion. To go further, we also tested the robustness of this perception based on how the stimulation is provided. Indeed, while historically the AHM is conveyed with vibrations, our previous study [10] suggested that the illusion can also be conveyed by intervals of mechanical pressure. To explore that possibility, we conducted a study with two main objectives. First, we determine and compare the threshold of velocity discrimination for the apparent motion using both vibrations and pressure intervals ("taps") on the skin. Secondly, we study the impact of these modes on the participants' confidence when answering.

2 User Study

This study aims to investigate the ability of discriminating a velocity change in the AHM illusion. The study has been approved by Inria's ethics committee (COERLE Dornell - Saisine 513).

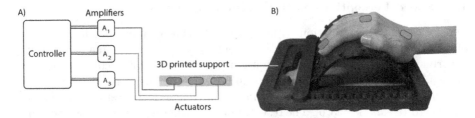

Fig. 1. Experimental set-up. A) The signals are generated via a controller and then amplified before being played by the custom-built actuators. B) Three electromagnetic actuators are placed on a curved hand-rest. The colored dots show the contact points of the actuators on the hand. (Color figure online)

2.1 Experimental Set-Up and Stimulation Modes

The experimental setup is shown in Fig. 1. It is composed of three custom actuators inspired by the work of Duvernoy et al. [5], with a coil as a stator and two magnets glued together in their repulsive position as a mover to increase the magnetic field. The actuators are mounted onto a curved 3D printed hand-rest, positioned in a comfortable bend for the participants. The signals for the three actuators are first created with Matlab and then processed through a National Instrument USB-6343 series controller, which sends them to three amplifiers enabling to deliver a 6.5 V signal to the motors, which corresponds to a force of approximately 0.4 N exerted on the hand. This last measure was recorded during a previous study, in which we characterized the force exerted by these actuators with a Nano17 force sensor (ATI, USA). The two magnets of the electromagnetic actuators move upward and downward along the center of the coil, depending on the electrical tension passing through it. This design enables to implement two stimulation modes: (i) a vibratory mode, where the actuators vibrate 120 Hz, and (ii) a "tap" mode, where the magnets elicit a single impact to the user's skin. The vibrating frequency for (i) was set based on previous studies investigating the apparent haptic motion illusion with vibrotactile stimuli, such as [17].

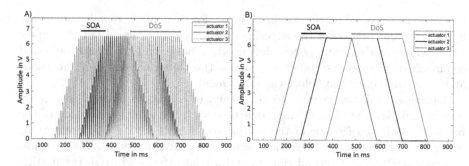

Fig. 2. Signals sent to the three actuators in the two actuation modes. A) Vibratory mode, made of sinusoïdal oscillations 120 Hz within ramp envelopes. B) "Tap" mode made of single ramp signals. In this Figure, we used DoS = 220 ms and SOA = 110 ms. (Color figure online)

Figure 2 shows the signals imparted to the three motors in the two stimulation modes. In both modes, asynchronous overlapping stimulations are sent to the same three locations on the hand (see Fig. 1). While the duration of activation of the actuators is fixed, we seek to change the time delay between the actuators activation, also called Stimuli Onset Asynchony (SOA). Based on pilot tests and [7], we set $DoS = 220$ ms and the reference $SOA = 110$ ms in both stimulation conditions. In the following experiment, we tested SOA values of 90%, 80%, 70%, 60% and 50% of the reference SOA, making the comparison SOA values [99, 88, 77, 66, 55] ms.

2.2 Experimental Design

Stimulations are conveyed between the middle finger and the proximal part of the palm, as shown in Fig. 1. We consider the two stimulation modes presented in Sect. 2.1, vibratory and tap, as well as two directions of motion, proximal-to-distal (orange-to-green in Fig. 1) and distal-to-proximal (green-to-orange).

The two modes (vibrations or taps) are tested in two blocks, carried out one after the other. A block is thus made of only vibratory or only tap trials. Each block is composed of 80 trials of which the changing parameters are the SOA and the direction of the motion. A trial is a sequence of the reference signal with a SOA = 110 ms and then a comparison signal with a different SOA, both having the same orientation (see also Sect. 2.3). The sequence of two signals is repeated a second time before the participants answer the questions. The order of the signal presentation is pseudo-randomized. Thus, blocks are only differentiated by the type of signal that is provided (vibratory or tap) and the order of the comparison, pseudo-randomized differently for each block and each participant.

2.3 Experimental Procedure

Ten persons participated in the experiment. They were all between twenty-one and thirty years old, of which two were women, and one was left-handed. Stimulations were delivered on the dominant hand. Participants were naive about the hypotheses and process of the experiment. Participants carried out the experiment while wearing headphones playing white noise, so as to mask the sound coming from the motors. Indications about the global number of stimulations and questions were given before the experiment.

During a trial, participants would receiv, in a random order (i) the reference stimulation ($SOA = 110$ ms), delivered with the stimulation mode of the block at hand, and (ii) one of the comparison signals having a different SOA, delivered with the same stimulation mode and same orientation. The identical sequence was played a second time to end a trial. After each trial, the participants answered two questions about what they perceived: (i) "Which one of the two motions was faster?" and (ii) "How certain are you of your answer?" The data collected from the participants were the index of the signal that seemed faster (1 or 2) and their confidence from 0 (no confidence at all) to 100 (total certainty). The mode of the starting block was counterbalanced between participants. At the end of the experiment, participants were also able to give open comments and feedback about their sensations and the experiment in general.

3   Results

Results are reported in Figs. 3 and 4. Figure 3 shows the rate of correct responses (score), while Fig. 4 shows the reported confidence. These two parameters are the dependent variables of our statistical analysis. As independent variables, we report the results for five SOA levels and four experimental conditions: two directions of motion (proximal-to-distal or distal-to-proximal) and two types of stimulation (vibrations or taps).

Results showed a significant decrease of correct answers when the time delay between actuators, the SOA, increases and thus, gets closer to the reference SOA. This performance trend is observable in all conditions, both in taps and in vibration mode as well as with proximal-to-distal or distal-to-proximal direction patterns. To confirm the visual perception, we performed a Friedman statistical test on the four experimental conditions. The test highlighted the effect of the changing value of the SOA on the participants' performance to discriminate the fastest stimulation they received for the conditions of proximal-to-distal taps, distal-to-proximal taps, proximal-to-distal vibrations and distal-to-proximal vibrations ($p < 0.01$).

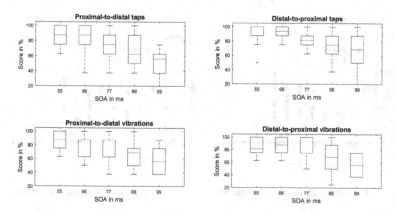

Fig. 3. Score when comparing five different SOAs vs. the reference one of 110 ms, across the two directions of motion (proximal-to-distal or distal-to-proximal) and type of stimulation (vibrations or taps). The boxplot gives the median, 25 and 75 percentiles with extrema values.

An identical Friedman test was performed on the effect the compared SOAs have on the confidence rates. A significant effect was also noted for the four experimental conditions ($p < 0.01$).

To interpret the effect of the direction (distal-to-proximal or proximal-to-distal) and the stimulation mode (tap or vibration), we performed matched-pairs Wilcoxon tests. The test showed no significant effect of the stimulation mode ($p > 0.05$) but it showed significant differences on the score between the proximal-to-distal and distal-to-proximal direction in tap stimulations ($p < 0.01$). However, a post-hoc test operated separately for each SOA did not show a significant difference for any of the comparisons.

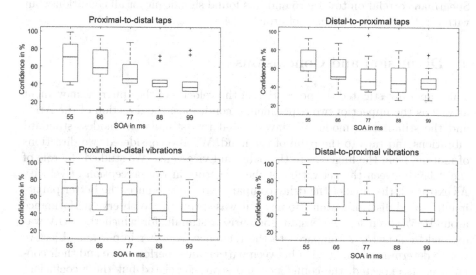

Fig. 4. Reported confidence of answer when comparing the SOAs, across the two direction of motion (proximal-to-distal or distal-to-proximal) and type of stimulation (vibrations or taps). The boxplot gives the median, 25 and 75 percentiles with extrema values.

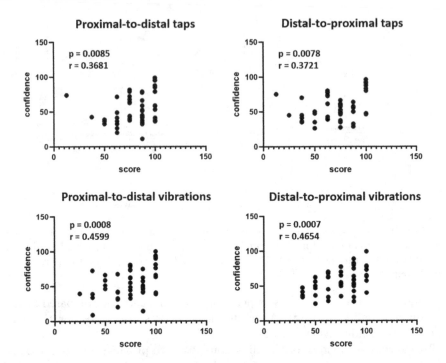

Fig. 5. Spearman correlation tests with the corresponding p-values and statistical dependence factors "r". We tested the correlation between the confidence and the score for the different conditions of mode and direction.

Finally, the matching between performance and confidence was tested by a Spearman correlation test Fig. 5 and was found significant for all conditions, but with a rather low r coefficient of around 0.4.

4   Discussion and Conclusions

This paper investigated the perception of the velocity of the apparent movement as well as the impact of two experimental conditions: the direction of the AHM and the stimulation mode. We investigated two stimulation modes, standard vibrations and taps to the palm of the hand. We also considered two directions of motion, from the fingertip to the palm and vice-versa. We studied the role of the delay between the activation of the actuators in the perception of velocity As expected, the smaller the delay compared to the reference, the better participants' speed discrimination. However, it was surprising to observe performance around 85% even for the easiest comparison stimuli, for which the SOA was divided by a factor 2 compared to the reference. Another important objective was to determine the matching between participants' performance and their confidence. As expected, the confidence and score correlated but the r coefficient was surprisingly low suggesting that participants struggled to assess their own performance. There was no significant influence of the mode of stimulation on the score or confidence, which showed a similar perception of both modes. Overall, the task was very challenging to participants and a few of them highlighted the difficulty of the task in their free comments. Thus, AHM illusions with different speeds might not be intuitive enough for people to use during everydat navigation tasks; the outcomes of the experiment were quite interesting in terms of haptic perception and confirm that human perception of tactile speed is inaccurate and prone to artefacts. The apparent haptic motion could still become a useful directional cue to integrate in navigation devices for impaired people, e.g., power wheelchairs, walkers, prewalkers [4] but modulating the speed might not be very informative. We wish to conduct further experiments, in which we let participants set what they perceive to be the best parameters for the AHM, e.g., duration, delay, intensity.

Acknowledgement. This work has received funding from the Inria Défi project "DORNELL". The authors thank Georges Daher, PhD student at the ISIR of Sorbonne Université, for his help in creating the experimental setup. They also thank Frédéric Marie, mechatronics engineer at the Laboratoire de Génie Civil et Génie Mécanique (LGCGM) of Rennes, for the design of the 3D printed hand-rest.

References

1. Burtt, H.E.: Tactual illusions of movement. J. Exp. Psychol. **2**(5), 371–385 (1917)
2. Dallmann, C.J., Ernst, M.O., Moscatelli, A.: The role of vibration in tactile speed perception. J. Neurophysiol. **114**(6), 3131–3139 (2015)
3. Delhaye, B.P., et al.: Feeling fooled: texture contaminates the neural code for tactile speed. PLoS Biol. **17**(8), e3000431 (2019)
4. Devigne, L., et al.: Power wheelchair navigation assistance using wearable vibrotactile haptics. IEEE Trans. Haptics **13**(1), 52–58 (2020)
5. Duvernoy, B., Farkhatdinov, I., Topp, S., Hayward, V.: Electromagnetic actuator for tactile communication. In: Prattichizzo, D., Shinoda, H., Tan, H.Z., Ruffaldi, E., Frisoli, A. (eds.) EuroHaptics 2018. LNCS, vol. 10894, pp. 14–24. Springer, Cham (2018). https://doi.org/10.1007/978-3-319-93399-3_2
6. Dépeault, A., Meftah, E.M., Chapman, C.E.: Tactile speed scaling: contributions of time and space. J. Neurophysiol. **99**(3), 1422–1434 (2008)
7. Gallo, S., et al.: Augmented white cane with multimodal haptic feedback. In: 2010 3rd IEEE RAS EMBS International Conference on Biomedical Robotics and Biomechatronics, pp. 149–155 (2010)
8. Israr, A., Poupyrev, I.: Control space of apparent haptic motion. In: 2011 IEEE World Haptics Conference, pp. 457–462 (2011)
9. Kirman, J.H.: Tactile apparent movement: the effects of interstimulus onset interval and stimulus duration. Percept. Psychophys. **15**(1), 1–6 (1974)
10. Lacôte, I., Pacchierotti, C., Babel, M., Marchal, M., Gueorguiev, D.: Tap stimulation: an alternative to vibrations to convey the apparent haptic motion illusion. In: 2022 IEEE Haptics Symposium (HAPTICS), pp. 1–6 (2022). hal-03551830
11. Lederman, J.: Tactile and haptic illusions. IEEE Trans. Haptics **4**(4), 273–294 (2011)

12. Lederman, S.J., Klatzky, R.L.: Haptic perception: a tutorial. Atten. Percept. Psychophys. **71**(7), 1439–1459 (2009)
13. Niwa, M., et al.: Determining appropriate parameters to elicit linear and circular apparent motion using vibrotactile cues. In: World Haptics 2009 - Third Joint EuroHaptics Conference and Symposium on Haptic Interfaces for Virtual Environment and Teleoperator Systems, pp. 75–78 (2009)
14. Park, J., Kim, J., Oh, Y., Tan, H.Z.: Rendering Moving Tactile Stroke on the Palm Using a Sparse 2D Array. In: Bello, F., Kajimoto, H., Visell, Y. (eds.) EuroHaptics 2016. LNCS, vol. 9774, pp. 47–56. Springer, Cham (2016). https://doi.org/10.1007/978-3-319-42321-0_5
15. Sherrick, C.E.: Bilateral apparent haptic movement. Percept. Psychophys. **4**(3), 159–160 (1968)
16. Sherrick, C.E., Rogers, R.: Apparent haptic movement. Percept. Psychophys. **1**(6), 175–180 (1966)
17. Zhao, S., Israr, A., Klatzky, R.: Intermanual apparent tactile motion on handheld tablets. In: 2015 IEEE World Haptics Conference (WHC), pp. 241–247 (2015)

Neutral Point in Haptic Perception of Softness

Anna Metzger[1,2(✉)], Anna Lotz[1], and Knut Drewing[1]

[1] Justus-Liebig University, Giessen, Germany
[2] Bournemouth University, Poole, UK
ametzger@bournemouth.ac.uk

Abstract. Haptic perception of objects' softness plays an important role in the identification and interaction with objects. How softness is represented in the brain is yet not clear. Here we investigated whether there is a neutral point in the perceptual representation of haptically perceived softness relative to which the objects are represented as being "soft" or "hard". We created a wide range of softness stimuli, varying from very hard (ceramic) to very soft foam with differently soft foam and silicone stimuli in between. Participants were assigned to one of three different stimulus set conditions: *full* set (18 stimuli), *soft* set (13 softest stimuli) or the *hard* set (13 hardest stimuli). They categorized each stimulus as "hard" or "soft" and we estimated the neutral point as the point of subjectively equal categorization as "hard" or "soft". We found that neutral points were different from the middle stimulus of each set. Furthermore, during the course of the experiment neutral points rather moved away from the middle of the stimulus set than towards it. Our results indicate that there might be a neutral point in the representation of haptically perceived softness, however range effects may play a role.

Keywords: Softness · Perception · Human

1 Introduction

How do we judge the quality of a mattress or the air pressure in bicycle tires? We need to touch these objects and judge their softness. In addition to warmth, roughness and friction, softness is a fundamental dimension in the haptic perception of objects [1]. It helps us to differentiate, classify and identify objects [2,3]. It is also important for the interaction with objects, e.g. adjusting contact forces [3]. Softness is a subjective measure of the ability of objects to deform under pressure [4]. It's physical correlate is compliance or Young's modulus, defined as the ratio between the displacement of the surface of an object and the force applied to the object [3] (but see [5]). To perceive the softness of an object humans usually press the object between two fingers or indented with a finger or a tool [6].

This work was supported by Deutsche Forschungsgemeinschaf (DFG, German Research Foundation) - project number 222641018 - SFB/TRR 135, A5.

H. Seifi et al. (Eds.): EuroHaptics 2022, LNCS 13235, pp. 57–65, 2022.
https://doi.org/10.1007/978-3-031-06249-0_7

This provides both tactile and kinesthetic information [7–10]. Also remembered softness of the object affects perceived softness [11]. Depending on the sensory and prior information, we mainly categorize objects as "hard" or "soft". It was also shown that we are quite good in discriminating different degrees of softness [3,7,8]. However, it is not yet clear how softness is represented in the brain.

Multiple studies have shown that changes in response to adaptation (after-effects) can provide information about how perceptual dimensions are coded in the brain. Also softness is susceptible to adaptation [12]. A silicone stimulus was perceived to be harder if the finger had previously adapted to a softer stimulus, and it was perceived as being softer if the finger pad had previously been adapted to a harder stimulus. In a general model of perception [13] it is assumed that a perceptual dimension is represented by the activity of several overlapping neural channels, each of which is finely tuned to a certain level along this dimension (e.g. orientation selective neurons). Perception depends on the overall activity of these neurons. When being exposed to a stimulus for a prolonged time, the channels that are most sensitive to it adapt, i.e. their responses are reduced. The relative activity of the adjacent channels then predominates and the overall response is shifted away from the adaption stimulus. Thus, the reported shift in the haptically perceived softness away from the softness of the adaptation stimulus suggests that there might be neural channels tuned to different softness values [12]. However, the authors also found a significant shift in the softness perception of a stimulus after adaptation to a stimulus which had approximately the same physical softness. This implies that a longer exploration of a stimulus leads to an altered perception of the softness of that stimulus. This finding cannot be explained by the above-mentioned model, in which perceived intensity of the adapted stimulus does not change [13]. However, such effects have been observed in visual perception. For instance, faces appear to be more average after prolonged viewing or colors fade after prolonged exposure [13]. It was proposed that this kind of after-effects arise in norm-based coding, i.e. the perceptual dimension is represented in the activity of two rather broadly tuned channels (matched to opposite poles of the perceptual dimension e.g. male vs. female in facial perception) and the information is encoded as the difference to a unique neutral point (equal activation of both channels) [13]. In such representation, if adaptation affects the sensitivity of one channel more than the other the neutral point shifts towards the adaptation stimulus. Consequently a stimulus with the same properties as the adaptation stimulus appears closer to the neutral point after adaptation [13]. In the case of softness perception, according to this model, it could be assumed that one channel codes the property "hard" and the other channel codes the property "soft". Adaptation to a hard stimulus would consequently shift the neutral point to a harder value. If the same stimulus was presented again, it would be closer to the shifted neutral point and thus be perceived to be softer, as found by [12].

Further, [8] showed that the point at which an object was categorized 50% "hard" and 50% "soft" was roughly the same as the measured compliance of the human fingertip-at least on average. The authors speculated, that the softness of one's own fingertip might be the neutral point in the perception of softness. However, the neutral point identified by [8] was roughly the middle of the presented stimulus set. Hence, it is not clear whether it actually reflects the neutral point or is created by task demands, i.e. the strategy to choose the middle of the stimulus range when being asked to assign the stimuli to two categories.

Here we test the speculation of [8] that there is a neutral point in the perception of softness. We created a wide range of stimuli, ranging from very hard (stone) to very soft (very soft foam) and presented three groups of participants each with a different stimulus set: the full set consisting of 18 stimuli; the soft set containing the softest 13 stimuli or the hard set containing the hardest 13 stimuli. This way, the stimuli overlap between the conditions, but each set had a different middle stimulus. In every trial, participants touched one stimulus and categorized it either as "hard" or "soft". From participants' responses we estimated the point of subjectively equal categorization as "hard" or "soft" (we refer to it as PSE in the following). We expected that if results are merely driven by participants' strategy, the PSEs would be each in the middle of the corresponding stimulus set. However, if participants would use the softness of their finger to decide whether a stimulus is hard or soft we could expect to find similar PSEs across set conditions. We also looked at how PSEs develop over the course of the experiment, as this could reflect learning. We found that PSEs change with the range of the stimulus set but they are different from the middle stimulus. Also, during the course of the experiment PSEs moved away from the middle of the stimulus set although participants had more and more information to identify the middle stimulus. Our results suggest that there is a perceptual neutral point of softness that does not fully depend on the range of the stimulus set.

2   Methods

2.1   Participants

24 volunteers (10 males, mean age 26 years, range 19–62 years) participated in the experiment. They were naive to the purpose of the experiment and were refunded. None of them reported any sensory or motor impairment of the dominant hand and arm. Written informed consent was obtained from each participant. The study was approved by the local ethics committee at Justus-Liebig University Giessen LEK FB06 (SFB-TPA5, 22/08/13) and was in line with the declaration of Helsinki from 2008.

2.2   Apparatus

The experiments were performed at a visuo-haptic workbench, which comprised a PHANToM 1.5A haptic force feedback device, a 22"-computer screen (120 Hz, 1280 × 1024 pixel), stereo glasses, a mirror and a force sensor consisting of a measuring beam (LCB 130) and a measuring amplifier (GSV-2AS, resolution 0.05 N, temporal 682 Hz). The participants sat in front of the workbench and the stimuli were placed in front of them on the force sensor. The mirror prevented direct sight on the experimental setup and the stimuli, and aligned them with their virtual visual representation. To guide the participants through the experiment, a virtual schematic 3D representation of the set up and the finger was displayed on the monitor, which was placed above the mirror. The index finger of the dominant hand was connected to the PHANToM arm using a custom-made adapter, leaving the finger pad uncovered and allowing free movement.

2.3 Stimuli

Stimuli were created from different materials to create a wide continuous elasticity range. We used differently soft foam for the two softest stimuli and fibreboard and ceramic for the two hardest stimuli. Stimuli with intermediate softness were produced by mixing a two-component silicone rubber solution (Alpa Sil EH 10:1) with varying amounts of a silicone oil (polydimethylsiloxane, viscosity 50 mPas). After mixing, the solutions were poured into cylindrical plastic dishes (75 mm diameter × 38 mm high) to harden. In order to prevent that the stimuli could be distinguished by their surface, all 18 stimuli were covered with pieces of cotton cloth. We measured the elasticity of the stimuli by following the standard methodology proposed by [14]. To obtain standardized samples of the same material for the measurement, a portion of the mixed solutions was poured into small cylinders (10 mm thick, 10 mm in diameter) and the same shapes were cut out from foam. The small stimulus samples were placed on the force sensor and intended by the force feedback device. For this purpose, instead of the fingertip adapter an aluminium plate (24 mm diameter) was attached on the PHANToM arm. The applied force was increased by 0.005 N every 3 ms until a minimum force of 1 N and a minimum displacement of 1 mm were detected. The resulting force and displacement data were converted into stress-strain data. A linear Young's modulus was fitted to each stress-strain curve (only for 0–0.1 strain) in MATLAB R2017. [4]. The elasticity of the two hardest stimuli could not be measured with the same method. However, it is reasonable to assume that their elasticity is highest and that they would constitute the upper asymptote in the psychometric function, thus a precise measurement of their elasticity could be neglected. The Young's moduli of the stimuli were: >200 for the hardest two stimuli, 199, 169, 141, 133, 127, 108, 102, 86, 81, 63, 57, 48, 43, 34, 3 and 0.8 kPa. The hard set ranged between >200 and 57 kPa and the soft one between 133 and 0.8 kPa.

2.4 Design and Procedure

The experiment was conducted using a between-subject design with three stimulus set conditions: *full* set containing all stimuli, *hard* set containing the hardest 13 stimuli, *soft* set, containing the softest 13 stimuli. Eight participants were assigned to each condition. The experiment consisted of 5 blocks. Each stimulus was presented twice in each block and overall 10 times in randomized order. This resulted in overall 180 trials in the full set condition, and 130 trials in the reduced set conditions. The independent variable was the presented stimulus set and the dependent variable was the PSE estimated from participants' judgments.

In each trial participants were presented with a stimulus. They could explore the softness of the stimulus as long as they wanted by using the index finger of their dominant hand. The task of the participants was to assign each stimulus either to the category "soft" or "hard". Participants indicated their response by moving their finger to the corresponding word in the virtual scene. Between the trials, the finger rested in the left corner of the work space. Each experimental session lasted about 30–45 min.

2.5   Analysis

Each participant's answers composed individual psychometric data, to which a cumulative Gaussian function was fitted using the psignifit 4 toolbox [15]. These functions represent the relationship between the Young's modulus and the likelihood of a "hard" response. The PSE (μ of the cumlulative function) was estimated as the Youngs's modulus at which the answers were 50% "hard" and 50% "soft".

3   Results

Figure 1 shows average PSEs for the three stimulus set conditions. A one-way ANOVA performed with the factor set condition and the dependent variable PSE showed a statistically significant influence of the set condition on the PSEs, $F(2,21) = 26.77$, $p < 0.001$. Post-hoc tests for pairwise comparisons (Bonferroni-corrected for 3 comparisons, $\alpha = .017$) revealed that the PSE in the hard condition significant differed from the other two conditions. The difference in PSEs between condition *full* and *soft* was not statistically significant.

In order to investigate the dependence of the responses on the stimulus set, we compared the average PSE in each condition to the Young's modulus of the middle stimulus of set in this condition (Fig. 1). The two-tailed t-tests performed for each condition showed a statistically significant difference between the average PSE and the Young's modulus of the middle stimulus in the *full* set condition, $t(7) = -6.42$, $p < 0.001$ and the *hard* set condition, $t(7) = -4.84$, $p = 0.002$. The difference in the soft set condition was not significant, $t(7) = 0.683$, $p = 0.517$.

In addition, we investigated how participants' answers changed over the course of the experiment. Figure 2 clearly shows that the PSEs shift towards softer values over the course of the experiment, which also implies a shift away from the middle stimulus of the set, in all three conditions. To check the statistical significance of this shift, we compared the average PSEs of the first half of the experiment and the last one in each condition with a two-tailed t-test. There was a statistically significant difference between the average PSEs of the two blocks in the *full* set condition, $t(7) = 2.84$, $p = 0.025$. In the *soft* set conditions the

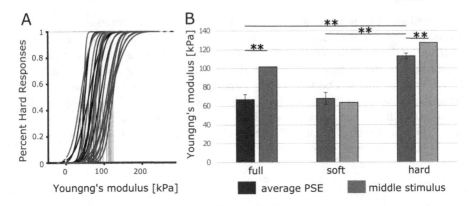

Fig. 1. A Individual psychometric functions. textbfB Average PSEs and the Youngs's modulus of the middle stimulus for each stimulus set. The error bars represent the standard error. ** $p < 0.01$

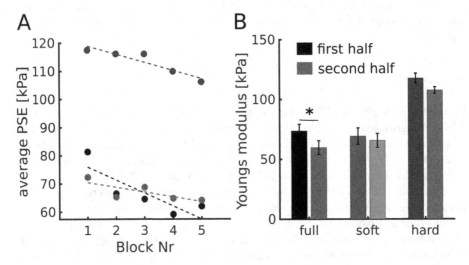

Fig. 2. A Development of the PSEs over the course of the experiment. Trials of each participant were divided into five blocks of equal size and PSEs were determined for each block by averaging the responses across participants for each block and each stimulus set condition. Dotted lines indicate regression lines. **B** Average PSEs in the first and second half of the experiment. For statistical analyses for each participant PSEs were estimated based on the trials in the first half of the experiment and the second half of the experiment. Error bars indicate the standard error, * $p < 0.05$

differences in average PSEs between the two halves was not significant, $t(7) = 1.00$, $p = 0.349$, and at the edge of significance in the *hard* set condition $t(7) = 2.19$, $p = 0.065$.

4   Discussion

Here we tested whether there is a neutral point in the perception of softness. We found that measured neutral points (i.e. PSEs) changed with the range of the stimulus set but they were different from the middle stimulus of the set even if participants had more time to learn the range of variability in each stimulus set. This showed that the results of [8] could not be explained just by the strategy of participants to choose the middle stimulus. However, the neutral points seems to change with the stimulus set. This could be due to range and adaptation effects.

Range effects describe distortions in people's reactions depending on the range of stimuli presented in the experiment [16]. This bias often includes a central tendency. However, it was shown that range effects can be avoided if the stimulus range corresponds to the range of lifelong experience [17]. Indeed, we found the largest deviation of the measured neutral point from the middle stimulus towards softer values in the full set condition, which included almost the entire reference system of lifelong perceived softness ranging from hard ceramic to very soft foam. This result is consistent with the idea that there is a unique neutral point in softness perception which is less distorted by range effects, if the stimulus range is wide enough. In future, a task in which the range of the stimulus set is less obvious, e.g. adjustment method, could be used to measure the neutral point.

It was shown that softness is susceptible to adaptation and adaptation effects suggest norm-based coding [12], similar to the representation of colours and faces. In such a representation adaptation would shift the neutral point towards the middle of the current stimulus range. However, we observed a drift away from the center of the set in the course of the experiment and towards softer values in each condition. This indicates that in the time range of our experiment adaptation played rather a minor role if any. Adaptation might be more important for short time ranges like in [12] or longer time ranges, e.g. adapting to the changing skin of one's own finger with increasing age (cf. adaptation to age-related changes in lens pigment density [13]). It is a possibility that the drift of the measured neutral points in the different conditions rather reflects convergence to the unique neutral point.

It was speculated that the softness of one's own finger might be the neutral point in the perception of softness [8]. Objects which deform the finger are classified as "hard" and objects which are deformed by the finger are classified as "soft". The elasticity of the fingertip differs between people [18]. It can be speculated that the neutral point correlated with finger elasticity. In future, this relationship could be tested. Additionally, it would be interesting to manipulate the elasticity of the finger experimentally and measure the effect on the neutral softness.

Taken together our results indicate that there might be a neutral point in the representation of haptically perceived softness, however range effects may play a role.

References

1. Okamoto, S., Nagano, H., Yamada, Y.: Psychophysical dimensions of tactile perception of textures. IEEE Trans. Haptics **6**(1), 83–91 (2012)
2. Tiest, W.M.B., Kappers, A.M.L.: Analysis of haptic perception of materials by multidimensional scaling and physical measurements of roughness and compressibility. Acta Psychol. **121**(1), 1–20 (2006)
3. Srinivasan, M.A., LaMotte, R.H.: Tactual discrimination of softness. J. Neurophysiol. **3**(1), 88–101 (1995)
4. Harper, R., Stevens, S.S.: Subjective hardness of compliant materials. Q. J. Exp. Psychol. **16**(768984023), 204–215 (1964)
5. Cavdan, M., Doerschner, K., Drewing, K.: Task and material properties interactively affect softness explorations along different dimensions. IEEE Trans. Haptics **14**(3), 603–614 (2021)
6. Lederman, S.J., Klatzky, R.L.: Hand movement: a window into haptic object recognition. Cogn. Psychol. **19**, 342–368 (1987)
7. Tiest, W.M.B., Kappers, A.M.L.: Cues for haptic perception of compliance. IEEE Trans. Haptics **2**(4), 189–199 (2009)
8. Friedman, R.M., Hester, K.D., Green, B.G., LaMotte, R.H.: Magnitude estimation of softness. Exp. Brain Res. **191**(2), 133–142 (2008). https://doi.org/10.1007/s00221-008-1507-5
9. Matsui, K., Okamoto, S., Yamada, Y.: Relative contribution ratios of skin and proprioceptive sensations in perception of force applied to fingertip. IEEE Trans. Haptics **7**(1), 78–85 (2014)
10. Metzger, A., Drewing, K.: Haptically perceived softness of deformable stimuli can be manipulated by applying external forces during the exploration. In: IEEE World Haptics Conference, WHC 2015, pp. 75–81. Institute of Electrical and Electronics Engineers Inc. (2015)
11. Metzger, A., Drewing, K.: Memory influences haptic perception of softness. Sci. Rep. **9**, 14383 (2019)
12. Metzger, A., Drewing, K.: Haptic aftereffect of softness. In: Bello, F., Kajimoto, H., Visell, Y. (eds.) Haptics: Perception, Devices, Control, and Applications. LNCS, vol. 9774, pp. 23–32. Springer, Cham (2016). https://doi.org/10.1007/978-3-319-42321-0_3
13. Webster, M.A.: Adaptation and visual coding. J. Vis. **11**(5), 1–23 (2011)
14. Gerling, G.J., Hauser, S.C., Soltis, B.R., Bowen, A.K., Fanta, K.D., Wang, Y.: A standard methodology to characterize the intrinsic material properties of compliant test stimuli. IEEE Trans. Haptics **11**(4), 498–508 (2018)
15. Schuett, H.H., Harmeling, S., Macke, J.H., Wichmann, F.A.: Painfree and accurate Bayesian estimation of psychometric functions for (potentially) overdispersed data. Vis. Res. **122**, 105–123 (2016)
16. Poulton, E.C.: Range effects in experiments on people. Am. J. Psychol. 3–32 (1975)
17. Müller, F., Giesecke, D.: Lautheitskonstanz oder Range-Effekt? Ein Experiment zur Differenzierung zwischen Wahrnehmung und Skalierungseffekt (2012)
18. Xu, C., Wang, Y., Gerling, G.J.: Individual performance in compliance discrimination is constrained by skin mechanics but improved under active control. In: IEEE World Haptics Conference, pp. 445–450 (2021)

Pilot Study on Presenting Pulling Sensation by Electro-Tactile Stimulation

Shota Nakayama$^{(\boxtimes)}$, Mitsuki Manabe, Keigo Ushiyama, Masahiro Miyakami, Akifumi Takahashi, and Hiroyuki Kajimoto

The University of Electro-Communications, 1-5-1 Chofugaoka, Chofu, Tokyo, Japan
{nakayama,manabe,ushiyama,miyakami,a.takahashi, kajimoto}@kaji-lab.jp

Abstract. When an object that is grasped with a finger is pulled by an external force, the traction force is perceived by cutaneous receptors and proprioception in the finger. Several attempts have been made to simulate the pulling sensation by using wearable devices, including mechanical asymmetric vibration and tightening by belt. In this study, we developed a new method that uses electrical simulation to generate an illusory force sensation by simulating the activity pattern of the cutaneous receptors. We validated our method through two experiments, one based on force direction judgment and the other on force magnitude adjustment.

Keywords: Electrical stimulation · Force sensation · Sensory illusion

1 Introduction

Compared to desktop type haptic displays, wearable type haptic displays are not particularly good at presenting external force. To solve this problem, numerous methods that create illusory phenomena by using skin sensation to present force sensation have been proposed.

A typical technique is the use of asymmetric vibration [1–4]. When a weight is vibrated such that it is driven quickly in the forward direction and slowly in the reverse direction, the illusion of being pulled is generated on the hand grasping the transducer. Another typical technique is the use of skin compression [5, 6]. A common method is belt tightening of the finger pad by two motors, which can present the sensation of the finger pad being pressed or the finger sliding sideways. The former technique involves a strong vibration sensation that spreads over the entire hand, whereas the latter requires a large mechanism to be attached around the finger.

We propose a method to overcome these problems by using a device that presents an illusory force sensation through electrical stimulation. It can be fabricated to be small and thin and does not involve transmitting a vibration sensation to the whole hand. We validated our method through two experiments, one based on force direction judgment and the other on force magnitude adjustment.

A. Takahashi—JSPS Research Fellow.

H. Seifi et al. (Eds.): EuroHaptics 2022, LNCS 13235, pp. 66–74, 2022.
https://doi.org/10.1007/978-3-031-06249-0_8

2 Method

2.1 Electrical Stimulation Device

Electrical stimulation was performed using the electrical stimulator developed by Kajimoto [7]. This stimulator is divided into a control unit that determines the current and stimulation pattern, and an electrode unit that consists of electrodes and switching circuits. The control unit is connected to a PC through a USB connection.

In the electrode unit (Fig. 1(a)), electrodes are attached to the top and bottom of a small box (4 cm × 3 cm × 1 cm, Fig. 1(b)). Sixty-three (7 × 9) circular electrodes (1.4 mm in diameter) are placed on one electrode board at 2 mm center-to-center intervals. The weight of the complete grasping part is 17 g. The maximum current for electrical stimulation is 6 mA.

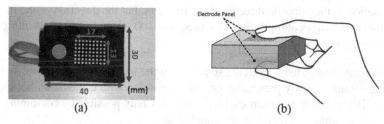

(a)   (b)

Fig. 1. (a) Electrode unit. (b) Structure of the grasping part.

2.2 Stimulus Pattern

The shallow part of finger skin contains mechanoreceptors called Meissner corpuscles and Merkel cells. Meissner corpuscles are involved in the recognition of object contact and surface texture, whereas Merkel cells identify the pressure sensation [8]. Anodic and cathodic stimulations are two types of electrical stimulations. Anodic stimulation mainly produces vibratory sensation, wherein the stimulating electrode is the anode, and the surrounding electrodes are the cathodes. In contrast, cathodic stimulation mainly produces pressure-like sensation. This is probably because the former tends to stimulate the nerves connected to Meissner corpuscles while the latter tends to stimulate the nerves connected to Merkel cells [9]. This suggests that when an external force is applied to a finger, continuous cathodic stimulation can present the sensation of the finger being pressed against an object. Furthermore, when the finger is in contact or detached, brief anodic stimulation can present the sensation of contact or detachment from the object.

Based on these considerations, we speculated that it was possible to present an illusory force sensation in the intended direction by applying cathodic stimulation to the electrode, and performing anodic or cathodic stimulation for a short period at the beginning and end of the stimulation. In addition, since both aforementioned methods produce stronger illusions at the beginning of stimulation than in steady state, we speculated that we could generate clearer illusory force sensation by repeatedly turning them on and off.

Through trial and error, we discovered a stimulus pattern, shown in Fig. 2, that can be expected to produce an illusory force in the intended direction from the back electrode to the front. The horizontal axis is the elapsed time, and the vertical axis is the value of command current. The discovered pattern consists of the following.

Cathodic Stimulation for 400 ms: The cathodic stimulation for 400 ms produces a pressure sensation on one finger. This is the main stimulus to generate the illusory force sensation.

Electrical Stimulation at the Beginning and End of Stimulation: For every 50 ms of stimulus onset and 50 ms of stimulus termination, anodic stimulation is performed on the front surface of the finger. This is expected to produce a situation wherein the front surface of the finger is tapped at the moment of traction. Conversely, cathodic stimulation is performed on the back side of the finger. This stimulation produces an illusory force sensation in the opposite direction for a moment, but the direction of the illusory force sensation changes abruptly in the subsequent 400 ms stimulation, resulting in a more enhanced illusory force sensation.

In this preliminary study, the force sensation in the intended direction was not sufficiently generated by only presenting pressure sensation with cathodic stimulation to one finger. Perceiving it as a clear external force was only possible by combining both stimuli at the beginning and end of the stimulation.

Sato et al. proposed and implemented a method for expressing the sense of contact, edge, and direction of force, by combining the cathodic and anodic stimuli [10]. Our proposed method can be considered as an attempt to generate illusory force sensation by applying this method to the action of pinching with two fingers.

The stimulating electrodes are shown in Fig. 3. The black points were stimulated on both sides 60 times per second (60 pulses per second (PPS)). We reduced the number of stimulation points owing to power and refresh-rate limitations.

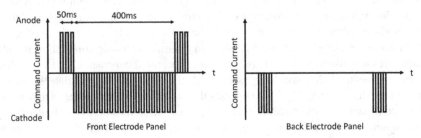

Fig. 2. Proposed stimulus pattern.

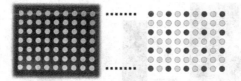

Fig. 3. Electrodes to be stimulated (black points were stimulated).

3   Experiment 1

The purpose of this experiment was to evaluate whether it is possible to create the illusion of traction force in the intended direction through the proposed method.

Ten subjects (including four authors), aged 21 through 27, participated in the experiment (Fig. 4(a)). Each subject sat at a desk and held the electrode unit in their right hand. They were instructed to pinch the electrode unit with their index finger on the front surface and the thumb on the back surface. To avoid the effect of moisture, they were instructed to wipe off the sweat from their fingers during the experiment [11]. The experiments were approved by the Ethics Committee of the University of Electro-Communications, Chofu, Tokyo, Japan.

3.1   Experimental Procedure

Subjects were instructed to pinch the electrode unit. While presenting the cathodic stimulus on both sides, the command current value was gradually increased until the subject felt pain. Thereafter, the command current value was lowered and adjusted to the maximum command current value at which the subject did not feel uncomfortable. In addition, we applied the stimulus pattern shown in Fig. 2, without divulging to the subject that it was an experimental pattern, and confirmed that it did not cause discomfort.

The stimulus pattern shown in Fig. 2 was intended to produce an illusory force sensation on the front side of the electrode where the index finger was placed (hereafter referred to as forward stimulus). By switching the stimulus pattern of the front and back electrodes, the illusory force sensation was produced on the back side of the electrode where the thumb was placed (hereafter called the backward stimulus). We presented either of these two stimuli patterns, and in a two-alternative forced choice asked the participants to choose the direction in which they felt the "traction force." The same stimulus pattern was repeated at a frequency of 1 Hz with an interval of 500 ms, until the participants answered. These trials were repeated ten times for each pattern in a random order, for a total of 20 trials. During the trials, the subjects were instructed to hear pink noise on headphones and close their eyes. They were asked to answer the following questions on a 5-point Likert scale (1: not at all, 5: very much).

- Did you feel as if you were being pulled from the outside? (Fig. 4(b) Pull)
- Did you feel as if you were being pushed from the inside? (Fig. 4(b) Push)
- Did you feel as if you were being sucked from the inside? (Fig. 4(b) Suck)
- Did you feel a clear difference between the two stimulus conditions? (Difference)

After the experiment, the participants were asked to voice their opinions freely.

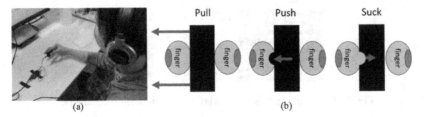

Fig. 4. (a) Experimental environment. (b) Definition of force sensation.

3.2 Experimental Result

The results of the experiment are shown in Fig. 5(a). The vertical axis shows the overall correct response rates for the forward and backward stimuli, and the error bars represent the standard errors among subjects. A t-test revealed that there was a significant difference from the chance rate (50%) at 5% level (p = 0.003 for the front side and p = 0.047 for the back side).

The swarm and violin plots of the answers to the questionnaire are shown in Fig. 5(b). The horizontal axis shows the questionnaire items, and the vertical axis shows the responses. The dashed lines indicate the quartiles.

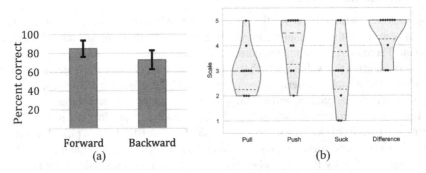

Fig. 5. (a) Experimental result. (b) Questionnaire.

3.3 Discussion

The force sensation was largely generated in the direction we intended. One subject interpreted both front and back stimuli in the opposite directions, whereas three subjects interpreted the forward stimulus almost correctly but tended to interpret the backward stimulus as forward. Furthermore, there was no subject who interpreted the backward stimulus correctly or the forward stimulus as backward. This is presumably because the current threshold was moderately higher in the thumb than in the index finger, and the electrode board used in this study could not cover the thumb completely. Therefore, some subjects might have answered without feeling a clear tactile sensation in the thumb.

Figure 5(b) shows that most subjects felt the force sensation of being pushed from inside but not being pulled. Because a typical asymmetric vibration imparts the sensation of being pulled from the outside, the quality of sensation appears to be different and might be insufficient. Considering that the asymmetric vibration incorporates not only skin surface vibration but also joint and deep tissue vibrations, it might be necessary to appropriately stimulate the muscle spindles and Golgi tendon organs related to the fingertips.

4 Experiment 2

The purpose of this experiment was to quantitatively measure the maximum illusory force generated by electrical stimulation, and to compare it with asymmetric vibrations, considering Rekimoto's method [2] as an example of a similar small device. Nine males and one female (including four of the authors), aged 21 through 27, participated in the experiment.

The experimental environment is shown in Fig. 6(a). The subject was seated, and electrical stimulus or vibration was imparted to the left hand, while a physical force was imparted to the right hand. As shown in Fig. 6(b), the physical pulling force was imparted by a string and a pulley with a suspended weight.

The asymmetric vibration was presented by using a short-vibration feedback device (Force Reactor, Alps Alpine). The vibration waveform was a square wave of 2 ms:6 ms, which was found to generate the strongest illusory force sensation by Rekimoto [2]. The drive voltage was 5 V, the absolute maximum rating, and the vibration was repeated for 500 ms with a period of 1 Hz to obtain a similar stimulation pattern as the electric stimulation.

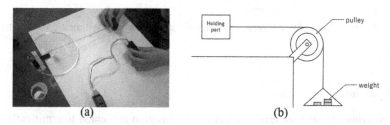

Fig. 6. (a) Experimental environment. (b) Physical force presentation mechanism.

4.1 Experimental Procedure

The subjects first experienced the vibration and the electric stimulus in order. They held the grasping part in one hand, and the stimulation was performed for approximately one minute. During the stimulation, subjects were told the intended direction of the illusory force sensation. After the experience, they were asked to confirm if the illusory force sensation was generated. This time, all the participants felt the illusory force. Then, the following two measurements were performed. The order of the measurements was counterbalanced.

During the measurement, the subjects were instructed to keep their arms in a floating position above the desk and not move them away from the desk. They then verbally instructed the experimenter to adjust the weight, to obtain the subjective point of equivalence (PSE). The weight was adjusted in 1 g increments. The measurements were repeated three times and the median value was considered as the measured value.

4.2 Experimental Result

Figure 7 is a slope chart of the measurement results for each subject. The average values of the electric stimulus and asymmetric vibration were 33.2 gf and 45.5 gf, respectively. A t-test revealed a significant difference between the two methods (p = 0.006).

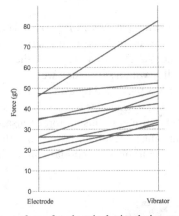

Fig. 7. Strength of illusory force for electrical stimulation and asymmetric vibration.

4.3 Discussion

In this experiment, we compared illusory force sensation presented intermittently with a physical force presented continuously. The illusory force sensation was presented intermittently because presently it is difficult to present a continuous illusory force sensation through electrical stimulation. As a result, a few subjects took a relatively long time to obtain the PSE; eventually all subjects were able to obtain it. In addition, since the illusory force sensation was presented intermittently with the same period for both the electric stimulus and vibration, we believe a that comparison between the two methods is fair.

Regarding the vibration presentation, Rekimoto [2] measured that the average illusory force sensation was 29.8 gf, while our measurement obtained a value of 45.5 gf. The reason for this difference might be that our stimulus pattern was intermittent; hence, the sensory adaptation did not occur. The vibration stimulus produced a larger force sensation than the electric stimulus for all the subjects. As mentioned in 3.3, this may be because of the fact that deep tissues were stimulated by the propagation of vibration but not by the electrical stimulation.

5   Conclusion

This paper examined the possibility of generating illusory force sensation by simulating the activity pattern of receptors, during the action of object grasping, through electrical stimulation. We used electrodes on two sides of a box, front and back, to provide electrical stimulation to the thumb and index finger when grasping the box with them.

As a result, it was established that the proposed method can generate illusory force sensation in a designated direction. However, the quality of the force sensation was different from the expected traction sensation, and the sensation of being pushed from the inside was dominant. Quantitative measurement of the presented force showed that the force sensation was close to that of asymmetric vibration.

In future work, by focusing on the spatiotemporal distribution of skin deformation, we intend to investigate stimulus patterns that produce stronger traction illusion by focusing on the spatiotemporal distribution of skin deformation.

Acknowledgements. This work was supported by JSPS KAKENHI Grant Number JP18H04110.

References

1. Amemiya, T., Ando, H., Maeda, T.: Virtual force display: direction guidance using asymmetric acceleration via periodic translational motion. In: First Joint Eurohaptics Conference and Symposium on Haptic Interfaces for Virtual Environment and Teleoperator Systems. World Haptics Conference, pp. 619–622, March 2005
2. Rekimoto, J.: Traxion: a tactile interaction device with virtual force sensation. In: ACM SIGGRAPH 2014 Emerging Technologies, New York, NY, USA, p. 1, July 2014
3. Culbertson, H., Walker, J.M., Raitor, M., Okamura, A.M.: WAVES: a wearable asymmetric vibration excitation system for presenting three-dimensional translation and rotation cues. In: Proceedings of the 2017 CHI Conference on Human Factors in Computing Systems, New York, NY, USA, pp. 4972–4982, May 2017
4. Choi, I., Culbertson, H., Miller, M.R., Olwal, A., Follmer, S.: Grabity: a wearable haptic interface for simulating weight and grasping in virtual reality. In: Proceedings of the 30th Annual ACM Symposium on User Interface Software and Technology, New York, NY, USA, pp. 119–130, October 2017
5. Minamizawa, K., Prattichizzo, D., Tachi, S.: Simplified design of haptic display by extending one-point kinesthetic feedback to multipoint tactile feedback. In: 2010 IEEE Haptics Symposium, pp. 257–260, March 2010
6. Chinello, F., Malvezzi, M., Pacchierotti, C., Prattichizzo, D.: A three DoFs wearable tactile display for exploration and manipulation of virtual objects. In: 2012 IEEE Haptics Symposium (HAPTICS), pp. 71–76, March 2012
7. Kajimoto, H.: Electro-tactile display kit for fingertip*. In: 2021 IEEE World Haptics Conference (WHC), p. 587, July 2021

8. Vallbo, A.B., Johansson, R.S.: Properties of cutaneous mechanoreceptors in the human hand related to touch sensation. Hum. Neurobiol. **3**(1), 3–14 (1984)
9. Yem, V., Kajimoto, H.: Comparative evaluation of tactile sensation by electrical and mechanical stimulation. IEEE Trans. Haptics **10**(1), 130–134 (2017)
10. Sato, K., Tachi, S.: Design of electrotactile stimulation to represent distribution of force vectors. In: 2010 IEEE Haptics Symposium, pp. 121–128, March 2010
11. Kaczmarek, K.A., Tyler, M.E., Bach-Y-Rita, P.: Electrotactile haptic display on the fingertips: preliminary results. In: Proceedings of 16th Annual International Conference of the IEEE Engineering in Medicine and Biology Society, vol. 2, pp. 940–941, November 1994

A Preliminary Study on the Perceptual Independence Between Vibrotactile and Thermal Senses

Jaejun Park[1], Jeongwoo Kim[1], Chaeyong Park[1], Seungjae Oh[1], Junseok Park[2], and Seungmoon Choi[1]([✉])

[1] Pohang University of Science and Technology, Pohang, Republic of Korea
{jjpark17,jwkim0417,pcy8201,oreo329,choism}@postech.ac.kr
[2] Electronics and Telecommunications Research Institute, Daejeon, Republic of Korea
parkjs@etri.re.kr

Abstract. We study whether the vibrotactile and thermal senses are independent in terms of information transmission. In Exp. 1, we estimated the respective information transmission capacities of vibrotactile and thermal stimuli. In Exp. 2, we measured the information transfer (IT) of vibrotactile-thermal multimodal stimuli. We compare the IT values obtained in the two experiments and demonstrate that approximately 90% of the information encoded through the two sensory modalities is preserved when they are combined. This result can contribute to the design of multimodal haptic stimuli for various user-interactive purposes.

Keywords: Information theory · Vibrotactile · Thermal · Multimodal

1 Introduction

The essential role of tactile devices is to deliver information, and the information is often abstract and categorical [2,6]. Research has sought effective methods in improving the amount of information that can be transferred by tactile communication. In this paper, we report a preliminary study that quantifies the extent to which the information transmission capability is preserved when two different types of tactile stimuli, vibrotactile and thermal, are factorially combined into multimodal stimuli.

A large number of previous studies measured the information transfer (IT) for vibrotactile stimuli. When a single vibration actuator is used, the highest IT value reported is 3.06 bits [17]. When multiple vibration actuators are used, the highest IT value reported is 7.02 bits [7]. In comparison, the IT values reported for thermal stimuli are rare. The highest IT value reported for the identification

This study was supported by a grant 21ZS1200, Fundamental Technology Research for Human-Centric Autonomous Intelligent Systems, from the Electronics and Telecommunications Research Institute (ETRI) of Korea.

J. Park and J. Kim—Equally contributed to this work.

H. Seifi et al. (Eds.): EuroHaptics 2022, LNCS 13235, pp. 75–83, 2022.
https://doi.org/10.1007/978-3-031-06249-0_9

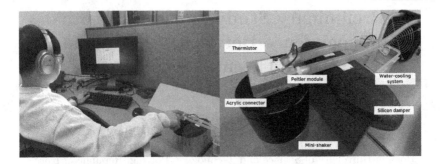

Fig. 1. Experiment setup (left) and apparatus for stimulation (right).

of thermal patterns is 2.13 bits [11]. Furthermore, there have been attempts to test the IT for multimodal stimuli. Shim et al. [9] measured the IT values for the simultaneous and sequential patterns of vibration and wind. The IT values were 1.70 bits (simultaneous) and 3.29 bits (sequential), where the IT values for individual vibration and wind were 1.76 bits and 1.23 bits, respectively. A recent comprehensive review on the IT of tactile displays is available in [13].

Presenting vibrotactile and thermal stimuli simultaneously is a promising approach to enriching tactile interaction [14, 16]. If the two tactile channels are sufficiently independent, vibrotactile-thermal stimuli may offer a substantially greater information transmission capacity (ITC). This research hypothesis was investigated by two perceptual experiments. In Exp. 1, we measured the individual ITCs of the vibrotactile and thermal channels. Only frequency was varied for the vibrotactile stimuli, and the number of stimuli was increased from 3 to 7. For the thermal stimuli, temperature change direction and rate were controlled. The number of stimuli was between 5 and 9. In Exp. 2, we factorially combined the optimum number of vibrotactile and thermal stimuli to make multimodal stimuli and obtained their estimated IT values (IT_{est}). Then, we examined whether the unimodal IT_{est}s and the multimodal IT_{est} satisfy the additivity, which means the complete perceptual independence of the two channels.

Our study can contribute to 1) the decision of optimal vibrotactile or thermal stimuli for user interaction purposes, 2) the design of vibrotactile-thermal stimuli that are highly recognizable for information and communication purposes.

2   Exp. 1: Unimodal Stimuli

We estimated the ITCs for both vibrotactile and thermal stimuli. The experiments reported in this paper were approved by the Institutional Review Board at POSTECH (PIRB-2021-E054).

2.1   Methods

Apparatus. We implemented a device for vibrotactile and thermal stimulation (Fig. 1). Its vibrotactile module consisted of a mini-shaker (Bruel & Kjær,

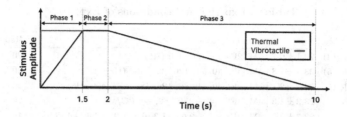

Fig. 2. Description of vibrotactile and thermal stimuli.

4810) and a power amplifier (Bruel & Kjær, 2718). The thermal module included a Peltier module ($20 \times 20 \times 3.1$ mm; MULTICOMP, MCPE1-07106NC-S), three DC motor drivers, a thermistor (SEMITEC, 223Fu3122-07U015), a microcontroller (Arduino Due), and a water-cooler. The temperature change rate of the Peltier module was controlled within ± 4 °C/s (0.1 °C average error) by PID control. The thermal module was put on the vibrotactile module using an acrylic connector. To block vibration propagation from the water-cooler, a silicon damper attenuating noise over 95% was placed between the tactile module and the cooler.

Experimental Conditions. Figure 2 depicts the time profiles of vibrotactile and thermal stimuli. We designed three conditions for vibrotactile stimuli with the number of stimuli $N = 3$, 5, and 7 (see Table 1). The frequencies of the vibrotactile stimuli were varied from 40 400 Hz in equidistant intervals on a logarithmic scale, stimulating RA 1 and PC channels [1]. The adjacent frequency differences were larger than the vibrotactile Just Noticeable Difference (JND) (approximately 18% [8]). The intensity of each stimulus was scaled to be 33 dB from the detection threshold on the finger at its frequency [3]. All of the vibration stimuli had the same duration of 1.5 s.

We also designed three conditions for thermal stimuli. Their baseline temperature was 32 °C. Considering the heat pain threshold, we varied the temperature change rate within ± 4 °C/s [15]. Given the number of stimuli $N = 5$, 7, and 9, the change rates were evenly spaced in a linear scale, as shown in Table 1. The temperature changes from the baseline were higher than the difference thresholds for both warm and cool stimuli ($+0.23$ and -0.14 °C at 33 °C [12]).

The thermal stimuli had three phases. In the beginning, the temperature changed to the target temperature at a given change rate over 1.5 s. In the second phase, the temperature remained for 0.5 s for stability. Then, the temperature returned slowly to the baseline temperature over 8 s.

Participants. The experiment had a between-subjects design. Each experimental condition was tested with seven participants each (see Table 1 for their demographics). No participants participated in the conditions of the same stimulus type. Before the experiment, the participants were informed of the experiment's goals and procedure via a written document, and then they signed a consent form. They were paid at the rate of KRW 10,000 ($\approx$ USD 9) per hour.

Table 1. Experimental conditions of Exp. 1

Condition	Age	Gender	Stimulus	Repetitions
Vibrotactile (3)	22.1 ± 0.7	5M, 2F	40, 126, 400 (Hz)	20
Vibrotactile (5)	23.4 ± 1.8	4M, 3F	40, 71, 126, 225, 400 (Hz)	20
Vibrotactile (7)	24.7 ± 2.1	5M, 2F	40, 59, 86, 126, 186, 273, 400 (Hz)	20
Thermal (5)	23.3 ± 1.3	4M, 3F	$-4, -2, 0, +2, +4$ (°C/s)	20
Thermal (7)	23.7 ± 3.0	4M, 3F	$-4, -2.6, -1.3, 0, +1.3, +2.6, +4$ (°C/s)	16
Thermal (9)	24.1 ± 2.4	7M	$-4, -3, -2, -1, 0, +1, +2, +3, +4$ (°C/s)	12

Procedure. Participants put their right index fingertip on the Peltier module. Note that vibration stimuli were presented through the Peltier module (Fig. 1). They manipulated a experiment program using their left hand. They wore noise-canceling headphones which played white noise to block environmental sounds.

The thermal conditions had additional constraints: 1) The room temperature was controlled to 25 °C; 2) The Peltier module maintained its temperature to 32 °C when there was no stimulus; 3) Participants were asked to put their fingertips on the Peltier module for 2 s before stimulus onset to make their contact skin temperature to 32 °C; and 4) They were also asked to take off their fingers from the Peltier module after receiving 1.5 s of thermal stimulus.

To complete each experimental condition, participants went through five sessions: training 1, practice 1, training 2, practice 2, and main session. In the training session, participants experienced the haptic stimuli assigned to the condition and wrote down their identification criteria on a sheet of paper for at least 5 min. In the practice session, participants perceived each stimulus and were asked to find the correct stimulus. Correct answer feedback was visually provided on the experiment program. Participants could freely revise their criteria notes. These training and practice sessions were repeated once again to ascertain and deepen their identification criteria. In the main session, participants' task was the same as that of the practice session, but there was no correct answer feedback. Participants could feel the stimulus as many times as they wanted.

Table 1 shows the numbers of repetition made for each stimulus. For each experimental condition, the stimuli were randomly distributed within four sub-sessions. A break of approximately 1 min was provided between sub-sessions. The experiment took 30 to 120 min, depending on the condition.

Data Analysis. For each experimental result, we computed the confusion matrix from the results of each participant and obtained its IT using the standard formulae in [13]. These individual IT_{est}s were used to compute the conditions' mean IT_{est} and additional statistical tests.

2.2 Results and Discussion

The confusion matrices obtained in the experiment are shown in Fig. 3 for the vibrotactile stimuli and Fig. 4 for the thermal stimuli. The IT_{est}s are presented in Fig. 5 along with the maximum possible values of IT.

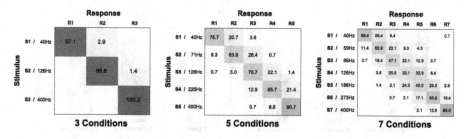

Fig. 3. Confusion matrices obtained with the vibrotactile stimuli.

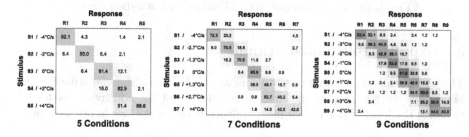

Fig. 4. Confusion matrices obtained with the thermal stimuli.

Vibrotactile Stimuli. The IT_{est}s were 1.51 bits (SD = 0.07 bits), 1.48 bits (SD = 0.40 bits), and 1.56 bits (SD = 0.23 bits) for $N = 3$, 5, and 7. We conducted one-way between-subjects ANOVA to analyze the effects of the number of stimuli on IT_{est}. The number of stimuli did not significantly affect IT_{est} ($F_{2,18} = 0.157, p = 0.856$). Thus, we can conclude that the IT_{est}s were saturated to 1.52 bits (the mean of the three IT_{est}s; 2.9 stimuli).

The vibrotactile stimuli had only one design variable of frequency. The IT_{est}s are comparable to those reported in the literature in similar settings [13].

Thermal Stimuli. The IT_{est}s were 1.72 bits (SD = 0.36 bits), 1.69 bits (SD = 0.26 bits), and 1.75 bits (SD = 0.36 bits). According to one-way between-subjects ANOVA, there was no significant difference in IT_{est} between the number of stimuli ($F_{2,18} = 0.068, p = 0.935$). Hence, the IT_{est}s were saturated to 1.72 bits (about 3.3 stimuli) regardless of the number of stimuli.

For further analysis, we calculated the percent-correct (PC) scores for different temperature change directions (cooling, no change, and warming). The results were 90.8, 88.5, and 84.3% for $N = 5$, 7, and 9, respectively. Thus, the participants could identify the direction of temperature change relatively well. Furthermore, the participants identified the cool stimuli with higher accuracies than the warm stimuli (see Fig. 4) because cold-sensitive thermoreceptors are distributed more densely than warm receptors in the human skin [5].

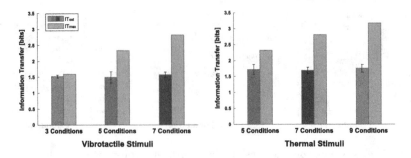

Fig. 5. Estimated and maximum IT values with standard errors.

3 Exp. 2: Multimodal Stimuli

The aim of this experiment was to estimate the IT of multimodal stimuli combining vibrotactile and thermal stimuli, and then test the additivity law for IT to assess the level of perceptual independence between the two haptic channels.

3.1 Methods

Experimental Conditions. The experimental conditions were designed by combining the vibrotactile and thermal stimuli selected from Exp. 1 (Fig. 2). For the vibrotactile stimuli, IT_{est}s were saturated to 1.52 bits (2.9 stimuli). Thus, we chose three stimuli (slightly larger than the vibrotactile channel capacity) with frequencies 40, 126, 400 Hz. For the thermal stimuli, IT_{est}s were saturated to 1.72 bits (3.3 stimuli). We selected four thermal stimuli, also slightly larger than the thermal channel capacity, as −4, −2, 0, and +4 °C/s (very cold, cold, no change, and warm). We included two cold stimuli because the cold stimuli were identified better than the warm stimuli in Exp. 1. Consequently, we used 12 experimental stimuli (3 vibrotactile and 4 thermal stimuli). The two types of stimuli were presented simultaneously with the same durations (1.5 s).

Participants and Procedure. Seven participants (3 males and 4 females; age $M = 22$ years and $SD = 0.58$) who did not participate in Exp. 1 took part in this experiment. The participants' task and experimental procedure were the same as those of Exp. 1. The number of repetitions was 12 for each multimodal stimulus. The experiment took up to 120 min.

3.2 Results and Discussion

The confusion matrix of the multimodal stimuli is shown in Fig. 6 (left). The IT_{est} was 3.06 bits (SD = 0.24 bits). This value is slightly less than the sum (3.24 bits) of the IT_{est}s of the vibrotactile and thermal stimuli obtained in Exp. 1.

For more detailed analysis, we show two confusion matrices for each modality Fig. 6 (right). Their IT_{est}s were 1.37 and 1.62 bits for the vibrotactile and

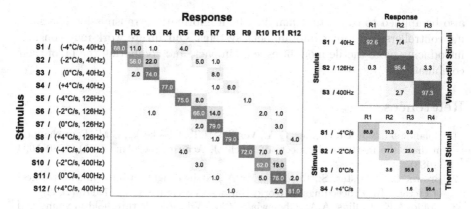

Fig. 6. Confusion matrices for the multimodal stimuli.

thermal stimuli. These numbers are reduced to 90.0% and 94.2%, respectively, compared to the IT_{est} of the individual modalities.

Singhal et al. [10] found that the perception of thermal stimuli was affected by concurrent vibrotactile cues in terms of IT. Green [4] also reported a multi-sensory effect between the tactile and thermal stimuli: a cold stimulus impairs the sensitivity of high-frequency vibration and a warm stimulus lowers the vibro-tactile sensitivity in a wide frequency range. Shim et al. [9] showed that the simultaneous presentation of wind and vibration stimuli remarkably reduce the accuracy and the IT of each unimodal stimuli compared to the sequentially com-bined ones. Despite such potential crossmodal interferences, the IT_{est} of the mul-timodal stimuli achieved over 90% of the information transmission performance of the individual stimuli. Therefore, the two modalities may not be completely perceptually independent, but combining them increases the identification abil-ity of tactile stimuli to a large extent.

4   Conclusions

In this paper, we evaluated the IT of vibrotactile and thermal stimuli in two cases. The first case was when the two stimuli were provided separately. The IT_{est}s were saturated to 1.52 bits (perfect recognition of 2.9 stimuli) and 1.72 bits (perfect recognition of 3.3 stimuli) for the vibrotactile and thermal stimuli. The second case was when the two types of stimuli were provided simultaneously to the same body location. The IT_{est} was 3.06 bits (perfect recognition of 8.3 stimuli), which is approximately 90% of the sum of the individual modality IT_{est}s (3.24 bits). This result indicates that the degree of crossmodal interference between vibrotactile and thermal senses is approximately 10% in terms of IT, suggesting a high potential of combined vibrotactile and thermal stimuli for effective information delivery.

This study was preliminary in nature. Its positive results encourage us to explore this research space with improved depth and breadth. Our results can also be applied to contexts demanding high information transmission like gaming controllers and in-car infotainment systems. Our future work may consider introducing more tactile modalities, e.g., impact, and comparing the IT of their combination.

References

1. Bolanowski, S.J., Gescheider, G.A., Verrillo, R.T., Checkosky, C.M.: Four channels mediate the mechanical aspects of touch. J. Acoust. Soc. Am. **84**(5), 1680–1694 (1988)
2. Brown, L., Brewster, S., Purchase, H.: A first investigation into the effectiveness of tactons. In: Proceedings of World Haptics Conference, pp. 167–176 (2005)
3. Goble, A.K., Collins, A.A., Cholewiak, R.W.: Vibrotactile threshold in young and old observers: the effects of spatial summation and the presence of a rigid surround. J. Acoust. Soc. Am. **99**(4), 2256–2269 (1996)
4. Green, B.G.: The effect of skin temperature on vibrotactile sensitivity. Percept. Psychophys. **21**(3), 243–248 (1977). https://doi.org/10.3758/BF03214234
5. Hensel, H.: Cutaneous thermoreceptors. In: Iggo, A. (ed.) Somatosensory System, vol. 2, pp. 79–110. Springer, Cham (1973). https://doi.org/10.1007/978-3-642-65438-1_4
6. MacLean, K.E.: Foundations of transparency in tactile information design. IEEE Trans. Haptics **1**(2), 84–95 (2008)
7. Park, G., Cha, H., Choi, S.: Haptic enchanters: attachable and detachable vibrotactile modules and their advantages. IEEE Trans. Haptics **12**(1), 43–55 (2019)
8. Pongrac, H.: Vibrotactile perception: examining the coding of vibrations and the just noticeable difference under various conditions. Multimedia Syst. **13**(4), 297–307 (2008). https://doi.org/10.1007/s00530-007-0105-x
9. Shim, Y.A., Lee, J., Lee, G.: Exploring multimodal watch-back tactile display using wind and vibration, pp. 1–12. ACM, New York (2018)
10. Singhal, A., Jones, L.A.: Perceptual interactions in thermo-tactile displays. In: Proceedings of World Haptics Conference, pp. 90–95 (2017)
11. Singhal, A., Jones, L.A.: Creating thermal icons—a model-based approach. ACM Trans. Appl. Percept. **15**(2), 1–22 (2018)
12. Stevens, J.C., Choo, K.K.: Temperature sensitivity of the body surface over the life span. Somatosens. Mot. Res. **15**(1), 13–28 (1998)
13. Tan, H.Z., Choi, S., Lau, F.W.Y., Abnousi, F.: Methodology for maximizing information transmission of haptic devices: a survey. Proc. IEEE **108**(6), 945–965 (2020)
14. Wilson, G., Brewster, S.A.: Multi-moji: combining thermal, vibrotactile, and visual stimuli to expand the affective range of feedback. In: Proceedings of the CHI Conference on Human Factors in Computing Systems, pp. 1743–1755 (2017)
15. Yarnitsky, D., Ochoa, J.L.: Studies of heat pain sensation in man: perception thresholds, rate of stimulus rise and reaction time. Pain **40**(1), 85–91 (1990)
16. Yoo, Y., Lee, H., Choi, H., Choi, S.: Emotional responses of vibrotactile-thermal stimuli: effects of constant-temperature thermal stimuli. In: Proceedings of International Conference on Affective Computing and Intelligent Interaction, pp. 273–278 (2017)
17. Yoo, Y., Regimbal, J., Cooperstock, J.R.: Identification and information transfer of multidimensional tactons presented by a single vibrotactile actuator. In: Proceedings of World Haptics Conference, pp. 7–12 (2021)

Spatial Compatibility of Visual and Tactile Stimulation in Shared Haptic Perception

Kimihiro Uemura[1], Hikari Yukawa[1](✉), Kota Kitamichi[1], Mina Shibasaki[2], Kouta Minamizawa[2], and Yoshihiro Tanaka[1]

[1] Nagoya Institute of Technology, Nagoya, Japan
k.uemura.377@nitech.jp, {yukawa.hikari,tanaka.yoshihiro}@nitech.ac.jp,
k.kitamichi.854@stn.nitech.ac.jp
[2] Keio University Graduate School of Media Design, Tokyo, Japan
{mina0415,kouta}@kmd.keio.ac.jp

Abstract. Tactile sharing with others facilitates improving communications and augmenting cooperative tasks. An increase of persons sharing tactile sensations increases the effectiveness whereas the area of tactile stimuli given should be investigated for intuitive perception. This study investigated the effect of spatial correspondence between tactile and visual stimuli in identifying tactile stimuli. In the experiment, participants viewed simultaneously two videos of other agents' hands each rubbing one of three textures and felt their vibrotactile stimuli in two locations. The videos were presented at different locations on the screen (Scene 1: left-right side or Scene 2 & 3: top-bottom) and the vibrotactile stimuli were presented either at the wrists of the left and right hand (Scene 1) or at the upper arm and the wrist of the right arm which either rested on the table (Scene 2) or was hanging down along the body (Scene 3). For each scene, visual and tactile stimuli were either spatially aligned (left and right video with tactile stimuli at the left and right wrist, and top and bottom video with top and bottom location on the right arm) or not. The result showed shorter response times for left-right spatial correspondence and for far (top) and close (bottom) visual stimuli corresponding to distal (wrist) and proximal (upper arm) locations on the body. This implied that the body schematic is an important factor for spatial compatibility of visual and tactile stimuli.

Keywords: Haptics · Communication · Body schematic · Wearable tactile display

1 Introduction

Tactile interfaces, such as tactile sensors and displays have advanced, thus making it possible to detect and transmit the tactile sensations that other people perceive. Tactile sensations are necessary for object identification and manipulation.

This work was supported by JST Moonshot R&D, JPMJMS2013 and JSPS 21H05071.

H. Seifi et al. (Eds.): EuroHaptics 2022, LNCS 13235, pp. 84–92, 2022.
https://doi.org/10.1007/978-3-031-06249-0_10

Sharing tactile sensations with others can improve communication and augment cooperative tasks with others, thus expanding the range of our object perception. Because of COVID-19, remote audio-visual communication has been spread rapidly. Tactile sharing can enrich the communication by allowing people to recognize what their partners are touching or how they move their fingers [1]. This may also induce simultaneously sharing experiences with multiple persons. Tactile sharing can be further employed for human-human/robot collaboration, where reciprocal awareness induces a smooth collaboration [2]. Casalino et al. [3] demonstrated that vibrotactile stimulation to an operator when a robot recognized his/her action improved the collaborative task.

For communication and cooperation, it is preferred that tactile stimulation from other people is presented to an area where the stimulation does not interfere with various operations of one's hands or fingertips. Wearable tactile displays are suitable for perceiving tactile stimulation in areas other than the fingers and hands [4]. An increase in the number of persons sharing tactile sensations improves communication and cooperation. However, perceiving the tactile sensations of multiple persons simultaneously disperses attention to the stimuli [5] and increases the cognitive load. Wang et al. [6] showed that the correct response rate decreased as the number of stimuli increased to estimate the stimulus position when multiple tactile stimuli were presented to both arms simultaneously. Hence, an intuitive presentation of visuo-tactile stimuli is important.

This study investigated spatial compatibility between visual stimuli and their corresponding tactile stimuli as shown in Fig. 1. A previous study demonstrated

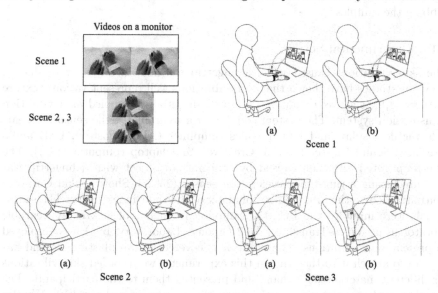

Fig. 1. Identification task of two tactile stimuli corresponding to two videos on a monitor with vibrotactile displays on the arm. Three scenes with different postures and tactile presentation positions were used and two spatial conditions between tactile and visual stimuli: (a) spatial correspondence (b) spatial non-correspondence were employed for each scene.

spatial compatibility in the cognitive processes of perception and action [7]. In the compatibility condition where the visual cue and action spatially correspond, such as responding to the visual stimuli presented on the right side with the button on the right side, the response is faster and more accurate than in the incongruent condition. When presenting the tactile sensations of others in the left and right positions to the body, correspondence between the left and right is be expected to facilitate recognition. However, it is unclear how the visual stimuli at the upper and lower positions correspond to the body area where the tactile stimuli are presented. Video conferencing systems divide the video of individuals into up, down, left, and right grids, and at the human-human/robot collaboration, the location of operators including robots can be arranged. In this study, we consider remote communication with multiple persons through a monitor, and as basic research, recognitions of the tactile sensations of two other persons are investigated. The three scenes shown in Fig. 1 were utilized as common postures of arms taken during video conferencing systems and possible locations of tactile stimulation. Wearable tactile displays were attached to the arm and presented vibrotactile stimuli corresponding to each person on the monitor. The videos of each person were placed on the left and right, or top and bottom of the screen.

2   Method

In the experiment, the videos of two persons' hands rubbing samples were given and participants identified each sample by perceiving tactile stimuli derived by rubbing the samples.

2.1   Experimental Setup

The skin vibrations caused when a fingertip touched an object were shared. Previous studies have shown that skin vibrations well represent various texture features [8,9], such as roughness. Figure 2 shows an assembled skin vibration transmission system. The system consists of a wearable tactile sensor, a wearable tactile display, and I/O modules (amplifier (AP05, Fostex), USB audio interface (Sound Blaster Play! 3, Creative), and laptop computer (PC)). The skin-propagated vibration caused by touching an object with a fingertip was measured using a ring-type acceleration sensor (2302B, Showa Sokki Corporation) attached to the intermediate phalanx of the right index finger. The sensor signals were input to the PC at a sampling frequency of 48 kHz. A wearable vibrator using Vibro-Transducer Vp2 (Acouve Laboratory, Inc.) was employed to present skin vibrations. The vibrator is covered with a plastic case and has a strap to attach it to the arm. In this experiment, we recorded skin vibrations for different materials beforehand and presented them to the participants. The signal from the sensor can be also sent to the vibrator in real time [10]. We also used a wide-angle web camera (Buffalo BSW200MBK) to capture the rubbing motion while collecting tactile information. The videos were synchronized with the sensor signals based on the moment that the finger contacted the object and presented to the participants along with the tactile stimuli.

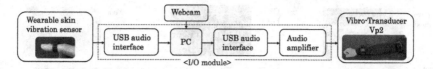

Fig. 2. Skin vibration transmission system. Skin vibrations induced in the fingertip are detected by a wearable acceleration sensor and presented on the arm of another person by a wearable vibrotactile display.

2.2 Conditions

Three natural postures of a person sitting in a chair and communicating were adopted (Fig. 1). For each scene, two conditions of spatial correspondence and non-correspondence were compared.

In Scene 1, the videos of two persons rubbing a sample were placed on the left and right panels of a monitor. Two tactile displays were attached to the left and right wrists of each participant. The tactile stimuli of the person in the video on the left and right panels on the monitor were presented to the left and right wrist, respectively (spatial correspondence), and vice versa (spatial non-correspondence). In Scenes 2 and 3, the videos of two persons were placed on the top and bottom panels. In Scene 2, the participants were asked to place their arms on the desk, while in Scene 3, participants were asked to place their arms hanging down. In the skin-propagated vibration with a distance of more than 80 mm for arm, the vibration intensities decrease below the perceptual detection threshold [11]. Therefore, two tactile displays were attached to the right wrist and the right upper arm of each participant, avoiding mechanical interference. For both Scene 2 and 3, the tactile stimuli of the person on the top and bottom panels on the monitor were presented to the right upper arm and the right wrist, respectively (spatial correspondence), and vice versa (spatial non-correspondence). In addition, for Scenes 1 and 2, participants were asked to place their both arms on the desk, with a distance of approximately 300 mm between the arms. During the experiment, the participant wore a black and white band beside the tactile displays, and correspondingly, the person in the video wore a black or white band. The colors of the bands of the two persons in the videos were randomly exchanged among trials.

2.3 Stimuli

Three different materials were prepared for the samples: wire mesh, glass beads, and wood as shown in Fig. 3. A preliminary test showed that they were easy to identify by presenting skin vibrations recorded for each sample. Figure 3 shows the skin vibration and its power spectrum density when each sample was rubbed with a fingertip. The duration for rubbing the sample once was approximately 1 s for all the samples. The samples in the video were processed to be blacked out so that they were visually indistinguishable.

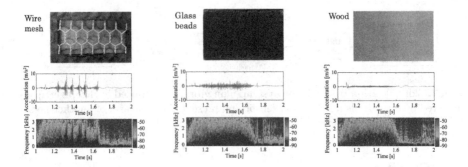

Fig. 3. Samples with different textures (wire mesh, glass beads, and wood) and skin vibration and its spectrogram when each sample was rubbed with the fingertip. The skin vibrations recorded were presented to participants.

2.4   Procedure and Analysis

Before the test, the intensity of the vibrotactile stimulation of the two tactile displays was adjusted to match the same in the subjective rating of each participant, and the participants practiced for about 2 min until they had sufficiently memorized all three samples while vibrotactile stimuli were simultaneously presented. Then, the participants were asked to recognize from the tactile stimuli which sample was explored in each video, and to say "Yes" as a signal of completing each trial after identifying the two samples. The responses were recorded in the identified sample and the required duration.

For each scene, 9 (combination of samples) × 2 (presentation conditions) × 2 (number of trials) = 36 trials were conducted for each participant. Different groups of eight volunteers participated in the experiment for each scene: 22–24 years old, 6 males and 2 females, 7 right handed and 1 left handed for Scene 1; 22–24 years old, 7 males and 1 females, all right handed for Scene 2; 22–24 years old, 7 males and 1 females, 7 right handed and 1 left handed for Scene 3. Informed consent was obtained from all of them. The experimental evaluation protocol followed the Declaration of Helsinki and was approved by the ethics committee of the Nagoya Institute of Technology. The participants wore headphones with white noise so they could not hear the sound of the vibrator.

The rate of correct responses and the mean response time were calculated for each participant and comparisons between each condition (spatial correspondence and spatial non-correspondence) were conducted for each scene. A Shapiro-Wilk test was conducted to confirm the assumption of normal distribution, and paired t-tests were conducted. When the normal distribution was denied, non-parametric Wilcoxon signed-rank tests were conducted. The significance level was set to $\alpha = 0.05$.

3   Results

The left panel of Fig. 4 shows the mean rate of correct responses for all participants and the standard deviation. The right panel of Fig. 4 shows the mean

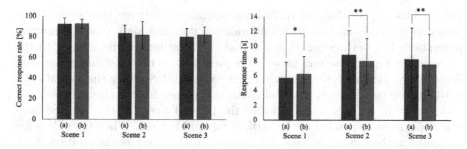

Fig. 4. Experimental results for each scene with two conditions. Left and right panels present correct response rate and response time, respectively. (a) and (b) denote the spatial correspondence and the spatial non-correspondence conditions, respectively. ** and * indicate $p < 0.01$ and $p < 0.05$, respectively.

response time and its standard deviation. The results for the two conditions were presented for each scene.

In Scene 1, the paired t-test between the two conditions showed no significant difference in the correct response rate ($t_7 = 0.158$, $p = 0.88$). However, the Wilcoxon signed-rank test showed a significant difference in the response time ($V_7 = 6373$, $p = 0.014$). This indicates that the response time in the spatial correspondence condition was significantly shorter than that in the spatial non-correspondence condition. In Scene 2, the paired t-test showed that there was no significant difference in the correct response rate between the spatial correspondence and the spatial non-correspondence conditions ($t_7 = 0.517$, $p = 0.62$). However, the Wilcoxon signed-rank test showed that the response time was significantly shorter in the spatial non-correspondence condition than that in the spatial correspondence condition ($V_7 = 6540$, $p = 0.0085$). In Scene 3, similar to the results in Scene 2, the paired t-test showed no significant difference in the correct response rate ($t_7 = 0.728$, $p = 0.49$). Wilcoxon signed-rank test showed that the response time in the spatial non-correspondence condition was significantly shorter than that in the spatial correspondence condition ($V_7 = 3707$, $p = 0.0025$).

4   Discussion

In the case where the videos were placed on the left and right panels, the results showed that the response time was improved under the spatial correspondence condition, where the spatial relationship of the tactile stimulation (right and left wrist) corresponded with the positions of the images on the monitor (right and left panels). This is consistent with the left-right spatial stimulus-response compatibility effect [7].

In contrast to the case of the left and right images using the left and right hands, the response time in the spatial non-correspondence condition was significantly shorter in the case where the videos were placed on the top and bottom panels. This implies that the spatial correspondence between the tactile stimu-lation on the body and the visual information does not always induce intuitive perception. For both Scenes 2 and 3, the tactile stimuli corresponding to the video on the top and bottom panels were presented on the wrist and upper arm, respectively, yielding a significantly faster response. Regarding the spatial com-patibility between the vertical and horizontal planes for the stimulus-response task, Vu et al. [12] demonstrated that the far and close locations on the hori-zontal plane corresponded to the vertical top and bottom. When the videos are presented at the top and bottom, participants might interpret the position of the videos on the screen along the vertical plane as depth, with the bottom video being perceived closer. These positions in depth are then mapped to the prox-imal and distal positions on the arm, independent of its pose. From the aspect of the body schematic, the wrist is distal and the upper arm is proximal.

In Scene 2, the participants took the posture placing the right arm on the desk, and the wrist and the upper arm were placed in the far and close posi-tion on the horizontal plane. Thus, considering the same manner of the spatial compatibility between the vertical and horizontal planes, it was considered that the wrist and the upper arm would correspond to the top and bottom positions in the video, respectively. The results supported this inference. The results of Scene 3, which showed the same trend as that of Scene 2, also indicate that body schematics affect spatial compatibility. In Scene 3, the right arm wearing tactile displays was down and the upper arm and the wrist were located above and below in the vertical plane, respectively. Therefore, both visual stimuli and tactile stimuli were located on the vertical plane, and the results showed that the correspondence of the body schematic had a significant effect (the wrist is distal and the upper arm is proximal). The spatial condition of tactile stimula-tion in Scene 1 also includes the correspondence of the body schematic. Thus, the present results indicate that the body schematic is an important factor for the correspondence between the spatial location of others and the tactile pre-sentation position of the body.

In this study, we examined the spatial compatibility between the visual stim-uli of other persons and the tactile stimuli on the body for three representative scenes in video communication. This finding is useful in arranging the location of other people in remote communication with others and in human-human/robot collaboration. Other viewpoints such as a face-to-face and third person perspec-tive will be investigated. In addition, the size of the video as a perseptive cue and various combinations of body presentation positions and arm postures are possible factors [13], and the arm can be moved during the work. Influence of the body schematic will investigated under various conditions.

For all scenes, the correct answer rates were not significantly different between the conditions. A possible reason is that the vibrations of the three types of sam-ples used in this study were easily identifiable. If we use samples that are difficult to discriminate or fix the time for discrimination, there may be a difference in the correct response rate between conditions. Different materials and an increase of materials will be tested for future work.

5   Conclusions

This study investigated the spatial compatibility between two visual stimuli and the corresponding two tactile stimuli presented on the arm to induce their accurate and fast perception. Although the combination was limited, the experimental results suggested the influence of the body schematic. The findings might be available for remote communication and human-human/robot collaborations. In the future, we would like to increase the number of tactile identifications and apply this method to interactive communication between multiple persons and cooperative work using shared tactile perception with other persons and robots.

References

1. Katagiri, T., Tanaka, Y., Sugiura, S., Minamizawa, K., Watanabe, J., Prattichizzo, D.: Operation identification by shared tactile perception based on skin vibration. In: 2020 IEEE International Conference on Robot and Human Interactive Communication (ROMAN), pp. 885–890 (2020)
2. Drury, J.L., Scholtz, J., Yanco, H.A.: Awareness in human-robot interactions. In: 2003 IEEE International Conference on Systems, Man and Cybernetics, Conference Theme-System Security and Assurance, vol. 1, pp. 912–918 (2003)
3. Casalino, A., Messeri, C., Pozzi, M., Zanchettin, A.M., Rocco, P., Prattichizzo, D.: Operator awareness in human-robot collaboration through wearable vibrotactile feedback. IEEE Robot. Autom. Lett. **3**(4), 4289–4296 (2018)
4. Pezent, E., Israr, A., Samad, M., Robinson, S., Agarwal, P., Benko, H., Colonnese, N.: Tasbi: Multisensory squeeze and vibrotactile wrist haptics for augmented and virtual reality. In: 2019 IEEE World Haptics Conference (WHC), pp. 1–6 (2019)
5. Connell, L., Lynott, D.: When does perception facilitate or interfere with conceptual processing? The effect of attentional modulation. Front. Psychol. **3**, 474 (2012)
6. Wang, D., Member, S., Peng, C., Afzal, N., Li, W., Wu, D., Zhang, Y.: Localization performance of multiple vibrotactile cues on both arms. IEEE Trans. Haptics **11**(1), 97–106 (2018)
7. Nishimura, A., Yokosawa, K.: Effects of visual cue and response assignment on spatial stimulus coding in stimulus-response compatibility. Q. J. Exper. Psychol. **65**(1), 55–72 (2012)
8. Bensmaia, S., Hollins, M.: Pacinian representations of fine surface texture. Percept. Psychophys. **67**(5), 842–54 (2005)
9. Wiertlewski, M., Lozada, J., Pissaloux, E., Hayward, V.: Causality inversion in the reproduction of roughness. In: Kappers, A.M.L., van Erp, J.B.F., Bergmann Tiest, W.M., van der Helm, F.C.T. (eds.) EuroHaptics 2010. LNCS, vol. 6192, pp. 17–24. Springer, Heidelberg (2010). https://doi.org/10.1007/978-3-642-14075-4_3
10. Fukuda, T., Tanaka, Y.: Skin vibration-based tactile tele-sharing. In: Kajimoto, H., Lee, D., Kim, S.-Y., Konyo, M., Kyung, K.-U. (eds.) AsiaHaptics 2018. LNEE, vol. 535, pp. 82–84. Springer, Singapore (2019). https://doi.org/10.1007/978-981-13-3194-7_17
11. Shah, A.V., Casadio, M., Scheidt, A.R., Mrotek, A.L.: Vibration propagation on the skin of the arm. Appl. Sci. **9**(20), 4329 (2019)

12. Vu, K.P.L., Proctor, R.W., Pick, D.F.: Vertical versus horizontal spatial compatibility: Right-left prevalence with bimanual responses. Psychol. Res. **64**(1), 25–40 (2000)
13. Nicoletti, R., Umiltà, C.: Right-left prevalence in spatial compatibility. Percept. Psychophys. **35**, 333–343 (1984)

Increasing Perceived Weight and Resistance by Applying Vibration to Tendons During Active Arm Movements

Keigo Ushiyama[⊠], Akifumi Takahashi, and Hiroyuki Kajimoto

The University of Electro-Communications, Tokyo, Japan
ushiyama@kaji-lab.jp

Abstract. We proposed to use kinesthetic illusion to achieve wearable/portable haptic devices for kinesthetic feedback in VR experiences. The kinesthetic illusion is the illusion of limb movement typically induced by vibratory stimulation. We investigated how the kinesthetic illusion affected the perceived weight and resistance of the handheld object. We designed vibration patterns that simulate constant gravity and velocity-related resistance. Two experiments were conducted to measure changes in perceiving weight and resistance when wielding cylindrical weights and hand fans. The results of the experiments indicated that the designed kinesthetic illusions enhanced these sensations; the real weight was perceived heavier, and the real resistance was perceived larger. However, we could not find the explicit difference between the two stimulation patterns, and the resistance sensation induced by the illusion differed from the actual sensation of using the hand fans.

Keywords: Heaviness · Kinesthetic illusion · Resistance · Tendon vibration

1 Introduction

Haptic feedback is essential for enhancing the quality of virtual reality (VR) experiences. By focusing on the physical interaction with an object, many ungrounded haptic devices have been proposed to provide force and tactile feedback in VR environment [1–3].

Conversely, haptic devices that simulate such forces, in general, tend to be complex and cumbersome. Many researchers have tackled this problem and proposed methods that employ haptic illusions induced by tactile or visual stimulation [4, 5]. Because the weight and length of a handheld object are related to the moment of inertia [6], Zenner et al. developed Shifty, which is a device that changes the position of the center of gravity to simulate the weight and length of handheld VR objects [7].

In the paradigm of kinesthetic feedback for VR objects, the illusion of proprioception, the sensation of force and movement of a body was rarely included. Because proprioception is known to contribute to the perception of the weight, length, and shape

A. Takahashi—JSPS Research Fellow.

H. Seifi et al. (Eds.): EuroHaptics 2022, LNCS 13235, pp. 93–100, 2022.
https://doi.org/10.1007/978-3-031-06249-0_11

of a handheld object [6, 8], such an illusion can also realize and enhance the kinesthetic feedback from a handheld VR object.

One such proprioceptive illusion is kinesthetic illusion, which is typically induced by applying low-frequency vibrations of approximately 100 Hz to tendons [9]. The illusion can induce a sensation of limb movement even when it is not moving and cause errors in the perception of movement velocity [9–11]. This effect has been studied since Goodwin et al. rigorously documented this phenomenon [12]. The basic mechanism of the kinesthetic illusion induced by tendon vibration is caused by changes in the firing rate of the muscle spindle afferents that are receptors in the muscles and contribute to the perception of limb position and movement [13].

The kinesthetic illusion is frequently used to generate the illusion of bodily movement. The illusion may also lead to the illusory deformation of the object being touched [14]. Though the kinesthetic illusion has the potential to modulate perception of the object's properties, the effects of the illusion on physical properties such as weights are still unclear. We previously reported that the heaviness of a handheld bar could be modulated through a preliminary experiment [15].

Following the previous report, this study investigates the possibility of modulating the virtual properties of a handheld object by the kinesthetic illusion. We conducted experiments to confirm the effect of tendon vibration on perceptions of heaviness and resistance.

2 Experiments

We conducted two experiments to investigate how much the kinesthetic illusion can independently increase the perceived weight and resistance through physical props. Twelve people (11 males, one female, ten right-handed, 21–26 years old) participated in the experiment. The experiment was conducted with the approval of the ethics committee of The University of Electro-Communications (No. 20067).

Prior to the detailed procedure for each experiment, common apparatus and stimulation conditions will be described.

2.1 Apparatus

Overview of the experimental environment is shown in Fig. 1 (a, b). An acrylic board was placed to hide the participant's arm. A glove was used to prevent perceiving the weight through tactile cues. Retroreflective markers for optical motion capture (OptiTrack V120 Duo) were fixed to the hand and elbow to track the arm's movements.

Four voice-coil-type vibrators (Acouve Lab, VP210) were mounted at the elbow and wrist by using supporters (Fig. 1 (b)). These vibrators were placed on the distal tendons of the biceps and triceps brachii and the radial side (abductor pollicis longus, extensor carpi radialis longus, and extensor carpi radialis brevis) and ulnar side (flexor and extensor carpi ulnaris) of the wrist. Although the loading force was not controlled, the supporters fixed vibrators with enough loading force to prevent them from slipping from the positions.

The vibrator was driven by a signal output from PC-based software (Cycling'74, Max 8) via an audio interface (Roland, OCTA-CAPTURE) and audio amplifiers (FX-AUDIO-FX202A/FX-36A PRO).

Fig. 1. (a) An overview of the environment as for experiment 1. (b) Positions of the vibrators and the markers. (c) Screen display of the sliders for movement control.

2.2 Tendon Vibration During Active Movements

We set three vibration conditions: Weight, Resistance, and Control (not applying stimuli). Figure 2 (c) illustrates the vibration patterns for the Weight and Resistance conditions during the lifting movement. Under the Weight condition, the tendons of the biceps and wrist abductors were constantly stimulated during the exercise. Under the Resistance condition, the tendons of the biceps and wrist abductors were stimulated during flexion, and the tendons of the triceps and wrist adductors were stimulated during extension.

The two stimulation conditions were selected to increase the apparent heaviness of a handheld object by the following mechanisms. The Resistance condition was set based on the previous report [15] that participants tended to perceive a handheld object as heavy when proprioceptive stimulation was applied to suppress velocity of the movement. The vibration pattern of this condition aims to represent resistance to the movement by evoking the movement illusion in the opposite direction. The Weight condition was set to induce the illusion of extension constantly, considering that the weight is affected by gravity.

The vibration frequency and amplitude were set to 70 Hz and from 70 m/s^2 to 100 m/s^2 for each vibrator based on a previous literature [9]. The amplitudes were adjusted within the range where the participants did not feel a tonic vibration reflex (the vibration-induced reflex of the muscle of which the direction is opposite to that of the illusion) and could still perceive the illusion strongly. The amplitude of the rise and fall was changed linearly to avoid vibration noise.

2.3 Experiment 1: Increasing the Weight of a Handheld Object

Figure 2 (a) shows the weight samples used in the experiment. The samples were created by using a cylindrical aluminum pipe 32 mm in diameter (29 mm inside diameter) and 150 mm in length, filled up by mixing salt and water, or sand (approx. 1.2 g/cm^3) and iron sand (approx. 2.8 g/cm^3) in several mixing rates. The samples weighed between 160 g and 330 g prepared in 10 g increments.

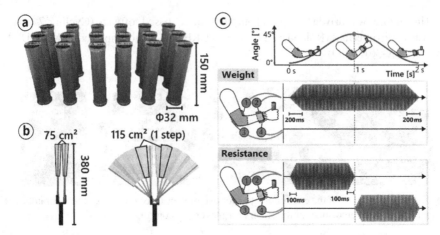

Fig. 2. (a) Weight samples used in experiment 1. (b) Hand fan samples used in experiment 2. (c) Vibration patterns of the weight and resistance conditions during the lifting movement.

Procedure. Two vibrators were placed at the participant's wrist and two at the elbow. When placing the vibrators, the experimenter confirmed the positions of the tendons of the targeted muscles by touching the participants' arms. The vibration amplitude was calibrated using an accelerometer (Sparkfun, LIS331). Whether participants could perceive the illusion was checked after calibration by applying the same vibratory stimulation as in the experiment.

The participants were asked to lift the objects following the sliders displayed on the screen (Fig. 1 (c)). One slider was for movement instruction, and the other was for informing participants of the forearm angle calculated from the hand and elbow positions. The vibration timing for each condition was pre-programmed based on the instruction slider (i.e., the vibration was applied regardless of the participant's movement). The timing to start the movement was notified with the counter. The participants braced their elbows on the desk, grasped the weight, and lifted it to 45° by a forearm movement (Fig. 1 (b)). The weight was then lowered to the desk again. This movement was practiced for each stimulation condition until the participants could confidently perceive heaviness while moving in response to the instruction slider.

The double staircase method using the Parameter Estimation by Sequential Testing (PEST) [16] was adopted to measure the subjective equivalent points of the perceived weight. PEST is a rule for deciding the step size for changing the compared stimuli (in this experiment, the weight of the sample). For each comparative trial, we randomly selected an ascending series from 160 g or a descending series from 330 g to prevent the participants from predicting the answers. The participants raised the reference sample (200 g) for each vibration condition and then lifted a sample for comparison without vibratory stimulation. They were then asked whether the comparison sample was lighter or heavier than the reference sample. Based on the answer, the subsequent step size was determined by PEST. The initial step size was set to 80 g for both series, and the series was completed when the step size was reduced to 10 g.

When participants answered beyond the range of samples prepared (for example, if the participant answered "heavy" even if the comparison sample was 330 g), the sample was presented again, and if the identical answer was given three consecutive times, the value was adopted as the series' result. The presentation order of the three conditions was counterbalanced between participants.

During the experiment, while lifting the weights, the participants wore headphones that emitted pink noise and metronome sounds. The metronome was presented at the lower and upper limits of the motion and was used as an additional cue in the exercise.

Result. Statistical analysis was carried out using SPSS (Statistics 24 Advanced, IBM). The results were the averages of the ascending and descending series for each stimulus condition for each participant. One-way repeated measures ANOVA (RM-ANOVA) was applied to analyze the perceived weight differences between the stimulus conditions. A post hoc test using the Bonferroni method was also conducted to investigate the differences between conditions.

Figure 3 (left) shows boxplots of the average subjective equivalent points of all participants for each stimulus condition. The main effect of the vibration conditions was significant ($F(2, 22) = 15.133$, $p < 0.001$). The post hoc test revealed significant differences between the Weight and Control conditions ($p = 0.001$) and the Resistance and Control conditions ($p = 0.011$). Conversely, there was no statistically significant difference between the Weight and Resistance conditions ($p = 0.246$).

The average weight in the Weight condition was 250 g, and the average weight in the Resistance condition was 235 g. The differences from a reference weight (200 g) were 25% and 17.5% respectively for the Weight and Resistance conditions.

2.4 Experiment 2: Increasing Resistance of a Handheld Object

Since the Resistance condition is intended to enhance resistance to the movement, the perception of a handheld object's resistance should be more affected than weight. Experiment 2 was carried out to investigate the effect of the kinesthetic illusion on the sensation of resistance. Two samples were prepared using hand fans based on Drag: on [17] to represent the sensation of resistance (Fig. 2 (b)).

This experiment was conducted with the same participants using an almost identical apparatus, experimental conditions, and procedure as in experiment 1, but the velocity of the movement was increased to facilitate perception of the hand fans' resistance. The participants flapped the hand fans three times at 1 Hz. Following the changes of the exercise, the duration of each vibration pattern was scaled to one-half time. The fan has ten width steps, and the area of the fan increases by 115 cm^2 per step. A fan opened four steps (535 cm^2) was used as the reference stimulus. The ascending series started from the completely closed state (75 cm^2), and the descending series was started from the opened state at the maximum (1225 cm^2). The initial step size in PEST was set to four steps, and the series ended when the step size was reduced to one. When changing the state, the hand fan was opened and closed symmetrically to avoid unnecessary torque.

In the preliminary experiment, we found that resistance was difficult to perceive when the same sample was continuously wielded. Therefore, the order of presentation of the reference and the comparison samples was randomized in this experiment. During the intervals between the trials, the participants wielded a completely closed sample to feel the lowest resistance to facilitate evaluation.

The sound of the metronome was not presented in this experiment because the sound tended to disturb participants' concentration during rapid movement.

Result. The results of the experiment are shown in Fig. 3 (right). Four participants reported greater than fully opened resistance with Weight and Resistance conditions: One participant for the Weight condition only, one participant for the Resistance condition only, and two participants for both conditions.

The RM-ANOVA showed the significant main effect of the vibration conditions $(F(2, 22) = 26.134, p < 0.001)$. A post hoc test revealed significant differences between Weight and Control $(p = 0.001)$ and Resistance and Control $(p < 0.001)$. Conversely, there was no statistically significant difference between Weight and Resistance $(p = 1.000)$.

The mean values of the Weight, Resistance, and Control conditions were 935 cm^2, 921 cm^2, and 572 cm^2, respectively. The difference from the reference area (535 cm^2) was 75% and 72% respectively for the Weight and Resistance conditions.

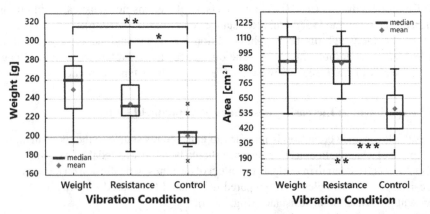

Fig. 3. Boxplots of all participants under each stimulus condition: weight (experiment 1) and resistance (experiment 2). The reference samples were 200 g and 535 cm^2 for each. ***: $p < 0.001$, **: $p < 0.01$, and *: $p < 0.05$.

3 Discussion

In experiment 1, although no significant difference was observed between the Weight and Resistance conditions, seven participants commented that it was easier to perceive weight under the Weight condition than under the Resistance condition. In addition, under the Weight condition, the sample was felt to be heavier than under the Resistance condition. Therefore, the Weight condition may enhance the sensation of weight more than the Resistance condition.

Conversely, in experiment 2, resistance was enhanced under both the Weight and Resistance conditions with no difference between them. Several participants commented that the resistance sensation of the hand fans and the kinesthetic illusion were different, and therefore the comparison was difficult. Some participants commented that they compared the sensation of resistance using the sensation only while lifting the sample, especially under the Weight condition. Therefore, there is a high possibility that the sensation evaluated to compare resistance differed among participants. On the other hand, there was a comment that the Resistance condition more closely represented a sensation of resistance than the Weight condition.

The main reason that the comparison became difficult was that the tendon vibration was applied to the forearm and upper arm while the resistance of the hand fan was mainly sensed in the wrist and forearm. Since the wrist and elbow vibrators were set using the same parameters, the vibration for the Resistance condition elicited a sensation that resistive force was being exerted on the forearm and upper arm as if the arm were moving under water. In addition, the strong vibration may have masked the resistance sensation from the hand fans and made the comparison difficult.

4 Conclusion

We investigated the effect of the kinesthetic illusion on the sensations of weight and resistance perceived while wielding handheld cylindrical weights and fans. Two vibration patterns that induce the kinesthetic illusion were designed to enhance the sense of weight and resistance; one was to simulate constant gravitational force, and the other was to simulate velocity-related resistance. The results of the experiments indicate that the designed kinesthetic illusions enhanced these sensations; the real weight was perceived heavier, and the real resistance was perceived larger. However, we could not find explicit difference between the two stimulation patterns. Our next step is to reconsider the position and parameters of the vibrators and to conduct the experiment using the haptic device that simulates the physical characteristics of the handheld object more accurately.

Acknowledgment. This research was supported by JSPS KAKENHI Grant Number JP18H04110.

References

1. Swindells, C., Unden, A., Sang, T.: TorqueBAR: an ungrounded haptic feedback device. In: Proceedings of the 5th International Conference on Multimodal Interfaces (2003)
2. Benko, H., Holz, C., Sinclair, M., Ofek, E.: Normaltouch and texturetouch: high-fidelity 3D haptic shape rendering on handheld virtual reality controllers. In: Proceedings of the 29th Annual Symposium on User Interface Software and Technology (2016)

3. Choi, I., Ofek, E., Benko, H., Sinclair, M., Holz, C.: CLAW: a multifunctional handheld haptic controller for grasping, touching, and triggering in virtual reality. In: Proceedings of the 2018 CHI Conference on Human Factors in Computing Systems (2018)
4. Choi, I., Culbertson, H., Miller, M.R., Olwal, A., Follmer, S.: Grabity: a wearable haptic interface for simulating weight and grasping in virtual reality. In: Proceedings of the 30th Annual ACM Symposium on User Interface Software and Technology (2017)
5. Heo, S., Lee, J., Wigdor, D.: Pseudobend: producing haptic illusions of stretching, bending, and twisting using grain vibrations. In: UIST (2019)
6. Turvey, M.T.: Dynamic touch. Am. Psychol. 1134–1152 (1996)
7. Zenner, A., Kruger, A.: Shifty: a weight-shifting dynamic passive haptic proxy to enhance object perception in virtual reality. IEEE Trans. Vis. Comput. Graph **23**, 1285–1294 (2017)
8. Proske, U., Allen, T.: The neural basis of the senses of effort, force and heaviness. Exp. Brain Res. **237**(3), 589–599 (2019). https://doi.org/10.1007/s00221-018-5460-7
9. Taylor, M.W., Taylor, J.L., Seizova-Cajic, T.: Muscle vibration-induced illusions: review of contributing factors, taxonomy of illusions and user's guide. Multisens. Res. **30**, 25–63 (2017)
10. Cordo, P.J., Gurfinkel, V.S., Brumagne, S., Flores-Vieira, C.: Effect of slow, small movement on the vibration-evoked kinesthetic illusion. Exp. Brain Res. **167**, 324–334 (2005)
11. Honda, K., Kiguchi, K.: Control of human elbow-joint-extension-motion change based on vibration stimulation for upper-limb perception-assist. IEEE Access **8**, 22697–22708 (2020)
12. Goodwin, G.M., McCloskey, D.I., Matthews, P.B.C.: The contribution of muscle afferents to kinaesthesia shown by vibration induced illusions of movement and by the effects of paralysing joint afferents. Brain **95**, 705–748 (1972)
13. Roll, J.P., Vedel, J.P.: Kinaesthetic role of muscle afferents in man, studied by tendon vibration and microneurography. Exp. Brain Res. **47**, 177–190 (1982)
14. Lackner, J.R.: Some proprioceptive influences on the perceptual representation of body shape and orientation. Brain **111**, 281–297 (1988)
15. Ushiyama, K., Takahashi, A., Kajimoto, H.: Modulation of a hand-held object's property through proprioceptive stimulation during active arm movement. In: Extended Abstracts of CHI (2021)
16. Taylor, M.M., Creelman, C.D.: PEST: efficient estimates on probability functions. J. Acoust. Soc. Am. **41**, 782–787 (1967)
17. Zenner, A., Krüger, A.: Drag: on: a virtual reality controller providing haptic feedback based on drag and weight shift. In: Proceedings of the 2019 CHI Conference on Human Factors in Computing Systems (2019)

A Comparison of Haptic and Auditory Feedback as a Warning Signal for Slip in Tele-Operation Scenarios

Femke E. van Beek[1]([⊠])[iD], Quinten Bisschop[1][iD], Kaj Gijsbertse[2][iD],
Pieter S. de Vries[2][iD], and Irene A. Kuling[1][iD]

[1] Department of Mechanical Engineering, Eindhoven University of Technology,
Eindhoven, The Netherlands
f.e.v.beek@tue.nl
[2] TNO, Soesterberg, The Netherlands

Abstract. Slip feedback is an important cue in everyday object manipulation, but it is generally missing in tele-operation systems. To test
the usefulness of simple, abstract types of feedback that warn the user
about slip events, we tested the effect of auditory and haptic vibration
feedback in a tele-operation task. Participants were asked to hold an
object in a remote robot hand, and the force profiles that they exerted
in response to slip events were measured. Haptic feedback did not significantly change the response characteristics, but auditory feedback did
significantly improve response latency. A small but significant difference
between haptic and auditory reaction times (60 ms) found in our control
experiment might explain the difference between the feedback types.

Keywords: Haptic feedback · Slip · Auditory feedback ·
Tele-operation

1 Introduction

Tele-operation is a technology in which a remote robot is controlled from a
distance by a human operator. This technology is helpful for performing tasks in
environments that are dangerous (e.g. nuclear power plants), unreachable (e.g.
space or deep sea), or require scaling (e.g. keyhole surgery) [8]. Many of these
tasks require high levels of dexterity, and are executed in environments that
cannot be fully predicted. Therefore, full automation is impossible, and thus the
human needs to be in the loop.

In daily life, humans perform dexterous tasks such as picking up an object
effortlessly and efficiently. A tight control is kept over the ratio between the
load force perpendicular to the object's surface and the grip force normal to the
object's surface, both during static holds [11], and during arm movements [2].
One of the sensory cues that helps to keep this safety margin small is slip force,
as micro-slips are acted upon reflexively to restore a proper safety margin [5].

© The Author(s) 2022
H. Seifi et al. (Eds.): EuroHaptics 2022, LNCS 13235, pp. 101–109, 2022.
https://doi.org/10.1007/978-3-031-06249-0_12

In tele-operation systems, slip force feedback is generally lacking, which could be a reason why dropping or crushing an object is much more likely in remote interactions than in direct interactions [3].

Several slip feedback displays have been developed (for instance [10]), and it is known that vibratory slip feedback improves task performance in virtual remote interactions [9]. Nonetheless, it is hard to integrate these systems in real-time tele-operation tasks with real remote environments. Most remote robots are only equipped with simple force sensors which are not able to measure slip. Pachierotti et al. do show that slip feedback improves performance in a real tele-operated setup [6], but they used grounded haptic devices to track only two finger tips. Many real tele-operation systems have complex input devices, such as gloves, which interfere with integrating large slip feedback devices. Since we did want to test real remote interactions with full hand tracking, we chose to focus on the effect of simple, abstract warning signals about slip on the user's behavior. We measured the participant's grip force profiles, and compared auditory and haptic vibration feedback to a condition with no slip force feedback. This allowed us to test if slip feedback affected the timing and the magnitude of the forces that participants used while holding a real object with a tele-operated arm.

2   Material and Methods

2.1   Participants

Fifteen healthy participants were recruited, of which two were authors (KG and PdV). All participants but the authors were naive to the purpose of the experiment. Four participants failed the initial stereo-vision test (30″ on the TNO test for stereoscopic vision, Laméris Instrumenten B.V), resulting in 7 men and 4 women, aged 35 ± 10 years (mean $\pm$ s.d.) completing the full experiment. All participants were right-handed, provided written informed consent, and were compensated for their time. Ethical approval for the experiment was provided by the TNO Institutional Review Board (#2021-036). One of the data sets was corrupted due to a technical failure, so 10 data sets were used for analysis.

2.2   Setup

A custom-designed tele-operation system was used in this experiment, as shown in Fig. 1. The remote robot consisted of a KUKA LBR iiwa robot arm (KUKA Aktiengesellschaft, Augsburg-Germany), to which a SHADOW robot hand (SHADOW ROBOT COMPANY LTD. London-United Kingdom) was attached. In this experiment, the Kuka arm was fixed in a convenient position, and only the robot hand was actuated. The standard SHADOW fingertips were replaced with BioTac sensors (SynTouch Inc, Montrose, CA-USA), which provide 3 DOF impedance measurements in Volts. These impedance values can be converted to forces by a per-sensor calibration, as shown in [7], but BioTac does not provide such a calibration. As we were not interested in absolute forces, but

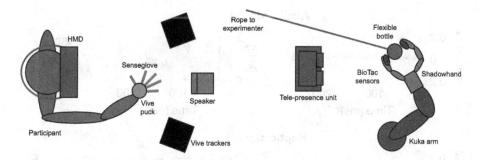

Fig. 1. Schematic overview of the setup, with equal colours indicating systems that were synchronized between participant on the left and remote robot on the right. The Kuka arm was set in a convenient fixed position, so only the Shadowhand reacted to the user's motion (for clarity, only 2 of the 5 fingers of the Shadowhand are shown). A rope, attached to the bottom of the water bottle, was lead through a pulley underneath the bottle, and lead back to the experimenter for slip induction.

only in differences, this was not a problem for our experiment. Nonetheless, it should be noted that whenever 'forces' are presented in this paper, these actually are impedances. A TNO-developed tele-presence unit housed a stereo-camera on a movable platform which filmed the remote robot. The movements of the platform were synchronized with the motion of the TNO-developed Head-Mounted Display (HMD), to which the filmed data was streamed. To control the remote robot, participants wore a Senseglove exoskeleton glove (Senseglove B.V., Delft-the Netherlands). The Senseglove's finger positions and orientations were mapped to the Shadowhand's positions and orientations. The Senseglove was used to provide passive kinesthetic feedback in all conditions using magnetic brakes on its tendon system, and active feedback in the haptic condition using the integrated ERM vibration motors on the back of the finger tips (C1026B002F, Vybronics, Inc., Shenzen-China. Motors rated 150 Hz, 1.5 mm).

2.3   Testing Auditory and Haptic Latencies

Because of the complexity of the full tele-operation system, we could not measure the communication latencies between all parts of the system. However, since a confounding factor for our specific research question could be a difference in system latency between the onsets of haptic and auditory feedback, we tested this prior to running our experiments. We used a stereo plug to record data from two contact microphones (AD-35 transducer, Otraki, Manila - Philippines) simultaneously. One microphone was taped to the Senseglove vibration motor, and the other to a speaker. We sent out two signals simultaneously, and measured the response at 41.1 kHz. From the recordings, spectrograms were calculated, as shown in Fig. 2. Auditory onsets were defined as the power spectrum 450 Hz exceeding −50 dB. Haptic onsets were defined as the absolute raw signal exceeding 0.035. Averaging the differences between these onsets across 10 repetitions

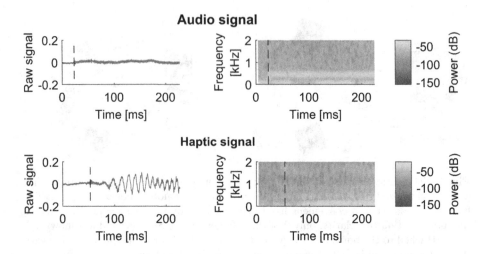

Fig. 2. Recordings from stereo contact microphones to test onset latencies between haptic and auditory stimuli. Left column: raw recordings, right column: spectrograms. The vertical dashed lines indicate determined onset times. For this recording, the measured haptic signal lagged the auditory one by 30 ms. Note that the measurement is only meant to determine the difference in onset times, so t = 0 has no specific meaning.

showed that the haptic signal onset lagged the auditory one by 44 ± 26 ms. To avoid this confound in the main experiment, we added a 44 ms delay in software between slip detection and onset of the auditory feedback.

2.4   Paradigm

Before commencing the experiment, all participants were asked to take a stereo-acuity test. Upon successful completion (score $< 30''$), participants were familiarized with the control of the tele-operation hand using the Senseglove by free exploration. When they felt comfortable with controlling the hand, they were asked to put on the HMD and continue exploration. Haptic and auditory feedback was demonstrated to avoid startle reactions, and whenever the participant felt ready to proceed, the main experiment was started.

In the main experiment, the participant's task was to hold a remote object with the tele-operated hand. The object was a compressible, partially-filled water bottle (total weight 250 g). Aluminum plates were attached to its sides to provide a flat surface for sensor readings from the robotic hand, and a rope was attached to its bottom. Participants were asked to only use their index finger and thumb to hold the object, while keeping the other fingers curled towards the palm. At the start of each trial, the experimenter placed the object in the tele-operated hand, and the participant adjusted their grip to a firm hold, without excessively squeezing the object. Next, the experimenter introduced a vertical disturbance on the object by pulling on the rope. The participant was asked to keep a firm grasp on the object, without it slipping and without excessive squeezing. The slip data gathered from the robot's sensorized fingers was used to provide slip feedback to the user. A derivative of the force tangential to the object's surface was calculated, as a quick change in vertical force would indicate slip. Whenever slip velocity exceeded a pilot-determined threshold in tangential force velocity (i.e., a sensor-specific impedance velocity), the object was considered to be slipping and feedback was initiated. Three slip feedback conditions were tested: (1) auditory feedback using 450 Hz pure tone, (2) haptic feedback using Senseglove's standard vibration signal at 50% of the maximum amplitude, and (3) no slip feedback. In all conditions, kinesthetic haptic feedback was provided by mapping normal forces detected on the remote robot's fingers to a gentle resistive force on the Senseglove tendons. In addition, visual feedback of the object and part of the rope were also present in all conditions. Although the visual feedback of the rope tension provided some feedback about the timing of the disturbance, the magnitude of the force disturbance was varied by the experimenter, so the time between disturbance start and start of the slip was hard to predict for participants. Whenever the object was dropped before the participant could respond to the disturbance, the trial was rejected and repeated.

All trials of each slip feedback type were presented as a block, with ten repetitions of the same feedback type making up one block. When a block was completed, the experimenter verbally administered the 6-item NASA TLX questionnaire to assess the participant's cognitive load on a 21-point scale (see [4] for a full description of the questions). The order of the blocks - and thus the feedback types - was counterbalanced between the participants.

After completing the main experiment, a control experiment was run to test the participant's baseline response times to auditory and haptic stimuli. In two sequences, participants were asked to press a key as soon as they perceived an auditory (450 Hz beep) or haptic (Senseglove vibration on thumb and index finger) stimulus. In each sequence, a single feedback type was presented 10 times in 0.5 s long stimuli, with random intervals (range: 1–5 s) between the previous key press and the presentation of the next stimulus. The full experiment never exceeded 1 h, and participants could take a break between blocks.

2.5   Data Analysis

To test our hypothesis, we looked at the differences in the grip force profiles between the different feedback conditions. To quantify grip force, we used the total of the force registered in the BioTac thumb and index finger tips that was normal to the object's surface. For each trial, all signals were resampled 100 Hz, and the time at which the slip force threshold was passed was set to time $= 0$. The squeeze force at time $= 0$ was subtracted from the signal to make responses comparable between trials. Resultant forces were low-pass filtered 10 Hz using a second-order Butterworth filter. Then, time traces were averaged across repetitions, resulting in one trace per participant-condition combination. For each of these traces, the first peak in the time series was determined, and the time and

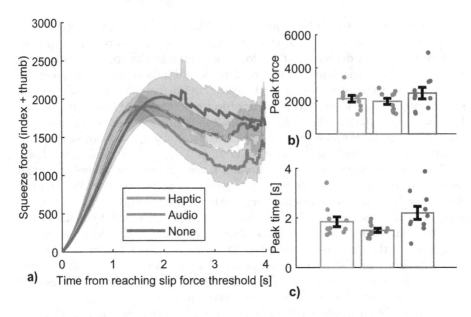

Fig. 3. Squeeze forces (mean ± 1 standard error), as measured at the remote robot's fingers. (a) Average time traces (solid lines) across participants. Time = 0 represents the time at which the slip threshold was passed. (b) and (c) Times and heights, respectively, of the squeeze force peaks, with one marker per participant.

height of the peaks were used as our outcome variables. Finally, squeeze force traces were averaged across participants for visualization purposes.

Prior to averaging across repetitions, and outlier analysis was performed by removing trials in which no squeeze force peak could be found. In this way, we selected trials in which participants reacted according to instructions. This outlier analysis removed 54 trials, which was 22% of the total number of trials.

For the control experiment, the time between the start of the stimulus and the participant's key press was calculated for each trial, and averaged across repetitions. To analyze the NASA TLX scores, data for each question were converted to a 0–100 score and averaged across participants.

3   Results

In our main experiment, we compared the effect of haptic and auditory feedback about object slip on the participant's squeeze behavior. The average time courses of the squeeze forces and the peak analyses are shown in Fig. 3. This figure suggest that feedback shifts the peak in squeeze force closer to initiation of slip (at time = 0). A repeated-measures ANOVA showed that there was indeed a significant effect of condition on peak timing ($F_{2,18} = 4.2, p = 0.032, \eta_p^2 = 0.32$). Bonferroni-corrected posthoc tests of peak timing showed a significant difference between auditory and no feedback ($t = -2.9, p = 0.029$), while neither auditory

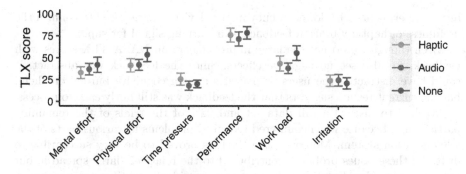

Fig. 4. Results (mean ± 1 S.E.) from NASA TLX questionnaire, which was administered after each feedback block. Physical effort was the only question that showed a significant effect of condition, with no significant posthoc comparisons.

and haptic nor haptic and no feedback differed significantly (all $t \leq 1.5$, $p \geq$ 0.48). For peak height, the repeated-measures ANOVA failed to find a significant difference between the conditions ($F_{2,18} = 0.90, p = 0.43, \eta_p^2 = 0.090$), so no posthocs were performed.

The control experiment showed that responses to auditory stimuli were 60 ms faster than responses to haptic stimuli, with haptic latency being $456 \pm$ 62 ms (mean $\pm$ standard error), and auditory latency being 396 ± 67 ms. A paired Student's t-test showed that this difference was significant ($t_8 = 2.8, p = 0.024$).

The NASA TLX scores, shown in Fig. 4, illustrate that the task required medium cognitive and physical load from participants, and that participants were satisfied with their own performance. Per-question repeated measures ANOVAs only found that Physical effort was exactly on the edge of significance ($F_{2,26} = 3.4, p = 0.050, \eta_p^2 = 0.21$), but in Bonferroni-corrected posthoc testing no significant comparisons remained. None of the other questions revealed significant differences between conditions (all $F_{2,26} \leq 2.5$, all $p \geq 0.10$).

4   Discussion

Auditory feedback helped participants to react significantly faster to slip, meaning that audio could be a useful warning signal for slip in tele-operation applications. While participants reacted faster, they did not significantly change their peak squeeze force. This shows that the auditory feedback did not induce a smaller safety margin, but also that users were still able to regulate their grip, while moving faster. Haptic vibration feedback failed to reach a significant effect on reaction time and squeeze force. This difference between the types of feedback could be caused by the small but significant (60 ms) difference in reaction times found in the control experiment. Another contributing factor could be the time it took the vibration motor to get to full vibration amplitude, which would add latency ($\sim$50 ms) to the onset latency which we already corrected for. Therefore, future experiments with lower-latency haptic devices are required to confirm the usefulness of haptic vibration feedback as a warning signal for slip.

Even though we do not see large improvements on NASA TLX scores with feedback, we also see no negative effects. Since the feedback was abstract, it could have distracted the users or placed an extra cognitive burden on them, but the current results suggests that the feedback was still fairly easy to process.

We had to discard a full data set and 22% of the trials of the remaining participants, because we encountered technical problems in various parts of the tele-operation system. Moreover, the BioTacs proved to be very susceptible to drift. All these issues probably contributed to the relatively large spread in our outcomes variables. Even though these were not ideal experimental conditions, we still were able to gather enough data to sketch an image of the usefulness of slip feedback. It also illustrates the complexity of full tele-operation systems, and thus underlines the need for testing feedback in these actual scenarios.

Future work on slip feedback could focus on developing a method to initiate slip more instantaneously, and with more control over its magnitude and duration. Such a method would allow for presenting realistic slip profiles, without providing additional visual cues. The slip detection procedure itself could be iterated on, by for instance adding information from the micro-vibration sensors in the BioTacs [1]. Another interesting avenue would be integrating more realistic, low latency slip feedback devices on the user's side. Ultimately, these developments might allow tele-operation users to recruit reflexive behavior that is present in normal interactions with physical objects.

References

1. Fishel, J., Loeb, G.: Sensing tactile microvibrations with the biotac comparison with human sensitivity. In: IEEE/RAS-EMBS BioRob, pp. 1122–1127 (2012)
2. Flanagan, J., Tresilian, J., Wing, A.M.: Coupling of grip force and load force during arm movements with grasped objects. Neurosci. Lett. **152**(1), 53–56 (1993)
3. Hannaford, B., Wood, L., McAffee, D.A., Zak, H.: Performance evaluation of a six-axis generalized force-reflecting teleoperator. IEEE Trans. Syst. Man Cybern. **21**(3), 620–633 (1991)
4. Hart, S.G.: Nasa-task load index (nasa-tlx); 20 years later. In: Proceedings of the Human Factors and Ergonomics, vol. 50, pp. 904–908. Sage publications: Los Angeles, CA (2006)
5. Johansson, R.S., Westling, G.: Signals in tactile afferents from the fingers eliciting adaptive motor responses during precision grip. Exp. Brain Res. **66**(1), 141–154 (1987)
6. Pacchierotti, C., Meli, L., Chinello, F., Malvezzi, M., Prattichizzo, D.: Cutaneous haptic feedback to ensure the stability of robotic teleoperation systems. Int. J. Robot. Res. **34**(14), 1773–1787 (2015)
7. Su, Z., Fishel, J., Yamamoto, T., Loeb, G.: Use of tactile feedback to control exploratory movements to characterize object compliance. Front. Neurorobot. **6**, 7 (2012)
8. Toet, A., Kuling, I.A., Krom, B.N., van Erp, J.B.F.: Toward enhanced teleoperation through embodiment. Front. Robot. AI **7**, 14 (2020)

9. Walker, J.M., Blank, A.A., Shewokis, P.A., O'Malley, M.K.: Tactile feedback of object slip facilitates virtual object manipulation. IEEE Trans. Haptics **8**(4), 454–466 (2015)

10. Webster, R.J., Murphy, T.E., Verner, L.N., Okamura, A.M.: A novel two-dimensional tactile slip display: Design, kinematics and perceptual experiments. ACM Trans. Appl. Percept. **2**(2), 150–165 (2005)

11. Westling, G., Johansson, R.S.: Factors influencing the force control during precision grip. Exp. Brain Res. **53**(2), 277–284 (1984)

Experiencing Touch by Technology

Judith Weda[1(✉)] , Dasha Kolesnyk[1] , Angelika Mader[1] ,
and Jan van Erp[1,2]

[1] University of Twente, Drienerlolaan 5, 7522 NB Enschede, The Netherlands
`j.weda@utwente.nl`
[2] TNO, Kampweg 55, 3769 DE Soesterberg, The Netherlands

Abstract. Touch technology can mediate social touch in situations when people cannot be physically close. Recent social touch technologies use haptic actuators capable of displaying pressure touch. We studied experience in two set-ups which use such actuators: a motorized ribbon and a McKibben sleeve. We investigated whether there is an inherent emotional and sensory experience attached to sensations produced by those set-ups. Participants were presented with pressure touches varying in rate of force change, peak force and contact area. Participants rated the sensory and emotional experience of each stimulus variation with a check-all-that-apply measure of 79 items in two sections and the Emoji-grid. We found that force has a major effect on the experience of a passive pressure touch. Speed and width also played a role, but to a lesser extent and only in one of the set-ups. The results inform the design of mediated social touch applications in making the technology more congruent with the context.

Keywords: Passive touch · Mediated social touch · Touch experience

1 Introduction

Social touch plays a key role in close social relationships. However, distance and social isolation create barriers for social touch. Touch deprivation might lead to loneliness [4]. The negative effects of touch deprivation can be partially mitigated by mediating social touch through technology [14]. Social touch technology (STT) is in continuous development. To inform this development, we studied user experience arising from technology-produced passive touch on the arm. Our findings can make mediated social touch more pleasant and acceptable.

The arm is one of the most comfortable and socially acceptable areas for receiving touch [12]. The arm is also suitable for mediated social touch, because it is easy to fix a device on the arm and to adjust fit to the user parameters. Therefore, a large proportion of STT is designed for use on the arm.

This project has received funding from the European Union's Horizon 2020 research and innovation programme under grant agreement No 825232.

H. Seifi et al. (Eds.): EuroHaptics 2022, LNCS 13235, pp. 110–118, 2022.
https://doi.org/10.1007/978-3-031-06249-0_13

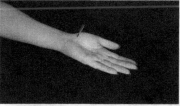

Fig. 1. Example stills of the videos shown during the interview (Videos 1, 5, and 9 from [7]. The custom videos are archived under DOI: 10.17632/9s6c2rzz8t.1)

STT can be based on vibration, force and temperature actuators [5]. More recent developments in STT focused on actuators that can produce pressure. Pressure actuators use different technologies: from tension bands [10], to pneumatics [17], to electroactive textile [8]. As pressure-based STT is being developed, more research is needed to determine what type of pressure would be most suitable to mediate social touch from the user perspective.

Limited guidance exists on factors that contribute to how users experience technology-produced sensation of pressure. Investigating the experience of touch with users is key to user-centred design. Tactile stimuli might have an inherent meaning or emotional associations. Although these associations will be modified by context [1], we must be aware of them to design a congruent experience.

In this study, we compared two actuators: a motorized ribbon, and a McKibben sleeve [16]. We investigated the effect of three actuator parameters - the peak force, the rate of force change, and the surface area with two levels each - on the sensory and emotional experience of the users. User experience was assessed through a multiscale experience profiling method (MEP method). The results can inform future STT design.

2   Study 1. Interviews for the Experience Profiling

The goal of this study was to identify a semantic field of expressions describing passive touch to develop the MEP method that will be applied in Study 2.

Participants. Since the experience of touch may depend on cultural background we performed semi-structured interviews with one participant from Sweden, one from Iran and one from China. The varying backgrounds provided a varied sample and starting point for the MEP method. All participants were interviewed in their own language. Interviews were conducted online. Before the interview started, participants read and agreed with an informed consent.[1] The participants received no payment.

Apparatus. Before the interview the participants were asked to do a brainstorm exercise designed to help them access their vocabulary on touch. The participants

---

[1] The research was reviewed and approved by the ethics committee of the EEMCS faculty of the University of Twente (reference RP 2020-104).

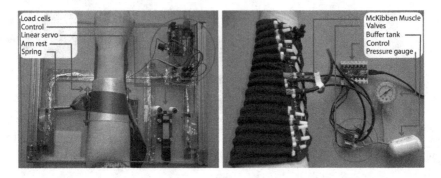

Fig. 2. The motorized ribbon on the left (the spring constant = 1.18 N/2 times the servo extension), the McKibben sleeve on the right

were shown touch videos from the socio-emotive touch database [7] and custom made videos of touch by objects [15] with the instruction to focus on the passive touch in the interaction. Figure 1 depicts stills taken from both types. Three videos of each category were shown and each video had a duration of two to eight seconds. The videos were viewed in the (online) presence of the interviewer. After each video, participants were asked eight open questions about the touch. The original interview questions were in English [15] and translated by interviewers to their native language. All conducted interviews were transcribed and translated to English for analysis[2]. The interview was semi-structured and took 45 min.

Results. We used grounded theory to analyze the interviews [6]. We found the following nine categories: Dynamic Properties (type, speed), Tactile properties (type, texture, localization, wet/dry), Comfort (mental/physical), Pain, Effect of the touch, Timbre, Affect, Bigger meaning and Context. For the development of the MEP method, we selected words in the categories most relevant to the emotional and sensory experience: Dynamic properties, Tactile properties, Effect of the touch, and Timbre. We limited the number of words to not overload the participants.

3   Study 2. Experience Profiling for Pressure Stimuli

Participants. A total of 52 students (local and international) and university staff members participated in study 2[3], 24 (age range 18–47, mean age 27.5, 9 females) experienced the ribbon set-up and 28 (age range 19–47, mean 25.5, 14 females) the McKibben set-up. Participants received a 5 Euro giftcard.

Apparatus. We used two set-ups to generate pressure stimuli on the lower arm: a motorized ribbon (Fig. 2, left) and a McKibben sleeve (Fig. 2, right) [16].

---

[2] Thanks Carin Backe, Hamid Souri and Fengdi Li for conducting the interviews.
[3] The research and Covid-19 precautions were reviewed and approved by the ethics committee of the EEMCS faculty of the University of Twente (reference RP 2021-200).

Table 1. The parameter settings for each set-up. The forces are for an arm circumference of 19 cm. AL denotes the length of the actuator in contact with the arm.

	Motorized Ribbon	McKibben Sleeve
Peak force (low, high)	6.3 Kpa, 24.8 Kpa (small area) 0.5 Kpa, 1.6 Kpa (large area)	5.6 Kpa, 8.9 Kpa
Rate of force change (slow, fast)	3027 ms, 1766 ms (low force), 4457 ms, 2724 ms (high force)	464 ms, 334 ms (low force), 496 ms, 453 ms (high force)
Surface area (small, large)	3 mm × AL, 40 mm × AL	13 mm × AL, 39 mm × AL

Motorized ribbon. A ribbon is wrapped around the arm on an armrest. A second ribbon is below the first to prevent the sensation of shifting. The ribbon is tightened around the arm using a linear motor that pushes the thread tied to the ribbon between two vertical rollers. The thread runs through ball bearings to reduce the amount of friction and lateral forces on the skin. The force on the arm is calculated by measuring the contraction force at two ends.

McKibben sleeve. The sleeve has 13 tunnels with McKibben actuators inserted with an average width of 13 mm each. McKibben actuators have a braided mesh outer sleeve and an elastic inner tube. When pressurized air enters the inner tube it expands. The longitudinal stiffness of the braided outer sleeve limits its increase in diameter causing linear contraction and creating pressure on the arm.

Stimulus parameters. In each set-up, we varied three stimulus parameters: the peak force, the rate of force change, and the surface area with two levels each. The eight stimuli for each set-up were presented three times to the participant. Table 1 summarizes the parameter settings for each set-up.

MEP Method. User interviews combined with literature research was our approach to create a check-all-that-apply (CATA) list [9]. We used the results of Study 1 and the work by [3] (lists with sensory and emotional properties of touch experience) to create two CATA lists: one for sensory and one for emotional qualities (understanding, happy, exciting, endearment, comforting, uplifting, thrilling, gentle, human, shocking, sexy, delicate, social, frightening, sensual, loving, mechanical, annoying, pleasurable, supportive, non-social, sad, desirable, friendly, surprising, upset, comfortable, aggressive, calming, irritating, relaxing, frustrated, dreadful, arousing, soothing, fleeting, pressing, smooth, dull, shaking, poking, rough, sharp, vibrating, slow, soft, hot, hitting, fast, flexible, cold, dragging, fluid, rubbery, lukewarm, tickling, itchy, tough, wet, stretching, friction, pointy, dry, patting, stinging, wrinkled, comfortable, squeezing, burning, textured, uncomfortable, embracing, flat, elastic, painful, tapping, hard, non-elastic, not painful). We added an option for not feeling touch, since sensitivity varies per person, and pilots showed that a slow rate of force change may be difficult to perceive. We also added the items human, mechanical, social and non-social, to measure if the touches are considered inherently social or human. In addition, we used the emojigrid [13] for the participant's rating of valence and

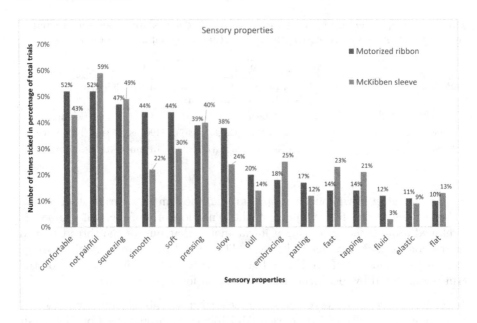

Fig. 3. Sensory properties for the motorized ribbon and the McKibben sleeve

arousal of each stimulus. These three parts combined allow us to broadly explore the touch qualities while keeping the effort for the participants acceptable.

Procedure. The participant's arm was placed in the set-up such that the middle between wrist bone and elbow was positioned under the ribbon, at this position the circumference was measured. The experiment was self-paced. After a signal by the participant a stimulus was presented followed by a part of the MEP method. The stimuli and the parts of the MEP method were presented in a randomized order. All three parts were completed before randomizing the next stimulus.

Results. Data of one participant from the McKibben sleeve experiment had to be deleted due to a technical error.

Sensory Properties. The frequencies of selecting each adjective across all trials are presented in Fig. 3 for all adjectives that were mentioned in at least 5% of the trials. As Fig. 3 shows, overall the participants experienced "comfortable", "non-painful", "squeezing", "smooth", "soft", and "pressing" sensations (mean percentage > 25%), which is in line with our expectations. Not one single trial felt "painful". The McKibben sleeve was less often characterized as "smooth", "soft", "slow", and more often as "fast", "tapping".

In order to assess whether the frequencies were affected by each of the stimulus parameters, independent t-tests with 1000 bootstraps were performed for frequencies as dependent variable and surface area, peak force, and rate of force change as independent variables. The bootstrapping was used because the

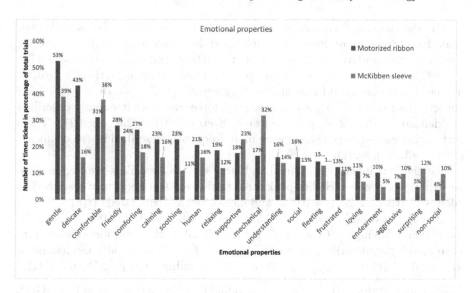

Fig. 4. Emotional properties for the motorized ribbon and the McKibben sleeve

assumption of observation independence does not hold for the trials, and the sample is very small. We did not correct for multiple testing because of low power and a high probability of Type II error. Since generalising to population is not the goal of the study, power concerns are favored over the risk of Type II error.

For the motorized ribbon, surface area showed no effects. Trials with higher peak force were perceived significantly more often as "squeezing" ($t(3.12) = -4.05$, p = .025), "embracing" ($t(4.09) = -5.17$, p = .006) and "rubbery" ($t(6) = -2.611$, p = .040), while trials with lower peak force were perceived more often as "flat" ($t(6) = 4.33$, p = .005) and "dull" ($t(6) = 2.53$, p = .044). Trials with a higher rate of force change were perceived more often as "fast" ($t(6) = -9.02$, p = .000).

For the McKibben sleeve, small surface area was more frequently described as "patting" ($t(6) = 4.92$, p = .003) compared to large surface area. Trials with high peak force were more often described as "slow" ($t(6) = -3.27$, p = .017), "itchy" ($t(3) = 5.0$, p = .015) and "uncomfortable" ($t(6) = -3.0$, p = .024) compared to lower peak force trials. Rate of force change showed no significant effects.

Emotional Properties. The frequencies of selecting each adjective across all trials are presented in Fig. 4 for adjectives mentioned in at least 5% of the trials. Overall the touch by the motorized ribbon elicited "gentle", "delicate", "comfortable", "friendly", "comforting" emotional experiences (mean percentage > 25%). McKibben sleeve seemed to elicit less delicate or gentle, and more mechanical emotional experiences.

The effects of surface area, peak force, and rate of force change were assessed in the same way as the sensory properties adjectives.

For the motorized ribbon, surface area showed no effects. Trials with high peak force were more frequently described as "endearing" (t(6) = −3.46, p = .013), "supportive" (t(6) = −3.14, p = .02), "comforting" (t(6) = −2.56, p = .043), "social" (t(6) = −2.53, p = .001) and "human" (t(6) = −6.48, p = .001), but also more frequently as "aggressive" (t(6) = −5.75, p = .001) and "upset" (t(6) = −3.27, p = .017). Low peak force was more frequently described as "delicate" (t(6) = 2.64, p = .038) and "fleeting" (t(6) = 2.9, p = .027). Trials with higher rate of force change were more frequently described as "mechanical" (t(6) = −3.69, p = .01).

For the McKibben sleeve, small and large surface area trials resulted in similar experience profiles. Higher peak force trials were less frequently described as "delicate" (t(6) = 5.29, p = .0.002) and more often as "frustrated" (t(6) = −3.38, p = .0015), "fleeting" (t(6) = −2.53, p = .0045), "exciting" (t(6) = −2.45, p = .05) and "loving" (t(6) = −2.45, p = .05). Trials with higher force change rate were more often described as "surprising" (t(3) = −7, p = .006), and less often as "calming" (t(6) = 2.65, p = .038) or "understanding" (t(6) = 2.78, p = .032).

Valence and Arousal. To access the emotional experience reflected in EmojiGrid, a 3 × 2 repeated measures MANOVA was performed with surface area, peak force, and rate of force change as within-subject independent variables, the scores for the dimensions of valence (x-axis) and arousal (y-axis) as two dependent variables. The scores varied between 0 and 220 for both axes.

For the motorized ribbon peak force had significant effects on emotional experience measured by the EmojiGrid, Wilk's lambda = .54, F(2, 22) =9.45, p = .001. Surface area had a marginally significant effect, Wilk's lambda = .76, F(2, 22) = 3.43, p = .051.

The univariate tests further revealed that arousal score was higher for larger surface area ribbon (M = 135.5) compared to smaller surface area ribbon (M = 118.9), F(1, 23) = 6.04, p = .022. The peak force also effected arousal so that the arousal score was lower for trials with higher peak force (M = 114.2) compared to trials with lower peak force (M = 140.2), F(1, 23) = 18.64, p < .001. There were no effects of surface area or peak force on the valence axis score.

For the McKibben sleeve, peak force, Wilk's lambda = .72, F(2, 25) = 4.96, p = .015, and rate of force change, Wilk's lambda = .65, F(2, 25) = 6.67, p = .005, had significant effects on emotional experience measured by the EmojiGrid. The univariate tests further revealed that peak force had an effect on both valence and arousal scores. Trials with higher peak force resulted in less positive emotional experience (M = 123.8) compared to trials with lower peak force (M = 141.4), F(1, 26) = 8.68, p = .007. Peak force also effected arousal so that the arousal score was lower for trials with higher peak force (M = 117.1) compared to trials with lower peak force (M = 129.8), F(1, 26) = 4.90, p = .036. The rate of force change effected the valence so that emotional score was more positive for lower rate of force change (M = 140.4) compared to higher rate of force change (M = 124.6), F(1, 26) = 13.80, p = .001. There were no effects of the rate of force change on the arousal axis score, F(1, 26) = 1.90, p = .176.

4   Discussion and Conclusion

We examined whether inherent meanings are associated with pressure presented to users' arms without any context. Several conclusions follow the results.

(1) We find that perceptions covary systematically with the properties of the stimuli. Therefore, there seems to be inherent meanings associated with different stimulus properties. (2) The two actuators seem to produce comparable experiences. Four of the top-5 sensory experiences for the two actuator types overlap (i.e., "comfortable", "not painful", "squeezing", and "soft"), and three of the top-5 emotional experiences ("gentle", "comfortable", and "friendly"). However, there are also differences in the profiles that may be related to the (confounding) parameter settings. Looking at the differences: the McKibben sleeve elicited less delicate, less gentle, and more mechanical emotional experiences in comparison to the motorized ribbon. The difference could be due to the fact that both rate of force change settings were faster than the motorized ribbon speed settings. This is supported by the findings that trials with higher rate of force change of the motorized ribbon were also more frequently described as mechanical. (3) In our studies for both set-ups higher peak force was more often perceived as comforting, social, and human. This is in line with previous findings that substantial pressure can have comforting and calming effects [2,11]. High peak forces tend to evoke more emotions and are more comfortable, thus are more suitable for STT. Pressure needs to be high enough to have a calming, comforting effect, but not become uncomfortable. The drawback is that valence of such pressure can vary depending on context, so the context must be chosen carefully.

Overall, the results suggest that it might be easier to produce a delicate, smooth, soft and calming sensation with the motorized ribbon than with the McKibben sleeve and thus be preferable for creating comforting social touch, at least with the conditions that we set for our investigation. More importantly, our findings suggest that every specific technology should be tested for inherent meanings and associations, so that stimuli are appropriately aligned with the context for the congruence of experience.

References

1. Askari, S.I., Haans, A., Bos, P., Eggink, M., Lu, E.M., Kwong, F., IJsselsteijn, W.: Context matters: The effect of textual tone on the evaluation of mediated social touch. In: Nisky, I., Hartcher-O'Brien, J., Wiertlewski, M., Smeets, J. (eds.) EuroHaptics 2020. LNCS, vol. 12272, pp. 131–139. Springer, Cham (2020). https://doi.org/10.1007/978-3-030-58147-3_15
2. Grandin, T.: Calming effects of deep touch pressure in patients with autistic disorder, college students, and animals. J. Child Adolesc. Psychopharmacol. 2(1), 63–72 (1992)
3. Guest, S., Dessirier, J.M., Mehrabyan, A., McGlone, F., Essick, G., Gescheider, G., Fontana, A., Xiong, R., Ackerley, R., Blot, K.: The development and validation of sensory and emotional scales of touch perception. Atten. Percept. Psychophys. 73(2), 531–550 (2011)

4. Heatley Tejada, A., Dunbar, R., Montero, M.: Physical contact and loneliness: Being touched reduces perceptions of loneliness. Adapt. Hum. Behav. Physiol. **6**, 292–306 (2020)

5. Huisman, G.: Social touch technology: A survey of haptic technology for social touch. IEEE Trans. Haptics **10**(3), 391–408 (2017)

6. Lazar, J., Feng, J.H., Hochheiser, H.: Research Methods in Human-Computer Interaction. Morgan Kaufmann, Cambridge, MA (2017)

7. Lee Masson, H., Op de Beeck, H.: Socio-affective touch expression database. PLoS One **13**(1), e0190921 (2018)

8. Melling, D., Martinez, J.G., Jager, E.W.: Conjugated polymer actuators and devices: Progress and opportunities. Adv. Mater. **31**, 1808210 (2019)

9. Ng, M., Chaya, C., Hort, J.: Beyond liking: Comparing the measurement of emotional response using essense profile and consumer defined check-all-that-apply methodologies. Food Qual. Prefer. **28**(1), 193–205 (2013)

10. Pezent, E., Israr, A., Samad, M., Robinson, S., Agarwal, P., Benko, H., Colonnese, N.: Tasbi: Multisensory squeeze and vibrotactile wrist haptics for augmented and virtual reality. In: 2019 IEEE World Haptics Conference, pp. 1–6. IEEE (2019)

11. Sato, W.: Inhibition of emotion-related autonomic arousal by skin pressure. Springerplus **4**(1), 1–4 (2015). https://doi.org/10.1186/s40064-015-1101-9

12. Suvilehto, J.T., Glerean, E., Dunbar, R.I., Hari, R., Nummenmaa, L.: Topography of social touching depends on emotional bonds between humans. Proc. Natl. Acad. Sci. **112**(45), 13811–13816 (2015)

13. Toet, A., van Erp, J.B.: The emojigrid as a rating tool for the affective appraisal of touch. PLoS One **15**(9), e0237873 (2020)

14. Van Erp, J.B., Toet, A.: Social touch in human-computer interaction. Front. Digit. Humanit. **2**, 2 (2015)

15. Weda, J., Mader, A., Kolesnyk, D., van Erp, J.: Experiencing touch by technology - videos and interview questions. https://doi.org/10.17632/9s6c2rzz8t.1

16. Weda, J., Henell, E., Kolesnyk, D., Mader, A., van Erp, J.: Perception and experience profiling, report H2020 WEAFING, Grant No 825232 (2021, in press)

17. Young, E.M., Memar, A.H., Agarwal, P., Colonnese, N.: Bellowband: A pneumatic wristband for delivering local pressure and vibration. In: 2019 IEEE World Haptics Conference (WHC), pp. 55–60. IEEE (2019)

Effect of Focus Direction and Agency on Tactile Perceptibility

Zane A. Zook[✉][ID] and Marcia K. O'Malley[ID]

Department of Mechanical Engineering, William Marsh Rice University,
Houston, TX 77005, USA
{gadzooks,omalley}@rice.edu

Abstract. Prior research has shown that the direction of a user's focus affects the perception of tactile cues. Additionally, user agency over touch stimulation has been shown to affect tactile perception. With the development of more complicated haptic and multi-sensory devices, simple tactile cues are rarely used in isolation and the effect of focus direction and of user agency on the perception of a sequence of tactile cues is unknown. In this study, we investigate the effect of both of these variables, focus direction and agency, on the perception of a cue sequence. We found that the direction of user focus and user sense of agency over tactile stimulation both had a significant effect on the accurate perception of a cue sequence. These results are presented in consideration for developing better haptic devices that account for users' focus on and control over these devices.

Keywords: Tactile perception · Tactile focus · Tactile agency

1 Introduction

Wearable haptic devices have prospered in the consumer world and become the subject of much recent haptics research aiming to expand their capabilities. These devices commonly leverage sequences of tactile cues such as vibration [11,17], skin stretch [1,5], and squeeze [2,13] in a perceptually distinct manner to the other senses, thus avoiding the negative consequences of sensory overload [10]. However, prior work has indicated the perception of tactile stimuli can change due to the direction of a user's focus (also referred to as attention in literature from other fields) [9,16] across the body and across multiple senses [3,8]. For brevity in this paper, we define focus direction as the aforementioned direction of a user's focus, either across the body spatially or across multiple

This work was supported by the National Science Foundation under Grant No. CMMI-1830146. This material is also based upon work supported by the National Science Foundation Graduate Research Fellowship Program under Grant No. 1842494. Any opinions, findings, and conclusions or recommendations expressed in this material are those of the author(s) and do not necessarily reflect the views of the National Science Foundation.

H. Seifi et al. (Eds.): EuroHaptics 2022, LNCS 13235, pp. 119–126, 2022.
https://doi.org/10.1007/978-3-031-06249-0_14

senses. To inform future research in developing sequences of tactile stimulation in increasingly complicated environments, one goal of this work is to understand the effect of focus direction on tactile perception.

In addition, there are differences in the perception of haptic cues that are "actively touched," where the user purposefully feels the environment as opposed to "passively touched," where the user reflexively senses a stimulus that has been applied to their skin [7]. These two "modes" of touch differ distinctly in the amount of agency, or control, the user has over the tactile stimulation. The effect of agency on tactile perception has not been well studied and is particularly relevant given new research directions, including studies on how to deliver rich information during active interaction with the environment, such as in responsive virtual reality [13,15], and motion feedback for prostheses [1,18]. Passive tactile interactions have also been investigated in prior work on delivering language cues through phonemes [6,17], and navigation cues [4]. The relevance of this type of research motivates the second goal of this work: to understand the role of agency on tactile perception.

To address these goals, we investigate the effect of focus direction and agency on the perception of tactile cues.

2   Methods

We tested participants' ability to accurately discriminate a single target vibration cue from a sequence of vibrations at different spatial locations on the forearm. We measured participant accuracy as the proportion of correct responses out of all trials presented in each condition while varying focus direction and agency. For this study, focus direction refers to the orientation of the participant's focus, either spatially across the skin or across two sensory modalities (touch and sound). Agency refers to the participant's ability to directly control the onset of stimulation or lack thereof. Our objective was to determine if these variations in focus direction and agency affected participants' ability to complete the task accurately.

2.1   Participants

A total of 18 participants (7 female, 14 right-handed, 19–28 years old, average age 24) took part in this study. All participants in the study were healthy adults and did not suffer any cognitive or motor impairment that would affect their ability to perform the experiment. All participants gave informed consent and all procedures and methods of the experimental protocol were approved by the Rice University Institutional Review Board (IRB-FY2019-49).

2.2   Experimental Hardware

This experiment used a modified version of the Vibro-Tactile Sleeve (VT-Sleeve) used in Macklin et al. [11]. The VT-Sleeve features 6 vibrotactile actuators (Compact Audio Exciter; Tectonic; Part No. TEAX14C02-8) embedded in a compression sleeve (Under Armour) with custom 3D-printed housings. The vibrotactors

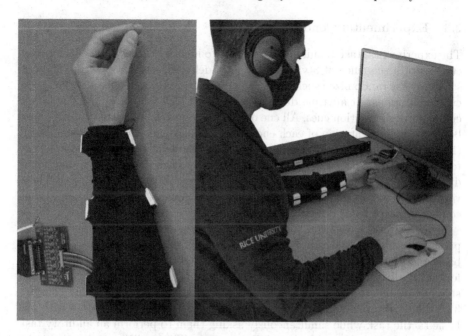

Fig. 1. Participant wearing the vibrotactile sleeve, positioned to complete the experimental task. (left) Vibrotactile sleeve used in experiments to deliver vibrotactile cues to the participant via 6 vibrotactors. Tactors were evenly spaced on the participant's forearm between their wrist and elbow with 3 vibrotactors on the ventral side of the arm and 3 vibrotactors on the dorsal side. (right) Participant wearing the vibrotactile sleeve while interacting with the experimental GUI on the monitor and listening to pink noise through headphones. (Color figure online)

were each press-fit into their respective 3D-printed housing, which clipped into slits in the compression sleeve. The vibrotactors were placed 60mm apart on the sleeve when relaxed with 3 vibrotactors on the ventral side of the arm and 3 vibrotactors on the dorsal side. The sleeve held each vibrotactor snugly against the skin as shown in Fig. 1. The sleeve was adjusted on all participants such that the vibrotactors on each side of the forearm were sufficiently separated to satisfy the two-point discrimination threshold reported for successive touch stimuli [12]. Wiring powering each vibrotactor was routed through another small slit in the compression sleeve directly next to the housing element, ensuring no direct contact between the wiring and skin. These wiring elements connected to a quick release ribbon cable connector, which in turn connected the tactors to a custom amplifier.

All tactile cues were 250 ms envelope sine waves presented 400 Hz for the target cue and 100 Hz for the distraction cues. Extensive pilot testing was done to verify that these characteristics resulted in clear and distinct vibrotactile cues. Tactile cues were rendered with Syntacts using a digital-to-analog converter (MOTU 24Ao USB) with signals amplified through the Syntacts V3.0 amplifier [14].

2.3 Experimental Conditions

The experimental task required participants to identify the location of the target cue out of a sequence of 6 vibrotactile cues delivered on the forearm. The cue sequence was randomized such that each vibrotactor vibrated once and the target cue could appear at anytime during the sequence. All sequences had one target cue and five distraction cues. All cue durations were 250 ms, with a 400 ms pause between the presentation of each cue. After each sequence, the participant used their mouse to indicate which vibrotactor they thought delivered the target cue to the forearm. This basic tactile task varied in each condition to shift focus direction or to adjust agency over sequence presentation in each trial. These variations made up the 6 conditions in this experiment to investigate the various combinations of focus direction and agency.

Focus direction was split into three categories: double focus, where participants were given a separate auditory task to perform simultaneously with the basic tactile task (as described above); wide focus, where the participant performed the basic task; and narrow focus, where the participant was told which side of the forearm the target would appear and therefore would only consider 3 of the 6 vibrotactors. The double focus task presented participants with the basic tactile task while simultaneously asking them to perform an auditory task that started in tandem with the tactile sequence. This auditory task presented a sequence of 6 auditory beeps 440 Hz each separated by 750 ms through the participant's headphones. These beeps were presented with two perceptually distinct volumes: one quiet and one loud. The participant was asked to count the number of loud beeps out of the 6 in the sequence while simultaneously performing the basic tactile task. While each participant was instructed to perform both tasks to the best of their ability, we report participant accuracy in completing the tactile task only, as the purpose of the auditory task was only to direct focus across two tasks. The wide focus category presented the basic tactile task with 6 cues in sequence to the participant and asked them to identify the location of the target cue. The narrow focus task informed the participant about which side of the forearm the target cue was going to appear via arrow icons presented in the GUI. This directed the participant's focus onto the 3 tactors on one side of the forearm. The participant then received the sequence of 6 cues and was asked where the target cue appeared.

Agency variations were split into only two categories: active, where the participants had a button to start the vibrotactile sequence; and passive, where the participants were allowed 30 s after each sequence was displayed to provide their response before the trial automatically advanced. The conditions in this experiment crossed these two categories such that there were 6 conditions for each participant presented in a random order: double focus passive, double focus active, wide focus passive, wide focus active, narrow focus passive, and narrow focus active.

2.4   Procedure

Participants were seated in front of a computer screen with their left hand placed on the table and their right hand using a computer mouse. The participant wore the vibrotactile sleeve on their left forearm as described in Sect. 2.2. The participant wore noise canceling headphones playing pink noise throughout the experiment to isolate them from any outside auditory stimulation. After receiving instructions from the experimenter, the participant interacted with a GUI on the computer to begin the experiment. Participants started with 5 practice trials and received feedback on the correct response in each trial. After practice, participants responded to 20 identification trials in a randomly selected condition as described in Sect. 2.3. Participants repeated this process for all six conditions.

3   Results

Participant accuracy was calculated for each condition based on the number of correct responses out of the 20 trials presented. For double focus conditions, accuracy scores were calculated solely on the participant's ability to complete the tactile task. Two participants were non-compliant with instructions and their accuracy scores were below random chance ($\approx 16.7\%$) for a majority of the experimental conditions. For these reasons the data from these two subjects were excluded from data analysis. The mean and standard deviation for each condition group are reported in Table 1. Figure 2 shows the spread of participant accuracy across the six conditions in this experiment. The narrow conditions showed the highest mean accuracy while the double conditions showed the lowest. Additionally while there was little variation in the means between the active agency and passive agency conditions, the average standard deviations in the active conditions were generally larger than the average standard deviations in the passive conditions.

A two way repeated measures ANOVA was performed to analyze the effect of focus direction and agency on participant accuracy. This ANOVA showed that both focus direction and agency had a statistically significant effect on participant accuracy after Greenhouse—Geisser correction (Focus Direction: $F(2, 30) = 4.05$, $p < .05$, Agency: $F(1, 15) = 5.40$, $p < .05$). There was no statistically significant interaction between focus direction and agency ($F(2, 30) = .60$, $p = .54$). Subsequent post-hoc pairwise t-tests performed between groups also showed no statistically significant difference between groups after Bonferroni correction.

Table 1. Participant mean and standard deviation accuracy scores for each condition

Agency focus direction	Active			Passive		
	Double	Wide	Narrow	Double	Wide	Narrow
Mean (%)	74.38	77.81	84.69	78.75	84.06	85.63
Std. dev. (%)	19.40	17.22	19.95	13.48	17.05	19.48

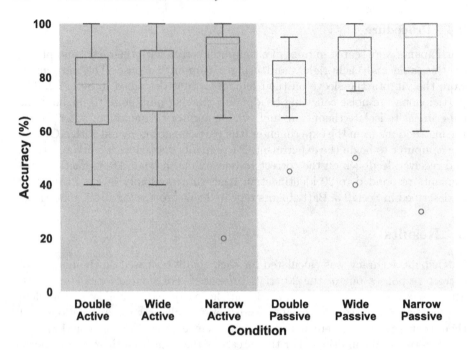

Fig. 2. Participant accuracy for each condition. On the left, the blue shaded region groups the three conditions with active agency and, on the right, the yellow shaded region groups the three conditions with passive agency. The labels double, wide, and narrow correspond to the double task conditions, the wide focus conditions, and the narrow focus conditions respectively. The red circles indicate a participant that varied by more than 3 standard deviations from the condition mean. (Color figure online)

4 Discussion

Results from this experiment confirm that there is a significant effect of focus direction and of agency on a participant's accuracy in perceiving a target tactile stimulus out of a randomly ordered sequence of five distraction stimuli and one target stimulus. This matches our expectations that the direction of participant focus on tactile stimuli and participant agency over tactile stimuli affect their ability to perceive the cue accurately. Contrary to our expectations however, the differences in accuracy between double, wide, and narrow focus and between active and passive agency were smaller than expected. After correction there were no significant differences in between condition groups, despite anecdotal increased confidence in the narrow focus conditions over the double focus conditions. Although between-group results were not shown to be significant in this initial investigation, these results suggest that between-group effects may exist for both focus direction and agency. Follow-up studies with a wider participant pool and designed to look more directly at focus direction and agency may show more granular differences between focus direction and agency groups.

5   Conclusion

We conducted a study to investigate the effect of focus direction and agency on participant accuracy in the perception of sequences of vibrotactile cues. Our experiment used a custom vibrotactile sleeve to deliver sequences of vibration stimuli to a user across six conditions that varied the direction of the participants' focus on and the participants' agency over each trial. Results from this experiment indicate focus direction and agency both have a significant effect on accuracy in perceiving a specific tactile stimulus out of a sequence. Further research is required to learn more comprehensively about specific differences in between levels of focus direction and levels of agency.

References

1. Battaglia, E., Clark, J.P., Bianchi, M., Catalano, M.G., Bicchi, A., O'Malley, M.K.: Skin stretch haptic feedback to convey closure information in anthropomorphic, under-actuated upper limb soft prostheses. IEEE Trans. Haptics **12**(4), 508–520 (2019). https://doi.org/10.1109/TOH.2019.2915075

2. Casini, S., Morvidoni, M., Bianchi, M., Catalano, M., Grioli, G., Bicchi, A.: Design and realization of the CUFF - clenching upper-limb force feedback wearable device for distributed mechano-tactile stimulation of normal and tangential skin forces. In: 2015 IEEE/RSJ International Conference on Intelligent Robots and Systems (IROS), pp. 1186–1193, September 2015. https://doi.org/10.1109/IROS.2015.7353520

3. Chica, A.B., Sanabria, D., Lupiáñez, J., Spence, C.: Comparing intramodal and crossmodal cuing in the endogenous orienting of spatial attention. Exp. Brain Res. **179**(3), 353–364 (2007). https://doi.org/10.1007/s00221-006-0798-7

4. Chinello, F., Pacchierotti, C., Bimbo, J., Tsagarakis, N.G., Prattichizzo, D.: Design and evaluation of a wearable skin stretch device for haptic guidance. IEEE Robot. Autom. Lett. **3**(1), 524–531 (2018). https://doi.org/10.1109/LRA.2017.2766244

5. Clark, J.P., Kim, S.Y., O'Malley, M.K.: The rice haptic rocker: altering the perception of skin stretch through mapping and geometric design. In: 2018 IEEE Haptics Symposium, pp. 192–197, March 2018. https://doi.org/10.1109/HAPTICS.2018.8357175

6. Dunkelberger, N., et al.: A multisensory approach to present phonemes as language through a wearable haptic device. IEEE Trans. Haptics **14**(1), 188–199 (2021). https://doi.org/10.1109/TOH.2020.3009581

7. Gibson, J.J.: Observations on active touch. Psychol. Rev. **69**(6), 477–491 (1962). https://doi.org/10.1037/h0046962

8. Gray, R., Mohebbi, R., Tan, H.Z.: The spatial resolution of crossmodal attention: implications for the design of multimodal interfaces. ACM Trans. Appl. Percept. **6**(1), 4:1–4:14 (2009). https://doi.org/10.1145/1462055.1462059

9. Halfen, E.J., Magnotti, J.F., Rahman, M.S., Yau, J.M.: Principles of tactile search over the body. J. Neurophysiol. **123**(5), 1955–1968 (2020). https://doi.org/10.1152/jn.00694.2019. https://journals.physiology.org/doi/full/10.1152/jn.00694.2019

10. Lipowski, Z.J.: Sensory and information inputs overload: behavioral effects. Compr. Psychiatry **16**(3), 199–221 (1975). https://doi.org/10.1016/0010-440X(75)90047-4. http://www.sciencedirect.com/science/article/pii/0010440X75900474

11. Macklin, A.S., Yau, J.M., O'Malley, M.K.: Evaluating the effect of stimulus duration on vibrotactile cue localizability with a tactile sleeve. IEEE Trans. Haptics **14**(2), 328–334 (2021). https://doi.org/10.1109/TOH.2021.3079727
12. Mancini, F., et al.: Whole-body mapping of spatial acuity for pain and touch. Ann. Neurol. **75**(6), 917–924 (2014). https://doi.org/10.1002/ana.24179. https://onlinelibrary.wiley.com/doi/abs/10.1002/ana.24179
13. Meli, L., Hussain, I., Aurilio, M., Malvezzi, M., O'Malley, M.K., Prattichizzo, D.: The hBracelet: a wearable haptic device for the distributed mechanotactile stimulation of the upper limb. IEEE Robot. Autom. Lett. **3**(3), 2198–2205 (2018). https://doi.org/10.1109/LRA.2018.2810958
14. Pezent, E., Cambio, B., O'Malley, M.K.: Syntacts: open-source software and hardware for audio-controlled haptics. IEEE Trans. Haptics **14**(1), 225–233 (2021). https://doi.org/10.1109/TOH.2020.3002696
15. Pezent, E., et al.: Tasbi: multisensory squeeze and vibrotactile wrist haptics for augmented and virtual reality. In: 2019 IEEE World Haptics Conference (WHC), pp. 1–6, July 2019. https://doi.org/10.1109/WHC.2019.8816098
16. Spence, C., Gallace, A.: Recent developments in the study of tactile attention. Can. J. Exp. Psychol./Revue canadienne de psychologie expérimentale **61**(3), 196–207 (2007). https://doi.org/10.1037/cjep2007021
17. Tan, H.Z., et al.: Acquisition of 500 English words through a TActile Phonemic Sleeve (TAPS). IEEE Trans. Haptics **13**(4), 745–760 (2020). https://doi.org/10.1109/TOH.2020.2973135
18. Witteveen, H.J.B., de Rond, L., Rietman, J.S., Veltink, P.H.: Hand-opening feedback for myoelectric forearm prostheses: performance in virtual grasping tasks influenced by different levels of distraction. J. Rehabil. Res. Dev. **49**(10), 1517–1526 (2012). https://doi.org/10.1682/jrrd.2011.12.0243

Haptic Technology

Haptic Guidance for Teleoperation: Optimizing Performance and User Experience

Leonie Becker[ID], Bernhard Weber[✉][ID], and Nicolai Bechtel

German Aerospace Center, Institute of Robotics and Mechatronics,
Muenchener Str. 20, 82234 Wessling, Germany
{leonie.becker,bernhard.weber,nicolai.bechtel}@dlr.de
https://www.dlr.de/rm/en

Abstract. Haptic guidance in teleoperation (e.g. of robotic systems) is a pioneering approach to successfully combine automation and human competencies. In the current user study, various forms of haptic guidance were evaluated in terms of user performance and experience. Twenty-six participants completed an obstacle avoidance task and a peg-in-hole task in a virtual environment using a seven DoF force feedback device. Three types of haptic guidance (translational, rotational, combination of both, i.e. 6 DoF) and three guidance forces and torques (stiffnesses) were compared. Moreover, a secondary task paradigm was utilized to explore the effects of additional cognitive load. The results show that haptic guidance significantly improves performance (i.e. completion times, collision forces). Best results were obtained when the guidance forces were set to a medium or high value. Additionally, feelings of control were significantly increased during higher cognitive load conditions when being supported by translational haptic guidance.

Keywords: Haptic shared control · Haptic guidance · Virtual fixtures · Teleoperation · Virtual reality

1 Introduction

Robotic teleoperation, enabled by technological advancements, offers a unique opportunity to complete tasks remotely, avoiding potential dangers or inconveniences for the operator in cases where human operation is still required. As a result, areas of application include space operations, surgical procedures and a variety of other domains [1]. However, manually operating a robot from a remote environment poses unique challenges for the human operator e.g. in terms of workload, situational awareness or operator well-being [2]. As one potential solution, the robotic system can assist the operator during task execution through the integration of various levels of automation [3]. However, it is well known that even though high levels of automation allow for higher task efficiency, they may come at additional costs, leading to e.g. low situation awareness, over-reliance

© The Author(s) 2022
H. Seifi et al. (Eds.): EuroHaptics 2022, LNCS 13235, pp. 129–137, 2022.
https://doi.org/10.1007/978-3-031-06249-0_15

and an erosion of skills [4]. Therefore, there is an urgent need to address issues caused by high levels of automation. Here, haptic shared control approaches have been associated with the beneficial effects of automated systems, without causing strong automation-induced challenges [5]. Haptic shared control has been described as allowing "(...) both the human and the [automation] to exert forces on a control interface, of which its output (its position) remains the direct input to the controlled system." ([6], p. 501). Hence, forces and movements applied by the operator and robotic system interact conjointly with each other during task completion [5]. Thus, haptic guidance forces can be used as virtual fixtures to guide the user along a predefined path or workspace [7]. Furthermore, redundant poses can be eliminated by limiting the operator's degrees of freedom, resulting in increased task performance. Additionally, the stiffness of guidance forces can be adjusted to provide adequate guidance forces for various task demands.

The use of haptic guidance has been implemented in a variety of applications. It has been demonstrated, e.g., that haptic guidance improves completion time, error rate and distance from the ideal trajectory in a surgical spiral path following task [8]. Studies also indicated that haptic guidance should be generally implemented with higher stiffness values. For instance, [9] utilized a haptic virtual guidance fixture for a curve following task. They distinguished between no guidance and three different degrees of stiffness ("soft", "medium", "complete"), whereas "complete" guidance was associated with best results in terms of performance accuracy and error reduction. Moreover, the benefits of assistance functions are particularly evident in situations with high workload. Yet, only little research has been conducted to investigate possible effects of cognitive load on haptic guidance. Here, [10], e.g., reported that when haptic guidance is applied to a virtual vehicle steering task via torques applied to the steering wheel, a reduction in deviation from the centerline can be achieved. Additionally, they concluded that haptic guidance can be used effectively to mitigate the performance-degrading effects of additional cognitive load when the latter is induced via a secondary task.

So far, no research has been conducted to investigate how different types of haptic guidance with different stiffnesses and cognitive load interact and how it affects task performance as well as user experience. In the current work, these factors were investigated in a user study, conducted in a virtual environment with a 7-DoF haptic device.

2   Methods

2.1   Apparatus

A force feedback device with seven actuated DoF and active gravity compensation (*lambda*.7, Force Dimension) was used as I/O device (see Fig. 1, left). The device can be used within a comparably large workspace and produces high maximal forces with a high resolution. A HTC VIVE Pro Eye head- mounted display (HMD) was used for displaying the experimental simulation to the user

(Resolution of 1440×1600 pixels per eye; 90 Hz refresh rate). The virtual environment was run in the *Unity* video game engine. The haptic rendering and physics simulation was performed via a combination of multiple asynchronous real time processes. Among them, the "Voxel Pointshell" (VPS) algorithm with an update rate of 1 kHz was used [11]. The haptic guidance paradigm was realized using a predefined hard-coded virtual guidance fixture that actively attracted the participant onto the ideal path independent of trajectory progress. The ideal path was constructed from 19 waypoints distributed in relation to the object positions and a hard-coded safety margin. A Catmull-Rom spline algorithm was then used to interpolate and smooth the final path. The delta transformation between the current haptic device transformation and the projected haptic device transformation on the ideal path was used to estimate the current deviation from the optimal trajectory. The resulting quadratic distance was multiplied with the chosen stiffness to produce the forces and/or torques of the virtual fixture. A quadratic increase of forces and/or torques was implemented, since [12] e.g. reported that this paradigm leads to best human performance compared to other paradigms.

2.2 Sample, Experimental Setup and Tasks

Sample. $N = 26$ employees of the DLR voluntarily participated in the study ($M_{Age} = 32.27$ [$SD = 10.72$]; 3 females, 23 males). All subjects were right-handed and had normal or corrected to normal vision.

Experimental Setup. The haptic device was positioned at a predefined position on ground. Participants were seated in such a way that the haptic input device could be operated optimally with the right hand (Fig. 1, left). Seat height and distance to the device were adjusted individually, so that the device's x-axis (with handle in null position) and the longitudinal forearm axis when holding the handle were aligned. A computer keyboard for responding to the secondary task was placed and fixated on a table next to the seat and was operated with the left hand. Finally, participants put on the HMD, individually adjusted head-strap and lens distance.

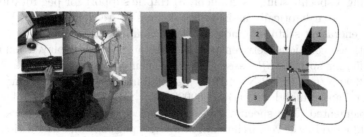

Fig. 1. Experimental Setup (left), VR Scene (middle), and Task Sequence (right)

Experimental Tasks. In the experimental simulation, various static objects as well as a user-controlled object were visually and haptically displayed. An obstacle avoidance and a peg-in-hole insertion task had to be performed in this scenario (Fig. 1, middle). A formation of four pegs (with square bases) in the simulation was used for the obstacle avoidance sub-task. Participants had to move a peg from a standardized starting position through the obstacles in a pre-defined chronology (Fig. 1, right) and were instructed to prioritize collisions avoidance while also completing the task as quickly as possible. The respective next obstacle that had to be circumnavigated was highlighted in red. The respective next passage between this obstacle and the neighboring obstacle was also highlighted by changing the latter object's color to blue. The completion of the task was indicated by a color change of the platform beneath the obstacles from grey to blue. No additional visual or auditory feedback was provided during task completion. After having completed the obstacle avoidance sub-task, participants were instructed to insert the controlled peg into a hole in the platform. A specific peg orientation was required for insertion due to the keyhole-like shape of the hole. Again, participants were asked to 1) prioritize contact force minimization during insertion and 2) to complete the task as quickly as possible. After having inserted the peg completely, the I/O device haptically guided the participants back to the starting position and the next trial was started. In some conditions, a *secondary task* had to be performed simultaneously with the above described primary task. There were four rectangular boxes displayed at the top, bottom, left and right sides of the experimental GUI. Randomly, one of these boxes changed its color from white to green and participants were instructed to respond as quickly and accurately as possible by pressing the corresponding arrow keys of the keyboard.

2.3   Experimental Design and Procedure

A secondary task paradigm was utilized to investigate the effects of additional *Cognitive Load*, i.e. there were conditions with vs. without secondary task. Four *Haptic Guidance* conditions were compared: 1) In a control condition, no haptic guidance was provided (C). 2) Haptic support for peg translation (T), i.e. the system guided movements along the trajectory in the dimensions X, Y, and Z by applying a specific stiffness gradient. 3) Haptic support for peg rotation (R). Thus, the system haptically assisted the degrees of freedom in X_{rot}, Y_{rot}, and Z_{rot} by generating stiffness gradients. 4) A combination of both ($T + R$). In all conditions, collisions between peg and obstacles were displayed haptically. Three different spring *Stiffnesses* were implemented for T and for R. A weak (K_1, T: 3600 N/m^2; R: 1.2 N $*$ m/rad^2), a medium (K_2, T: 33420 N/m^2; R: 6.6 N $*$ m/rad^2) and a high level (K_3, T: 60000 N/m^2; R: 12 N $*$ m/rad^2) of stiffness were implemented. The specific values were selected based on a series of test trials conducted by the research team. Moreover, peg sizes and user's viewpoints were varied, to explore whether findings generalize across different scenarios. For each condition, subjects started with a small peg (1.5 cm $\times$ 1.5 cm $\times$ 8 cm) and then performed the task with a large peg (2.55 cm $\times$ 2.55 cm $\times$ 12 cm). Also, there were two different viewpoints on the virtual scene, which were exactly mirroring each other. The perspective was randomly assigned to the first trial (small peg) and then mirrored for subsequent trial (large peg).

Experimental Design. A 2 (Cognitive Load, C) × 4 (Haptic Guidance, H) × 3 (Stiffness, K_i) within-subject experimental design was utilized. Within each H condition, the three different degrees of Stiffness (K_1, K_2 & K_3) were implemented. The orders of C, H and K_i conditions were counterbalanced across subjects. Each participant performed two trials for each type of Stiffness (1. Small, 2. Large Peg). Thus, 24 experimental trials had to be completed for both Cognitive Load conditions (=48 trials).

Procedure. Firstly, subjects completed a demographic questionnaire, read and signed an informed consent form. Secondly, participants were briefed on the study's background, experimental tasks and hardware. Thirdly, the seating position and HMD were adjusted and the experiment was started. Prior to each Haptic Guidance condition, subjects completed a test trial. Subsequent to each Stiffness condition, subjects verbally indicated the usefulness of the Haptic Guidance. When having completed a Haptic Guidance condition, the HMD was taken off and an additional questionnaire was completed. Between the two Cognitive Load conditions, subjects had a 5-minute break.

2.4   Measures and Statistical Analysis

Time to Complete (TTC) and *Collision Forces* were recorded as objective performance measures. After each Stiffness trial, participants rated the *Usefulness* of the respective Haptic Guidance ("The haptic guidance was useful: Rate on a 5-point scale from disagree (1) to agree (5)"). Subsequent to each Haptic Guidance condition, subjects filled the NASA-TLX *Workload* questionnaire [13]. Furthermore, subjects' *Perceived Control* was rated by two self-constructed items ("I felt in control of the system" and "I felt controlled by the system"; 5-point Likert-type scales ranging from disagree (1) to agree (5)).

Repeated measure ANOVAs (rmANOVA) with Cognitive Load, Haptic Guidance and Stiffness as within factors were performed on all measures. In case of non-sphericity, Greenhouse-Geisser (GG.) adjustments were made. Post-hoc comparisons were conducted with Bonferroni correction.

3   Results

3.1   Objective Performance Measures

Time to Complete (TTC). For the *Obstacle Avoidance Task*, a marginally significant main effect of Cognitive Load was evident ($F(1, 25) = 4.03$, $p < .10$), i.e. TTC tended to be longer for conditions with secondary task ($M = 3.58$ s; $SD = 1.26$ s) than for tasks without secondary task ($M = 3.27$ s; $SD = 1.42$ s). Furthermore, a significant Haptic Guidance × Stiffness interaction effect was found

($F(6, 150) = 7.46$, $p < .001$). For all Stiffness conditions, $T + R$ was associated with shorter TTCs than in the control group and R guidance. However, for K_2 and K_3, T guidance also yielded shorter TTCs than in the control group (see Fig. 2, left).

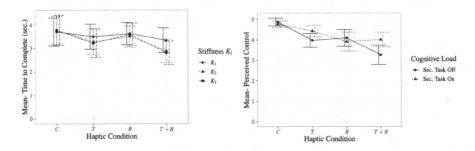

Fig. 2. Interaction effects between Haptic Condition and Stiffness for the time to complete during the obstacle avoidance task (left) and for the interaction effect between Haptic Condition and Cognitive Load for subjective control ratings (right). "T" = translational guidance, "R" = rotational guidance, "$T + R$" = translational + rotational guidance, "C" = control group. Error bars indicate 95% confidence interval.

For the *Peg-In-Hole Task*, a significant Haptic Condition × Stiffness interaction effect ($F(2.66, 66.56) = 5.00$, GG., $p < .01$) was found. For K_1 no significant Haptic Condition effects could be identified; for K_2, $T + R$ led to shorter TTCs compared to the control group and T guidance. For K_3, R guidance additionally reduced TTCs when compared to the control group.

Collision Force. For the *Obstacle Avoidance Task* rmANOVA indicated a significant interaction between Haptic Condition and Stiffness Condition ($F(6, 150) = 8.67$, $p < .001$). While no Haptic Guidance effects were evident with K_1, $T + R$ and T guidance resulted in lower collision forces than R guidance and the control group for K_2 and K_3. Regarding the *Peg-in-Hole Task*, no significant effects could be observed.

3.2   Subjective Ratings

Workload. A significant main effect of Cognitive Load on the raw NASA-TLX sum scores was found ($F(1, 23) = 17.88$, $p < .001$). As expected, Cognitive Load resulted in significantly higher ratings ($M = 9.29$; $SD = 2.25$; scale range: 1(very low) - 20 (very high)) compared to conditions without Cognitive Load ($M = 7.95$; $SD = 2.22$). Also, Haptic Guidance yielded a significant main effect ($F(3, 69) = 10.70$, $p < .001$): $T + R$ guidance ($M = 8.00$; $SD = 2.40$) was associated with significantly lower ratings than R ($M = 8.88$; $SD = 2.29$) and the control condition ($M = 9.24$; $SD = 2.03$). Also, ratings in T ($M = 8.34$; $SD = 2.13$) were lower than in the control condition.

Perceived Control. RmANOVA indicated a significant interaction between Cognitive Load and Haptic Guidance ($F(3, 69) = 6.32$, $p = .001$), which revealed that in conditions with secondary task, the ratings for $T + R$ as well as T were significantly higher compared to the conditions without secondary task (both ps = .001), see Fig. 2 (right).

Usefulness. RmANOVA revealed that there was a significant main effect of Haptic Guidance ($F(2, 46) = 25.76$, $p < .001$) and Stiffness ($F(1.16, 26.76) = 23.91$, GG., $p < .001$) on usefulness ratings. Overall, $T + R$ guidance ($M = 3.92$; SD = .50) was rated as being most useful, followed by T guidance ($M = 3.49$; SD = .77) and lastly by R guidance ($M = 2.99$; $SD = .50$). For Stiffness, K_3 (M = 3.71; $SD = .47$) and K_2 ($M = 3.65$; $SD = .50$) were associated with the significantly highest usefulness scores compared to K_1 ($M = 3.02$; $SD = .85$).

4   Discussion

In the present study, different implementations of haptic shared control were evaluated in a virtual environment with basic obstacle avoidance and insertion tasks. Specifically, human task performance and subjective experience when working with haptic support providing guidance for translational, rotational, or both motions and various stiffnesses were investigated in a user study. Altogether, clear evidence for improved task performance was found when haptic guidance for translations was provided. Given that stiffness was sufficiently high ($\geq 33420 \, \text{N/m}^2$) obstacle avoidance tasks were completed faster and with significantly lower collision forces. During the peg-in-hole task, there were two subtasks: moving the peg into the correct pose for insertion and then insertion in contact itself. Here, higher stiffnesses (Translations: $\geq 33420 \, \text{N/m}^2$; Rotations: $\geq 6.6 \, \text{N} * \text{m/rad}^2$), allowed faster task completion and the haptic guidance for all DoF was particularly beneficial. Rotational guidance only yielded improved completion times if a high stiffness was implemented ($12 \, \text{N} * \text{m/rad}^2$). There was a trend that cognitive load (introduced by a secondary task) led to increased completion times, indicating that participants tried to save cognitive resources by slowing down. Interestingly, however, the overall result pattern for haptic guidance and stiffness was evident independently from cognitive resource availability. The positive effects of haptic guidance were also reflected in the subjective ratings of participants: haptic guidance for all DoF was rated best in terms of workload and usefulness, followed by translational guidance and then by rotational guidance. Moreover, participants' ratings of usefulness was more positive for medium or high stiffnesses. Not surprisingly, participants felt that they had less control over the system when haptic guidance was activated. However, under conditions of additional cognitive load, the feelings of control increased when being supported by haptic guidance with translational support. Subjects seemingly had the impression of having the dual task situation more under control with these types of assistance.

Conclusions. In this work, it has been shown that haptic guidance improves task performance and reduces workload. Medium or high levels of stiffness are preferable in terms of task performance and user experience. Furthermore, the study provides evidence that translational guidance leads to a stronger sense of control under conditions of higher cognitive load.

Additional research should be conducted to determine how the proposed haptic guidance system could be made more adaptive. For instance, it may be advantageous to combine haptic guidance with the concept of a haptic wall or significantly stronger guidance when approaching obstacles. Here, varying degrees of stiffness could be implemented at different points throughout the parkour. In narrow passages, a high degree of stiffness may be implemented, while the user retains greater flexibility when navigating around the obstacles. This may have a beneficial effect on subjective feelings of control, as the operator retains more control and the haptic guidance only intervenes when necessary.

References

1. Sheridan, T.: Teleoperation, telerobotics and telepresence: a progress report. Control. Eng. Pract. **3**, 205–214 (1995)
2. Kaber, D., Onal, E., Endsley, M.: Design of automation for telerobots and the effect on performance, operator situation awareness, and subjective workload. Hum. Factors Ergon. Manuf. Serv. Ind. **10**(4), 409–430 (2000)
3. Boessenkool, H., Abbink, D., Heemskerk, C., Helm, F., Wildenbeest, J.A.: Task-specific analysis of the benefit of haptic shared control during telemanipulation. IEEE Trans. Haptics **6**, 2–12 (2013)
4. Endsley, M., Kiris, E.: The out-of-the-loop performance problem and level of control in automation. In: Human Factors, vol. 37, pp. 381–394. SAGE Publications, Los Angeles (1995)
5. Abbink, D., Mulder, M., Boer, E.: Haptic shared control: smoothly shifting control authority? Cogn. Technol. Work **14**, 19–28 (2012)
6. Abbink, D., Mulder, M.: Neuromuscular analysis as a guideline in designing shared control. In: Advances in Haptics. InTech (2010)
7. Rosenberg, L.: Virtual fixtures: perceptual tools for telerobotic manipulation. In: Proceedings of IEEE Virtual Reality Annual International Symposium, pp. 76–82 (1993)
8. Xiong, L., Chng, C.B., Chui, C.K., Yu, P., Li, Y.: Shared control of a medical robot with haptic guidance. Int. J. Comput. Assist. Radiol. Surg. **12**(1), 137–147 (2016). https://doi.org/10.1007/s11548-016-1425-0
9. Marayong, P., Bettini, A., Okamura, A.: Effect of virtual fixture compliance on human-machine cooperative manipulation. In: IEEE/RSJ International Conference on Intelligent Robots and Systems, vol. 2, pp. 1089–1095 (2002)
10. Griffiths, P., Gillespie, R.: Sharing control between humans and automation using haptic interface: primary and secondary task performance benefits. Hum. Factors **47**, 574–590 (2005)
11. Sagardia, M., Hulin, T.: A fast and robust Six-DoF god object heuristic for haptic rendering of complex models with friction. In: Proceedings of the 22nd ACM Conference on Virtual Reality Software and Technology, pp. 163–172 (2016)

12. Rognon, C., Wu, A., Mintchev, S., Ijspeert, A., Floreano, D.: Haptic guidance with a soft exoskeleton reduces error in drone teleoperation. In: International Conference on Human Haptic Sensing and Touch Enabled Computer Applications, pp. 404–415 (2018)
13. Hart, S., Staveland, L.: Development of NASA-TLX (Task Load Index): results of empirical and theoretical research. Adv. Psychol. **52**, 139–183 (1988)

A Multi-modal Haptic Armband for Finger-Level Sensory Feedback from a Prosthetic Hand

Alexandre Berkovic[1], Colin Laganier[1], Digby Chappell[1,2(✉)],
Thrishantha Nanayakkara[1], Petar Kormushev[1], Fernando Bello[2],
and Nicolas Rojas[1]

[1] Dyson School of Design Engineering, Imperial College London, London, UK
`d.chappell19@imperial.ac.uk`
[2] Department of Surgery and Cancer, Imperial College London, London, UK

Abstract. This paper presents the implementation and evaluation of three specific, yet complementary, mechanisms of haptic feedback—namely, normal displacement, tangential position, and vibration—to render, at a finger-level, aspects of touch and proprioception from a prosthetic hand without specialised sensors. This feedback is executed by an armband worn around the upper arm divided into five somatotopic modules, one per each finger. To evaluate the system, just-noticeable difference experiments for normal displacement and tangential position were carried out, validating that users are most sensitive to feedback from modules located on glabrous (hairless) skin regions of the upper arm. Moreover, users identifying finger-level contact using multi-modal feedback of vibration followed by normal displacement performed significantly better than those using vibration feedback alone, particularly when reporting exact combinations of fingers. Finally, the point of subjective equality of tangential position feedback was measured simultaneously for all modules, which showed promising results, but indicated that further development is required to achieve full finger-level position rendering.

Keywords: Multi-modal haptics · Prosthetic hands · Proprioception

1 Introduction

Sensory feedback is essential for dexterous manipulation and its absence in hand prostheses hinders the correct achievement of numerous fine-motor tasks, such as grasping, stroking or throwing [1]. Human sensory dexterity mechanisms are inherently dependent on both complex nervous arrangements and two sensory

Digby Chappell was supported by the UKRI CDT in AI for Healthcare http://ai4health.io (Grant No. P/S023283/1)

A. Berkovic and C. Laganier—Equal contribution.

H. Seifi et al. (Eds.): EuroHaptics 2022, LNCS 13235, pp. 138–146, 2022.
https://doi.org/10.1007/978-3-031-06249-0_16

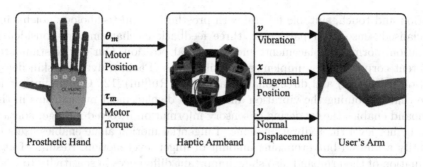

Fig. 1. Overview of the introduced multi-modal haptic armband. Signals are read from a prosthetic hand and translated to vibration, tangential position, and normal position feedback.

systems: namely proprioception, which tracks the motion of our hand and its muscular effort; and touch, which communicates key information on environment interactions [2]. With visual guidance being an imperfect surrogate for its highly intricate somatosensory complement and the fact that our motor systems are inherently coupled with our dexterity [3,4] the underlying necessity of a sensory feedback system for prosthetic hands is clear.

The vast majority of works in the literature has focused on creating haptic feedback systems to interpret different aspects of touch, such as tactile sensations or contact forces. To achieve this, the prosthetic hand used is typically equipped with some form of touch sensor [5]. However, the sensing capabilities of pros- thetic hands are currently still limited, as the underlying technology lacks the sophistication of the natural counterpart. Nonetheless, vibration and pressure- based feedback have shown promising results in rendering touch, to the point where individual finger contacts can be discerned by users [5–7].

When rendering proprioception, both the position and force applied by the hand must be encoded into the feedback applied to the user. This is consid- erably more applicable to hand prostheses, given that the position and applied force of each controlling motor can be monitored without any specialised sensing technology. Despite this, very little work exists reproducing elements of propri- oception to prosthetic hand users. Of note, [8] utilised a rolling mechanism to encode the position of the whole hand, resulting in users being able to identify small changes in grasped object size.

Regarding modes of haptic feedback for prosthetic hands, previous works have mostly focused on a single form of information [9], thus inherently limiting the level of perception available to the user—often aiming to represent the state of the entire hand [10,11], rather than doing so at a finger level. An ideal haptic feedback system should indeed utilise mutli-modal feedback to encode differ- ent aspects of proprioception and touch. Multiple feedback sites should also be utilised to allow the user to differentiate feedback from individual fingers.

The aim of this work is to develop a multi-modal haptic feedback armband, shown in Fig. 1, which is able to encode key aspects of somatotopic proprioception and touch, suitable for use with prosthetic hand technology not having specialised sensors. In particular, three feedback mechanisms are considered: vibration, normal displacement, and tangential position. Our device stimulates different corpuscules by implementing vibrations (Pacini), dynamic skin deformation (Meissner) and directional shear forces (Ruffini) [12]. Our findings indicate that combining the vibration and normal displacement mechanisms in the armband enables the rendering of sensory information involved in finger contact to a higher level than vibration alone. Tangential motion alone enables users to identify a limited but promising amount of finger-level motion. Results of user evaluation of the armband also show important differences in sensitivity to feedback of locations around the upper arm, which should be taken into consideration in future research and development.

2   Multi-modal Haptic Armband

2.1   Rendering of Feedback

The multi-modal haptic armband is used to render feedback from a prosthetic hand using the position, θ_m measured by the encoders, and torque, τ_m, of the motors driving each finger. In this work, we specifically consider the OLYMPIC hand [13], in which each finger is driven in the flexion direction by an individual motor, while springs mounted on the dorsal side of each finger drive extension. Contact can be detected as the hand closes around an object, and relayed to the user using the feedback mechanisms of the armband; vibration for indicating touch sensations (from motor position and torque), normal displacement for rendering contact force and force elements of proprioception (from motor torque), and tangential position for rendering position elements of proprioception (from motor position).

Vibration. Vibration is used to render the tactile element of contact. Without a direct sensor mounted on each fingertip, contact must instead be inferred from motor position and torque. This can be achieved by monitoring the acceleration of each finger motor; large deceleration indicates that the finger is making contact with an object. The vibration of finger i is applied as a pulse of fixed duration T_v, and intensity k_v, upon acceleration falling below a negative threshold, $\ddot{\theta}_v$:

$$v_i(t) = \begin{cases} k_v & \ddot{\theta}_{m,i} < \ddot{\theta}_v \quad \text{for } 0 < t < T_v \\ 0 & \text{Otherwise} \end{cases} \tag{1}$$

Normal Displacement. When each finger is stationary, the motor torque required to hold the finger at its current position is equal to the combined torque of the extension springs acting on each joint [14]. A linear estimation of contact-free motor torque, $\hat{\tau}_{s,i}$, can therefore be calculated. Motor torque is therefore approximately linear to position when each finger of the hand is contact-free. Assuming the skin behaves according to a visco-elastic model, the

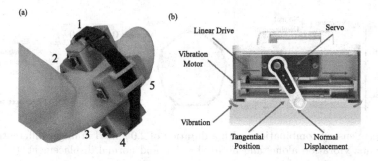

Fig. 2. (a) User wearing the armband, modules labelled. (b) Schematic of the feedback mechanisms.

contact is stable, and displacement is small, then contact force can be rendered by applying a proportional displacement normal to the surface of the skin. The displacement applied to the user is proportional to the difference in current motor torque and the contact-free stationary torque, scaled by k_y:

$$y_i = k_y(\tau_{m,i} - \hat{\tau}_{s,i}). \tag{2}$$

Tangential Position. Proprioceptive position feedback is achieved by relaying the current motor position to the position of a linear drive, x_i, moving tangential to the surface of the skin, with linear scaling k_x:

$$x_i = k_x\theta_{m,i}. \tag{3}$$

2.2   Mechanisms of Feedback

The multi-modal haptic armband is composed of five modules equally spaced around the arm, each corresponding to one of the user's fingers. Each module, shown in Fig. 2, consists of three feedback mechanisms: a vibration motor rendering tactile contact sensations, a servo-motor to apply displacement normal to the skin expressing contact force and force elements of proprioception, and a linear drive that moves tangentially to the skin rendering position elements of proprioception. Each individual module is self-contained and has a footprint of 68 × 40 mm, a height of 33 mm, and a mass of 56 g, which makes the armband suitable for daily use due to its relatively compact size and small weight. Studies have shown that the upper-arm region in which our research is being implemented has a sensitivity threshold distance of around 30 mm [15] which validates the 40 mm width of the design.

The vibration motor is placed in the upper inside wall of the casing such that vibrations are distributed throughout its shell, as to render non-position-specific feedback. The servo is mounted on a carriage attached to both the linear drive and two support rods allowing for smooth travel and the application of normal displacement at a desired position. To keep module size to a minimum, a small 10 mm diameter stepper motor was used, coupled with a 38 mm threaded rod to

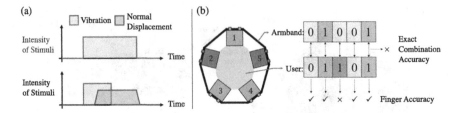

Fig. 3. (a) Sensory combinations with a duration of 2.0 s used in finger differentiation experiments; vibration alone, and both vibration and normal displacement. (b) Summary of finger differentiation experimental setup and performance metrics calculated. For any trial, each module has a probability of 0.5 that it will apply feedback.

form a linear drive. An extension to the servo motor (Tower Pro, SG90) horn was also designed to create better contact with the user's skin using a spherical tip.

3   Experimental Evaluation

In order to evaluate the haptic armband, 5 users wore it on their right arm, with the first module aligned with the centre of the bicep, and subsequent modules located around the upper arm ordered in the direction of the outside of the arm. By design, the armband blocks the user's view of each feedback mechanism, and headphones were used to block audio cues.

Just-Noticeable Difference. The just-noticeable difference (JND) of the servo-based force feedback applying a nominal normal displacement of 3.0 mm and the linear position feedback at a nominal tangential position of 17.5 mm were individually evaluated. One feedback module was used to deliver paired stimuli, using a staircase procedure to find the JND. This was repeated for each module of the armband, evaluating the sensitivity to feedback of 5 points around the upper arm. JND is important in this case as it allows us to gauge how small changes in stimuli are perceived. When grasping fragile or delicate objects, slight variations in position and force can be of great consequence, so understanding the perceptual limits of these feedback mechanisms is critical.

Finger Differentiation. These experiments were performed to evaluate the ability of the haptic armband to provide finger-level feedback that is differentiable to the user. First, using a digital twin of the OLYMPIC hand [14], users were asked to identify which fingers of the prosthetic hand made contact with a virtual object based on vibration feedback alone, then the experiment was repeated with normal displacement feedback also present. At each trial, each finger randomly contacted the virtual object with a probability of 0.5. In both scenarios, 100 trials were used and feedback was applied for a total of 2.0 s;

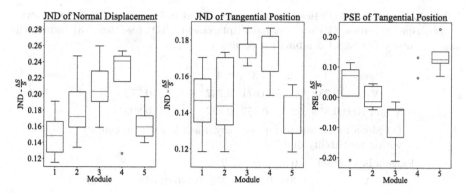

Fig. 4. Left: JND of normal displacement for each module, measured with a nominal displacement of 3.0 mm. Middle: JND of tangential position for each module, measured with a nominal position of 17.5 mm. Right: PSE of tangential position for five modules used simultaneously (note that module 4 suffered an electrical fault and only 2 user results are presented).

for vibration and normal displacement, vibration was applied for 0.75 s, and a normal displacement of 3.0 mm applied for 1.75 s, such that the two feedback mechanisms overlapped in application Fig. 3(a).

Two metrics are calculated from users results; individual finger accuracy and exact combination accuracy. As summarised in Fig. 3(b), individual finger accuracy allows us to monitor how successful the user is at detecting contact on each finger. Exact combination accuracy is a harsher metric, only counting when the exact combination of fingers has been correctly identified. This may be a more appropriate metric when considering prosthetic hands; users should be able to identify precisely which fingers are making contact with objects if they are to achieve dexterous control of the prosthesis.

Point of Subjective Equality. Measuring the point of subjective equality (PSE) of position feedback is crucial. If this armband is to be used during prosthetic hand control, the user must be able to control their hand to a desired pose and be confident that, based on the position feedback they receive, the hand has indeed reached that pose. In this experiment, users received position feedback from a control pose, then were asked to adjust 5 sliders on a user interface in 0.24 mm increments until they produced a position feedback perceived as equal to that of the control pose.

4   Results and Discussion

To measure the efficacy of the system, JND was measured for normal displacement and tangential position feedback, which measures the amount of change in a stimulus that allows it to be detectable at least 75% of the time, as estimated by fitting a psychometric curve to user responses. As seen in Fig. 4(left and middle), JND of modules located at the rear of the upper arm (modules 3, 4) was

Table 1. Finger differentiation accuracy for users identifying contact of digital twin fingers using vibration (V) and normal displacement (ND) feedback. (a) Individual finger accuracy. (b) Exact combination accuracy.

Finger	1	2	3	4	5
V	0.782	0.768	0.671	0.737	0.772
V and ND	0.858	0.811	0.757	0.722	0.809

(a) Mean individual finger accuracy. Each finger is in contact with a probability of 0.5.

Fingers in Contact	0	1	2	3	4	5
V	1.000	0.735	0.272	0.146	0.048	0.222
V and ND	1.000	0.875	0.371	0.181	0.130	0.421

(b) Mean exact combination accuracy. Each trial is randomly sampled, so the probability of f fingers being in contact is $\binom{5}{f}$. The probability of randomly guessing correctly is $1/32$ (0.031).

significantly lower than modules at the front of the upper arm (modules 1, 2, 5), with median values of 0.72 mm versus 0.45 mm for normal displacement, and 3.06 mm versus 2.53 mm for tangential position. This is expected, considering the high sensitivity of glabrous skin [2], which is located primarily on the anterior and medial regions of the upper arm. The difference in JND around the upper arm can be used in future to inform module placement and specific feedback mapping that takes this into account when rendering sensory cues.

Finger differentiation results reveal that multi-modal feedback generally improves user ability to identify whether a finger has made contact (see Table 1a), although user performance with vibration alone is still high. The ability of users to recognise the exact combination of fingers in contact at any point is heavily dependent on the number of fingers in contact (see Table 1b). Combining vibration with normal displacement feedback improves exact combination accuracy considerably. With vibration alone, exact combination accuracy when four fingers are in contact is little over random chance $(1/32 = 0.031)$, at 0.048, whereas when normal displacement is included, this rises to 0.130. Although each module is larger than the sensitivity threshold distance of the upper arm (30 mm), vibration transmits through the skin and can 'blur' feedback [6], which may account for the poor finger differentiation performance of vibration alone.

Point of subjective equality results, shown in Fig. 4(right), reveal that user ability to interpret tangential position is accurate to within ±20% when multiple modules are active. Unfortunately, module 4 suffered electronic failure during evaluation, so results for it are only present for two users. PSE for remaining modules follows a similar pattern to previous results; the anterior region of the forearm is more sensitive to stimuli, which may account for user over-estimation of tangential position in modules located in these regions, while the reverse is true for modules located toward the less sensitive posterior regions. It is hypothesised that PSE results could be improved by increasing the travel of the linear drive, and improving the geometry of the servo arm to contact more skin.

5 Conclusion and Future Work

In this work, we have presented a haptic feedback armband that is capable of rendering sensory aspects of touch and proprioception of a prosthetic hand. User evaluation revealed that JND of normal displacement feedback and tangential position feedback is highest on the anterior regions of the upper arm, dropping significantly in non-glabrous regions at the posterior of the upper arm. Using normal displacement to translate contact force supplements vibration-rendered touch sensation and greatly improves user ability to identify exact fingers in contact with objects. Users were able to resolve tangential position to within a noticeable difference as little as 2.0 mm for a single finger, but PSE results indicate that improvements are required to relay accurate position information to the user. Alternative designs investigating a larger range of motion, improved reliability of the linear drive, and different roller geometries should be considered.

References

1. Kim, K., Colgate, J.E.: Haptic feedback enhances grip force control of sEMG-controlled prosthetic hands in targeted reinnervation amputees. IEEE Trans. Neural Syst. Rehabil. Eng. **20**(6), 798–805 (2012)
2. Sobinov, A.R., Bensmaia, S.J.: The neural mechanisms of manual dexterity. Nat. Rev. Neurosci. **22**, 741–757 (2021)
3. Jiang, W., Tremblay, F., Chapman, C.E.: Context-dependent tactile texture-sensitivity in monkey M1 and S1 cortex. J. Neurophysiol. **120**(5), 2334–2350 (2018). https://doi.org/10.1152/jn.00081.2018
4. Umeda, T., Isa, T., Nishimura, Y.: The somatosensory cortex receives information about motor output. Sci. Adv. **5**(7), eaaw5388 (2019)
5. Yunus, R., et al.: Development and testing of a wearable vibrotactile haptic feedback system for proprioceptive rehabilitation. IEEE Access **8**, 35172–35184 (2020)
6. Ranasinghe, A., Althoefer, K., Dasgupta, P., Nagar, A., Nanayakkara, T.: Wearable haptic based pattern feedback sleeve system. Adv. Intell. Syst. Comput. **547**, 302–312 (2017)
7. Antfolk, C., et al.: Transfer of tactile input from an artificial hand to the forearm: experiments in amputees and able-bodied volunteers. Disab. Rehabil. Assistive Technol. **8**(3), 249–254 (2013)
8. Rossi, M., Bianchi, M., Battaglia, E., Catalano, M.G., Bicchi, A.: HapPro: a wearable haptic device for proprioceptive feedback. IEEE Trans. Biomed. Eng. **66**(1), 138–149 (2019)
9. Svensson, P., Wijk, U., Björkman, A., Antfolk, C.: A review of invasive and non-invasive sensory feedback in upper limb prostheses. Expert Rev. Med. Dev. **14**(6), 439–447 (2017)
10. Motamedi, M.R., Florant, D., Duchaine, V.: A wearable haptic device based on twisting wire actuators for feedback of tactile pressure information. J. Robot. Mechatron. **27**(4), 419–429 (2015)
11. Huaroto, J.J., Suarez, E., Krebs, H.I., Marasco, P.D., Vela, E.A.: A soft pneumatic actuator as a haptic wearable device for upper limb amputees: toward a soft robotic liner. IEEE Robot. Autom. Lett. **4**(1), 17–24 (2019)

12. Hayward, V.: A brief overview of the human somatosensory system, pp. 29–48 (2018)
13. Liow, L., Clark, A.B., Rojas, N.: OLYMPIC: a modular, tendon-driven prosthetic hand with novel finger and wrist coupling mechanisms. IEEE Robot. Autom. Lett. **5**(2), 299–306 (2020)
14. Chappell, D., et al.: Virtual reality pre-prosthetic hand training with physics simulation and robotic force interaction. IEEE Robot. Autom. Lett. **7**, 4550–4557 (2022)
15. Koo, J.P., et al.: Two-point discrimination of the upper extremities of healthy Koreans in their 20's. J. Phys. Ther. Sci. **28**(3), 870–874 (2016)

Sound Pressure Field Reconstruction for Ultrasound Phased Array by Linear Synthesis Scheme Optimization

Jianyu Chen[1]([⊠])[iD], Shun Suzuki[1][iD], Tao Morisaki[1][iD], Yutaro Toide[1][iD], Masahiro Fujiwara[1,2][iD], Yasutoshi Makino[1,2][iD], and Hiroyuki Shinoda[1,2][iD]

[1] Graduate School of Frontier Sciences, The University of Tokyo, Kashiwa-shi, Japan
`7707189878@edu.k.u-tokyo.ac.jp`

[2] Graduate School of Information Science and Technology, The University of Tokyo, Tokyo, Japan

Abstract. Ultrasound phased array is a device that is usually used to provide mid air tactile sensations like three-dimensional shape haptics images by generating various specific sound fields. Forming foci for the ultrasound phased array using the linear synthesis scheme (LSS) is a straightforward technique to induce tactile feeling. The matching phase set for each focal point is calculated separately in LSS, and then they are linearly superimposed to generate multiple focal points. Due to the fact that adding an arbitrary offset to the entire phase pattern has no effect on the generated focus patterns, adjusting the offset in linear summation may result in a superior sound field. In our study, we propose that optimize the offset before linear superposition. These offsets are determined based on the number of focal points, which means it will not cause an explosive increase in computing cost with the increase of transducers. To optimize the offset of each focus pressure generated by LSS, we used a greedy algorithm with a brute-force search optimization method. The computing cost of our proposed method is dictated by the number of foci after calculating the phase sets of LSS once. We demonstrate the proposed method's optimum performance in varied numbers of foci and transducers in this study.

Keywords: Foci field · Linear synthesis scheme · Optimization

1 Introduction

Ultrasound phased array, such as airborne ultrasound tactile display (AUTD) [1], can produce various tactile sensations remotely on a human skin surface. Monnai et al. used an ultrasound focus generated by AUTD to develop a mid-air interaction system that allows users to touch a floating virtual screen with non-contact tactile feedback [2]. Moreover, the foci field has been applied to the

Supported by organization JST CREST JPMJCR18A2.

H. Seifi et al. (Eds.): EuroHaptics 2022, LNCS 13235, pp. 147–154, 2022.
https://doi.org/10.1007/978-3-031-06249-0_17

formation of three-dimensional (3D) haptic images, allowing people to perceive static 3D shapes by touching them with their fingers and hands [3]. As a result, the question of how to generate a specialized sound field with stronger sound pressure has been widely studied to provide a better tactile experience.

Many methods for optimizing the foci sound field have been proposed. Long et al. optimized the transmission matrix to achieve a 3D tactile sensation like virtual objects in the air [4]. GS-PAT, which can optimize the foci field using both CPU and GPU [5]. The preceding optimization approaches can all obtain a better sound field, but the computing cost rises dramatically as the transducers increases, and the calculation becomes more difficult. As a result, we anticipate discovering a strategy that is simple to calculate, while also optimizing the sound field effectively.

One of the simplest ways to generate a foci field is a linear synthesis scheme (LSS). When designing a foci sound field with LSS, the phases set for each focus are calculated separately, and then those sets are linearly synthesized. The phase pattern that forms each focal point has a degree of freedom in terms of phase offset. Because even if the phase pattern is given an arbitrary offset, the spatial pattern of the generated sound pressure will be the same, it is vital to give an appropriate offset for each focus when linear addition is conducted. We present a greedy algorithm with brute-force search in this study for generating a stronger sound field by optimizing the offset when synthesizing the phase patterns of multiple focal points.

Suzuki el al. have proven the superiority of the greedy algorithm in optimizing individual transducer phase in the ultrasonic phased array [6]. In this paper, we applied a greedy algorithm to the offsets optimization to improve the performance of LSS.

After calculating the phases sets of each focus field based on the LSS for once, we only have to explore the optimal offsets, and add them to the sets, which means that the computing cost for optimization is linearly proportional to the number of foci. As a result, the computing cost of creating a foci sound field will be lowered even when a huge number of transducers are used. Furthermore, the pressure of the sound field generated by LSS can be strengthened. When using ultrasound phased array to provide tactile sensation, this can deliver a better experience. In our study, all the experiments are based on simulations and the final phases are normalized while the amplitude of the transducers are set to be in the range of $(0, 1]$.

2   Methods

2.1   Linear Synthesis Scheme Optimization

Firstly, we introduce how LSS works. The linear synthesis scheme, as its name implies, generates the foci field by linearly synthesizing each focus signal. Thus, we must calculate the phase set which generates each focus at first.

Let the phase set be $\boldsymbol{q}(\phi_1, ..., \phi_M)$, defined as,

$$\boldsymbol{q}(\phi_1, ..., \phi_M) = [e^{-j\phi_1}, ..., e^{-j\phi_M}]. \tag{1}$$

where, ϕ_i is the phase of i-th transducer. Assuming the sound emitted from the transducer is a spherical wave, the acoustic field $p(\boldsymbol{r})$ generated by M transducers is expressed as:

$$p(\boldsymbol{r}; \boldsymbol{q}) = \sum_{i=1}^{M} \frac{1}{4\pi |\boldsymbol{r} - \boldsymbol{r}_i|} e^{j(k|\boldsymbol{r}-\boldsymbol{r}_i|)} q_i, \tag{2}$$

where k is a wavenumber, and $\boldsymbol{r}_i$ is the position of i-th transducer. Note that, the amplitude of the transducers are set to be 1 in the study. From Eq. (2), when creating a focus at a position $\boldsymbol{r}_f$, we should set ϕ_i as follows:

$$\phi_i = k |\boldsymbol{r}_f - \boldsymbol{r}_i|. \tag{3}$$

Here, let the phase sets of the transducers that generates the focus at $\boldsymbol{r}_f$ be $\boldsymbol{q}(\boldsymbol{r}_f)$, i.e.,

$$\boldsymbol{q}(\boldsymbol{r}_f) = [e^{jk|\boldsymbol{r}_f-\boldsymbol{r}_1|}, ..., e^{jk|\boldsymbol{r}_f-\boldsymbol{r}_M|}]. \tag{4}$$

By linearly synthesizing the phase set that generates each focus, we can generate N foci located at $\boldsymbol{r}_{f_1}, ..., \boldsymbol{r}_{f_N}$. The human sense of touch cannot distinguish a phase difference of 40 kHz vibration at the focus, and thus, there is room to optimize the phase of each focus. Therefore, the phase set $\boldsymbol{q}$ which generate N foci is represented as,

$$\boldsymbol{q} = \boldsymbol{q}(\boldsymbol{r}_{f_1})e^{jo_1} + \cdots + \boldsymbol{q}(\boldsymbol{r}_{f_N})e^{jo_N}, \tag{5}$$

where o_n is the phase offset of n-th focus (Fig. 1).

With Eq. (5), the calculation of LSS has been done, but we have to set the amplitude in the range of $(0, 1]$, we have to do the normalization in the final:

$$\boldsymbol{q}_{\text{normal}} = \frac{\boldsymbol{q}}{\max\{|q_i|\}} \tag{6}$$

In theory, setting all the offsets to 0 is the fastest way to form a foci field, but the sound pressure may become weaker. GS-PAT uses an iterative method to optimize the offsets, and we suggest a faster method: greedy algorithm with brute-force search, to deal with the optimization.

For each focus, we use the greedy algorithm to explore the optimal offset that generates the strongest sound pressure. As a result, the calculation cost of such LSS optimization is the order of the number of foci.

2.2   Greedy Algorithm with Brute-Force Search

Greedy algorithm is an intuitive algorithm that is used in optimization problems. The algorithm breaks the problem into sub problems, and searches for the best solution for each one to arrive at the overall best solution for the entire problem. Suzuki et al. treated each transducer in the ultrasonic phased array as a separate problem, sampling from the original continuous $[0, 2\pi)$ phase space in an equal and discrete manner. In the study, we focus on the offsets of the phase set from each focus and discretize the offset from $[0, 2\pi)$ as L division. Then we calculate

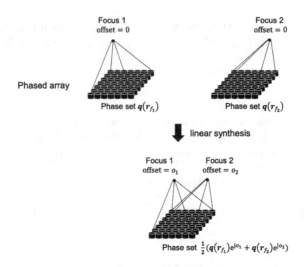

Fig. 1. Linear synthesis scheme

phase set $q(r_{f_n})$ of each focus and obtain p_n, where p_n means the sound field vector of foci which is generated by the transducers with phase set $q(r_{f_n})$. Unlike Suzuki, we optimize the phase set offset of the foci in each simulation at random, which means we may receive different outcomes each time. In the loops of N foci and equal division L offsets, we explore the optimal phase offset o_n one by one. That is, we search optimal phase offset o_1 which obtained the maximum $\|p^l\|$ on the focus position, and then keep its optimum and search next optimal offset o_2, and we continue this process for all phase offsets. The optimal phase set and sound field can be retrieved when the loop is completed. The algorithm we suggest in this study is depicted in the diagram below:

Algorithm 1. Greedy Pressure Field Reconstruction with Brute-force Search

---

Input: $q_{r_1}, ..., q_{r_N}$
Output: q
1: discretize offset o as $\left\{ o^l \middle| o^l \in [0, 2\pi), l = 1, \cdots, L \right\}$
2: calculate $p_n = [p(r_{f_1}; q(r_{f_n})), ..., p(r_{f_N}; q(r_{f_n}))]$ for $n = 1, ..., N$
3: Set $p_t = 0$
4: **for** $n = 1, ..., N$ **do**
5:      **for** $l = 1, ..., L$ **do**
6:          obtain $p^l = p_t + p_n e^{jo^l}$
7:      **end for**
8:      $o_n \leftarrow o^{l^*}$ s.t $l^* = \text{argmax}_l \left\{ \|p^l\| \right\}$
9:      $p_t \leftarrow p_t + p_n e^{jo_n}$
10: **end for**
11: Substitute optimal offset $o_1, ..., o_N$ into Eq. (5)

---

3 Experiments

3.1 Outline

The experiments are run by a desktop computer that has an Intel(R) Core(TM) i9-9900X CPU @3.50 GHz CPU. And we run the code in C++ base on Eigen library. It should be noted that the computing cost of GS-PAT depends on the matrix calculation. Therefore, the calculation of GS-PAT would be faster by using a library or computational resources that can perform faster matrix calculations. The experiments mainly simulate the performance of each approach for generating sound fields with various foci number using a phased array of 18×14 transducers. The x-axis is the side with 18 transducers, and the y-axis is the side with 14 transducers. The number of x-axis and y-axis transducers is multiplied by K to generate a larger phased arrays, as the number of phased arrays grows, implying that the number of phased arrays will expand by K^2. We calculated the field in the plane of $x = (-50, 50)$ mm, $y = (-50, 50)$ mm at 150 mm above the center of the phased array vertically. The speed of sound is set to 340 m/s. Furthermore, the frequency of the ultrasound was set to 40 kHz, and thus the wavelength was 8.5 mm. In this study, we will evaluate our proposed method compared to the LSS without offset optimization and GS-PAT (CPU).

First, we will sequentially generate (2, 4, 6, 8, 10, 12)-foci fields on the periphery of a circle with a radius of 30 mm (Fig. 2a) based on one phased array unit, and evaluate the average sound pressure of each focus generated by the methods. Then, in three experiments, we will compare the computing cost, with the field of these experiments being the 12 foci field on the circle as shown in Fig. 2a. The first shows the computing cost of varying numbers of foci with one phased array unit with 252 transducers, whereas the second one increases the number of foci like the first one but with 16 phased array units which have 4032 transducers. The third experiment increases the number of phased array units in the situation of generating 12 foci field. The phase space of the offset in the greedy algorithm was divided into $L = 16$ equal parts and discretely sampled, the each offset of focus will be optimized in random order. To calculate the mean and standard deviation of the foci pressure, we ran ten simulations for each sound field.

Then, to evaluate the versatility of the proposed method, we also designed the other foci sound field that can be applied to the sense of tactile. A sound field of a five-pointed star similar to Inoue et al. [3] (Fig. 2b) with 11 foci based on one phased array. The outer foci of the five-pointed star are based on a concentric circle with a radius of 43 mm, the inner foci are based on a circle with a radius of 16 mm, and the center of the circle is the last point. Our proposed method, like the prior experiment, calculates each offset in random order.

3.2 Evaluations

Due to space limitations, we only list the result field of the 12 foci situation in Fig. 3 (the result of LSS+Greedy is picked one of the ten simulations), and the evaluation of the average foci pressure in the other situations is listed in Fig. 4a.

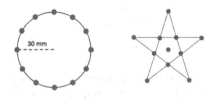

(a) Foci on a circle   (b) Five-point star

Fig. 2. Objective fields

The computing cost by the three methods to form a different number of foci fields with the 252 transducers in Fig. 4b, and 16 phased array units with 4032 transducers in Fig. 4c. The computing cost of increasing number of transducers while generating 12 foci is shown in Fig. 4d.

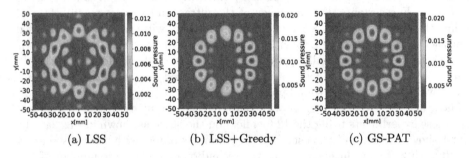

(a) LSS   (b) LSS+Greedy   (c) GS-PAT

Fig. 3. Average pressure of the 12 foci field on a circle

From the results, we can know Greedy algorithm has strengthened the sound field pressure generated by LSS without offset optimization, although the sound pressure is lower than GS-PAT while the foci is lager than 8, in some cases our proposed method obtained the strongest sound pressure.

In terms of computing cost, we can see that LSS without offset optimization took the least computing cost in any case. The computing cost of our proposed method is faster than GS-PAT under all the situations.

As Fig. 5 shows, LSS without offset optimization forms the field that the side lobe has other peaks as well, which may influence the tactile experience. After optimizing the field by our proposed method, it formed an 11 foci sound field clearly.

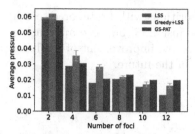

(a) Average pressure for the number of foci

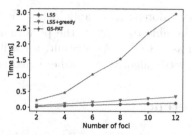

(b) Computing cost for the number of foci with 252 transducers

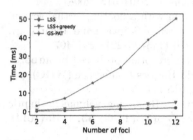

(c) Computing cost for the number of foci with 4032 transducers

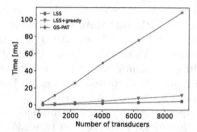

(d) Computing cost for the number of transducers

Fig. 4. Evaluation of computing cost

(a) LSS   (b) LSS+Greedy   (c) GS-PAT

Fig. 5. Fields of five-point star

4   Conclusion

To use the ultrasonic phased array to generate a foci sound field with stronger sound pressure to provide a tactile experience in a shorter time, we applied greedy algorithms with brute-force search to optimize the foci obtained by LSS. The solution we proposed has increased the sound pressure while also optimizing the side lobe. Even the computing cost is larger than LSS without offset optimization, our proposed method will not cause an explosive increase in computational cost with the increase of transducers. Furthermore, we will use the proposed method in the actual world to detect and evaluate pressure using the sensor. Moreover, we anticipate that greedy algorithms with brute-force search will be able to improve the performance of not just LSS, but also approaches such as GS-PAT. As a result, we will investigate the improvement of applying the greedy algorithm to other existing methods in the future.

References

1. Iwamoto, T., Tatezono, M., Shinoda, H.: Non-contact method for producing tactile sensation using airborne ultrasound. In: International Conference on Human Haptic Sensing and Touch Enabled Computer Applications, pp. 504–513 (2008)
2. Monnai, Y.: HaptoMime: mid-air haptic interaction with a floating virtual screen. In: The 27th Annual ACM Symposium on User Interface Software and Technology, UIST 2014, pp. 663–667 (2014). https://doi.org/10.1145/2642918.2647407
3. Inoue, S., Makino, Y., Shinoda, H.: Active touch perception produced by airborne ultrasonic haptic hologram. In: 2015 IEEE World Haptics Conference (WHC), pp. 362–367 (2015). https://doi.org/10.1109/WHC.2015.7177739
4. Long, B., Seah, S.A., Carter, T., Subramanian, S.: Rendering volumetric haptic shapes in mid-air using ultrasound. ACM Trans. Graph. (TOG) **33**, 1–10 (2014). https://doi.org/10.1145/2661229.2661257
5. Plasencia, D.M., Hirayama, R., Montano-Murillo, R., Subramanian, S.: GS-PAT: high-speed multi-point sound-fields for phased arrays of transducers. ACM Trans. Graph. Article no. 138 (2020). https://doi.org/10.1145/3386569.3392492
6. Suzuki, S., Fujiwara, M., Makino, Y., Shinoda, H.: Radiation pressure field reconstruction for ultrasound midair haptics by Greedy algorithm with brute-force search. IEEE Trans. Haptics (Open Access, Early Access). https://doi.org/10.1109/TOH.2021.3076489

A Rotary Induction Actuator
for Kinesthetic and Tactile Rendering

Georges Daher$^{(\boxtimes)}$, Stéphane Régnier$^{(\boxtimes)}$, and Sinan Haliyo$^{(\boxtimes)}$

Sorbonne Université, CNRS, Institut des Systèmes Intelligents et de Robotique (ISIR), 75005 Paris, France
{georges.daher,stephane.regnier,sinan.haliyo}@sorbonne-universite.fr

Abstract. Actuators with low inertia and high bandwidth are of great interest for haptic devices, as they improve the quality of force rendering and transparency. This paper describes, as a proof of concept, a new design in rotary induction motors, the Axial-DSIM (Axial Double-Sided Induction Motor). This motor has a simple design construction that consists of a thin and lightweight disc-shaped moving secondary (rotor) surrounded by fixed primaries on both sides that generate a rotating magnetic field that induces a force on the disc. The low inertia of this motor and its principle of operation make it possible to render high-fidelity torques with high dynamics.

Keywords: Design of haptic interfaces · Axial-DSIM (Axial Double-Sided Induction Motor) · Kinesthetic and tactile device

1 Introduction

Nowadays, the field of haptics is experiencing substantial growth; interest in haptic interfaces has increased considerably given their wide range of applications, including teleoperation, rehabilitation, education, games, arts, sciences, etc. However, the mechanical structure of most existing interfaces limits the transparency and rendering of haptic interactions [2]. The type of motor-drive system in a haptic interface is the predominant factor behind its selection. For instance, haptic devices that can produce high forces will not usually be capable of producing tiny and precise ones. The actuator, in this case, will be relatively large and massive; as a result, its large inertia will mask the perception of small forces. A small motor with low inertia, on the other hand, will be capable of producing weak and precise forces, but not large ones due to its small size [5].

An ideal haptic interface should be designed to cover the bandwidth of human haptic perception, and to faithfully render and scale forces without any structural distortion. The fidelity of a haptic interface is expressed as its transparency, which is obtained when the haptic signals rendered by the device are not distracted or scrambled by its mechanical dynamics [4]. Consequently, the ideal

This work was supported by the French National Research Agency through the ANR-Colamir project (ANR-16-CE10-0009).

H. Seifi et al. (Eds.): EuroHaptics 2022, LNCS 13235, pp. 155–163, 2022.
https://doi.org/10.1007/978-3-031-06249-0_18

haptic interface should exhibit low inertia, low mass, low friction, and a high structure stiffness while generating force with high dynamic range [8]. Thus, a lot of attention should be dedicated to the design, choice, and supply/control of the actuators used in the design of haptic interfaces.

This paper introduces the Axial-DSIM (Fig. 1) and details its design and some of its parameters and characteristics, and its advantages over other types of Eddy-current actuators previously introduced in the literature.

The Axial-DSIM seems to be an excellent candidate because the forcer (rotor, moving part, or secondary) is simply a thin sheet of conductor that could weigh a few grams only and upon which a force is exerted when placed in a traveling magnetic field.

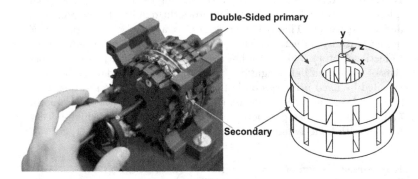

Fig. 1. The Axial-DSIM (Axial Double-Sided Induction Motor). The rotor's axis is coupled to a handle to transmit torque to the human operator.

2 Eddy-Current Based Haptic Devices

Eddy-current based actuators are appealing for haptics because of their low inertia. They have been employed in designing haptic interfaces but have not yet been commercialized. The primary goal behind the design of these interfaces is to reduce inertia and improve transparency [3,8,9], and [10].

In [8,9], and [3], rotary Eddy-current clutches are used in which a motor is used to rotate magnet-carrying discs around a non-ferromagnetic low inertia conductor to induce force on it; this complicates the control and makes the response time slower because to change the direction of the torque, it is necessary to change the direction of the motor that spins the magnet-carrying disc. And because the inertia of the motor and the magnet-carrying disc is high, the change in direction will be slower, resulting in a slower response time. In addition, when the magnets do not rotate and the handle is moved, viscosity is felt.

To have more flexibility in the control and a simpler mechanical design, we decided to replace the rotating magnet-carrying discs with a set of two electro-magnets that surround the rotor. Thus, instead of having a fixed magnetic field that is moved mechanically, we will have a magnetic field generated by fixed electromagnets whose amplitude and speed are determined by the amplitude and frequency of the windings supply, which should result in a faster response time and a finer torque due to the higher actuation frequency.

3   Human Haptic Perception and the Design Requirements of an Ideal Haptic Interface

Knowledge gained on human haptic perception helped to understand how to develop and improve the design of haptic interfaces. The human haptic perception relates to two cognitive senses: the tactile sense and the kinesthetic sense. Both senses are very important for to manipulation and locomotion [1,4,12].

Haptic Interfaces are divided into two: Kinesthetic and tactile. Kinesthetic interfaces produce force feedback, and tactile interfaces deliver tactile feedback. Both types have progressed in recent years, but the two types are commonly addressed separately. That's mainly because kinesthetic actuators that render force feedback aren't usually capable of rendering tactile information through vibration over a large bandwidth. And vibrators that render tactile information naturally aren't capable of generating force feedback. Coupling the two displays is an essential feature to have in haptic interfaces, especially if they are intended to be used for teleoperation tasks. Obtaining both stimuli simultaneously is usually accomplished by mixing both types of haptic interfaces [6,7,9,11], and [12].

4   The Principle of Operation of the Axial-DSIM

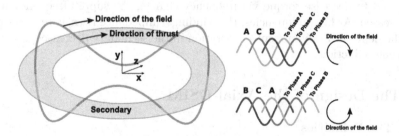

Fig. 2. The principle of operation of the Axial-DSIM.

When the primaries of the motors are fed with a three phased supply, a time traveling magnetic field will flow in the air gap between the two primaries and causes the secondary to move in the same direction as the field as illustrated in Fig. 2. When the time-traveling field passes through the non-ferromagnetic conductor, a current is induced in the conductor to oppose the change in the field by creating another magnetic field that opposes the change in the original magnetic field generated by the primaries. The induced current and the magnetic field of the primaries generate the force on the conductive plate. This is given by Laplace's law, which states that "A conductor through which a current flows and placed in a magnetic field is subject to a force." The frequency of the three-phase input supply determines the speed of the magnetic field, and the amplitude of the supply voltage determines the amplitude of the field. To reverse the direction of the magnetic field, i.e., the direction of the torque, the phase sequence must be altered from ACB to BCA by swapping the power supply of phase A with the power supply of phase B (Fig. 2).

5   Important Parameters and Considerations

The width of the air gap, i.e., the distance that separates the two sides of the primary, has a significant effect on the thrust and efficiency of the motor. When the length of the air gap increases, the efficiency and thrust decrease. Thus, the air gap must be as small as is mechanically possible. The smaller the air gap, the better the performance and thrust. The secondary thickness and conductivity are also important parameters to consider. The higher the electrical conductivity of the secondary, the higher the thrust produced. Here, aluminum was prioritized because it has the best weight to conductivity ratio. Keeping a small thickness secondary with a small air gap width is recommended. That falls to our advantage in minimizing the weight and inertia of the secondary. The input frequency is an important parameter as well. The frequency must be chosen in accordance with the magnetic characteristics of the primary core of the motor. The Axial-DSIM uses the SMC (Soft Magnetic Composite) "Somaloy 700HR 5P" that has fewer core losses at 60 Hz. Consequently, to output the maximum torque, the input supply frequency must be set to near the rated frequency of the primary core. And to decrease torque via frequency change, the supply frequency must be increased. At high frequencies, the windings' impedance rises, core losses rise, and the depth of Eddy-current penetration in the secondary decreases, resulting in torque reduction.

6   The Design of the Axial-DSIM

6.1   The Primaries

Aside from the primary core's material considerations, its geometry is also quite important. In the initial design that was intended to be sent to fabrication (Fig. 3 (a)), we wanted the primary to have an outer diameter of no more than 8 cm to avoid increasing the inertia of the secondary by increasing its diameter, and a deep slot with a small opening to reduce the slotting effect while yet allowing for the placement of coils with a sufficient number of turns. With SMC materials, complex shapes could be formed and fabricated at a lower cost than traditional silicon steel laminations. Making the mold is the most expensive aspect of the production process. Manufacturing a tiny amount is not cost-effective, which is why in this proof of concept, we have chosen as an alternative to use the

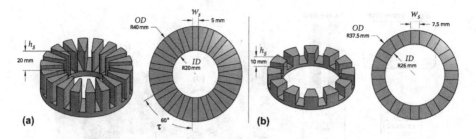

Fig. 3. (a) the initial intended design, (b) the design used in the experiments (primary used usually for single sided axial brushless DC motors).

primaries that are already used in the manufacture of single-sided axial brushless DC motors (Fig. 3 (b)). Although this primary core isn't perfect for an induction motor, it could be used to demonstrate the principle of actuation and its benefits.

6.2 The Winding Design

There are mainly two types of windings design: the one-layer planar (concentrated) non-overlapping windings and the overlapping double-layer windings. Induction motors often use overlapping double-layer windings, which provide a traveling field with fewer harmonic content as compared to the one-layer planar windings design. Here we have chosen to combine the two by adopting a three-layer planar, double-layer winding design (Fig. 4). This winding design is not recommended for single-sided configurations because the magnitude of the magnetic field generated by the phase closer to the air gap (third layer) is stronger than the amplitude of the field generated by the phase of the first layer. However, in a double-sided configuration that could be accounted for in the design by reversing the order of layers in the second side (Fig. 4). Each phase on each side consists of four sets of 60-turn coils connected in series (Fig. 5).

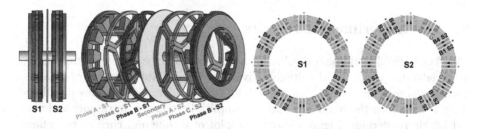

Fig. 4. The winding design of the Axial-DSIM (The three-layer planar, double-layer winding design).

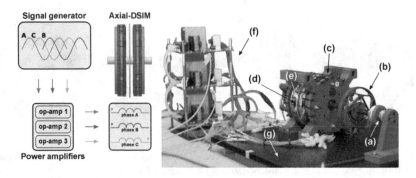

Fig. 5. The experimental setup. (a) the "ATI nano-17 force sensor", (b) the handle, (c) the first side of the stator, (d) the second side of the stator (3 mm apart from the first side), (e) the rotor (aluminum disc 80 mm in diameter and 1 mm in width), (f) the linear power amplifiers, and (g) the anti-vibration plate.

7 The Experimental Setup and the Shape of the Supply

Since the main interest is the quality of the rendered torque, the axis of the rotor has been attached to a high-resolution 6-DoF force sensor, the "ATI nano-17 force sensor". The motor and sensor are mounted on an anti-vibration plate to limit noise and vibrations coming from nearby equipment. The motor was supplied by three 120° phase shifted sine waves. The three sines were generated by a signal generator and amplified by linear power amplifiers. The shape of the power supply voltage greatly influences the quality of the rendered force and other factors such as noise and heating. The secondary disc is significantly less likely to overheat when the three phases are supplied with purely sinusoidal voltages. Therefore, as in HI-FI audio systems, we used linear power amplifiers fed with symmetrical linear power supplies. Linear power supplies have very low noise and ripple levels and react quickly to changes in voltage, resulting in a faster response time.

8 The Experimental Results

The Axial-DSIM could be controlled by a fixed-frequency/variable-voltage drive, a variable-frequency/fixed-voltage drive, or by variable-frequency/variable-voltage drive. This control flexibility falls to our advantage because it allows us to act on both the frequency and the amplitude to create intriguing and varied haptic renderings. Figure 7 shows the plot of torque and current per phase when the frequencies vary while the voltage remains constant. Hence, if a fixed-frequency/variable-voltage drive is to be used, the frequency should be adjusted at the primary core's rated frequency, which is 75 Hz, as seen in the plot of Fig. 7. Figure 8 shows the reaction time the motor takes to change the direction of torque. The change in direction was simply performed by swapping the power supply of phase A with the power supply of phase B. Given the Axial-DSIM's compelling dynamics, we chose to modulate the amplitude of the input supply to add vibrations to the constant force (Fig. 6).

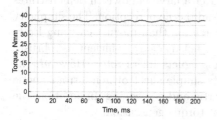

Fig. 6. Shape of constant torque generation as a function of time.

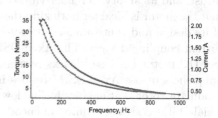

Fig. 7. Measured torque and current per phase as a function of frequency.

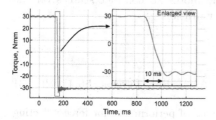

Fig. 8. Axial-DSIM reaction time to changes in direction of torque, the change from 30 Nmm to −30 Nmm took about 10 ms.

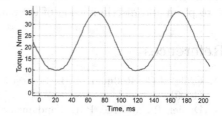

Fig. 9. Shape of the torque generated by the Axial-DSIM when supplied with a three-phase amplitude modulated signal; here a vibration 10 Hz was added.

In Fig. 9, we can see the shape of the torque exerted on the rotor when the Axial-DSIM is supplied with a three-phase supply 100 Hz amplitude modulated by a sinusoidal wave 10 Hz. The motor's vibration bandwidth is quite large; vibrations up to 1000 Hz Hz can be rendered and perceived. The maximum torque that can be generated with this design is about 100 Nmm. The maximum torque could be significantly increased if the primary cores were designed in the manner depicted in Fig. 3 (a).

9   Conclusion

The Axial-DSIM, a prototype haptic actuator with a low inertia rotor, was designed and built. The basic principle of the motor's operation was presented to familiarize the reader with the physics behind its operation. The experiments performed have proved that the Axial-DSIM is indeed interesting for haptics due to its wide range of rendering abilities that no other type of actuator of only one kind could provide. The Axial-DSIM's response time to change is the direction of thrust, and its ability to render vibrations over a large bandwidth demonstrates that the motor is ideal for both kinesthetic and tactile perception. The simplicity of its design and control makes it more interesting than magnet-carrying Eddy current coupling devices. The Axial-DSIM is not as energy efficient as a brushless DC motor because it lacks permanent magnets in its rotor. However, the uniqueness of the Axial-DSIM resides in its simple sheet of non-ferromagnetic conductor rotor that enables having low inertia and no cogging. In addition, the Axial-DSIM does not require an encoder for torque generation.

The main goals for future work include the further optimization of the motor and its design, the coupling of the Axial-DSIM with a slave micro-manipulator to perform micro teleoperated tasks (the specificities of the motor make it ideal for use in applications such as micro-teleoperation where high dynamics and vibrations are present), and the use of the motor to conduct studies on the perception and discrimination of force and vibration.

References

1. Craig, J.C., Rollman, G.B.: Somesthesis. Annu. Rev. Psychol. **50**(1), 305–331 (1999)
2. Daniel, R., McAree, P.R.: Fundamental limits of performance for force reflecting teleoperation. Int. J. Robot. Res. **17**(8), 811–830 (1998)
3. Ge, X., Peng, S.K., Wang, B., Shapiro, J., Gillespie, B., Salisbury, C.: A high bandwidth low inertia motor for haptic rendering based on clutched eddy current effects. In: 2012 IEEE Haptics Symposium (HAPTICS), pp. 83–89. IEEE (2012)
4. Hayward, V., Astley, O.R., Cruz-Hernandez, M., Grant, D., Robles-De-La-Torre, G.: Haptic interfaces and devices. Sens. Rev. (2004)
5. Hayward, V., Maclean, K.E.: Do it yourself haptics: part I. IEEE Robot. Autom. Mag. **14**(4), 88–104 (2007). https://doi.org/10.1109/M-RA.2007.907921
6. Lim, S.C., Lee, H.K., Park, J.: Role of combined tactile and kinesthetic feedback in minimally invasive surgery. Int. J. Med. Robot. Comput. Assisted Surg. **11**(3), 360–374 (2015)
7. Lu, T., Pacoret, C., Hériban, D., Mohand-Ousaid, A., Regnier, S., Hayward, V.: Kilohertz bandwidth, dual-stage haptic device lets you touch brownian motion. IEEE Trans. Haptics **10**(3), 382–390 (2016)
8. Millet, G., Haliyo, S., Regnier, S., Hayward, V.: The ultimate haptic device: first step. In: World Haptics 2009-Third Joint EuroHaptics Conference and Symposium on Haptic Interfaces for Virtual Environment and Teleoperator Systems, pp. 273–278. IEEE (2009)
9. Mohand-Ousaid, A., Millet, G., Régnier, S., Haliyo, S., Hayward, V.: Haptic interface transparency achieved through viscous coupling. Int. J. Robot. Res. **31**(3), 319–329 (2012)
10. Ortega, A., Weill-Duflos, A., Haliyo, S., Regnier, S., Hayward, V.: Linear induction actuators for a haptic interface: a quasi-perfect transparent mechanism. In: 2017 IEEE World Haptics Conference (WHC), pp. 575–580. IEEE (2017)

11. Ousaid, A.M., Haliyo, D.S., Régnier, S., Hayward, V.: A stable and transparent microscale force feedback teleoperation system. IEEE/ASME Trans. Mechatron. **20**(5), 2593–2603 (2015)
12. Pérez Ariza, V.Z., Santís-Chaves, M.: Haptic interfaces: kinesthetic vs. tactile systems. Revista EIA (26), 13–29 (2016)

Haptic Feedback for Wrist Angle Adjustment

Michiel den Daas⬭, Femke E. van Beek⬭, and Irene A. Kuling(✉)⬭

Eindhoven University of Technology, 5612 AZ Eindhoven, The Netherlands
i.a.kuling@tue.nl

Abstract. Haptic feedback is envisioned to be a powerful tool in (digital) orthosis fitment procedures. In context of a larger research project on digital molding and developing a glove for orthopedic experts, we explored the use of vibrotactile feedback on the wrist for wrist angle adjustments. Five different patterns are presented on both the inside and outside of the wrist as well as crossing signals. Participants were asked to indicate whether the pattern was communicating that the wrist angle had to be increased or decreased by moving the hand up or down. The results show that the vibrotactile stimuli are being interpreted consistently by the participants, provided the patterns are presented on one side of the arm. Although the interpretations were consistent within participants, there were individual differences in the reported directions of the signals, which makes it important to take into account personal preferences and calibration when implementing haptic feedback.

Keywords: Haptic feedback · Orthopedic fitment · Vibrotactile

1 Introduction

In the development of orthoses, typically, a physical mock-up is made to base the design of the aid on. To make this mock-up, plaster is used in order to create a mold around the patient's affected body part. Although the method has proven to be successful, there are some downsides to this method. First, it produces an excess of waste, and second, there is a rather small amount of objective data on the fitment process of orthoses which limits further research and development. In order to solve this, SmartScan is being developed [1]. SmartScan is a glove, which is meant to provide the practitioner digital support to develop orthopedic aids. This glove will allow the user to move their hands over a target body part to generate a digital 3D model of it, including the locally applied pressure patterns. To further improve the functionalities of SmartScan the possibilities of adding haptic feedback to help the practitioner in the fitment procedure are being explored.

One of the use cases envisioned for haptic feedback is the fitment of an ankle-foot orthosis. An ankle-foot orthosis is typically used for correcting drop-foot [2]; a condition in which the patient is unable to effectively control their foot. This results in an uncontrolled drop of the foot during loading, and the foot might also be dragged during the swing. In this case, it can prove effective to constrain the foot in a certain direction, where obtaining the right angle is vital [3]. Therefore, it would be useful to give haptic feedback on this angle to practitioners during the fitment procedure, which they perform with their hands (see Fig. 1A). This feedback information needs to be easy to understand

© The Author(s) 2022

H. Seifi et al. (Eds.): EuroHaptics 2022, LNCS 13235, pp. 164–170, 2022.

https://doi.org/10.1007/978-3-031-06249-0_19

by the practitioners to be effective. In this study, feedback about directionality of desired movement direction of the wrist (and thereby changing the wrist angle) is given by a wristband with vibrotactile motors. The four vibrotactile motors can be activated at different times creating different spatial and temporal patterns, which allows to study the consistency of the perceptual interpretation of vibrotactile feedback patterns about directionality on the angle of the wrist.

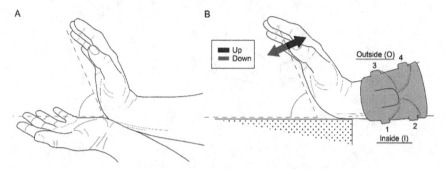

Fig. 1. A. Angle between the two hands used by practitioners during fitment of an ankle-foot orthosis. **B.** Schematic representation of the experimental set-up. Vibrotactile patterns were presented on the inside (actuator 1 and 2) and outside (actuator 3 and 4) of the arm. Participants were asked to indicate whether the presented pattern was communicating to move the hand up or down by moving their hand in the perceived direction and verbally answering 'up' or 'down'.

2   Vibrotactile Feedback on the Wrist Through a Wristband

Vibrotactile stimuli can be used for haptic feedback (e.g. [4, 5]) as well as for haptic communication e.g. [6, 7] or social touch (e.g. see [8] for a review). Typical findings in these studies are that spatiotemporal patterns created by vibrotactile motors can be used for communication, but the results are not always veridical (e.g. [9]). In this study, we use vibrotactile patterns as an indicator for the desired wrist angle adjustments. The actuators (Arduino vibration motor modules) are placed in a sweatband on the wrist. Two actuators are placed on both the dorsal (outside) and the ventral (inside) side of the arm, with one being on the proximal side (actuators 1 and 3 in Fig. 1B) of the wrist joint and one being on the distal side (actuators 2 and 4 in Fig. 1B) of the wrist joint. Two patterns were designed as a warning signal with either simultaneous or alternating actuation on/off actuation of the two motors on one side of the arm. The three other patterns were designed to create an illusion of movement (e.g. [10, 11]) by having a short overlap in the actuation of the subsequent motors. Actuation consisted of two motors vibrating subsequently on one side of the arm or switching between both sides. The five spatiotemporal vibration patterns that were designed for this experiment can be found in Fig. 2 (upper and middle row) and Table 1.

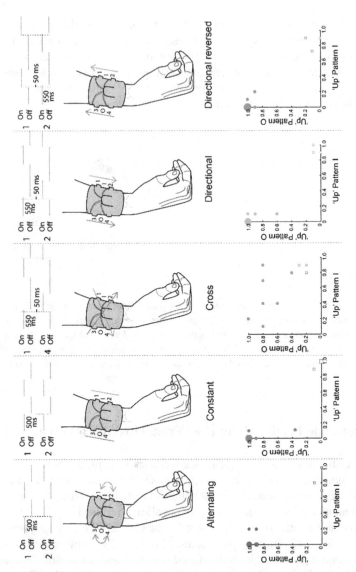

Fig. 2. The top row and middle row show the different vibrotactile patterns used in the experiment. At the wrists (middle row) the patterns are indicated with arrows, while on the top row the timing (two cycles) of the stimuli is schematically shown for the patterns on the inside of the arm (I, congruent with blue arrows in the middle row). Note that there was no overlap between the first and the second stimulus in the signals for the Alternating and Constant patterns, and 50 ms overlap for those in the Cross, Directional and Directional reversed patterns (which also lead to stimuli being 50 ms longer). The latter three signal types also had a 500 ms break in between the repetitions of the directional stimuli. The graphs at the bottom show the interpretation of each signal for all participants. Larger markers indicate more participants having the exact same results. Note that the data is clustered around the top left and bottom right corners (except for the Cross pattern), indicating consistent interpretations within the participants. The seven participants that interpreted the Alternating signal on the outside of the arm as 'Up' are indicated with filled circles in the graphs of all patterns, and the three participants that interpreted the Alternating signal on the inside of the arm as 'Up' are presented as open squares in the graphs of all patterns.

Table 1. Overview of the spatiotemporal vibrotactile patterns used in the experiment as well as the type of signal and the motors involved in the pattern.

Pattern	Abbreviation	Type	Motors pattern inside (I)	Motors pattern outside (O)
Alternating	Alt	Warning	1 & 2	3 & 4
Constant	Cons	Warning	1 & 2	3 & 4
Cross	Cross	Directional	1 & 4	2 & 3
Directional	Dir	Directional	1 & 2	3 & 4
Directional reversed	Dir-Rev	Directional	1 & 2	3 & 4

3 Experimental Design

3.1 Participants

Ten right-handed participants (seven female, 18–28 years, mean age 24.6 ± 6.5 years) volunteered to take part in the experiment, including two of the authors (FB, IK). None of the participants had known haptics deficits and except for the authors all participants were naive about the details of the experiment. All participants gave their written informed consent prior to the experiment. The study was approved by the Ethics Review Board of the TU/e.

3.2 Procedure

Participants received verbal instructions about the task. They were asked to sit in front of a table with their wrist on the table while wearing the sweatband with the actuators (Fig. 1B). Five different patterns were presented, once starting from the inside (I) and once starting from the outside (O). An overview of the patterns can be seen in Fig. 2. In each trial, a single pattern was presented and participants were asked to move their hand and answer 'up' or 'down' based on how they interpreted the presented signal. The pattern was repeated until the choice was made. All 10 stimuli (five patterns on two sides) were presented to the participant in a random sequence. After completing a sequence, a new random sequence was presented, for a total of 10 sequences, resulting in 100 trials per participant. The experiment took about 15 min to complete.

3.3 Analysis

For each participant, pattern and side the percentage of responses in which 'up' was chosen was calculated and compared for the inside (I) and outside (O) versions of the patterns. The consistency of the interpretation of the pattern was defined as the absolute difference between the fractions chosen 'up' in the inside and outside versions of the patterns. The consistency was analyzed with a one-way repeated measures ANOVA. Bonferroni corrections were used for post-hoc comparisons.

4 Results

The results showed large individual differences in the interpretations of the signals, as can be seen in Fig. 2, bottom row. For each participant a value between 0 and 1 for the inside (I) and (O) outside pattern is presented, which represents the fraction of 'up' responses for each pattern. For example, for the Alternating pattern, seven of the participants interpreted the pattern on the outside of the wrist as a cue to move towards the vibrated side (i.e. a value close to 1 for O and close to 0 for I), while the other three participants interpreted the signal as a cue to move away from that side (i.e. a value close to 0 for O and close to 1 for I). To assess the within-participant consistency of pattern interpretations, the three participants who had a value close to 1 for O for the Alternating pattern were indicated with a square marker in all subfigures. The same grouping of circles and squares arises in all subfigures except Cross, which suggests that there are two different (between participants), but consistent (within participants) interpretations for the same patterns.

The consistency values of all participants can be seen in Fig. 3. All participants gave consistent answers for all patterns that were on the same side of the arm (Alt, Cons, Dir, Dir-Rev), but not for the Cross pattern which gave vibrations on both sides of the arm (Fig. 3). The one-way repeated measures ANOVA showed a significant effect of pattern on the consistency ($F_{4,36} = 14.95$, $p < 0.001$, $\eta_p^2 = 0.624$). Post-hoc comparisons showed a significant difference between pattern Cross and all other patterns (all p's < 0.02) and no other significant differences.

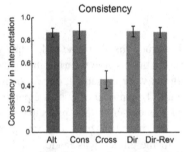

Fig. 3. Results. All patterns on one side of the arm show high consistency, while the Cross pattern is significantly less consistently interpreted. There is no difference between the other four patterns, indicating that the pattern itself might not be relevant.

5 Discussion and Conclusion

In this study we explored whether different vibrotactile patterns on the wrist could be used to give the user feedback about the desired change in wrist angle. The results show that the information can be consistently and efficiently sent by all patterns that have the vibrotactile actuation on one side of the arm. Only for the pattern in which the actuation switched sides, the consistency and therefore efficiency of the communication was low. This suggests that the side of the actuation seems to be more relevant than the actual pattern. Since no difference was found between directional patterns and warning patterns, it might be concluded that a single actuator on both sides of the arm could be enough for effective communication of the wrist angle.

An interesting finding is that there were individual differences in the interpretation of the actuations on the different sides of the arm. Seven participants interpreted the actuated side as the direction they had to move towards, while three participants interpreted the signal oppositely; they moved away from the actuated side. This means that for future application of such haptic feedback systems in practice, individual calibration will be needed, both on interpretation (similar to the preferred scrolling direction on a mousepad). Another aspect that should be individually calibrated is the intensity of the signal. Some participants indicated that the vibration was too intense, while others felt it was at the right intensity.

A next step for this project is to make the feedback signal dynamic and responsive to the real-time angle of the wrist to test whether the found solution would be suitable for online angle adjustment while developing an orthopedic aid.

To conclude, a simple vibration on one side of the wrist is an effective way to give information about the required direction of change of the angle of the wrist, but only if the interpretation of the signal can be individually calibrated.

References

1. Fontys SmartScan webpage. https://fontys.nl/Over-Fontys/Fontys-Paramedische-Hogesc hool/Onderzoek/SmartScan.htm. Accessed 03 Dec 2021
2. Alam, M., Choudhury, I.A., Mamat, A.B.: Mechanism and design analysis of articulated ankle foot orthoses for drop-foot. Sci. World J. **2014**, 14 (2014). https://doi.org/10.1155/2014/867869. Article ID 867869
3. Bregman, D.J.J., Rozumalski, A., Koops, D., De Groot, V., Schwartz, M., Harlaar, J.: A new method for evaluating ankle foot orthosis characteristics: BRUCE. Gait Posture **30**(2), 144–149 (2009)
4. Israr, A., Poupyrev, I.: Tactile brush: drawing on skin with a tactile grid display. In: CHI 2011, Vancouver, BC, Canada, pp. 2019–2028 (2011)
5. Kuling, I.A., Gijsbertse, K., Krom, B.N., van Teeffelen, K.J., van Erp, J.B.F.: Haptic feedback in a teleoperated box & blocks task. In: Nisky, I., Hartcher-O'Brien, J., Wiertlewski, M., Smeets, J. (eds.) Haptics: Science, Technology, Applications. LNCS, vol. 12272, pp. 96–104. Springer, Cham (2020). https://doi.org/10.1007/978-3-030-58147-3_11
6. Jones, L.A., Kunkel, J., Piateski, E.: Vibrotactile pattern recognition on the arm and back. Perception **38**(1), 52–68 (2009)
7. Plaisier, M.A., Kappers, A.M.L.: Social haptic communication mimicked with vibrotactile patterns-an evaluation by users with deafblindness. In: ASSETS 2021: The 23rd International ACM SIGACCESS Conference on Computers and Accessibility Proceedings, pp. 1–3. Association for Computing Machinery (2021)
8. Van Erp, J.B., Toet, A.: Social touch in human–computer interaction. Front. Digit. Humanit. **2**, 2 (2015)
9. Plaisier, M.A., Sap, L.I., Kappers, A.M.: Perception of vibrotactile distance on the back. Sci. Rep. **10**(1), 1–7 (2020)

10. Geldard, F.A., Sherrick, C.E.: The cutaneous "rabbit": a perceptual illusion. Science **178**(4057), 178–179 (1972)
11. Wu, W., Culbertson, H.: Wearable haptic pneumatic device for creating the illusion of lateral motion on the arm. In: 2019 IEEE World Haptics Conference (WHC), Tokyo, Japan, pp. 193–198 (2019)

Larger Skin-Surface Contact Through a Fingertip Wearable Improves Roughness Perception

David Gueorguiev[1,2]([✉]), Bernard Javot[1], Adam Spiers[1,3],
and Katherine J. Kuchenbecker[1]

[1] Haptic Intelligence Department, Max Planck Institute for Intelligent Systems,
70569 Stuttgart, Germany
`dgueorguiev@is.mpg.de`
[2] CNRS, Institut des systèmes intelligents et de robotique, 75005 Paris, France
[3] Manipulation and Touch Lab, Department of Electrical and Electronic Engineering,
Imperial College London, London SW7 2BX, UK

Abstract. With the aim of creating wearable haptic interfaces that allow the performance of everyday tasks, we explore how differently designed fingertip wearables change the sensory threshold for tactile roughness perception. Study participants performed the same two-alternative forced-choice roughness task with a bare finger and wearing three flexible fingertip covers: two with a square opening (64 and 36 mm^2, respectively) and the third with no opening. The results showed that adding the large opening improved the 75% JND by a factor of 2 times compared to the fully covered finger: the higher the skin-surface contact area, the better the roughness perception. Overall, the results show that even partial skin-surface contact through a fingertip wearable improves roughness perception, which opens design opportunities for haptic wearables that preserve natural touch.

Keywords: Roughness perception · Psychophysics · Tactile devices

1 Introduction

An increasing number of wearable devices, sensing gloves, and video game controllers now provide haptic feedback due to strong interest in tactile communication for extended reality (XR), which includes augmented reality (AR), and virtual reality (VR) [10]. These devices provide many advantages compared to desktop haptic interfaces, such as freedom of motion, transportability, and efficient XR integration. However, wearing haptic devices on the finger also has drawbacks. One major challenge is that devices that provide force or vibrotactile feedback often cover the fingertip with a plastic platform or belt, e.g., [6,10,11]. Thus, whenever a user wears the device, they cannot directly feel objects and surfaces that are part of the real environment surrounding them.

© The Author(s) 2022
H. Seifi et al. (Eds.): EuroHaptics 2022, LNCS 13235, pp. 171–179, 2022.
https://doi.org/10.1007/978-3-031-06249-0_20

This constraint is limiting because it prevents the use of mice, keyboards, and touchscreens, and because it interferes during contact with real materials that are increasingly used to complement and enrich virtual objects [2,12]. For example, a novel research trend moves a small number of physical objects to render interaction with many virtual items [5]. Therefore, it seems that future XR applications will need to mix purely virtual haptic rendering generated by a device with the efficient use of tangible objects and surfaces.

The use of a probe, glove, or finger sheath is known to impair roughness discrimination [4]. Previous studies have also shown that wearing medical gloves impairs dexterity and tactile sensitivity during medical tasks such as determining pulse location [7]. Perception of vibration magnitude is mostly preserved during the use of a probing tool, though acuity depends on how the tool is held in the hand [16]. The size of the probe also impacts the tactile perception of vibration, which points to a role of the skin area in contact [14]; this process is probably mediated by the specific receptive fields of the tactile afferents [1].

Recent research has already begun creating devices that generate haptic feedback while preserving the user's tactile and kinesthetic capabilities. For example, a vibrotactile actuator can be placed on a proximal segment of the finger rather than the fingertip; to some extent, the brain relocates the stimulus to the tip of the finger, where the interaction is believed to take place [9]. In another recent design, the fingertip device folds itself up when the user interacts with real objects [13]. A third approach is to build very thin actuators that partially preserve haptic perception. Thin soft layers with tiny embedded actuators preserve most of the tactile acuity [8] but can provide only specific electrical or vibrotactile cues that aim more at tactile communication rather than rich rendering [15].

The current study explores how preserving direct skin-surface contact in a fingertip band (as used in several wearables [6,11]) through openings of different sizes impacts roughness perception. To that end, we have designed three flexible fingertip wearables that are either solid or have a central square opening. Since roughness perception relies on either skin deformations or vibration, the size of the skin contact is likely to affect the amount of preserved tactile accuracy. For small openings, the occurring deformation patterns might be limited, and skin vibrations might be distorted. Results from the experiment suggest that larger preserved skin-surface contact improves roughness perception.

2   Materials and Methods

Data were collected from eight right-handed volunteers and one left-handed volunteer. They were healthy and aged between 21 and 59. This study was approved by the ethics committee on human research of Université catholique de Louvain. All participants gave written informed consent.

Experimental Setup. Three wearable interfaces were custom-built for the experiment (Fig. 1). They are made of Dycem sheet that is 0.7 mm thick, and they are fixed to the finger with two elastic bands. Dycem was chosen to be

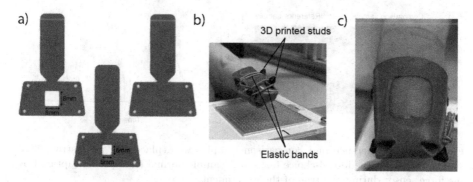

Fig. 1. a) The three wearable interfaces that were used in the study included square fingerpad openings that are $64\,mm^2$, $36\,mm^2$, and $0\,mm^2$. b) Each one is held on the finger by two elastic bands, which are retained by four 3D printed studs mounted at the corners of the surface covering the fingerpad. c) Underside of the interface with the $64\,mm^2$ opening while it is worn by a participant.

flexible but not extensible, and to have a high coefficient of friction; such characteristics are compatible with the creation of a wearable device that would provide haptic cues such as shear force, normal force, or vibration. Three types of fingertip wearables were laser-cut with respective openings of $64\,mm^2$, $36\,mm^2$, and $0\,mm^2$ (no opening); the corresponding horizontal and vertical aperture dimensions are $8\,mm$ and $6\,mm$. The apertures featured $0.5\,mm$ fillets on the corners to prevent tearing of the material during use, given that these are the points of highest stress concentration. The wearables were fixed to the finger rather tightly but with special care not to have a bump of skin protruding through the opening due to excessive pressure. The interfaces also included a rectangular section that wrapped around the distal part of the fingertip and was secured over the nail to prevent surface contact by other regions of the participant's skin.

Square $40\,mm \times 40\,mm$ surfaces covered by grids of rigid raised dots were used in the psychophysical experiments (Fig. 2a). These surfaces were 3D-printed (Objet Connex 260) from a design that has already been used in the literature to estimate the human sensory threshold for roughness [8]. The reference surface had a center-to-center dot spacing of $1.00\,mm$, and the comparison surfaces had dots spacings of $1.05\,mm$, $1.08\,mm$, $1.20\,mm$, $1.30\,mm$, $1.50\,mm$, $1.70\,mm$, $1.80\,mm$, and $2.00\,mm$. These intervals are identical to the 3D-printed grids that were used by Nittala et al. [8]. All surfaces had a dot height of $0.65\,mm$.

The experimental setup consisted in a sample holder mounted on a Nano17 force/torque sensor (ATI, USA). The holder had two square recesses in which the square surfaces could be inserted and interchanged between trials. A curtain placed between the sample holder and the participant prevented the use of visual cues. A MATLAB script and an NI USB-6343 X series acquisition card were used to run the psychophysical experiment and record the force data (Fig. 2b).

Psychophysical Procedure. Participants had to perform a roughness discrimination task in four conditions: with their bare finger and while wearing

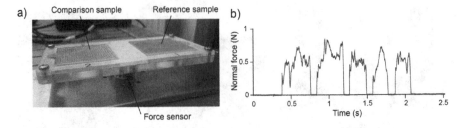

Fig. 2. a) The reference and comparison samples were placed on an instrumented sample holder for the forced-choice task. b) A sample normal force profile captured by the force sensor during one trial of the experiment.

the prepared Dycem interfaces with openings of $64\,\mathrm{mm}^2$, $36\,\mathrm{mm}^2$, and $0\,\mathrm{mm}^2$. The order of the conditions was counterbalanced across participants. However, it has to be noted that the $64\,\mathrm{mm}^2$ opening was added after the sixth participant because performance with the $36\,\mathrm{mm}^2$ opening was poorer than expected. The $64\,\mathrm{mm}^2$ condition was performed last by the first six participants; hence, the presentation order was not fully counterbalanced.

For each condition, participants performed the same forced-choice roughness task, which consisted in freely exploring the two samples and reporting whether the left or right one felt rougher. We used a Zwislocki staircase procedure that targets the 75% just noticeable difference (JND), which is common for threshold estimation [17]. We started with a comparison square grid whose dots have a center-to-center dot spacing of 1.70 mm, which is a difference of 70% compared to the 1.00 mm spacing of the reference grid. During the procedure, the difference with the reference grid was decreased by switching to a grid with smaller spacing after three non-consecutive correct answers and increased in the same manner after each error. The staircase was ended after either five reversals or 30 trials. To reduce left/right confusion, the samples were identified with the words "one" and "two", and each answer was double-checked by the experimenter before being recorded in the MATLAB script. The surfaces and interfaces were washed with soap, and volunteers thoroughly cleaned their hands with alcohol disinfectant.

Data Analysis. The 75% JND was computed by averaging the center-to-center spacing of dots across the last ten trials before the end of the psychophysical staircase for each participant in each condition. Several times during the experiment, especially with the bare finger, participants reached the minimum difference that was possible with the square grids. In those cases, a sensory threshold of 5% was recorded, corresponding to the smallest tested grid. In addition, the collected force measurements were used to compute the average normal force during contact as well as the duration of the participant's tactile interaction with the samples. The statistical analyses were performed with the Graphpad prism software; the D'Agostino & Pearson normality test was used to decide whether a parametric or non-parametric statistical analysis should be performed.

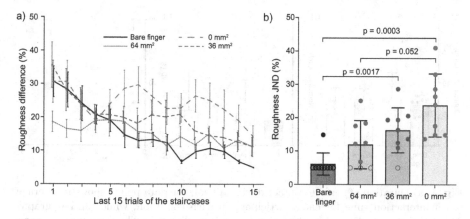

Fig. 3. a) Last 15 trials of the participants' staircases averaged for each condition (mean ± SEM). b) 75% just noticeable difference (JND). Each bar graph represents the mean value for an experimental condition, and the error bars show the standard deviation across participants. The individual sensory thresholds appear as solid dots when the staircase ended due to the number of errors and as open circles when the participant successfully discriminated the two most similar roughness samples.

3   Results

The aim of the study was to test whether varying the conditions of skin-surface contact significantly affects tactile sensory performance. For each experimental condition, we computed and plotted the mean staircases across participants for the last 15 trials before the end of the procedure (Fig. 3a). This interval was chosen because 15 trials was the lowest possible number of trials. The results suggest that convergence occurred for the conditions in which the skin directly touched the surface. The curve is noisier when the fingertip was fully covered, perhaps due to artefacts such as random slips or a higher cognitive load during this indirect exploration. Thus, we performed a Friedman non-parametric statistical test between the four conditions (Fig. 3b), which showed a significant impact of the condition on the 75% JND (n = 9, Q(4) = 19.64, p = 0.0002). The post-hoc analysis that was performed with a Dunn's multiple comparisons test showed significant differences between the bare finger and the $0\,\mathrm{mm}^2$ conditions (p = 0.0003) and between the bare finger and the $36\,\mathrm{mm}^2$ conditions (p = 0.0017), as well as a marginally significant difference between the $64\,\mathrm{mm}^2$ and the $0\,\mathrm{mm}^2$ conditions (p = 0.052). Overall, the data indicate that the amount of skin-surface contact matters for roughness discrimination.

We also used the force recordings to quantify how participants interacted with the samples during each trial. A threshold of 0.05 N, which is around five times higher than the noise level of the sensor, was used to determine when tactile exploration was occurring. By this method, we could obtain the average duration of tactile interaction across the final ten trials that were used to compute the 75% JND (Fig. 4a). A Friedman statistical analysis showed no influence on the experimental condition on the duration of tactile exploration.

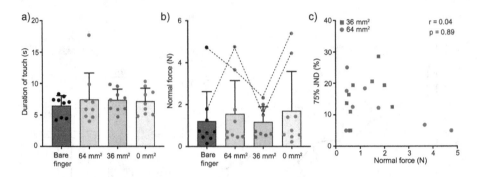

Fig. 4. a) Duration of surface contact during a trial; the data points show the average touch interaction time for each participant in each condition, while the bar graphs depict the mean and standard deviation for each condition. b) The same type of graph for the mean normal force exerted by participants. The dashed lines connect the surprisingly large values from two participants. c) Relationship between the mean normal force and the 75% JND for the conditions with $64\,\text{mm}^2$ and $36\,\text{mm}^2$ openings in the interface; no significant correlation is observed.

We used the same threshold to compute the average normal force exerted during the final ten trials that contributed to the JND (Fig. 4b). As for the duration, the Friedman statistical test showed no significant impact of the experimental conditions. Surprisingly, some individual data points showed average normal force values as high as $5\,\text{N}$; we found that all of the average normal force values over $2\,\text{N}$ came from two participants. Thus, these values resulted from the general strategy of these participants rather than because of a specific condition.

Finally, we tested the possibility that pushing harder could impact the protrusion of skin through the opening in the wearable interface and hence the participant's sensitivity to roughness. However, as seen in Fig. 4c, a Spearman analysis between the exerted normal force and the sensory threshold for the conditions with an opening in the interface did not show any significant correlation.

4   Discussion

The results have shown that adding an opening in a flexible skin cover can improve the wearer's tactile acuity compared to a solid cover of the finger. Three of the nine participants achieved the smallest threshold when the cover had a large opening ($64\,\text{mm}^2$), while none of them could distinguish this small differences in surface roughness when the wearable had no opening ($0\,\text{mm}^2$). Interestingly, we did not find an effect of the condition on the duration of tactile exploration or the applied normal force. It seems that participants chose a duration and normal force strategy at the start and did not deviate much from it during the experiment, unlike in [3], where participants pushed much harder when the button detection task was more challenging. However, we did not find an influence of the duration of tactile exploration or the normal force on the participant's 75% JND. Thus, it seems that performance at the task was predominantly driven by the tactile sensitivity of participants and that exploration behaviour played little role. We were especially interested in whether an increased amount of skin put into contact with the surface by pushing harder could improve roughness perception for conditions that included an opening, but we found no evidence of a correlation between average normal force and the sensory threshold.

The main result of the experiment is that a rather small contact area between the skin and external surfaces (through a 64 mm^2 opening) is sufficient to improve the JND from 23.8% to 12.0%, which is a two-times-better sensory threshold. Overall, our results are in line with existing literature on perception with a solid finger covering [4,8] and the importance of skin-surface contact area [14], and they provide a new insight about the potential of partial skin-surface contact. Therefore, we imagine the opportunity to design wearable haptic interfaces that achieve contact force rendering through bands or a cover containing openings while largely preserving natural tactile sensation.

Although we carefully controlled their fit, the wearable interfaces used in our study were the same for all participants. Using constant opening sizes probably impacted our results since finger sizes are variable; the ratio between the preserved contact area and the contact area of the bare finger thus varied across people. Some participants commented that skin-surface contact was intermittent with the small opening since the cover was rather thick. Contact in this condition might also have been impacted by the elasticity of each participant's skin. Future research will determine whether openings in a fingertip cover also improve other dimensions of tactile acuity such as texture and softness perception. We are also curious whether it is possible to superpose accurate rendering of virtual haptic features with preserved natural sensation of the touched surface or object.

Acknowledgement. This work was funded by the Max Planck Society, CNRS, and the ANR grant Maptics. The authors thank the MPI-IS Robotics ZWE for helping to create the interfaces.

References

1. Abraira, V., Ginty, D.: The sensory neurons of touch. Neuron **79**(4), 618–639 (2013). https://doi.org/10.1016/j.neuron.2013.07.051
2. Bouzbib, E., Bailly, G., Haliyo, S., Frey, P.: CoVR: a large-scale force-feedback robotic interface for non-deterministic scenarios in VR. In: Proceedings of the ACM Symposium on User Interface Software and Technology (UIST), pp. 209–222 (2020). https://doi.org/10.1145/3379337.3415891
3. Gueorguiev, D., Kaci, A., Amberg, M., Giraud, F., Lemaire-Semail, B.: Travelling ultrasonic wave enhances keyclick sensation. In: Prattichizzo, D., Shinoda, H., Tan, H.Z., Ruffaldi, E., Frisoli, A. (eds.) EuroHaptics 2018. LNCS, vol. 10894, pp. 302–312. Springer, Cham (2018). https://doi.org/10.1007/978-3-319-93399-3_27
4. Klatzky, R.L., Lederman, S.J.: Tactile roughness perception with a rigid link interposed between skin and surface. Percept. Psychophys. **61**(4), 591–607 (1999)

5. Mercado, V., Marchai, M., Lécuyer, A.: Design and evaluation of interaction techniques dedicated to integrate encountered-type haptic displays in virtual environments. In: 2020 IEEE Conference on Virtual Reality and 3D User Interfaces (VR), pp. 230–238 (2020). https://doi.org/10.1109/VR46266.2020.00042
6. Minamizawa, K., Fukamachi, S., Kajimoto, H., Kawakami, N., Tachi, S.: Gravity grabber: wearable haptic display to present virtual mass sensation. In: Proceedings of ACM SIGGRAPH Emerging Technologies, SIGGRAPH 2007, p. 8-es (2007). https://doi.org/10.1145/1278280.1278289
7. Mylon, P., Lewis, R., Carré, M.J., Martin, N.: Evaluation of the effect of medical gloves on dexterity and tactile sensibility using simulated clinical practice tests. Int. J. Ind. Ergon. **53**, 115–123 (2016). https://doi.org/10.1016/j.ergon.2015.11.007
8. Nittala, A.S., Kruttwig, K., Lee, J., Bennewitz, R., Arzt, E., Steimle, J.: Like a second skin: understanding how epidermal devices affect human tactile perception. In: Proceedings of the CHI Conference on Human Factors in Computing Systems, pp. 1–16. ACM (2019). https://doi.org/10.1145/3290605.3300610
9. Pacchierotti, C., Salvietti, G., Hussain, I., Meli, L., Prattichizzo, D.: The hRing: a wearable haptic device to avoid occlusions in hand tracking. In: Proceedings of the IEEE Haptics Symposium (HAPTICS), pp. 134–139 (2016). https://doi.org/10.1109/HAPTICS.2016.7463167
10. Pacchierotti, C., Sinclair, S., Solazzi, M., Frisoli, A., Hayward, V., Prattichizzo, D.: Wearable haptic systems for the fingertip and the hand: taxonomy, review, and perspectives. IEEE Trans. Haptics **10**(4), 580–600 (2017). https://doi.org/10.1109/TOH.2017.2689006
11. Prattichizzo, D., Chinello, F., Pacchierotti, C., Minamizawa, K.: Remotouch: a system for remote touch experience. In: Proceedings of the IEEE International Symposium on Robot and Human Interactive Communication (RO-MAN), pp. 676–679 (2010). https://doi.org/10.1109/ROMAN.2010.5598606
12. Roo, J.S., Hachet, M.: One reality: augmenting how the physical world is experienced by combining multiple mixed reality modalities. In: Proceedings of the ACM Symposium on User Interface Software and Technology (UIST), pp. 787–795 (2017). https://doi.org/10.1145/3126594.3126638
13. Teng, S.Y., Li, P., Nith, R., Fonseca, J., Lopes, P.: Touch&Fold: a foldable haptic actuator for rendering touch in mixed reality. In: Proceedings of the ACM CHI Conference on Human Factors in Computing Systems, New York, NY, USA (2021). https://doi.org/10.1145/3411764.3445099
14. Verrillo, R.T.: Investigation of some parameters of the cutaneous threshold for vibration. J. Acoust. Soc. Am. **34**(11), 1768–1773 (1962). https://doi.org/10.1121/1.1909124
15. Withana, A., Groeger, D., Steimle, J.: Tacttoo: a thin and feel-through tattoo for on-skin tactile output. In: Proceedings of the ACM Symposium on User Interface Software and Technology (UIST), pp. 365–378 (2018). https://doi.org/10.1145/3242587.3242645
16. Zamani, N., Culbertson, H.: Effects of dental glove thickness on tactile perception through a tool. In: Proceedings of the IEEE World Haptics Conference (WHC), pp. 187–192 (2019). https://doi.org/10.1109/WHC.2019.8816166
17. Zwislocki, J.J., Relkin, E.M.: On a psychophysical transformed-rule up and down method converging on a 75% level of correct responses. Proc. Natl. Acad. Sci. **98**(8), 4811–4814 (2001). https://doi.org/10.1073/pnas.081082598

Expanding Dynamic Range of Electrical Stimulation Using Anesthetic Cream

Takumi Hamazaki[1](✉), Taiga Saito[1], Seitaro Kaneko[1,2], and Hiroyuki Kajimoto[1]

[1] The University of Electro-Communications, 1-5-1 Chofugaoka, Chofu, Tokyo, Japan
{hamazaki,saito,kaneko,kajimoto}@kaji-lab.jp
[2] Tokyo, Japan

Abstract. Electrical stimulation is one of the methods to stimulate skin sensation, and can provide sensations such as vibration and pressure by changing the polarity of the stimulus. These stimuli can be combined to design a variety of tactile sensations. However, there is a major problem with electrical stimulation: As the amount of electric current is increased, itching or pain sensation is elicited. This study aims to suppress the itching and pain caused by electrical stimulation, and to present strong, clear, and stable, pressure and vibration sensations. We applied an anesthetic cream containing lidocaine, which is one of the most used local anesthetics, to reduce the induced pain and itching. Therefore, we specifically examine the applicability of lidocaine toward a desirable situation, in which pain thresholds are increased and tactile thresholds are not significantly affected. The results showed a significant relationship between the application of the cream and the dynamic range of stimulating current, and subsequently the quality of experience by human participants.

Keywords: Chemical haptics · Local anesthetic cream · Electrical stimulation

1 Introduction

The methods of stimulating cutaneous sensations can be approximately classified into two categories: mechanical stimulation by physical deformation of the skin, and electrical stimulation through the direct generation of nerve activity. The mechanical stimulation can generate a sense of texture [1] and unevenness [2] by vibrating the skin or presenting a spatial skin distortion pattern, whereas electrical stimulation [3] directly stimulates the nerve axons extending from mechanoreceptors using electrodes placed on the skin surface. Electrical stimulation exhibits the following advantages over mechanical stimulation: low thickness and weight, low power consumption, and the absence of mechanical moving parts. Devices such as visual-tactile conversion devices that use these

---

S. Kaneko—JSPS Research Fellow.

H. Seifi et al. (Eds.): EuroHaptics 2022, LNCS 13235, pp. 180–188, 2022.
https://doi.org/10.1007/978-3-031-06249-0_21

advantages of electrical stimulation have been previously proposed [4], for the visually impaired and for large area tactile displays [5] that present tactile sensations of virtual object surfaces to the entire palm.

There are two types of electrical stimulations: anodic stimulation, in which current flows from a single electrode to a group of surrounding electrodes; and cathodic stimulation, in which current flows from a group of surrounding electrodes to a single electrode. There are differences in thresholds and sensations between these two modes [6], with cathodic stimulation characteristically producing pressure sensations, believed to originate primarily from Merkel cells; and anodic stimulation characteristically producing vibration sensations, believed to originate primarily from Meissner corpuscles [7]. These stimuli can be combined to design a variety of tactile sensations [8].

However, there is a major problem with electrical stimulation: As the amount of electric current is increased, itching or pain sensation is elicited. This problem is exaggerated in the case of multi-point stimulation, in which each electrode induces pain under different thresholds, and pain from just a single electrode degrades the entire experience.

This study aims to suppress the itching and pain caused by electrical stimulation, and to present strong, clear, and stable, pressure and vibration sensations. We applied an anesthetic cream containing lidocaine, which is one of the most used local anesthetics, to reduce the induced pain and itching. However, it is undesirable for lidocaine to reduce the target sensations such as vibration and pain. Therefore, we specifically examine the applicability of lidocaine toward a desirable situation, in which pain thresholds are increased (pain becomes less perceptible) and tactile thresholds are not significantly affected (tactile sensation is fully perceptible). Dynamic range was used to investigate the expansion of the electric stimulus presentation range.

In recent years, many efforts have been made to use chemical substances for tactile displays. Lu et al. [9] proposed a method of providing numbness and other sensations in VR space, through a series of efforts called Chemical Haptics by applying solutions such as sansho, capsaicin, and lidocaine. Based on their study, the current method can be considered as an attempt to combine electrical stimulation with Chemical Haptics.

2   Experiments and Result

We conducted three experiments: Experiment 1, in which we applied electrical stimulation to three locations (fingertip, forearm, and forehead), with different skin thicknesses, under three conditions direct contact (C1), castor oil application (C2), and anesthetic cream application (C3); Experiment 2, in which we performed an experiment similar to Experiment 1, under a different polarity of the electrical stimuli; and Experiment 3, in which we focused on the intensity of subjective tactile sensation, and examined whether the anesthetic cream affected the perceived intensity of target tactile sensation.

All experiments were approved by the Ethics Committee of the University of Electro-Communications, Japan.

2.1   Experiment 1

Conditions. The anesthetic cream commercially available in Japan (Daiichi Sankyo Healthcare, lidocaine concentration 2%) was used. As a control condition, castor oil cream (Casoda, Heritage Products, USA), which comprises the same base material as the anesthetic cream, was used.

The forearm was chosen imitating a wristwatch-type wearable device, and the forehead was chosen assuming a HMD-embedded device (Fig. 1a).

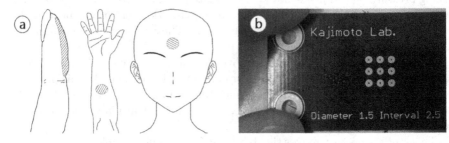

Fig. 1. Experimental conditions: (a) Application location, Finger pad; center of the forearm; and center of the forehead. (b) is the electrode used in this experiment.

Apparatus. The electrical stimulation device consists of nine electrodes with a diameter of 1.5 mm and a center-to-center distance of 2.5 mm (Fig. 1b). The central electrode was used as the stimulating electrode, and the 8 surrounding electrodes operated as the returning current electrodes. In anodic stimulation, the center electrode becomes the anode, and in cathodic stimulation, the center electrode becomes the cathode. In this experiment, we monitored the current flowing through the skin by measuring the voltage across a series-connected 1 kΩ resistor using an oscilloscope.

Procedure. The order of the experiments was counterbalanced among the subjects to overcome the effect of the order of application substances. Six male participants between 21–27 years of age were tested. The experiment was conducted over 3 days, with one condition at each location measured each day.

We applied 1.0 g of the ointment per 10 cm^2, on the skin of the fingers, forearm, and forehead, sealing the area with plastic wrap and masking tape, for 1 h to allow adequate penetration of the ointment. This procedure was skipped for condition C1. The areas under cream application were wiped off with gauze and we started the measurement. Anodic current stimulation with a pulse width of 200 us was applied at 30 pulses per sec (pps), and the participants adjusted the amplitude by interacting with the up and down keys of the keyboard, to find the threshold value for the slight perception of the stimulus (herein referred to as the "tactile threshold"), and the threshold amplitude for the perception of pain (herein referred to as the "pain threshold"). Three trials were performed for each condition and location, at an interval of 30 s between each trial.

Results. The tactile and pain thresholds, recorded under each condition, are shown in Fig. 2. The variance of the values can be observed to be large and the individual differences were large, especially for the pain threshold. This agrees with the results of a previous research [10].

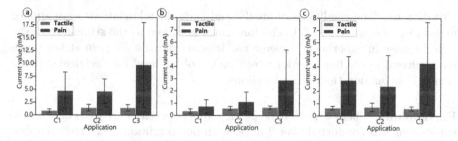

Fig. 2. Distribution of tactile and pain thresholds in the (a) finger, (b) forearm, and (c) forehead. Only (a) has a different value on the vertical axis.

Subsequently, we focused on the ratio of pain threshold to tactile threshold. This ratio is believed to indicate the ease of adjustment to the stimulated tactile sensation without pain, and can be referred to as a dynamic range in electro-tactile sensation. The dynamic range was calculated and normalized through Condition C1 (Fig. 3). A two-way ANOVA with correspondence was performed on these results. The main effect was observed only for the change in application condition ($F = 12.995$, $p < 0.05$), and not for that of the application location ($F = 2.412$, n.s.). No interaction was observed between the application conditions and locations ($F = 2.703$, n.s.). The Bonferroni corrected t-test for the results of the change of application condition showed a significant difference between the Conditions C2 and C3 ($p < 0.05$), and a marginally significant trend between Conditions C1 and C3 ($p < 0.1$). Therefore, the application Condition C3 can be concluded to possess the widest dynamic range. The dynamic range change in the arms appear to be larger and that in the fingers was smaller, however with no significant difference.

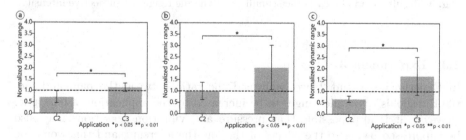

Fig. 3. Dynamic ranges at each location: (a) finger, (b) forearm, and (c) forehead.

2.2   Experiment 2

Experiment 2 was based on the variation of electrical stimulation. We measured and compared the tactile and pain thresholds of anodic and cathodic stimuli for Conditions C1 and C3.

Conditions. The conditions of anodic and cathodic stimulations were applied under Conditions C1 and C3. The forearm was chosen as the stimulation location, because in experiment 1, some participants did not fell pain at the maximum current at the fingers. Moreover, the use of forehead for electrical stimulation is less common than other locations.

Participants. There were 12 male participants between 21–27 years of age. The experiment was conducted over 2 days, with one condition measured each day. The same procedure as Experiment 1 was adopted.

Results. Similar to Experiment 1, the dynamic range was calculated for each participant, normalized through Condition C1, and a two-way ANOVA with correspondence was performed (Fig. 4a). The main effect was observed under application conditions ($F = 10.39$, $p < 0.01$), and not for the variation of stimulus type ($F = 0.014$, n.s.). No interaction was detected between application conditions and stimulus type ($F = 0.014$, n.s.). The dynamic range was shown to be significantly higher under Condition C3, than under Condition C1.

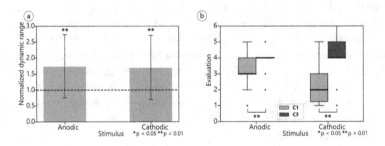

Fig. 4. Results of anodic-cathodic stimuli: (a) dynamic range (b) subjective intensity.

2.3   Experiment 3

In Experiments 1 and 2, we confirmed the ratio between the pain and tactile thresholds (dynamic range) to be increased by the application of lidocaine. However, it is also possible that pressure and vibration sensations may also be suppressed owing to the effects of the anesthetic cream, and this would be detrimental to study objective of the application of lidocaine. Experiment 3 was conducted to confirm this, in which we compared the subjective intensity of two types of stimuli with and without application of the local anesthetic cream.

Conditions. The same stimulation and application conditions as Experiment 2 were adopted.

Procedure. Fourteen participants (13 males and 1 female) between 21–27 years of age were included in the study. The experiment was conducted over 2 days, with one condition measured per day.

The same procedure as Experiments 1 and 2 were followed until the electrical stimulation. First, the pain threshold was measured by adjusting the current across the electrode, after which the participants rated the subjective intensity of the stimulus at the threshold on a 7-point Likert scale from 0–6 (0: "very weak stimulus", 6: "very strong stimulus"). At the pain threshold, pain is just barely perceived, and the strength of the sensation is significantly related to vibratory and pressure sensations.

Results. The responses to the Likert scale were analyzed through ART-ANOVA (Fig. 4b). Similar to Experiment 2, the main effect was observed under the application condition ($F = 23.119$, $p < 0.01$), and not under the stimulus conditions ($F = 2.323$, n.s). No interaction was observed between the application conditions and stimuli ($F = 10.39$, n.s). The responses obtained for Condition C3 were significantly higher than those under Condition C1, suggesting that the application of anesthetic cream allowed a strong electrical current, and enabled a strong perception of target tactile sensation.

After the experiment, positive comments such as "the stimulus was stronger than in the previous application" and "I felt pure vibration and pressure sensation", were obtained. Comments such as "I did not feel any significant change", "I could observe that the range of stimulation became wider, but I did not feel any change in the intensity of stimulation" were also obtained.

3   Discussion

The experiments confirmed that the application of an anesthetic cream prior to electrical stimulation increased the dynamic range, both with respect to the magnitude of electrical current and the subjective intensity of stimulation. The comparison of the results of conditions C2 and C3 with those of castor oil (control condition) attributed this phenomenon to the anesthetic effect, and eliminated the influence of sweating from the application of the cream. Through these experiments, the effectiveness of the proposed method of applying local anesthetic cream was demonstrated.

When local anesthetics are applied, C fibers, which are unmyelinated fibers that control pain and itching, are anesthetized, followed by the anesthetization of thin myelinated fibers (sensory nerves: $A\delta$ fibers control warmth and pain, $A\gamma$ fibers control intrinsic sensation and muscle tone, and $A\beta$ fibers control touch and pressure), and finally $A\alpha$ fibers (motor nerves), which are thick myelinated fibers [11]. The results of the current experiments are with existing literature.

Although the results of Experiment 1 did not show the main effect under changing application locations, the effect of lidocaine on fingers were observed to be marginally smaller than that of the other two locations. The transdermal absorption of the chemicals in the palm of the hand was approximately 0.83 times lower than that on the back of the forearm [12], suggesting the anesthetic effect to be weaker in the fingertips. In contrast, although the transdermal absorption rate of the forehead was approximately 6 times greater than that of the back of the forearm, the dynamic range was approximately equal to that of the forearm. Since electrical stimulation is easily affected by skin conditions such as sweat, and as the forehead comprises concentrated sweat glands, the anesthetic effect might be underestimated owing to sweat. The details of the effect of sweat and transdermal absorption rate require to be focused in future studies.

In Experiment 3, there was a significant difference between the subjective perception intensity of the subjects with and without the application of local anesthetic cream, but it was not a dramatic change. Furthermore, the increase of dynamic range with respect to the increase in the magnitude of current, as shown in experiments 1 and 2, the volume adjustment of stimulation current to become easy.

The limitation of the proposed method in this experiment is that it requires an application time of 1 h, prior to the main stimulation. Increased concentration of the cream might shorten the time, which can be a separate research topic in the future. However, since we do not need to turn off and on the effect quickly, we consider it is not a very significant practical limitation.

4   Conclusion

In this study, we proposed a method to expand the dynamic range of electrical stimulation for tactile display by applying a local anesthetic cream containing lidocaine to reduce itching and pain. The results suggested that the application of local anesthetic creams can increase the dynamic range with respect to the magnitude of electrical current and subjective perceptional intensity, enabling the perception of strong stimuli. The experimental results, conducted based on multiple locations, anodic stimulation, and cathodic stimulation, revealed no significant difference caused by the difference in locations, and the dynamic range of the electrical stimulation to be expanded by the local anesthetic cream, independent of the type of stimulation.

Although this experiment was conducted with a single electrode, the problem of pain perception under electrical stimulation is exaggerated with multi-point electrodes (during electrical stimulation through multiple points, if even single electrode causes pain, it will be an unpleasant experience). In the future, we will verify the results under multi-point electrical stimulation. We will also conduct study to stimulate more immediate and vivid electrical stimuli by changing the concentration of lidocaine, to elucidate the relationship between anesthetic effects and electrical stimuli.

Acknowledgement. This research was supported by JSPS KAKENHI Grant Number JP20H05957.

References

1. Tanaka, Y., Nguyen, D.P., Fukuda, T., Sano, A.: Wearable skin vibration sensor using a PVDF film. In: 2015 IEEE World Haptics Conference (WHC), pp. 146–151 (2015)
2. Hayward, V., Terekhov, A.V., Wong, S.C., Geborek, P., Bengtsson, F., Jörntell, H.: Spatio-temporal skin strain distributions evoke low variability spike responses in cuneate neurons. J. Roy. Soc. Interface **11**(93), 20131015 (2014)
3. Kaczmarek, K.A., Webster, J.G., Bach-y-Rita, P., Tompkins, W.J.: Electrotactile and vibrotactile displays for sensory substitution systems. IEEE Trans. Biomed. Eng. **38**(1), 1–16 (1991)
4. Bach-y-Rita, P., Kaczmarek, K.A., Tyler, M., Garcia-Lara, J.: Form perception with a 49-point electrotactile stimulus array on the tongue: a technical note. J. Rehabili. Res. Dev. **35**, 427–430 (1998)
5. Kajimoto, H.: Design of cylindrical whole-hand haptic interface using electrocutaneous display. In: Isokoski, P., Springare, J. (eds.) EuroHaptics 2012. LNCS, vol. 7283, pp. 67–72. Springer, Heidelberg (2012). https://doi.org/10.1007/978-3-642-31404-9_12
6. Kaczmarek, K.A., Tyler, M.E., Bach-Y-Rita, P.: Electrotactile haptic display on the fingertips: preliminary results. In: Proceedings of 16th Annual International Conference of the IEEE Engineering in Medicine and Biology Society, vol. 2, pp. 940–941 (1994)
7. Yem, V., Kajimoto, H.: Comparative evaluation of tactile sensation by electrical and mechanical stimulation. IEEE Trans. Haptics **10**(1), 130–134 (2017)
8. Sato, K., Tachi, S.: Design of electrotactile stimulation to represent distribution of force vectors. In: 2010 IEEE Haptics Symposium, pp. 121–128 (2010)
9. Lu, J., Liu, Z., Brooks, J., Lopes, P.: Chemical haptics: rendering haptic sensations via topical stimulants, pp. 239–257 (2021)
10. Mason, J.L., MacKay, N.A.M.: Pain sensations associated with electrocutaneous stimulation. IEEE Trans. Biomed. Eng. BME **23**(5), 405–409 (1976)
11. Liu, S., Kopacz, D.J., Carpenter, R.L.: Quantitative assessment of differential sensory nerve block after lidocaine spinal anesthesia. J. Am. Soc. Anesthesiol. **82**(1), 60–63 (1995)
12. Feldman, R.J.: Regional variation in percutaneous penetration of 14C cortisol in man. J. Inves. Dermatol. **48**, 181–183 (1967)

Haptic Rattle: Multi-modal Rendering of Virtual Objects Inside a Hollow Container

Emilie Hummel[1]([✉]), Claudio Pacchierotti[2], Valérie Gouranton[1],
Ronan Gaugne[3], Theophane Nicolas[4], and Anatole Lécuyer[5]

[1] Univ Rennes, INSA Rennes, IRISA, Inria, CNRS, Rennes, France
emilie.hummel@irisa.fr
[2] CNRS, Univ Rennes, Inria, IRISA, Rennes, France
[3] Univ Rennes, Inria, CNRS, IRISA, Rennes, France
[4] Inrap, UMR Trajectoires, Rennes, France
[5] Inria, Univ Rennes, CNRS, IRISA, Rennes, France

Abstract. The sense of touch plays a strong role in the perception of
the properties and characteristics of hollow objects. The action of shaking a hollow container to get an insight of its content is a natural and
common interaction. In this paper, we present a multi-modal rendering
approach for the simulation of virtual moving objects inside a hollow
container, based on the combination of haptic and audio cues generated
by voice-coils actuators and high-fidelity headphones, respectively. We
conducted a user study. Thirty participants were asked to interact with
a target cylindrical hollow object and estimate the number of moving
objects inside, relying on haptic feedback only, audio feedback only, or
a combination of both. Results indicate that the combination of various
senses is important in the perception of the content of a container.

Keywords: Haptics · Multi-modal rendering · Interaction

1 Introduction

The action of shaking a hollow container to get an insight of its content is a
natural reflex [6]. This way of interacting can be of great value in Virtual and
Augmented Reality applications, e.g., for the manipulation of virtual instruments
such as maracas or to get insights on the number of incoming messages on a
smartphone [11]. When interacting with a hollow container that houses one or
multiple objects inside, we take advantage of multiple senses to identify various
characteristics of the hidden objects, e.g., their shape, weight, size, material, all
of them affecting the way they interact with the hollow object when shacked.

This paper introduces a multi-modal method for the rendering of multiple
virtual moving objects inside a hollow container, combining audio and haptic

This project has received funding from the European Union's Horizon 2020 program
under grant agreement No. 801413; project "H-Reality".

H. Seifi et al. (Eds.): EuroHaptics 2022, LNCS 13235, pp. 189–197, 2022.
https://doi.org/10.1007/978-3-031-06249-0_22

feedback for applications in Virtual and Augmented Reality. Starting from an acceleration-based interaction model of the moving objects with respect to the hollow object, we generate the interactions to be rendered through the audio and/or haptic output systems. We called the proposed approach "Haptic Rattle", and performed a user study to assess this approach.

2   Related Work

The perceptual and interaction aspects of manipulating a hollow container filled with small moving objects have been investigated in the literature. Sekiguchi et al. [7] evaluated the perception of feeling a moving object inside a larger container using two solenoids. They assessed the user's ability to discriminate between four interaction models representing a box with an object inside, with only one model faithfully simulating such interaction. Results showed that users were able to effectively discriminate each interaction model [8]. Tanaka et al. [10] assessed the ability of users to discriminate changes in the inner parameters of their model to render the presence of moving objects in a container, using voice coil motors. Similarly, Yamamoto et al. [12] tested the perception between a real and virtual model for solid and liquid content inside a container, with no significant difference between the two models. On another line, the Gravity Grabber of Minamiza et al. used two motors to actuate the motion of a belt onto the fingertip of the user, providing the sensation of holding a glass with liquid inside [1]. Such approach ignores the proprioceptive/kinesthetic aspects of holding a mass, but still offers a reliable feeling of weight and the ability to discriminate between objects with different levels of liquid [2]. Other works aim at improving the realism of simulating the presence of one or multiple objects inside a container using the combination of a voice coil motor and impact actuators [4], the high-speed change of the rotational inertia [9], physics-based haptic vibrations interface [11] or library [5].

While effective, these works focus on rendering the motion of single objects inside a hollow container. On the other hand, Plaisier and Smeets [6] performed an experiment on the ability to perceive the numerosity of objects inside a container, using a tangible box filled with real spheres. Their experiment assessed the number of spheres the participants could accurately detect using audio and haptic or only audio cues, considering between one and five spheres. Results showed that participants accurately detected up to three spheres but underestimated their numerosity when presented with four or five. Finally, results were significantly more accurate when the participants had audio and haptic feedbacks with respect to when they were provided only with audio feedback.

With respect to the abovementioned works, we present a simplified physic-interaction virtual model that can be easily evaluated at runtime for an arbitrary number of virtual moving hidden objects. To assess the viability of our rendering model, we carry out a user study inspired by the work of Plaisier and Smeets [6].

3   Methods

When interacting with a container filled with one or more moving objects, the manipulation of the hollow container provides various feedbacks, such as the feeling of the objects hitting or rolling on the inner surface of the container. If the container is rigid, this feedback has audio and haptic components that are characterized by many factors, such as the magnitude of the container's movement and the materials and size of the moving objects. The user's imparted movement on the container is essential in this perception as it directly causes the interactions of the moving objects and thus the resulting audiohaptic sensations. To render such interactions, we propose an interactive acceleration-based model combined with a multi-modal audio-haptic rendering.

3.1   Apparatus

As a proof-of-concept, we use a 3D-printed hollow cylinder which dimensions allow comfortable grasping, containing two haptic voice coil actuators on each base (HapCoil-One, Actronika, FR) and an IMU MPU6050 with a sampling rate of 1 kHz, a 3-axis gyroscope and a 3-axis accelerometer in the middle (see Fig. 1a). The voice coils are connected to a TDA3116D2 dual-channel amplifier, powered by a 5 V power supply, and a stereo 24-bit 96 kHz MOTU sound card.

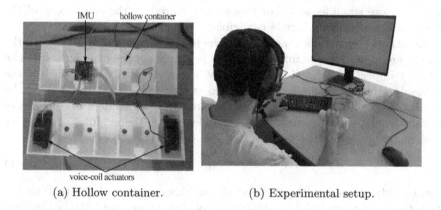

(a) Hollow container.   (b) Experimental setup.

Fig. 1. (a) Our cylindrical prop, housing an IMU and two voice-coil actuators (b) Experimental setup for the manipulation of the device.

The actuators are used to simulate the interaction of virtual spheres inside the container, while the IMU registers the cylinder accelerations when it is shaken by the user. The haptic actuators are inserted symmetrically inside the cylinder, to avoid favoring one handedness over the other. The IMU is set up in a plane position to minimize the impact of rotational movements.

3.2 Acceleration-Based Rendering Model

We chose a simple acceleration-based model to enable the fast simulation of an arbitrary number of moving objects inside the container. It registers the data from the IMU and then provides the acceleration parameters for the multi-modal rendering. The acceleration data is filtered using a low-pass filter at frequency $\omega_0 = 60$ Hz and then integrated to output the movement velocity. When the acceleration is higher than a threshold $a_t = 0.5$ m/s^2, the magnitude and direction of the movement are transmitted to the rendering algorithm. The frequency ω_0 and the threshold a_t are tuned based on empirical data coming from pilot experiments, so as to correctly detect when the user is shaking the cylinder. In the world reference frame, the forces acting on the system are the gravity, the inertia of each element, and the force applied by the user. The force applied on each sphere is induced by the acceleration induced by the cylinder movement. Each sphere follows the simple equation: $\vec{F} + \vec{g} + \vec{i} = m\vec{a}$, with $\vec{F}$ being the acceleration applied to the cylinder by the user, $\vec{g}$ the acceleration gravity, $\vec{i}$ the inertia component, m the mass and $\vec{a}$ the acceleration of the sphere. The gravity and inertial physic interactions are handled through the physics engine of Unity 3D, in which we add the forces induced by the user movements. In the cylinder's reference frame, the sphere follows the movement applied by the user with a small delay, induced by the inertia, which depends on the mass of the object. The inertia has an impact on the realism of the interaction [4], however, the simulated spheres are small objects with a small mass. The induced inertia is therefore small as well. For this reason, to reduce the complexity of the algorithm, we considered inertia negligible.

Vibrations and sounds generated by the interaction of the spheres with the cylinder are highly dependant on the inner properties of the cylinder itself. The vibratory impact of a small object against a surface can be simplified as the combination of a sinusoidal signal and an exponential decay, inspired by the model of [3] (see Eq. (1)). The resulting low-frequency signal depends on a decay constant B, and the frequency ω. B and ω depend on the modulus of elasticity of the material, and on the density and geometry of the object,

$$Q(t) = e^{-Bt} sin(\omega \cdot t) \tag{1}$$

If we consider objects with a regular shape, this equation allows to tune the interaction for any material by recording repeated impacts and fitting the sinusoidal waveform and decay to match the record. This signal corresponds to the output feedback and only needs to be modulated by the object velocity $\vec{v}$ before being displayed through the audio and/or haptic output systems. In our implementation, we used voice-coil actuators for haptic feedback and hi-fi headphones for audio feedback. The signal is displayed through two audio channels, which are independently controlled depending on the direction/location of the impact. Doing so, when the user shakes the container, the system is able to detect its movement and effectively simulate the presence of an arbitrary number of spheres moving inside. Figure 2 shows a diagram of the rendering process. The IMU acceleration data are computed into an estimated velocity vector $\vec{v}$. $f(v)$ is a linear

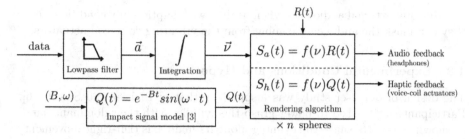

Fig. 2. Diagram of the rendering process: from the data sensed by the IMU to the feedback output.

function computing a gain factor from each sphere acceleration, based on ν value and the configuration of the spheres inside the container. This factor is used to modulate the audio and haptic signals. Audio signal $S_a(t)$ is then computed based on signal $R(t)$ of recorded impact sounds from the desired objects, e.g. wooden spheres, with multiple frequencies and intensities of impact,

$$S_a(t) = f(\nu)R(t) \tag{2}$$

On the other hand, the parameters (B, ω) are chosen according to a wood material ($B = 154 \text{ s}^{-1}$; $\omega = 67 \text{ rad/s}$) to compute $Q(t)$ in Eq. (1), on which is based the haptic signal $S_h(t)$,

$$S_h(t) = f(\nu)Q(t) \tag{3}$$

These signals are then duplicated according to the number of spheres n and displayed according to the feedback condition considered, with a randomized delay between two empirical values (0 ms and 100 ms) to give a natural effect of collisions.

4 User Study

We conducted a user study to assess the effectiveness of our approach in simulating the presence of an arbitrary number of virtual moving objects inside a hollow container. A video is available at https://youtu.be/cMXTvAOvQtc.

4.1 Population, Materials and Setup

Thirty participants volunteered to participate in this experiment, aged between twenty and fifty-five years old. Among the participants were eight women and twenty-two men, five left-handed and twenty-five right-handed.

The setup is that described in Sect. 3.1 and shown in Fig. 1b. Users are asked to sit comfortably in front of a computer screen while manipulating the hollow cylinder. The screen shows the instructions of the experiment and enables the user to answer the related questions. Users also wear a pair of headphones for providing audio feedback when this type of feedback is considered, as well as for masking any external audio cues when dealing with haptics-only conditions, e.g., they can mask the audio cues coming from the vibrating voice-coil actuators.

4.2 Experimental Conditions and Hypotheses

The design of our user study was inspired by that of Plaisier and Smeets [6]. Participants were asked to power grasp the cylinder with their dominant hand, as shown in Fig. 1b, shake the cylinder along its main axis (left-right movement) for five seconds, and answer the question: "How many spheres do you feel moving inside the container?". As the cylinder was shaken, the rendering algorithm described in Sect. 3 provided the user with compelling audio or/and haptic feedback to simulate the presence of multiple spheres. We considered three feedback conditions: combined haptics and audio feedback (H+A), audio-only feedback (A), and haptics-only feedback (H). For each condition, the number of simulated spheres inside the hollow container varies from one to five. Each participant carried out 15 repetitions of the trial per condition and per number of spheres, yielding 15 (repetitions) × 3 (feedback conditions) × 5 (number of simulated spheres) = 225 trials, that were randomized. Following the results of Plaisier and Smeets [6] (see Sect. 2), our hypotheses (HP) are as follows:

HP1: Participants can effectively discriminate the presence of up to three spheres inside the container.

HP2: Participants underestimate the number of spheres when there are four or five spheres in the container.

HP3: Participants show a better recognition score with composite audiohaptic (H+A) feedback than with audio (A) or haptic (H) only feedback.

Our objective is to achieve a performance as close as possible to that that humans achieve when interacting with a hollow container filled with real objects.

4.3 Results

The following results are the analysis of the experiment data. We observed a learning effect on the five first trials for each participant and therefore removed the related answers, leaving seventy trials for each condition. Two subjects were considered outliers and removed from this data analysis, as they were unable to follow the experiment indications and did not understand the experimental task. To avoid any bias, we did not provide them with additional information with respect to the other participants. Figure 3 shows the mean and standard deviation of the user's answer in discriminating the number of spheres inside the container, for each actual rendered number of spheres and conditions.

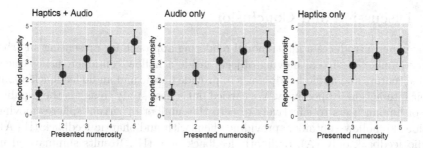

Fig. 3. Presented vs. Reported numerosity of the spheres inside the container. Mean and standard deviation are plotted.

We evaluated the error between the user's reported numerosity and the actual rendered one, computed simply as the absolute value of the difference between the actual and reported numerosity. Then, we carried out statistical analysis to see whether there is a difference with respect to the feedback condition (H+A, A, H) or the number of rendered spheres (1, 2, 3, 4, 5). Boxplots of the reported number of spheres and error per condition and number of presented spheres are reported in Fig. 4. We used non-parametric tests because the user's response is not on a continuous scale. The Kruskal-Wallis rank sum test for error according to the conditions shows a significant difference between at least two conditions ($p < 0.001$, $\chi^2(2) = 33.279$). The pairwise Wilcoxon rank sum test with continuity correction confirms these results, with a significant difference between conditions (H) vs. (A), $z = 3.83$ and $p < 0.001$, and (H) vs. (H+A), $z = 5.62$ and $p < 0.001$, but no significant difference between (A) vs. (H+A), $z = 1.83$ and $p > 0.1$.

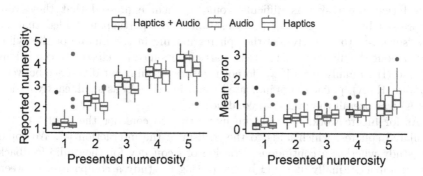

Fig. 4. (Left) Presented vs. Reported numerosity of the spheres inside the container, per condition. (Right) Presented numerosity vs. error, per condition.

5   Discussion and Conclusion

This paper introduces a multi-modal rendering approach for presenting an arbitrary number of virtual moving objects inside a hollow container, using voice-coil actuators and audio headphones to provide haptic and audio feedback about their interaction, respectively. We carried out a user study to assess the capability of human users to effectively discern the number of virtual spheres moving inside the hollow container, providing them with audiohaptic feedback (H+A), audio feedback only (A), or haptic feedback only (H). Results summarized in Figs. 3 and 4 show that, in general, subjects were quite good at estimating the number of spheres rendered inside the hollow container, proving the effectiveness of the proposed rendering techniques. This is especially true when 1 to 3 spheres were rendered, confirming the results of [6] and **HP1**. As also seen in [6], as the number of rendered spheres augments, users tend to underestimate their number, confirming also **HP2**. Indeed, after 3 rendered spheres the curves shown in Fig. 3 flatten and never reach 5. This phenomenon is stronger for condition (H), and it is very similar in conditions (A) and (H+A). Also, the standard deviation increases as the number of rendered spheres increases. The statistical analysis of the results reveals that performance under (H) is significantly different (worse) than the two other conditions, but no significant difference between (H+A) and (A) was found, partially disproving hypothesis **HP3**.

Asking the users about their strategies revealed a certain similarity. They fall into two main categories: the first one consists of repeatedly and energetically shaking the container and comparing the intensity of the feedback with previous trials; the second one consists of counting the impacts on one extremity of the cylinder after a single energetic stroke. These two strategies can be of course mixed together and carried out one after the other.

After the experiment, participants shared their impressions. While they agreed that the task was difficult, some of them expressed that they were "impressed by the realism of the sensation". One participant added that he was "surprised to actually feel the spheres moving in the middle of the cylinder". About the difficulty of the task, participants mostly referred to conditions (A) and (H). Finally, even if we do not measure a significant difference between (A) and (H+A), participants indicated condition (H+A) as significantly easier. (H+A) was overall the preferred condition.

As for future work, it would be interesting to consider the inertia of the simulated spheres, which adds to the realism of the interaction [4]. We could also study whether a continuous variation of frequency for the audio feedback has an impact. Finally, it could be interesting to study a comparison between simulated and real interactions and the impact of the number of actuators inside.

References

1. Minamizawa, K., Fukamachi, S., Kajimoto, H., Kawakami, N., Tachi, S.: Gravity grabber: wearable haptic display to present virtual mass sensation. In: ACM SIGGRAPH 2007 Emerging Technologies, pp. 8–11 (2007)

2. Minamizawa, K., Fukamachi, S., Kawakami, N., Tachi, S.: Interactive representation of virtual object in hand-held box by finger-worn haptic display. In: IEEE Symposium on Haptic Interfaces for Virtual Environment and Teleoperator Systems, pp. 367–368 (2008)
3. Okamura, A.M., Dennerlein, J.T., Howe, R.D.: Vibration feedback models for virtual environments. In: IEEE International Conference on Robotics and Automation, vol. 1, pp. 674–679 (1998)
4. Park, C., Park, J., Oh, S., Choi, S.: Realistic haptic rendering of collision effects using multimodal vibrotactile and impact feedback. In: IEEE WHC, pp. 449–454 (2019)
5. Park, G., Choi, S.: A physics-based vibrotactile feedback library for collision events. IEEE Trans. Haptics 10(3), 325–337 (2018)
6. Plaisier, M.A., Smeets, J.B.: How many objects are inside this box? In: IEEE WHC, pp. 240–244 (2017)
7. Sekiguchi, Y., Hirota, K., Hirose, M.: Haptic interface using estimation of box contents metaphor. In: ICAT, vol. 203 (2003)
8. Sekiguchi, Y., Matsuoka, S., Hirota, K.: Inertial force display to represent content inside the box. In: Kappers, A.M.L., van Erp, J.B.F., Bergmann Tiest, W.M., van der Helm, F.C.T. (eds.) EuroHaptics 2010. LNCS, vol. 6191, pp. 81–86. Springer, Heidelberg (2010). https://doi.org/10.1007/978-3-642-14064-8_12
9. Shimizu, S., Hashimoto, T., Yoshida, S., Matsumura, R., Narumi, T., Kuzuoka, H.: Unident: providing impact sensations on handheld objects via high-speed change of the rotational inertia. In: IEEE VR, pp. 11–20 (2021)
10. Tanaka, Y., Hirota, K.: Shaking a box to estimate the property of content. In: Isokoski, P., Springare, J. (eds.) EuroHaptics 2012. LNCS, vol. 7282, pp. 564–576. Springer, Heidelberg (2012). https://doi.org/10.1007/978-3-642-31401-8_50
11. Williamson, J., Murray-Smith, R., Hughes, S.: Shoogle: excitatory multimodal interaction on mobile devices. In: SIGCHI Conference on Human Factors in Computing Systems, pp. 121–124 (2007)
12. Yamamoto, T., Hirota, K.: Recognition of weight through shaking interaction. In: IEEE WHC, pp. 451–456 (2015)

Design of a 2-DoF Haptic Device for Motion Guidance

Lisheng Kuang[1(✉)], Maud Marchal[2], Marco Aggravi[1],
Paolo Robuffo Giordano[1], and Claudio Pacchierotti[1]

[1] Univ Rennes, CNRS, Inria, IRISA, Rennes, France
{lisheng.kuang,marco.aggravi,paolo.giordano,
claudio.pacchierotti}@irisa.fr
[2] Univ Rennes, INSA Rennes, CNRS, Inria, IRISA – France and IUF, Rennes, France
maud.marchal@irisa.fr

Abstract. We present a 2-degrees-of-freedom (2-DoF) haptic device, which can be either used as a grounded or a hand-held device. It is composed of two platforms moving with respect to each other, actuated by two servomotors housed in one of structures. The device implements a rigid coupling mechanism between the two platforms, based on a three-legged 3-4R constrained parallel linkage, with the two servomotors actuating two of these legs. The device can apply position/kinesthetic haptic feedback to the user hand(s). This paper presents the device and its kinematics, together with a human subjects experiment where we evaluate its capabilities to provide meaningful directional information.

Keyword: Motion guidance

1 Introduction

Kinesthetic haptic interfaces have been very popular in the past, for applications ranging from industrial to surgical robotics. Researchers have designed many different types of such interfaces, focusing on improving their, e.g., peak force, bandwidth, workspace, and/or price, according to the target field of application. In this respect, we can identify two main categories of kinesthetic interfaces: grounded and ungrounded. Grounded devices have their base placed on an external support, such as a table, while ungrounded devices have their base on the user's body [8]. Grounded kinesthetic devices include popular commercial kinesthetic systems such as the Virtuose (Haption, FR), Omega.x (Force Dimension, CH), Falcon (Novint Tech., USA) and the Phantom (Geomagic, 3D Systems, USA) series. More recently, in research, Jang et al. [3] presented a grounded isometric interaction device to induce whole-body interaction; Okui et al. [7] designed a delta-type 4-degrees-of-freedom (4-DoF) grounded haptic

This work has received funding from the Inria Défi project "DORNELL" and the China Scholarship Council No. 201908440309.

H. Seifi et al. (Eds.): EuroHaptics 2022, LNCS 13235, pp. 198–206, 2022.
https://doi.org/10.1007/978-3-031-06249-0_23

device actuated with a magnetorheological clutch, thus able to adjust its stiffness and viscosity; and Satler et al. [10] devised a portable interface composed of controlled wheel torques to render forces to a user handle placed on the top of the device. On the other hand, ungrounded kinesthetic devices come in very different forms, spanning from hand-held devices for gaming and VR interaction [5,12] or guidance [11,13] to body-worn exoskeletons [1,9].

This paper presents a 2-degrees-of-freedom (2-DoF) kinesthetic device that can be either attached onto an external support or held between the two hands. Its design is inspired by the anti-parallelogram mechanism [6] featuring a quaternion joint [4]. It is composed of two structures, connected by three articulated legs which are in turn actuated by two servo motors. By moving with respect to each other, they can provide the user with directional information.

2   Device Design and Actuation

The proposed device is shown in Fig. 1. A video is available at https://youtu.be/vc6B-OOj590.

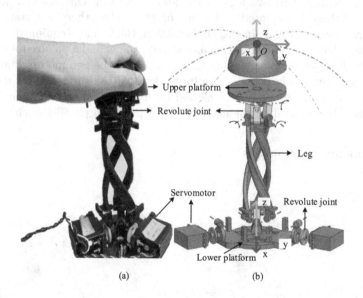

Fig. 1. The device is composed of two servo motors actuating an anti-parallelogram mechanism moving two platforms with respect to each other. (a) Device in its grounded configuration, with the lower platform secured to an external support and the user hand posed on the upper one. The device can be either used as a grounded interface, like in this figure and in our experiment of Sect. 3, or it can be held between the two hands (see the video for this configuration). (b) CAD of the device. The grey dotted lines show the surface of the sphere on which the upper platform moves. The spherical cap posed on the upper platform makes the interaction with the user hand more comfortable. Of course, it can be changed with other shapes according to the task at hand. (Color figure online)

2.1   Mechanism and Structure

The design of the proposed device is inspired by the principle of the anti-parallelogram mechanism [6], which consists of three identical supporting link-ages forming an interlaced structure with no interference between them [4]. Compared to standard serial mechanisms, parallel mechanisms enable fast dynamics and high payload with relatively small size and low weight. Moreover, a large range of motion and uniform manipulability can be obtained by choosing appropriate dimensional parameters and actuation.

As shown in Fig. 1, the device consists of two platforms, a lower and an upper platforms, connected by three legs each having a spiral curved link. The device has dimensions of $15 \times 15 \times 23$ cm and weighs 150 g. The two ends of each link are connected with two serial revolute joints to the platforms, forming a 3-4R coupling parallel mechanism which can move freely with 2 degrees of freedom, according to the orthogonal rolling motion on each leg [4]. If the lower platform is fixed on an external support (as in Fig. 1(a)), the motion of the upper platform is confined on the surface of a sphere centered in the center of the lower platform (see grey dotted lines in Fig. 1(b)). Two Hitech-625MG servo motors are attached to two legs on the lower platform. One revolute joint on the short linkage is mounted to the motor shaft, the other joint connects to the leg, as shown in Fig. 1. In this configuration, the two legs equipped with the servomotors have an active rotation on the lower joints and a passive rotation on the other three joints. By changing the actuation of the two motors, the upper platform can reach any point within its workspace (the surface of a sphere, as mentioned above). The graphical method using reciprocal screw system theory can be used [14] to analyze the mobility of the proposed parallel mechanism.

2.2   Kinematics Analysis

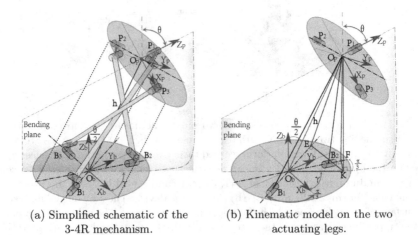

(a) Simplified schematic of the   (b) Kinematic model on the two
        3-4R mechanism.   actuating legs.

Fig. 2. Schematic of the proposed device.

We study the kinematics of the device so as to evaluate the relationship between the motion of the servomotors and that of the upper platform with respect to the lower one. As illustrated in Figs. 1 and 2, three legs are regarded as individual open-chain manipulators. The connection points between the upper platform and legs 1, 2, 3 are defined as P_1, P_2, P_3, respectively, while the connection points between the lower platform and the legs is denoted by points B_1, B_2, B_3. Let us consider the device in its grounded configuration, with the base attached to an external support. The coordinate frames of the upper (mobile) and lower (static) platforms, $C_p(O_p - x_p y_p z_p)$ and $C_b(O_b - x_b y_b z_b)$, are fixed on their geometric centers, O_p and O_b, respectively. The frames are defined as indicated in Figs. 1 and 2.

The plane where the z_b and z_p axes coexist is defined as the bending plane, which is always perpendicular to the surfaces of the two platforms. Here, the orientation angle θ and bending angle γ denote the angle between z_b and z_p on the bending plane and the angle from the x_b axis to the bending plane, respectively. Considering the bending and orientation angles, the homogeneous transformation from C_b to C_p is as follows:

$$
\begin{aligned}
{}_p^b T &= R(Z,\gamma)R(Y,\frac{\theta}{2})T(h)R(Y,\frac{\theta}{2})R(Z,-\gamma) \\
&= \begin{bmatrix}
1 - 2C^2(\gamma)S^2(\frac{\theta}{2}) & -S(2\gamma)S^2(\frac{\theta}{2}) & C(\gamma)S(\theta) & hC(\gamma)S(\frac{\theta}{2}) \\
-S(2\gamma)S^2(\frac{\theta}{2}) & 1 - 2S^2(\gamma)S^2(\frac{\theta}{2}) & S(\gamma)S(\theta) & hS(\gamma)S(\frac{\theta}{2}) \\
-C(\gamma)S(\theta) & -S(\gamma)S(\theta) & C(\theta) & hC(\frac{\theta}{2}) \\
0 & 0 & 0 & 1
\end{bmatrix}
\end{aligned}
\tag{1}
$$

where $S(x) = sin(x)$ and $C(x) = cos(x)$, $R(.,.)$ denotes a rotation, T an homogeneous transformation around the z axis, and h the distance between O_b and O_p, which is also the diameter of the sphere onto which the upper platform moves.

Geometrically, we can also derive the relationship (forward and inverse kinematics) between γ and θ with respect to the motor's inputs α_1 and α_2. As shown in Fig. 2b, motor 1 is mounted at B_1 with the motor shaft rotating along $B_1 O_b$, while motor 2 is mounted at B_2, with its shaft rotating along $B_2 O_b$. The motor's rotating angle is limited within $(-\pi/2, \pi/2)$. For any bending pose of the upper platform, O_p has a projection point K in the plane $B_1 B_2 B_3$. Obviously, $O_p K$ is perpendicular to $B_1 O_b$ and $B_2 O_b$, and it is in the bending plane. Moving perpendicularly from K to $B_1 O_b$ and $B_2 O_b$, we obtain two pedal points, E, F, which lie in the shaft axis of motor 1 and motor 2, respectively. Right-angled triangle $\triangle O_p K E$ and $\triangle O_p K F$ share the same edge $O_p K$. As $O_p E$ and $O_p B_1$ lie in the same plane $O_p O_b B_1$, then the rotation angle of motor 1, α_1, is equal to $\angle O_p E K$, and that of motor 2, α_2, is equal to $\angle O_p F K$. From a simple geometrical derivation, we obtain

$$\tan(\alpha_1) = \tan(\angle O_p E K) = \frac{O_p K}{EK} = \frac{O_p K}{O_b K \sin(\gamma + \frac{\pi}{3})}$$
$$= \frac{h \cos \frac{\theta}{2}}{h \sin \frac{\theta}{2} \sin(\gamma + \frac{\pi}{3})} = \frac{1}{\tan \frac{\theta}{2} \sin(\gamma + \frac{\pi}{3})} \qquad (2)$$

$$\tan(\alpha_2) = \tan(\angle O_p F K) = \frac{O_p K}{FK} = \frac{1}{\tan \frac{\theta}{2} \sin(\gamma - \frac{\pi}{3})} \qquad (3)$$

3   Experimental Evaluation

We carried out an experiment evaluating the capabilities of the device in providing directional/motion information. The device renders a set of shapes with its end-effector, i.e., the moving upper platform, that users are asked to recognize.

3.1   Setup

The experimental setup is shown in Fig. 1. The device is used in its grounded configuration, with the lower platform attached to a table. Users are seated in front of the device and are asked to place their dominant hand on the device upper platform. A cardboard prevents the user from seeing his or her hand on the device during the experiment. A computer screen is also placed in front of the user, from which he or she can receive information about the experiment and answer the related questions.

3.2   Participants

Fourteen subjects (1 female and 13 males, aged from 23 to 33 years) participated in the experiment. Six are left-handed, eight are right-handed. Participants received an information sheet with the experiment details and signed a consent form. The study has been approved by Inria's ethics committee (Saisine 513).

3.3   Procedure

The experiment is divided in two blocks, carried out one after the other.

In the first one, the device moves along eight linear patterns, shown in Figs. 3a and 3c. We recall that the center of the upper platform O_p moves across the surface of a sphere having diameter h and centered at the center of the lower platform O_b (see Sect. 2.2). Each linear pattern starts from the resting position of O_p, then moves towards its designated direction and back, along a circular arc with chord of 20 cm (see also the video). For example, the linear pattern referred to as "South" in Fig. 3, starts from the resting position of O_p, then moves towards the S (South) direction (see Fig. 3a), then back to the resting

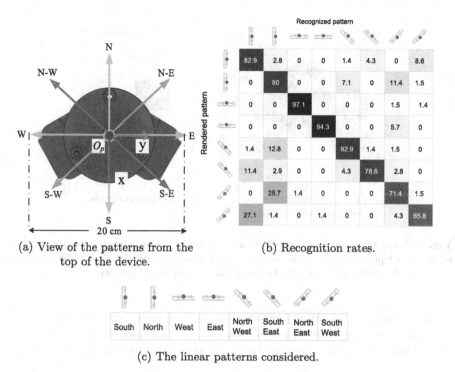

(a) View of the patterns from the top of the device.

(b) Recognition rates.

South	North	West	East	North West	South East	North East	South West

(c) The linear patterns considered.

Fig. 3. Linear patterns. (a) The patterns as rendered by the device, view from the top of the upper platform; (b) recognition rates of linear patterns; (c) the linear patterns. All linear patterns start and finish in the resting position of O_p, moving along a circular arc having a chord of 20 cm.

position of O_p, then moves towards the N (North) direction, and finishes in the resting position of O_p. All linear patterns start and finish in the resting position of O_p. The user sits on the South side of the device.

In the second block, the device moves along three shape patterns, shown in Figs. 4a and 4b. Similarly as before, these shapes are actuated over the surface of the sphere that is the workspace of our device. As indicated in Fig. 4a, they all start from a point located at the North of the resting position of O_p (named P in the Figure). Before starting the experiment, the experimenter explained the procedure to the user and spent about two minutes adjusting the chair armrest. Subjects placed their palm on the upper platform of the device and lean on the chair armrest to keep the forearm at the same level of the upper platform, ensuring maximum comfort. After each pattern was rendered, users were asked to select which pattern, in their opinion, the device just rendered. This choice was made through a GUI on the computer screen in front of the user. The device actuating all the patterns is shown at https://youtu.be/vc6B-OOj590.

Each pattern was provided five times, yielding (8 linear patterns + 3 shape patterns) × 5 = 55 repetitions of this pattern recognition task.

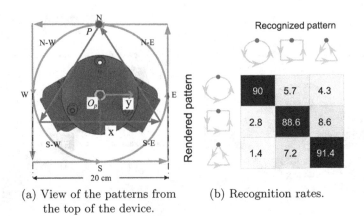

(a) View of the patterns from   (b) Recognition rates.
the top of the device.

Fig. 4. Shape patterns. (a) The patterns as rendered by the device, view from the top of the upper platform. All shape patterns start and finish north of the resting position of O_p, in the point named P. (b) Recognition rates of shape patterns.

Immediately after the experiment, participants were asked to fill in a questionnaire where we asked which pattern was the harder/easier to recognize and if they had any further comment about the rendering and the experiment.

3.4   Results

Figures 3b and 4b report the results of the experiment, in the form of two confusion matrices showing the percentage of recognition of the rendered vs. the recognized patterns. For the linear patterns the chance level is 1/8 (12.5%), while for the shape patterns the chance level is 1/3 (33.3%). The questionnaire showed that "East" and "West" linear patterns were considered as the easiest to identify, while the square shape pattern resulted to be the hardest. Diagonal linear patterns were reported to be harder to identify than other linear patterns.

4   Discussion and Conclusions

This paper presented a kinesthetic haptic device composed of two platforms connected by three interleaved legs. Two servo motors, actuating two of the three legs, can move the upper platform on the surface of a sphere centered in the lower platform. While the device can be used both in a grounded or handheld configuration, this paper focused on the former. In this case, the lower base is firmly attached to an external support, and the upper platform moves so as to provide the user with haptic feedback.

We carried out an experimental evaluation of our device. 14 human subjects were asked to place their hand on the upper platform of the device and recognize the pattern being displayed. The device was commanded to render 8 linear and 3 shape patterns, which can be used to provide, e.g., navigational information.

Results show a very high recognition rate for all patterns, with the highest and lowest recognition rate being 97.1% and 65.8%, respectively, with a chance level of 12.5%. For the linear patterns recognition, it was easier to recognize horizontal patterns (lateral-median motions with respect to the user, e.g., West and East patterns), while diagonal patterns were the hardest to recognize. The sense of motion along a certain pattern did not seem to affect much performance. The most common mistake was to identify diagonal patterns as anterior-posterior ones (North and South patterns). For the shape patterns recognition, recognition rates are very high in all conditions. In this case, subjects reported to be mostly relying on the number of edges being rendered by the device.

The design of this device shows also some limitations. First, in the current configuration, as we are using servo motors, the device is not backdrivable and cannot be used as a standard impedance-type haptic interface. Second, the motors are both placed on the same platform, which is a good idea if the device is used in a grounded configuration, but not ideal if held between the two hands. In a hand-held configuration, it is better to distribute the weight equally between the two platform, which should be straightforward to do.

Given such promising results, in the next future we plan to test a wider range of patterns, evaluating, e.g., the smallest change in orientation between two linear patterns and the smallest change in the number of edges between two shape patterns users can discriminate. We will also test the device when held between two hands, as well as the performance of the device in terms of resolution, precision, repeatability, and maximum output force. This device was developed within a larger effort to design a multi-modal haptic handle for various mobility aids, e.g., power wheelchairs, walkers, prewalkers. For this reason, we also plan to test this device effectiveness in rendering directional information when mounted on one of these mobility aids, similarly to [2].

References

1. Chinello, F., Malvezzi, M., Prattichizzo, D., Pacchierotti, C.: A modular wearable finger interface for cutaneous and kinesthetic interaction: control and evaluation. IEEE Trans. Ind. Electron. **67**(1), 706–716 (2019)
2. Devigne, L., et al.: Power wheelchair navigation assistance using wearable vibrotactile haptics. IEEE Trans. Haptics **13**(1), 52–58 (2020)
3. Jang, B.G., Kim, G.J.: Evaluation of grounded isometric interface for whole-body navigation in virtual environments. Comput. Anim. Virtual Worlds **25**(5–6), 561–575 (2014)
4. Kim, Y.J., Kim, J.I., Jang, W.: Quaternion joint: dexterous 3-DOF joint representing quaternion motion for high-speed safe interaction. In: IEEE/RSJ International Conference on Intelligent Robots and Systems (IROS), pp. 935–942 (2018)
5. Kovacs, R., et al.: Haptic PIVOT: on-demand handhelds in VR. In: Proceedings of Annual ACM Symposium on User Interface Software and Technology, pp. 1046–1059 (2020)
6. Okada, M., Nakamura, Y.: Development of a cybernetic shoulder-a 3-DOF mechanism that imitates biological shoulder motion. IEEE Trans. Robot. **21**(3), 438–444 (2005)

7. Okui, M., Kobayashi, M., Yamada, Y., Nakamura, T.: Delta-type four-DOF force-feedback device composed of pneumatic artificial muscles and magnetorheological clutch and its application to lid opening. Smart Mater. Struct. **28**(6), 064003 (2019)
8. Pacchierotti, C., Sinclair, S., Solazzi, M., Frisoli, A., Hayward, V., Prattichizzo, D.: Wearable haptic systems for the fingertip and the hand: taxonomy, review, and perspectives. IEEE Trans. Haptics **10**(4), 580–600 (2017)
9. Sarac, M., Solazzi, M., Frisoli, A.: Design requirements of generic hand exoskeletons and survey of hand exoskeletons for rehabilitation, assistive, or haptic use. IEEE Trans. Haptics **12**(4), 400–413 (2019)
10. Satler, M., Avizzano, C.A., Ruffaldi, E.: Control of a desktop mobile haptic interface. In: IEEE World Haptics Conference (WHC), pp. 415–420 (2011)
11. Spiers, A.J., Dollar, A.M.: Design and evaluation of shape-changing haptic interfaces for pedestrian navigation assistance. IEEE Trans. Haptics **10**(1), 17–28 (2016)
12. de Tinguy, X., Howard, T., Pacchierotti, C., Marchal, M., Lécuyer, A.: WeATaViX: wearable actuated tangibles for virtual reality experiences. In: Nisky, I., Hartcher-O'Brien, J., Wiertlewski, M., Smeets, J. (eds.) EuroHaptics 2020. LNCS, vol. 12272, pp. 262–270. Springer, Cham (2020). https://doi.org/10.1007/978-3-030-58147-3_29
13. Walker, J.M., Zemiti, N., Poignet, P., Okamura, A.M.: Holdable haptic device for 4-DOF motion guidance. In: IEEE World Haptics Conference (WHC), pp. 109–114 (2019)
14. Yu, J., Dong, X., Pei, X., Kong, X.: Mobility and singularity analysis of a class of two degrees of freedom rotational parallel mechanisms using a visual graphic approach. J. Mech. Robot. **4**(4) (2012)

Preliminary Design of a Flexible Haptic Surface

Romain Le Magueresse[1,2](✉), Frédéric Giraud[2]⑩, Fabrice Casset[1],
Anis Kaci[2], Brigitte Desloges[1], and Mikael Colin[1]

[1] Univ. Grenoble Alpes, CEA, Leti, 38000 Grenoble, France
`romain.lemagueresse@cea.fr`
[2] Univ. Lille, Arts et Metiers Institute of Technology, Centrale Lille,
Junia, ULR 2697 - L2EP, 59000 Lille, France

Abstract. This paper presents the preliminary development of a flexible haptic surface in order to produce texture rendering on a large conformable area. For this purpose, Haptic Pixels vibrating at ultrasonic frequencies are actuated by piezoelectric elements and implanted on a flexible matrix. The design leads to square glass plates of $10 \times 10\,mm^2$ with a thickness of $500\,\mu m$, actuated by PZT ceramics with a thickness of $200\,\mu m$ and a radius of $2.5\,mm$ bonded on a $100\,\mu m$ thick PEEK film. Electromechanical characterizations validate the design. The PEEK film between two pixels is exploited to separate them, to obtain the flexibility of the surface and to create an area of friction reduction with a stationary wave. Haptic evaluations are carried out to confirm the performances of the approach on a Haptic Pixel.

Keywords: Haptic · Flexible haptic · Friction reduction

1 Introduction

A Haptic Surface displays to a user's fingertip a tactile stimulation during active touch [1]. Researchers have proposed several actuation solutions to create the interaction forces, taking advantage of wave propagation into the Haptic Surface. For that purpose, they use localized vibrations thanks to inverse filtering [12], time reversal [8], stimuli confinement [6] or friction modulation [7] techniques. To be efficient, these solutions require specific properties from the surface to be actuated, and their principle have been demonstrated with flat and rigid material, with a simple geometry.

However, the trend in user interfaces is a demand toward "smart-surfaces" [5] that incorporate several functionalities into a wide variety of material and shapes. Moreover, the emergence of new flexible technologies such as flexible phones, rollable screens or wearable devices push us to think about conformable haptic devices. A current solution for that goal is to bury an array of actuators in a polymer substrate, as in [14] with electromagnetic actuators, with Shape Memory Alloy actuators [3] which need a pneumatic circuit or with Dielectric Polymer

© The Author(s) 2022
H. Seifi et al. (Eds.): EuroHaptics 2022, LNCS 13235, pp. 207–215, 2022.
https://doi.org/10.1007/978-3-031-06249-0_24

Actuators [15]. In [11], the authors use a gel, that can change its stiffness with a temperature increased produced by an electrical current. These devices have demonstrated their capability to produce vibrotactile stimulation, but they cannot be used to create programmable texture rendering by friction modulation, because elastic waves doesn't propagate in the soft substrate. In [2], electroadhesion is used to create friction modulation, but it requires the substrate to be in metal.

This paper presents the preliminary design of a flexible haptic device that can produce texture rendering over a large bendable surface. The principle is to create vibrating plates at ultrasonic frequency, actuated by piezoelectric actuators (named *Haptic Pixels* in the remaining of the paper), and to embed them onto a flexible substrate. This solution combine the characteristics of rigid haptics devices with the flexibility of a polymer to obtain a solution with friction modulation, as depicted Fig. 1 when used as a wristband. This interface can be developed for any tactile application where the surface must be conformable or foldable. Haptic Pixels could be controlled independently to allow localized effect and multiple finger interaction or their actions could be combined to create areas on the polymer for friction modulation.

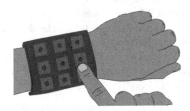

Fig. 1. Application of the proposed design to a Haptic Wristband. Each Haptic Pixel produce friction modulation while the whole device is bendable and conformable to the user's body.

The paper is organized as follows: the design and the evaluation of an elementary Haptic Pixel are first presented. Then, two pixels are combined together in a matrix in order to create an area of friction modulation and an overview of the complete surface is shown; an electromechanical characterization validates the design. Finally, a tribological and psychophysical studies are presented to confirm the results for an elementary Haptic Pixel.

2   Design of the Device

2.1   Design of the Haptic Pixel

A Haptic Pixel is built with a rigid glass plate, actuated by a piezoelectric actuator and covered by a polymer film.

The first step is the design of the glass plate with the PZT actuator. The size of the glass plate should be as small as possible to make the device more bendable. However, it should be large enough to allow friction reduction by ultra-sonic lubrication; we have considered that the pixel could not be smaller than the fingerpulpe. Indeed the spatial resolution is about a few mms [9], thus the finger cannot detect the amplitude variations at the surface if the pixel is too small. As a result, we set in our case a width of 10 mm with a square shape and a thickness of 500 μm. A piezoelectric disk actuator is added to the center of the glass plate. Simulations are carried out in order to determine the vibration mode of the plate and the size of the actuator. The values found are a radius of 2.5 mm, which corresponds approximately to the central antinode of vibration of this mode, and a thickness of 200 μm which maximise the displacement for this frequency. On Fig. 2, the pixel is detailed with the CAD drawing, the vibration mode in FEM and its realization. For the fabrication, a PZT ceramic (PI 255, 5 mm diameter and 0.2 mm thick) is bonded to the pixel with Vitralit 6128VT UV adhesive. Then two wires are welded on the ceramic, which have a wrap-around electrode. The first vibration mode is measured after realization at 28.6 kHz. This measurement, in good agreement with simulations, validates the design.

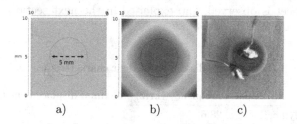

Fig. 2. a) Pixel CAD drawing; b) 1st vibration mode in FEM; c) Realization

The polymer that is placed on the plate influences its resonant behaviour. To evaluate this change, new simulations are carried out with several types of polymer films, presented in Table 1, used for flexible electronic circuits. For each film the same propagation behavior is observed as presented Fig. 3a). Furthermore, the simulations results show a significant influence of the polymer on the resonant frequency and amplitude. The polymer decreases the resonant frequency and damps the vibration. The KAPTON and the PEEK film have the lowest vibration amplitude reduction; so they seem the most efficient for a haptic used. These results are confirmed by an experimental study presented in Table 1. For that purpose, 80×80 mm^2 polymer sheets of several types and thickness were bonded on plates with an epoxy glue (EPOTEK OG116-31).

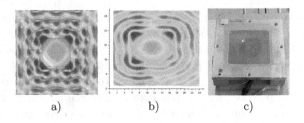

a)   b)   c)

Fig. 3. Displacement field with a PEEK layer bonded on a Haptic Pixel: a) Simulation results; b) Measurements; c) Realization

Among all the tested material, the PEEK (polyetheretherketone) film has the lowest vibration amplitude reduction compared with the uncovered plate, with the same voltage (0.4 μm at the center of the Haptic Pixel for 40 Vpp). Therefore, the PEEK film is used in this work. Finally, the cartography of the elementary Haptic Pixel is carried out, with a Polytec OFV-5000 modular vibrometer base with a sensor head OFV-505, and a resolution of 1 mm, as presented on Fig. 3b), which is close to the simulation. In particular, the propagation of the wave in the polymer sheet is observed, and exploited in the next section to create the flexible haptic surface.

Table 1. Realization and their characteristics

	Frequency (kHz)	Maximal displacement (μm)
No polymer	28.6	1.2
PC - 175 μm	25.7	0.2
PET - 100 μm	26.8	0.2
PEEK - 100 μm	25.8	0.4
KAPTON - 75 μm	25.9	0.3

2.2 Design of the Surface

The flexible haptic surface is built by spreading ultrasonic plates over a 100 μm thick sheet of PEEK polymer. The further apart the plates, the more flexible the surface. But, in order to create a friction reduction between two plates, we take advantage of the wave propagation in the polymer sheet. Indeed, the simulation and experimental results of the Fig. 3 show a stationary bending wave with a wavelength of 3.5 mm around the square plate.

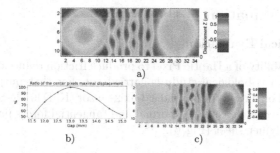

a)

b)

c)

Fig. 4. The 2 Haptic Pixels prototype: a) Measured vibration field when the 2 pixels are energized for a distance of 13 mm; b) Simulated vibration amplitude at the center of the 2 pixels as a function of the distance between them; c) Measured vibration field when right pixel is energized.

To precisely set the distance between two plates, a study with 2 pixels is carried out. By simulation, we obtain the vibration amplitude at the center of the two pixels as a function of the distance; the results are depicted on Fig. 4b). As it can be seen, there is an optimal distance between the two plates, estimated at 13 mm, that allows a maximum vibration amplitude at a given voltage on the two pixels. This distance corresponds to an integer number of wavelength of the standing wave that occur in the gap. This creates the conditions of an acoustic impedance matching between the plates and the polymer.

A new prototype has been built, in which the plates was placed with a gap of 13 mm. Laser interferometer measurement clearly demonstrates that the polymer sheet vibrates in the gap as expected by the FEM model (Fig. 4a). Interestingly, when only one pixel is energized, the vibration doesn't propagate to the other one as seen on Fig. 4c). Hence, every Haptic Pixel can produce a haptic feedback independently of the other. This property is validated through psychophysical evaluation in the next section.

Then by combining several times this elementary pattern of two Haptic Pixels on the surface, we obtain a haptic interface of the desired size: for example three by three as presented on Fig. 1 and on Fig. 5. The realization on the bottom right of Fig. 5, on a cylindrical support with a diameter of 40 mm, is the one obtained after a few steps of realization in clean room.

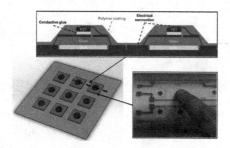

Fig. 5. Schematic views of an interface of three by three pixels and photography of a realization after a few steps of manufacturing

3   Haptic Evaluation of the Device

3.1   Tribological Evaluation

To validate the ability of a Haptic Pixel to produce friction reduction in condition of active touch, a tribological study is carried out. For that purpose, a Haptic Pixel is mounted on a rigid frame, and a 3-axis sensor (K3D40 Me-System) placed under the surface can measure the normal and tangential forces produced by the finger on the polymer sheet (Fig. 6).

Fig. 6. Set up for the tribological test

A sinusoidal voltage at the first resonant frequency of the Haptic Pixel is modulated by a 5 Hz square signal in order to create the friction modulation. The voltage amplitude is set between 20 V peak to peak (20 Vpp) to 140 Vpp. During the tests, the participants explore the surface laterally during 20 s at a velocity of 50 mm/s, while applying a normal force of 0.2 N. To ensure the velocity and the normal force, a training phase is performed for each participant. The typical response for 1 participant is depicted in Fig. 7a) with a voltage of 100 Vpp and which corresponds to one swipe across the surface. It should be emphasized here that the friction reduction produced by the ultrasonic vibration occurs all over the pixel, and not only on top of the plate, as expected from the results of Fig. 3.

The friction contrast [10] is then calculated and plotted for the 8 subjects on Fig. 7b) as a function of the applied voltage. On each graph a cross represents a value of the friction contrast on a swipe of the surface. The median of these values is then plotted. As expected, the friction contrast increases as the applied voltage increases. However, depending on the participants, the friction contrast evolves in a more or less important way. It hardly reaches 0.2 for some participants while for others it reaches 0.4 at 140 Vpp. The sweep velocity may have been too high. Indeed, the sweep velocity plays a very important role in the performance of friction reduction [13].

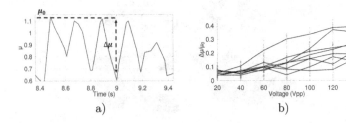

Fig. 7. a) Friction coefficient during 1 s for one subject at 100 Vpp; b) Friction contrast at different applied voltage for each subject

3.2   Psychophysical Evaluation

It has been shown in Sect. 2 that if not energized, a Haptic Pixel produces no ultrasonic vibration. We want to check in this part if this property can result in independent zones, that can produce a tactile feedback or not. For that purpose, we have conducted a psychophysical study on the 2 Haptic Pixels prototype of Fig. 4 and Fig. 8a).

The protocol is as follows. We energize the left (L) pixel, or the right (R) pixel or none (∅). The solution where both pixels are energized is not evaluated since we aim to validate the perception of a single Haptic Pixel. The voltage supplied to the plates is at a voltage amplitude V = {20, 30, 40, 60, 80} Vpp and is modulated by a square signal at a frequency f = {50, 250, 500} Hz. The participants are presented with each condition, in a random order, and each set is repeated 10 times, leading to $3 \times 3 \times 5 \times 10 = 450$ tests. The experiment duration is approximately 60 min. Data were collected from 9 consenting and inexperienced volunteers (3 females and 6 males) between 22 and 28 years old. The participants were asked to slide their index finger over the whole surface, were phonically isolated with earphones playing white noise and blindfolded. They were asked to answer 'L', 'R' or 'None', depending on where they perceived the stimulus.

As expected, the mean correct ratio answer increases when the voltage applied to the PZT ceramics increases, this voltage being directly related to the vibration amplitude of the pixel. This rate reaches 1 above 80 Vpp. The confusion matrix is given for a voltage of 40 Vpp in Fig. 9. The Pixel R seems less detectable. This might be due to a lower vibration amplitude than for L pixel, which can be due to an incorrect driving frequency. The modulation at 250 Hz allows to have the best rates of correct answers. This was also expected since this frequency is the optimal frequency for detection with the haptic sensation finger [4]. Thus for an applied voltage of 30 Vpp, the mean correct answer ratio reaches 92%.

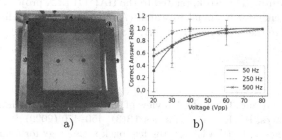

a)   b)

Fig. 8. a) Surface used for the psychophysical test; b) Mean Correct answer ratio

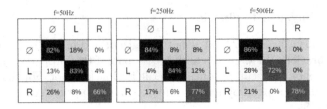

Fig. 9. Confusion matrix at 40 Vpp

Hence this study validates the concept of Haptic Pixel for a flexible haptic surface. The sensation can be localized around a pixel and is well detectable by the users. Different signals allow to modulate the haptic feeling.

4   Conclusion

A flexible haptic surface has been designed in this paper. This surface is composed of elementary Haptic Pixels assembled on a PEEK film polymer. The Haptic Pixel and the surface are developed and validated electromechanically. A tribological study is carried out and allows to validate the capacity of a Haptic Pixel to produce the illusion of texture thanks to the friction reduction. Similarly, a psychophysical study validates the detection of the Haptic Pixels.

Future work will validate the friction reduction on a complete surface and on bending conditions. Complementary psychophysical studies will also be performed. The surface designed will be manufactured in a cleanroom and will be adapted for uses where the flexibility of the surface is necessary: Haptic Wristband for example. A specific electronic will be developed to drive each Haptic Pixel independently or to combine them to create large area of friction modulation.

Acknowledgements. This work, related to the HAPTIFLEX project, was supported by the French Agency for National Research (ANR) via Carnot funding. The second author acknowledge the support of IRCICA, USR CNRS 3380.

References

1. Basdogan, C., Giraud, F., et al.: A review of surface haptics: enabling tactile effects on touch surfaces. IEEE Trans. Haptics **13**(3), 450–470 (2020)
2. Bau, O., Poupyrev, I.: REVEL: tactile feedback technology for augmented reality. ACM Trans. Graph. **31**(4), 1–11 (2012)
3. Besse, N., Rosset, S., et al.: Flexible active skin: large reconfigurable arrays of individually addressed shape memory polymer actuators. Adv. Mater. Technol. **2**(10), 1700102 (2017)
4. Bolanowski, S.J., Gescheider, G.A., et al.: Four channels mediate the mechanical aspects of touch. J. Acoust. Soc. Am. **84**(5), 1680–1694 (1988)

5. Breitschaft, S.J., Clarke, S., et al.: A theoretical framework of haptic processing in automotive user interfaces and its implications on design and engineering. Front. Psychol. **10**, 1470 (2019)
6. Dhiab, A.B., Hudin, C.: Confinement of vibrotactile stimuli in narrow plates: principle and effect of finger loading. IEEE Trans. Haptics **13**(3), 471–482 (2020)
7. Giraud, F., Hara, T., et al.: Evaluation of a friction reduction based haptic surface at high frequency. In: 2018 IEEE Haptics Symposium (HAPTICS), pp. 210–215 (2018)
8. Hudin, C., Lozada, J., et al.: Localized tactile feedback on a transparent surface through time-reversal wave focusing. IEEE Trans. Haptics **8**(2), 188–198 (2015)
9. Lederman, S.J., Klatzky, R.L.: Haptic perception: a tutorial. Attention Perception Psychophys. **71**(7), 1439–1459 (2009)
10. Messaoud, W.B., Bueno, M.A., et al.: Relation between human perceived friction and finger friction characteristics. Tribol. Int. **98**, 261–269 (2016)
11. Miruchna, V., Walter, R., et al.: GelTouch: localized tactile feedback through thin, programmable gel. In: Proceedings of the 28th Annual ACM Symposium on User Interface Software & Technology, pp. 3–10. ACM, Charlotte, November 2015
12. Pantera, L., Hudin, C.: Multitouch vibrotactile feedback on a tactile screen by the inverse filter technique: vibration amplitude and spatial resolution. IEEE Trans. Haptics **13**(3), 493–503 (2020)
13. Vezzoli, E., Vidrih, Z., et al.: Friction reduction through ultrasonic vibration part 1: modelling intermittent contact. IEEE Trans. Haptics **10**(2), 196–207 (2017)
14. Yu, X., Xie, Z., et al.: Skin-integrated wireless haptic interfaces for virtual and augmented reality. Nature **575**(7783), 473–479 (2019)
15. Yun, S., Park, S., et al.: A soft and transparent visuo-haptic interface pursuing wearable devices. IEEE Trans. Industr. Electron. **67**(1), 717–724 (2020)

Human Self-touch vs Other-Touch Resolved by Machine Learning

Aruna Ramasamy[1,2]([⊠]) [iD], Damien Faux[1] [iD], Vincent Hayward[1,3] [iD],
Malika Auvray[3] [iD], Xavier Job[4] [iD], and Louise Kirsch[5] [iD]

[1] Actronika SAS, 68 boulevard de Courcelles, 75017 Paris, France
aruna.ramasamy@actronika.com
[2] École Normale Supérieure, CNRS, Laboratoire des Systèmes Perceptifs,
24 rue Lhomond, 75005 Paris, France
[3] Sorbonne Université, CNRS, Institut des Systèmes Intelligents et de Robotique,
4 Place Jussieu, 75005 Paris, France
[4] Department of Neuroscience, Karolinska Institutet,
C4 Neurovetenskap, 171 77 Stockholm, Sweden
[5] University of Paris, CNRS, Integrative Neuroscience and Cognition Center,
45 Rue des Saint-Pères, 75006 Paris, France

Abstract. Using a database of vibratory signals captured from the
index finger of participants performing self-touch or touching another
person, we wondered whether these signals contained information that
enabled the automatic classification into categories of self-touch and
other-touch. The database included signals where the tactile pressure
was varied systematically, where the sliding speed was varied systematically, and also where the touching posture were varied systematically.
We found that using standard sound feature-extraction, a random forest
classifier was able to predict with an accuracy greater than 90% that a
signal came from self-touch or from other-touch regardless of the variation of the other factors. This result demonstrates that tactile signals
produced during active touch contain latent cues that could play a role
in the distinction between touching and being touched and which could
have important applications in the creation of artificial worlds, in the
study of social interactions, of sensory deficits, or cognitive conditions.

Keywords: Self-touch · Touchant-touché · Social tactile interactions ·
Machine learning

1 Introduction

Skin-to-skin touch is an important tactile interaction. This type of touch has
attracted the attention of many authors (e.g. [9,24,31]) and motivated research
across many fields; from philosophy [16,23], cognitive neuroscience [2,5,10,20,28],

Supported by Skłodowska-Curie Actions Programme (Horizon 2020), Innovative Training Network INTUITIVE.

H. Seifi et al. (Eds.): EuroHaptics 2022, LNCS 13235, pp. 216–224, 2022.
https://doi.org/10.1007/978-3-031-06249-0_25

to human development and well-being [1,4,8,25]. All these works are based on introspection or on behavioural observations since, by necessity, they cannot rely on the objectification of the mechanical consequences, hence of the sensory consequences of skin touching skin. It was however recently been realised that the objectification of tactile interactions is possible when hands actively interact with inanimate objects [12,29,30]. Motivated by this observation, some of us collected a database of vibration signals collected from a index finger interacting with another finger, or a forearm, with a view to provide objective data produced during skin-to-skin interactions [17]. This latter study demonstrated that the "tactile waves" measured on a touching finger bore features related to the interaction, to wit, the pressure applied (a tonic characteristic) and the sliding speed (a kinematic characteristic). The signals were shown to be relatively independent from the posture with which the interaction was effected, making this technique potentially useful for analyses about tactile behaviour [17].

1.1 Present Study

In the present study we advanced the hypothesis that information contained in single-channel vibration signals recorded from a finger in sliding contact with another finger, or with a forearm, contained information that would enable the discrimination between self-touch and touching another person. In the foregoing, we show that certain supervised machine classifiers can achieve a very high level of success in deciding whether tactile vibrations came from *self-touch* or from *other-touch* (touching another person). Supervised machine classifiers trained models through ground-truth labels which are indicated here by SELF and OTHER. Our findings could contribute to the study of behaviour in many domains, chief among them is the study of the role of touch in social interactions and investigations related to cognitive conditions such as autism or schizophrenia where self/other touch discrimination might be impaired [8,31]. It also bears the intriguing conclusion that the determining factors differentiating self from ordinary touch are not limited to a unique convergence of sensory and motor signals [3,14,15,23] but that the tactile inputs *per se* contain cues that are special to self-touch.

1.2 Signal Database

A key attribute of the database is that it included signals recorded in similar conditions of pressure and speed during self-touch but also when touching another person. The vibrations recorded from a finger sliding on skin clearly depended on the pressure applied and on the sliding speed [17]. On this account, it would be surprising if machine learning classifiers were not able to discriminate between categories of intensity or speed. In fact, the multichannel whole-hand recordings described in [29] contained sufficient information to enable a support vector machine classifier to categorise twelve different tactile gestures, three types of materials, as well as the shape of the objects being touched.

1.3 Feature Extraction

Since the data at hand were available in form of time-dependent signals arising from mechanical interactions between objects in contact, it stands to reason that techniques developed to classify sounds would also be appropriate to classify signals arising from sliding fingers. A first option was to train and then to test classifiers using raw data. Another possibility was to extract domain knowledge features from the signals.

Features frequently used in the processing of sound include: Maximum Mel Frequency Cepstral Coefficients (MFCC), a quantification technique for vibratory signals [11], minimum MFCC, mean MFCC, Zero-Crossing Rate which capture the rhythmic features of a signal [13], Chromograms which are commonly used for the analysis of musical sounds [27], Spectral Roll-Off which measures the right skewness of a spectrum [18], Spectral Flux which describes rate of change a time-varying spectrum arising from a non-stationary process [21], and Pitch which is a well-known instantaneous attribute of sounds [26]. Features do not contribute equally significant to a given classification problem. Their significance can be assessed through the decrease of accuracy in classification when a feature is dropped. To this end, GINI importance, or mean decrease in impurity (MIDI), may be used to evaluate the importance of each feature [6].

1.4 Performance Measures

The performance of a classifier is relative to a `test` dataset used to examine the model trained with a `train` dataset. Here we use standard performance metrics. Accuracy is measured by

$$\text{Accuracy} = (N_{\text{tp}} + N_{\text{tn}})/(N_{\text{tp}} + N_{\text{fp}} + N_{\text{fn}} + N_{\text{tn}}),$$

which accounts for the number, N, of predictions labeled as true positives (tp), true negatives (tn), false positives (fp), and false negatives (fn), commonly expressed in percent, while precision is defined by the proportion of true positive predictions to the total number of positive predictions, and recall which reaches one when there are no false negatives. We also used a statistical measure termed, F_1-Score, which is the harmonic mean of precision and recall. These metrics are recalled below.

$$\text{Precision} = N_{\text{tp}}/(N_{\text{tp}} + N_{\text{fp}}), \quad \text{Recall} = N_{\text{tp}}/(N_{\text{tp}} + N_{\text{fn}}),$$
$$F_1\text{-Score} = 2(\text{Recall} \cdot \text{Precision})/(\text{Recall} + \text{Precision}).$$

We used a graphical representation borrowed from Signal Detection Theory [22]. Here, a Receiver Operating Characteristics (ROC) curve plots the false positives vs. true positives. It can be interpreted as a plot of 1-sensitivity vs. sensitivity. Even though these curves may cross, any curve clearly above another is better. A single-number separability measure, the area under the ROC curve (AUC), follows from this representation. An AUC of 0.7 is said to be acceptable, excellent if it around 0.8, and outstanding above 0.9.

1.5   Ambiguity and Abstention

In the present study we employed a recently introduced technique which proposes that in case of ambiguity it is better to abstain rather than to make a prediction [7]. This technique can be implemented, for example, in a three-way random forest algorithm which can extract probabilities for each class. Given two probability thresholds, α and β, having value 1.0 for the ground-truth class and 0 for the incorrect class, predictions are declared positive when the score of the positive class is greater than α and the score the positive class is greater than the score of the negative class. Conversely, predictions are declared negative when the score of the negative class is greater than β and the score of the negative class is greater than the score of the positive class. When neither of these conditions are met, then there is an abstention. Typical values for α and β are 0.75.

2   Results

The skin-to-skin touch datasets described in [17] contained signal recorded with eighteen participants of balanced gender and hence captured a reasonable diversity of individual behaviours. The signals were recorded at audio-rate and downsampled 10-fold. The initial one second interval of each recording was edited out to eliminate the energy burst due to the stick-to-slip transition. This deletion potentially eliminating useful information for the purpose of this study. The **pressure** dataset comprised ten 10-s recordings where participant touched ten times for each condition their own or the other participant's index finger with a gentle or firm touch, resulting in 720 trials. The **speed** dataset comprised similar recordings but the participants touched their own or the other participant's forearm at three different speeds giving rise to 1080 trials. The **posture** dataset comprised similar recordings but the participants touched their own or the other participant's index finger in two different orientation to vary the relationship between the sensor and the regions of skin contact, resulting in 720 trials.

2.1   Relative Performance of Classification Techniques

Table 1 shows the performance of various classification techniques using the **pressure** dataset suggesting that the random forest classifier performed best compared to other classifiers. It produced negligible Mean Squared Error (MSE) for the **test** dataset with a classification accuracy of 81% greatly surpassing logistic regression, decision tree, Gaussian, and support vector classification.

2.2   Importance of Feature Extraction

Tests conducted with the combined datasets to evaluate the contribution of extracted features vs raw data unequivocally confirmed the importance of providing the algorithms with extracted domain knowledge features. Most metrics,

Table 1. Classifier performance & Importance of feature extraction. Accuracy in %.

Classifier w pressure	Accuracy	Precision	Recall	F_1-score	train-MSE	test-MSE	AUC
logistic regression	52.5	0.52	0.52	0.52	0.41	0.47	0.56
decision tree	65.0	0.65	0.65	0.65	0.00	0.35	0.65
Gaussian	55.0	0.56	0.46	0.50	0.42	0.45	0.57
support vector	53.4	0.53	0.61	0.56	0.40	0.47	0.53
random forest	81.0	0.81	0.79	0.80	0.00	0.19	0.84
Raw/features w dataset							
Raw w pressure	61.0	0.58	0.63	0.60	0.02	0.39	0.64
Raw w speed	62.0	0.57	0.67	0.62	0.17	0.38	0.64
Raw w posture	53.0	0.54	0.56	0.55	0.10	0.47	0.54
Features w pressure	81.0	0.81	0.79	0.80	0.00	0.19	0.83
Features w speed	78.0	0.76	0.76	0.76	0.00	0.22	0.85
Features w posture	76.0	0.78	0.75	0.76	0.01	0.24	0.81

Table 1, show low to unacceptable values when raw data was used. Figures 1a, b further indicate that classification models become skilful with the introduction of domain knowledge features since the figure shows across-the-board reduction in false positives rate.

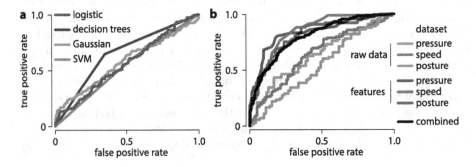

Fig. 1. ROC curves. a, Weak performance of classifiers on the raw data of combined datasets. **b,** Overall effect of domain knowledge feature extraction on different datasets.

The results shown in Table 2 indicate that a three-way classification algorithm with abstention greatly improved discrimination between the labels SELF and OTHER. Accuracy was increased from 81% to 92% with the **pressure** dataset, from 78% to 97% with the **speed** dataset, and from 76% to 84% with the **posture** dataset. For the case of the combined dataset accuracy was increased from 73% to 90%, which is very significant.

Figure 2 summarizes the GINI importance of the different domain knowledge features used relatively to the datasets.

Table 2. Three-way classification of SELF and OTHER using abstention.

	Accuracy (%)	Precision	Recall	F1-score	Train-MSE	Test-MSE	AUC
pressure	92.2	0.96	0.89	0.92	0.00	0.08	0.92
speed	97.6	1.00	0.95	0.87	0.00	0.02	0.98
posture	84.8	0.88	0.85	0.86	0.00	0.15	0.84
combined	90.2	0.87	0.97	0.91	0.00	0.01	0.90

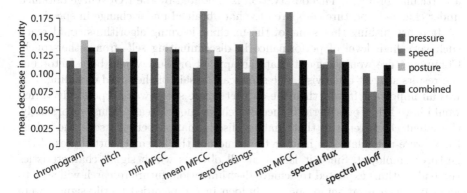

Fig. 2. GINI importance of features. This measure is plotted for each features during three-way random forest classification of different datasets. Features are ranked by order of decreasing importance with combined data.

2.3   Discussion and Conclusion

Overall, the best performance was achieved by the three-way random forest classification algorithm which makes use of an ensemble learning method though a multitude of decision trees. The mutually exclusive branches represent subcategories of the input features. The random forest classifier overcomes overfitting though voting to predict an output. The technique known as bootstrap aggregating de-correlates the decision trees corresponding to different training sets. Noise in single tree affects the performance of the model but not the average of many trees. This strategy was very successful in the classification of self-touch and other-touch (labels SELF and OTHER) from single-channel recordings of tactile waves in the index finger.

As a whole, the eight domain knowledge features showed little relative advantages over the others in the task of discriminating self-touch from other-touch. The mean and max MFCC features which made important contributions when the speed of sliding contact varied could be considered as exceptions. Also, the zero-crossing feature was important when the posture was changed. Chromogram and pitch had the highest importance in terms of classification with combined data. These findings indicates that none of the commonly used audio features preferentially revealed the latent characteristics of skin-to-skin friction-induced vibrations that can be used to distinguish self-touch from touching other people. It is possible that the removal of the stick-to-slip frictional transitions was responsible for this general lack of sensitivity. These findings therefore suggest that further research is needed to discover better domain knowledge features for this type of data.

The AUC measure was found to be a preferred performance indicator over accuracy, an observation that was commented in [19] although it is recommended to consider different evaluation metrics to discuss performance of a classifier on a particular problem. This observation is validated by the ROC curve obtained under change in posture compared to that obtained under change in speed.

It is astonishing that some of the machine learning algorithms could reach such very high level of performance in discriminating self- from other-touch. Common sense would suggest that the applied pressure would be a factor but the results suggest otherwise. The same can be said of the speed of sliding. If it was an important factor, then the dataset where speed was purposefully varied would have led to poor performance, which was not the case. Thus, surprisingly, the latent characteristics that enabled discrimination between self- and other-touch were not related to neither the tonic nor the kinematic attributes of the gestures employed. Since it is not at all obvious which signal characteristics these algorithms exploited to achieve discrimination, future research will seek to identify which invariant properties hidden in the recorded tactile signals were used by the classifier to discriminate self- from other-touch.

References

1. Ardiel, E.L., Rankin, C.H.: The importance of touch in development. Paediatr. Child Health **15**(3), 153–156 (2010)
2. Bays, P.M., Wolpert, D.M.: Predictive attenuation in the perception of touch. In: Haggard, P., Rosetti, Y., Kawato, M. (eds.) Sensorimotor Foundations of Higher Cognition, vol. 22, pp. 339–358. Oxford University Press, Oxford (2008)
3. Bermúdez, J.L.: The Paradox of Self-consciousness. MIT Press, Cambridge (2000)
4. Blackwell, P.L.: The influence of touch on child development: implications for intervention. Infants Young Child. **13**(1), 25–39 (2000)
5. Blakemore, S.J., Wolpert, D.M., Frith, C.D.: Why can't you tickle yourself? NeuroReport **11**(11), R11–R16 (2000)
6. Breiman, L.: Random forests. Mach. Learn. **45**(1), 5–32 (2001)
7. Campagner, A., Cabitza, F., Ciucci, D.: Three–way classification: ambiguity and abstention in machine learning. In: Mihálydeák, T., et al. (eds.) IJCRS 2019. LNCS (LNAI), vol. 11499, pp. 280–294. Springer, Cham (2019). https://doi.org/10.1007/978-3-030-22815-6_22
8. Cascio, C.J., Moore, D., McGlone, F.: Social touch and human development. Dev. Cogn. Neurosci. **35**, 5–11 (2019)
9. Classen, C.: The Book of Touch. Routledge (2020)
10. Crucianelli, L., Metcalf, N.K., Fotopoulou, A.K., Jenkinson, P.M.: Bodily pleasure matters: velocity of touch modulates body ownership during the rubber hand illusion. Front. Psychol. **4**, 703 (2013)
11. Davis, S., Mermelstein, P.: Comparison of parametric representations for monosyllabic word recognition in continuously spoken sentences. IEEE Trans. Acoust. Speech Signal Process. **28**(4), 357–366 (1980)

12. Delhaye, B., Hayward, V., Lefèvre, P., Thonnard, J.L.: Texture-induced vibrations in the forearm during tactile exploration. Front. Behav. Neurosci. **6**(37), 1–10 (2012)
13. Gouyon, F., Pachet, F., Delerue, O.: On the use of zero-crossing rate for an application of classification of percussive sounds. In: Proceedings of the COST G-6 Conference on Digital Audio Effects (DAFX-00), Verona, Italy, vol. 5 (2000)
14. Haggard, P., Clark, S., Kalogeras, J.: Voluntary action and conscious awareness. Nat. Neurosci. **5**(4), 382–385 (2002)
15. Hara, M., et al.: Voluntary self-touch increases body ownership. Front. Psychol. **6**, 1509 (2015)
16. Husserl, E.: The constitution of psychic reality through the body. In: Ideas Pertaining to a Pure Phenomenology and to a Phenomenological Philosophy, pp. 151–169. Springer, Heidelberg (1989)
17. Kirsch, L.P., Job, X.E., Auvray, M., Hayward, V.: Harnessing tactile waves to measure skin-to-skin interactions. Behav. Res. Methods **53**(4), 1469–1477 (2020). https://doi.org/10.3758/s13428-020-01492-3
18. Kos, M., Kačič, Z., Vlaj, D.: Acoustic classification and segmentation using modified spectral roll-off and variance-based features. Digit. Signal Process. **23**(2), 659–674 (2013)
19. Ling, C., Huang, J., Zhang, H.: AUC: a statistically consistent and more discriminating measure than accuracy. In: Proceedings of 18th International Joint Conference on Artificial Intelligence (IJCAI) (2003)
20. Löken, L.S., Olausson, H.: The skin as a social organ. Exp. Brain Res. **204**(3), 305–314 (2010)
21. Lu, L., Jiang, H., Zhang, H.J.: A robust audio classification and segmentation method. In: Proceedings of the Ninth ACM International Conference on Multimedia, pp. 203–211 (2001)
22. Marcum, J.I.: A statistical theory of target detection by pulsed radar. Technical report, Rand Corp Santa Monica, CA (1947)
23. Merleau-Ponty, M.: Phenomenology of Perception. Routledge (1962)
24. Morrison, I.: Keep calm and cuddle on: social touch as a stress buffer. Adapt. Hum. Behav. Physiol. **2**, 344–362 (2016)
25. Moscatelli, A., Nimbi, F.M., Ciotti, S., Jannini, E.A.: Haptic and somesthetic communication in sexual medicine. Sex. Med. Rev. **9**(2), 267–279 (2021)
26. Nielsen, A.B., Hansen, L.K., Kjems, U.: Pitch based sound classification. In: 2006 IEEE International Conference on Acoustics Speech and Signal Processing Proceedings, vol. 3, p. III. IEEE (2006)
27. Pelkowitz, L.: A generalization of the spectrogram for colored displays. IEEE Trans. Acoust. Speech Signal Process. **31**(1), 222–225 (1983)
28. Schütz-Bosbach, S., Musil, J.J., Haggard, P.: Touchant-touché: the role of self-touch in the representation of body structure. Conscious. Cogn. **18**(1), 2–11 (2009)
29. Shao, Y., Hayward, V., Visell, Y.: Spatial patterns of cutaneous vibration during whole-hand haptic interactions. Proc. Natl. Acad. Sci. **113**(15), 4188–4193 (2016)
30. Tanaka, Y., Horita, Y., Sano, A.: Finger-mounted skin vibration sensor for active touch. In: Isokoski, P., Springare, J. (eds.) EuroHaptics 2012. LNCS, vol. 7283, pp. 169–174. Springer, Heidelberg (2012). https://doi.org/10.1007/978-3-642-31404-9_29
31. Van Erp, J.B.F., Toet, A.: Social touch in human-computer interaction. Front. Digit. Humanit. **2**, 2 (2015)

Investigating Movement-Related Tactile Suppression Using Commercial VR Controllers

Immo Schuetz$^{(\boxtimes)}$ (ID), Meaghan McManus (ID), Katja Fiehler,
and Dimitris Voudouris (ID)

Justus-Liebig University Giessen, Giessen, Germany
schuetz.immo@gmail.com

Abstract. When we perform a goal-directed movement, tactile sensitivity on the moving limb is reduced compared to during rest. This well established finding of movement-related tactile suppression is often investigated with psychophysical paradigms, using custom haptic actuators and highly constrained movement tasks. However, studying more naturalistic movement scenarios is becoming more accessible due to increased availability of affordable, off-the-shelf virtual reality (VR) hardware. Here, we present a first evaluation of consumer VR controllers (HTC Vive and Valve Index) for psychophysical testing using the built-in vibrotactile actuators. We show that participants' tactile perceptual thresholds can generally be estimated through manipulation of controller vibration amplitude and frequency. When participants performed a goal-directed movement using the controller, vibrotactile perceptual thresholds increased compared to rest, in agreement with previous work and confirming the suitability of unmodified VR controllers for tactile suppression research. Our findings will facilitate investigations of tactile perception in dynamic virtual scenarios.

1 Introduction

When we perform a movement using a specific body part, sensitivity to external tactile stimuli on this body part is substantially reduced [6,7]. This *tactile suppression* effect has now been reliably established for a variety of self-generated movements, such as single finger abductions [6,7], goal-directed reaching [5,9,11,18,19], and grasping [15]. It is commonly attributed to central mechanisms that predict sensory action outcomes using a feed-forward model and suppress corresponding afferent sensory inputs [1,20], although peripheral mechanisms such as masking may also be involved [6]. Tactile suppression phenomena are a useful avenue to study predictive processing in the human sensorimotor system [10] and in understanding the informational value of tactile feedback during active movement.

Tactile sensitivity is usually investigated using specialized haptic actuators ("tactors", cf. [8,14] for review), which can be expensive and require special

H. Seifi et al. (Eds.): EuroHaptics 2022, LNCS 13235, pp. 225–233, 2022.
https://doi.org/10.1007/978-3-031-06249-0_26

expertise to use. At the same time, many research labs now have virtual reality (VR) setups at their disposal due to advances in consumer hardware, and research is shifting towards freely moving participants in virtual environments. Consumer VR hardware bundles typically include wireless game controllers, most of which now feature a built-in vibrotactile actuator for haptic feedback. Consequently, leveraging built-in vibrotactors for behavioral research would facilitate studying tactile sensitivity in freely moving participants and without the need for custom tactile stimulation devices. However, the exact haptic properties of these controllers are largely undocumented. Here, we evaluate two commercially available VR controllers (HTC Vive and Valve Index, cf. Fig. 1) to determine whether a controller's vibration amplitude (stimulus intensity) can be controlled finely enough to measure tactile perceptual thresholds using psychophysics. Additionally, we compare tactile detection thresholds between a resting baseline and a simple goal-directed movement task to assess the suitability of this setup for the detection of movement-related tactile suppression. Moreover, the tested controllers allow for specifying vibration frequency. Human tactile perception integrates a combination of four main sensory channels with different temporal and spatial sensitivity profiles [2,12,13]. Because of spatial summation effects due to the relatively large physical size of the controllers and differences in hand posture, it is unclear how different actuator frequencies propagate to the participant's skin surface. We thus further compare perceptual thresholds for three different frequencies (100, 250, and 400 Hz) to investigate possible frequency-dependent differences in haptic perception when using VR controllers.

2   Methods

2.1   Participants

A total of 19 participants took part in the experiment. Data from one participant were excluded due to excessive movement in the baseline condition, leading to a total sample of 18 participants (13 female, 5 male; mean age 25.1 years ±4.9 years, range 19–37 years). All participants were right handed as confirmed using the Edinburgh Handedness Inventory (EHI [16]; mean score 83.2, range 33–100) and had normal vision or corrected-to-normal vision using contact lenses. Participants gave written informed consent and received course credits or 8€ per hour for their participation. The experiment was approved by the research ethics board at Justus Liebig University Giessen (protocol number 2019-0003) and was performed in accordance with the Declaration of Helsinki (2008).

2.2   Apparatus

Participants sat on a desk chair without armrests, with both arms hanging at their sides in a relaxed position and holding a VR game controller in each hand. Two different controllers were used, as shown in Fig. 1 (left): the HTC Vive Pro controllers[1] (hereafter: *Vive*; HTC Corp., Xindian, New Taipei, Taiwan) and

---

[1] https://www.vive.com/eu/accessory/controller2018/.

Fig. 1. *Left*: Controllers used in the study (HTC Vive Pro and Valve Index). *Middle*: Target sphere in VR (fixation or reach target, depending on the condition). *Right*: Gaze response procedure. The participant selected the right cube ("Yes") in this trial.

Valve Index controllers[2] (hereafter: *Index*; Valve Corp., Bellevue, WA, USA). In each experimental session, participants held controllers of the same type in both hands to keep muscle activation and somatosensory feedback across both hands comparable. Only the right controller was used for movement and vibrotactile stimuli. Participants wore an HTC Vive Pro Eye HMD, used to present task instructions and collect participants' responses using the integrated eye tracker.

Participants were presented with a minimal visual scene consisting of a tiled floor and sky (Fig. 1). Instructions were presented as light gray text on a dark background panel which floated in front of the participant. No 3D models of the controllers were rendered in the VR environment, because pilot testing determined that stronger vibrotactile stimuli could cause visible jitter in the controller model, potentially yielding a visual cue for whether a vibrotactile stimulus was present or absent[3]. Auditory pink noise was played over the HMD's headphones to ensure that participants based their responses only on the vibrotactile stimuli and not on audible noise from the haptic actuator. The experiment was implemented using Unity (version 2019.4.16f1; Unity Technologies, Inc., San Francisco, CA, USA), SteamVR (version 1.20.1), the Vive SRanipal eye tracking framework (version 1.1.2.0), and the Unity Experiment Framework [4]. It was run on an Alienware desktop PC (Intel Core i9-7980XE CPU at 2.6 GHz, 32 GB RAM, Dual NVidia GeForce GTX1080 Ti GPU).

2.3 Experimental Task

Participants performed a *movement* condition, in which they had to execute a reaching movement towards a visual target using the controller in their right hand, and a *baseline* condition, in which they had to simply hold the controllers at their side without moving. Movement and baseline conditions were run as separate consecutive sessions. Each trial started with the presentation of a light gray sphere (distance 1.5 m, radius 5 cm; cf Fig. 1, middle) at eye level and at a

---

[2] https://www.valvesoftware.com/de/index/controllers.

[3] Presumably, this is due to the inertial measurement unit within each controller, which is used for positional tracking and can become overwhelmed by noise when vibrotactile feedback is activated.

random angle ($\pm 20°$) from the participant's body midline. In the movement conditions, a vibrotactile stimulus of 200 ms duration and with varying amplitude and frequency was presented using the right controller at movement onset, which was determined online using a movement speed criterion ($0.3\,\text{m/s}$). If a participant failed to move within 3 s after sphere onset, a reminder text appeared and the trial was excluded from analysis. In the baseline condition, the vibrotactile stimulus was always applied after a fixed delay of 500 ms after sphere onset and participants had to simply fixate the sphere. They were instructed to let their arms hang at their sides instead of resting them on their leg to avoid additional tactile perception on their thigh or knee. After each trial, participants were asked whether they felt a vibration of the right controller. To avoid confounding their response with the stimulated limb, such as by using a controller button press, responses were collected using the eye tracker in the HMD. Two light gray cubes (distance 1.5 m, side length 20 cm, Fig. 1, right) were shown at eye level, labeled with the text "Yes" and "No" alongside the question "Did you notice any vibration?". Participants responded by looking at the cube corresponding to their chosen response for 1 s, after which the selected cube changed color to indicate the chosen answer. The eye tracker was calibrated at the start of each session using the built-in calibration routine.

Vibrotactile stimuli were presented at three possible frequencies (100, 250, and 400 Hz) and at seven possible intensities (0.00001, 0.0001, 0.001, 0.01, 0.1, 1.0, and no stimulation)[4]. Each frequency and stimulus intensity combination was repeated eight times, leading to a total number of 168 trials per session (3 frequencies × 7 intensities × 8 repetitions). Each session took around 12 min to complete, and each participant performed four sessions in total (one movement and one baseline session per controller type). Sessions using the same controller type were performed consecutively. The order of controllers was counterbalanced across participants, as was the order of baseline and movement sessions within each controller type. The full experiment took around one hour.

2.4 Data Analysis

Data analysis was performed using Python (version 3.8) and jamovi (version 2.2.3). Trials in which participants did not move in the movement condition, or performed a movement in the baseline condition, were excluded from the dataset (57 trials or 0.47% total). Psychometric functions were then fit to each participant's rate of "yes" responses (detection rate) per stimulus intensity using the psignifit 4 toolbox [17]. Initially, functions were fit to all response data per condition, independent of the presented frequency to gain a robust estimate of individual response behavior. This *combined analysis* thus contained 24 responses per stimulus intensity (3 frequencies × 8 repetitions). As stimulus intensity levels ranged from 10^{-5} to 1 in steps of one order of magnitude, data were fitted with

---

[4] Intensity values are provided to SteamVR on a scale of zero to one and do not directly correspond to any physical property such as peak-to-peak displacement.

a Weibull psychometric function and perceptual thresholds (stimulus intensities at which a participant would have a detection rate of 0.5) are thus returned on a log scale. Since a stimulus intensity of zero is not defined on a log scale, response rates in trials without a stimulus (false alarms) were set as a fixed lower bound (gamma parameter) when fitting each participant's psychometric function for each condition. Individual false alarm rates ranged from 0–0.33 (mean: 0.07) in the baseline and 0–0.21 (mean: 0.05) in the movement condition.

To assess tactile suppression, we calculated threshold difference values by subtracting each participant's baseline detection threshold from their movement threshold, separately for each controller type. Due to the log-scaled intensity values, a threshold difference value of 1 indicates a difference of one order of magnitude in stimulus intensity. To test for suppression, we then used one-sample t-tests to compare these threshold differences against zero. To compare suppression effects and baseline detection thresholds between controllers, we used paired t-tests where data were normally distributed (Shapiro-Wilk test), otherwise the Wilcoxon signed-rank test was used. To further investigate effects of controller vibration frequency on tactile perception, we also fit individual psychometric functions for each participant, frequency, and condition (*frequency analysis*, 8 repetitions per intensity). These data were analyzed using a linear mixed effects model with factors controller × frequency and a random intercept coefficient for participant. Bonferroni-Holm correction was used for all multiple comparisons.

3 Results

We first investigated baseline and movement tactile thresholds independent of vibration frequency (24 responses per intensity). Figure 2 plots individual and averaged detection thresholds for the movement and baseline conditions for the Vive (left) and Index controller (middle). Average thresholds were larger in the movement compared to baseline condition for both controllers (Vive: $t_{17} = 3.25$, p = 0.002, d = 0.77; Index: $t_{17} = 4.17$, p < .001, d = 0.98), indicating that participants experienced tactile suppression. Baseline thresholds were greater in the Index compared to the Vive controller ($W_{17} = 137$, p = 0.024, r = 0.60).

For a more direct comparison of tactile suppression effects, threshold difference values for each controller type are shown in Fig. 2 (right). Suppression magnitude was not significantly different between both controllers ($t_{17} = 0.13$, p = 0.90). Moreover, individual threshold differences for both devices were strongly correlated (r = 0.82, $R^2 = 0.67$, $F_{16,1} = 32.7$, p < .001), suggesting similar individual levels of suppression regardless of controller type.

Results for the frequency analysis are shown in Fig. 3, plotting the data similar to Fig. 2 but when fit separately for each vibration frequency (8 responses per intensity). Here, baseline thresholds differed significantly between controllers ($F_{85,2} = 11.8$, p < .001, $\eta_p^2 = .20$) and frequencies ($F_{85,1} = 19.7$, p < .001, $\eta_p^2 = .18$). In post-hoc tests, 400 Hz had higher thresholds than both 100 Hz ($t_{85} = -3.86$, p < .001, $\eta_p^2 = .19$) and 250 Hz ($t_{85} = -4.49$, p < .001, $\eta_p^2 = .18$), but 100 Hz and 250 Hz did not differ significantly ($t_{85} = 0.64$, p = 0.526). Threshold differences per frequency (Fig. 3, right) all indicate tactile suppression (all

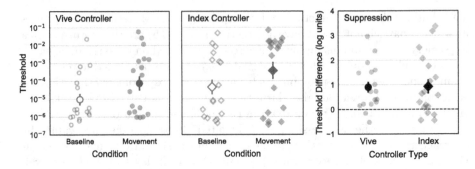

Fig. 2. *Left, Middle*: Detection thresholds in the baseline (blue, open markers) and movement condition (red, filled markers) for the two tested controller types. Functions were fit independent of frequency (24 responses per intensity). *Right*: Threshold differences (movement - baseline) for each controller. Values above zero (dashed line) indicate tactile suppression. Error bars show ±1 SEM. (Color figure online)

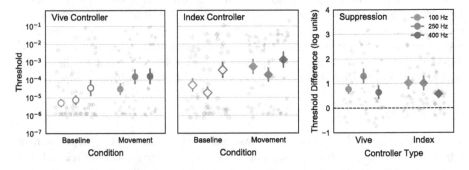

Fig. 3. *Left, Middle*: Thresholds in the baseline (open markers) and movement condition (filled markers) for each controller type, fit separately for each tested vibration frequency (8 responses per frequency × intensity). *Right*: Threshold differences for each controller and frequency. Error bars indicate ±1 SEM.

$W_{17} > 127$, all $p < 0.027$, all $r > .52$), but suppression strength significantly varied between frequencies ($F_{85,2} = 4.9$, $p = 0.01$, $\eta_p^2 = .08$), with the only significant post-hoc comparison between 250 and 400 Hz ($t_{85} = 3.13$, $p = 0.007$, $\eta_p^2 = .09$).

4 Discussion

Here, we present the first evaluation of consumer VR controllers for tactile psychophysics research. Baseline (resting) perceptual thresholds differed between controller types, indicating that devices may differ in vibrotactile presentation. Nevertheless, we found consistent tactile suppression during movement, and individual suppression was of similar magnitude and correlated between devices.

Baseline thresholds significantly differed between controllers, with the Vive controller apparently presenting the "stronger" tactile stimulus (lower threshold). This might be explained by a stronger actuator in the Vive controller or differences in actuator mounting or placement. Another explanation could relate to possible mass differences between controllers. In this case, participants would need to apply greater grip forces to hold the heavier controller, leading to stronger afferent input masking the vibrotactile stimulus. Yet, recent work does not suggest stronger suppression for greater forces [3] and mass was very similar when measured (Vive: 205 g, Index: 196 g), making this explanation unlikely. It is further unknown whether all controllers of the same type produce similar physical stimuli for a given intensity value. In any case, different controller types may be more or less suitable for perceptual threshold estimation, and care should be taken to evaluate devices before comparing results across studies, which was beyond the scope of this conference paper.

We found significant suppression during movement compared to rest in both controllers, in good agreement with previous work that found similar effects with specialized tactors [5,6,11,18,19]. The magnitude of this suppression effect was comparable for both controllers and individual threshold differences were highly correlated between devices. This suggests that we indeed measured suppression as a participant-specific variable, further supported by significant suppression effects in all tested vibration frequencies. While the fixed stimulus onset times in the baseline condition might have facilitated prediction of the stimulus and thus lower thresholds, movement-related tactile suppression is evident also with jittered baseline stimulus onset times [11]. Additionally, the larger number of trials with stimulus compared to without could have biased participants towards responding "yes" more often, but such a bias would affect all conditions equally as trial counts were identical.

The frequencies tested here predominantly activate the Pacinian corpuscles, which have a receptive range of 40–800 Hz and a peak sensitivity around 250 Hz [2,12,13]. Based on prior work, 250 Hz should therefore have the lowest baseline thresholds, with lower sensitivity and thus higher thresholds for 100 Hz and 400 Hz [2,14]. In the Index controller, baseline and movement thresholds indeed generally follow this pattern. In the Vive controller, thresholds increased with frequency, with 100 Hz being the most detectable. This might be explained by hardware differences as well, e.g. higher force at 100 Hz for the Vive actuator or resonance with the controller case. Future work could investigate the force output of each controller at different frequencies. Nonetheless, the 250 Hz widely used in prior work (e.g., [11,18,19]) appear to work well for the tested devices.

These findings also open up questions for future work. Notably, the lowest intensity tested in this study was 10^{-5}, yet quite a few participants had estimated thresholds below this level (cf. Fig. 2), indicating that they perceived some or all stimuli at the lowest level(s) presented. True thresholds thus might lie even lower for some participants. Another possibility is that steps in intensity near the lower end of the tested range might produce only little difference in the physical stimulus amplitude, or that the range of stimuli produced by the controller is limited by hardware capabilities. To address these questions, we are currently examining the physical stimulus produced across a large range of intensities to determine a usable span of values. Notably, the suppression effects found here are unlikely to be affected by this as we compared within-subject baseline and movement thresholds. Taken together, our findings pave the way for future tactile psychophysics studies in VR using wireless, off-the-shelf hardware.

References

1. Bays, P.M., Flanagan, J.R., Wolpert, D.M.: Attenuation of self-generated tactile sensations is predictive, not postdictive. PLoS Biol. 4(2), e28 (2006)
2. Bolanowski, S.J., Jr., Gescheider, G.A., Verrillo, R.T., Checkosky, C.M.: Four channels mediate the mechanical aspects of touch. J. Acoust. Soc. Am. 84(5), 1680–1694 (1988)
3. Broda, M.D., Fiehler, K., Voudouris, D.: The influence of afferent input on somatosensory suppression during grasping. Sci. Rep. 10(1), 1–11 (2020)
4. Brookes, J., Warburton, M., Alghadier, M., Mon-Williams, M., Mushtaq, F.: Studying human behavior with virtual reality: the unity experiment framework. Behav. Res. Methods 52, 1–9 (2019)
5. Buckingham, G., Carey, D.P., Colino, F.L., Degrosbois, J., Binsted, G.: Gating of vibrotactile detection during visually guided bimanual reaches. Exp. Brain Res. 201(3), 411–419 (2010)
6. Chapman, C.E., Beauchamp, E.: Differential controls over tactile detection in humans by motor commands and peripheral reafference. J. Neurophysiol. 96(3), 1664–1675 (2006)
7. Chapman, C., Bushnell, M., Miron, D., Duncan, G., Lund, J.: Sensory perception during movement in man. Exp. Brain Res. 68(3), 516–524 (1987)
8. Choi, S., Kuchenbecker, K.J.: Vibrotactile display: perception, technology, and applications. Proc. IEEE 101(9), 2093–2104 (2012)
9. Colino, F.L., Buckingham, G., Cheng, D.T., van Donkelaar, P., Binsted, G.: Tactile gating in a reaching and grasping task. Physiol. Rep. 2(3), e00267 (2014)
10. Fiehler, K., Brenner, E., Spering, M.: Prediction in goal-directed action. J. Vis. 19(9), 10 (2019)
11. Gertz, H., Voudouris, D., Fiehler, K.: Reach-relevant somatosensory signals modulate tactile suppression. J. Neurophysiol. 117(6), 2262–2268 (2017)
12. Gescheider, G.A., Wright, J.H., Verrillo, R.T.: Information-Processing Channels in the Tactile Sensory System: A Psychophysical and Physiological Analysis. Psychology press (2010)
13. Johnson, K.O., Yoshioka, T., Vega-Bermudez, F.: Tactile functions of mechanoreceptive afferents innervating the hand. J. Clin. Neurophysiol. 17(6), 539–558 (2000)
14. Jones, L.A., Sarter, N.B.: Tactile displays: guidance for their design and application. Hum. Factors 50(1), 90–111 (2008)
15. Manzone, D.M., Inglis, J.T., Franks, I.M., Chua, R.: Relevance-dependent modulation of tactile suppression during active, passive and pantomime reach-to-grasp movements. Behav. Brain Res. 339, 93–105 (2018)
16. Oldfield, R.C.: The assessment and analysis of handedness: the Edinburgh inventory. Neuropsychologia 9(1), 97–113 (1971)
17. Schütt, H.H., Harmeling, S., Macke, J.H., Wichmann, F.A.: Painfree and accurate Bayesian estimation of psychometric functions for (potentially) overdispersed data. Vis. Res. 122, 105–123 (2016)

18. Voudouris, D., Broda, M.D., Fiehler, K.: Anticipatory grasping control modulates somatosensory perception. J. Vis. **19**(5), 4 (2019)
19. Voudouris, D., Fiehler, K.: Dynamic temporal modulation of somatosensory processing during reaching. Sci. Rep. **11**(1), 1–12 (2021)
20. Wolpert, D.M., Flanagan, J.R.: Motor prediction. Curr. Biol. **11**(18), R729–R732 (2001)

Estimation of Frictional Force Using the Thermal Images of Target Surface During Stroking

Mitsuhiko Shimomura[1]($\boxtimes$) ⓘ, Masahiro Fujiwara[1] ⓘ, Yasutoshi Makino[1,2] ⓘ,
and Hiroyuki Shinoda[1] ⓘ

[1] The University of Tokyo, 7-3-1 Hongo, Bunkyo-ku, Tokyo 113-0033, Japan
shimomura@hapis.k.u-tokyo.ac.jp, Masahiro_Fujiwara@ipc.i.u-tokyo.ac.jp,
{yasutoshi_makino,hiroyuki_shinoda}@k.u-tokyo.ac.jp
[2] JST PRESTO, 7 Gobancho, Chiyoda-ku, Tokyo 102-0076, Japan

Abstract. We propose a method for estimating the frictional force between a contacted surface and the human touch using thermal video images captured using an infrared thermographic camera. Because this method can estimate force remotely, its application to various situations, in which the measurement is difficult to obtain using conventional contact-based methods, is expected. Furthermore, thermal images have the advantage of measuring physical quantities directly related to frictional force. As a result of machine learning using the measured data from multiple subjects and materials, we succeeded in estimating the frictional force with a high accuracy from the information of the temperature change on the surface. In addition, we account for both the frictional and direct heat transferred between the finger and object affecting the temperature change; therefore, we attempted to improve the accuracy by extracting only frictional heat. Consequently, our method succeeded in improving the accuracy.

Keywords: Friction · Thermal image · Machine learning

1 Introduction

We propose a method for estimating the frictional force between the surface of an object and a human finger using thermal video recorded by an infrared thermographic camera. The frictional force is estimated using machine learning, which accounts for the temperature of the surface changing because of the frictional heat present when a human strokes an object. Because the surface temperature of an object is measurable with a high spatial resolution using a non-contact method, such as an infrared thermographic camera, remotely estimating the force acting on various surfaces is possible.

Supported by JST PRESTO 17939983.

H. Seifi et al. (Eds.): EuroHaptics 2022, LNCS 13235, pp. 234–242, 2022.
https://doi.org/10.1007/978-3-031-06249-0_27

Specialized equipment is typically used to measure physical quantities related to tactile sense. For example, a strain gauge sensor and durometer are used for force and hardness measurements, respectively. Recently, methods have been proposed for estimating tactile information remotely based on visible images. Previous studies [3,4,7,8] have proposed a method for estimating the applied pressing force from images of fingernail using the change in fingernail color caused by the changing blood flow when a finger presses an object. Unlike conventional methods that use measuring instruments, these image-based methods only require cameras and are expected to be applied in various situations [6].

Previous studies have used thermal images instead of visible images to estimate tactile events. For example, Dunn et al. proposed a method of classifying the strength of pressure as strong or weak using a random forest based on thermal images [2]. Several studies have also attempted to extract tactile histories from thermal images [1,5].

Infrared thermographic camera-based methods have the following advantages over general cameras: (1) Because the strength of the friction force appears as a change in temperature from friction heat, an infrared thermographic camera can measure physical quantities near the force. (2) Sequential information is useable as spatial data because the touch history remains as the temperature changes. Hence, although superficial changes are unobserved in a visible image, the touch history can be captured in a thermal image.

This study propose a new technique for measuring tactile information remotely in a non-contact manner, that is, a method that estimates frictional force regressively using machine learning and thermal images. To verify the suitability of this method for general use, we examined the accuracy within a simple environment. Unlike previous methods, our estimates frictional force as a continuous quantity, not as a classification of force strength. This method can improve the means of information input. Moreover, the proposed method has the potential to measure force where conventional tactile sensors are difficult to install. For example, if we use a contact-type sensor to measure the force between touching skin, the tactile experience may change. However, we can measure the force without interference using a non-contact method.

2   Proposed Method

2.1   Problem Definition

Our method estimates frictional force using a thermal video in which the temperature change of a surface is recorded when touched by humans. We use the system shown in Fig. 1 to measure the training and validation data for machine learning. Each material is fixed to a force sensor that measures the frictional force when humans stroke it. While the contacted object is undeformed, a force sensor placed underneath the object can measure the frictional force exerted on the target surface. The thermographic camera is set above the target such that the lens and surface are parallel. The subjects draw a straight line left-to-right on the surface while changing the applied force. A thermographic camera and force sensor then simultaneously measure the temperature of the surface and frictional force, respectively.

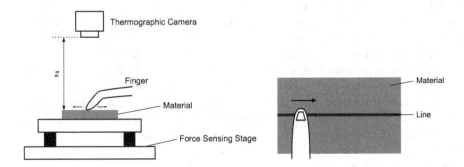

Fig. 1. An overview of the system. When subjects touch each material fixed on a force sensor, a thermographic camera and force sensor measure the temperature of the surface and frictional force, respectively.

2.2   Making Dataset

The measured thermal image sequence is used as input, and the force at the time as output; thus, creating a training dataset that associates the thermal image with force is necessary.

To begin, we preprocess each of the measured data. We crop thermal videos using a fixed window to retain only the area around the straight line traced by subjects. Because the spatial resolution of the thermographic camera we used is 640×480, the input thermal video was cut to the range of ± 64 pixels from the straight line; thus, 640×128 was used. Next, we smooth the force data to equalize the frame number of the force data and thermal video. The force sensor now measures at 120 fps, and the thermographic camera at 30 fps. Thus, the force sensor data is averaged every 4 frames.

We then extract the periods in which subjects touch objects. These are periods in which the vertical force is larger than the threshold of 0.1 N for longer than 6 frames (200 ms). We denote each period as I_n (see Fig. 2). In addition, each period I_n is divided every 6 frames (200 ms), and each is used as a segment S_i. We generate input and output data from each segment, as shown in Fig. 3. The input data are the six differential images in each segment obtained by subtracting the thermal image of the previous frame, and the corresponding label data are defined as the average value of the frictional force in the segment. The use of the subtracted images between frames reduces the effect of variations in the initial temperature of the surface. We show some examples of input and label data in Fig. 4.

When the frictional force is measured, the thermal image does not contain information beneath the finger. We assume that the tracing speed changes

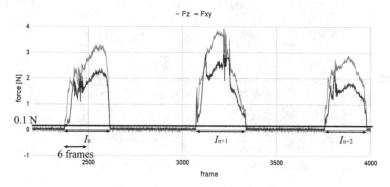

Fig. 2. An example of intervals $\{I_n\}_n$. Each period is longer than 6 frames (200 ms), and the vertical force is continuously larger than 0.1 N.

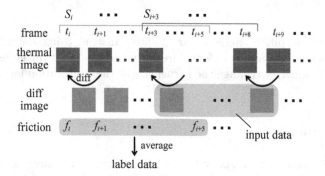

Fig. 3. An overview of dataset generation. Label data are the average value of friction force in a segment, and the corresponding input data are a set of differential thermal images in a time-delayed segment.

insignificantly, and, by setting a constant delay time, the thermal image and frictional force are made to correspond. That is, the thermal image information after 3 frames (100 ms) is used when the frictional force at frame t is estimated. This relies on the thermal image containing the contact history, and the frictional force before 3 frames (100 ms) can be estimated from the image after the finger moves.

2.3 Machine Learning Model

We use an efficient convolutional network model (ECO: the extended model of a basic convolutional neural network (CNN) model [9]) to estimate the frictional force from thermal images. Because the original ECO model is used for the classification problem, we modify the output layer for the regression problem. That is, in classification tasks, the probability of each class is output in the last

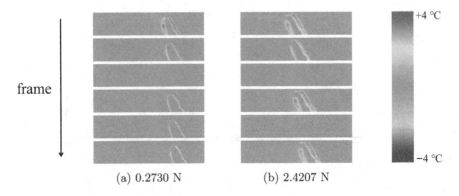

Fig. 4. These figures show examples of the input and label data used for learning: (a) small force and (b) large force. The input images represent a heat map of the line traced by the subject. Because the time resolution of the thermographic camera is not exactly 30 fps, some frames without change are green. (Color figure online)

layer; thus, we output a single scalar quantity instead using the affine layer. Moreover, because the size is different from that of the input image used in the original ECO model, we change the kernel size of the convolution layer near the input layer.

3   Experiment

Data were obtained from six subjects (three men and three women). We experimented with the system illustrated in Fig. 1 and adopted two materials: polystyrene foam and cardboard. We used a tactile force plate (Tech-Gihan, TF-2020) and infrared thermographic camera (Testo, testo 883) to measure the frictional force and temperature of the object surface, respectively. The force plate had an A/D converter with a 16 bit resolution that can measure up to 10 N. In addition, the camera resolution was 640 × 480 px. We set the thermographic camera 30 cm above the target surface. To prevent the average surface temperature from increasing when the subjects trace the surface, a fan blows air such that the surface temperature remains constant while uncontacted.

We synchronize the time between the thermal video captured by the thermographic camera and the force data measured by the tactile force plate using the following method. First, we placed a finger on an object and quickly swiped in a horizontal direction by momentarily adding force at the stationary state. Because the time resolution of the thermographic camera was 30 fps at most, a frame exists in which the position of the finger changed discontinuously. In contrast, the tactile force plate detected a horizontal impulse input. Hence, we treat these as the same time data and as synchronized. Because the tactile force plate measured forces at 120 fps, our dataset included time synchronization errors of up to 33 ms.

Subjects traced a 20 cm straight line left-to-right on each material while changing the applied pressing force, and we measured the frictional force and temperature change. Although the subjects were not required to use a precise stroke speed, they attempted to stroke each trace in about a second. All subjects traced each piece of material for 10 min, and the sensor output was recorded. We used 80% of the data obtained for training and 20% for testing. Three datasets of patterns were used for training and testing: (a) polystyrene foam and cardboard, (b) polystyrene foam, and (c) cardboard. The data was obtained only for patterns (b) and (c), whereas we mixed and used pattern (a).

In addition, we conducted another experiment to determine the effect of direct heat transfer between the finger and object. When humans touch, two factors change the temperature: frictional heat and heat transfer. However, the temperature change owing to heat transfer is primarily caused by finger contact time, which is independent of frictional force. Thus, we conducted the following experiment to determine if the accuracy improved by removing the heat transfer effect. Subjects wore a glove made of cotton and polyester to touch the object instead of touching it with their fingers. Wearing a glove creates a layer that insulates and reduces the effect of heat transfer between the finger and object. We performed the same experiment as without the glove and compared the results of both experiments.

4   Results and Discussion

The estimation results for each pattern in the skin and glove experiments are shown in Fig. 5 for all patterns. The horizontal axis shows the actual value of the frictional force, and the vertical axis the estimated value. This means that a point closer to the 45° diagonal line can be estimated with high accuracy. Table 1 lists the root mean square error (RMSE).

Compared to the results without gloves, the accuracy improved when wearing gloves for all patterns except (c), in which the accuracy deteriorates Therefore, this improvement may be owed to the removed heat transfer when wearing gloves. Possible reasons for accuracy deteriorating in (c) is the small friction coefficient between the cardboard and glove and the temperature change owing to frictional heat possibly being insufficiently large. Increasing the length of the segment may solve this. Because the heat transfer is slower than finger motions, further improvements in accuracy are expected by extending the segment length and observing the heat transfer more carefully.

In our experiment, the method is limited to a fixed path, but the same method may be applied to a path of arbitrary shape. For example, if we track the position of a finger and crop a thermal image around it, the subsequent procedure is the same as that of the present method. Furthermore, by modifying the model to simultaneously estimate the dynamic friction coefficient, the frictional and normal forces can be estimated, and the spatial pressure distribution is obtainable.

In both experiments, the results of pattern (a) are as accurate as those of patterns (b) and (c), which used only one material. This result suggests that

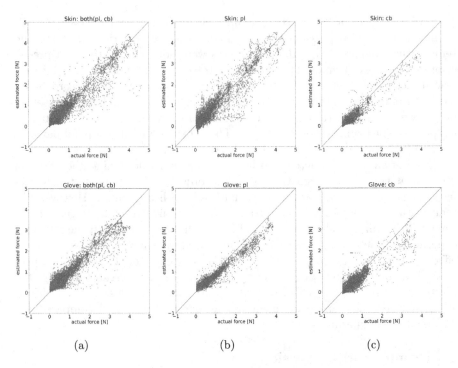

(a)   (b)   (c)

Fig. 5. Estimation results of the frictional force estimated from thermal images using machine learning: (a) polystyrene foam and cardboard, (b) polystyrene foam, and (c) cardboard.

the proposed method has a generalization performance with respect to tactile objects.

Points with a lower estimation accuracy farther from the straight line, are primarily distributed below the line, which means that they are estimated to be smaller than the actual values. This result is attributable to the biased distribution of the label data used for training. More data for smaller forces than larger ones exist in any given dataset; therefore, the model is likelier to estimate smaller forces than the actual values. To address this problem, we may improve our method in various ways, such as performing data augmentation to unify the distribution of data and weighing a loss function to equalize the influence from the data.

Table 1. RMSEs of each pattern in the skin and glove experiments.

	(a)	(b)	(c)
RMSE(Skin) [N]	0.2676	0.3679	0.1941
RMSE(Glove) [N]	0.2480	0.2562	0.2988

5   Conclusion

This paper proposes a method for estimating the frictional force between a finger and object using the temperature change of the surface when humans stroke it. As a result of the experiment using the data when 6 subjects touch two types of materials, we succeeded in estimating with an error of less than 10% of the data range. Because the number of materials used for the experiment was insufficient for general applications; thus, more objects with varying heat capacities must be tested. However, as a basic test of our proposed method for estimating the frictional force using thermal images, we showed this possibility.

In addition, we confirmed that the estimation accuracy improved in many cases by reducing the effect of direct heat transfer between the finger and object.

These results show the possibility of inputting information through finer force changes, contrasting the classification of force strength demonstrated by a previous study. In addition, we showed the possibility of measuring tactile information in locations where installing tactile sensors is challenging.

References

1. Brahmbhatt, S., Ham, C., Kemp, C., Hays, J.: Contactdb: analyzing and predicting grasp contact via thermal imaging. In: Proceedings of the IEEE/CVF Conference on Computer Vision and Pattern Recognition (2019)
2. Dunn, T., Banerjee, S., Banerjee, N.K.: User-independent detection of swipe pressure using a thermal camera for natural surface interaction. In: 2018 IEEE 20th International Workshop on Multimedia Signal Processing (MMSP), pp. 1–6. IEEE (2018)
3. Grieve, T., Lincoln, L., Sun, Y., Hollerbach, J.M., Mascaro, S.A.: 3D force prediction using fingernail imaging with automated calibration. In: 2010 IEEE Haptics Symposium, pp. 113–120. IEEE (2010)
4. Grieve, T.R., Hollerbach, J.M., Mascaro, S.A.: 3-D fingertip touch force prediction using fingernail imaging with automated calibration. IEEE Trans. Rob. **31**(5), 1116–1129 (2015)
5. Kishino, Y., Shirai, Y., Yanagisawa, Y., Ohara, K., Mizutani, S., Suyama, T.: Identifying human contact points on environmental surfaces using heat traces to support disinfect activities: poster abstract. In: Proceedings of the 18th Conference on Embedded Networked Sensor Systems, SenSys 2020, pp. 768–769. Association for Computing Machinery, New York (2020)
6. Marban, A., Srinivasan, V., Samek, W., Fernández, J., Casals, A.: A recurrent convolutional neural network approach for sensorless force estimation in robotic surgery. Biomed. Signal Process. Control **50**, 134–150 (2019)
7. Mascaro, S.A., Asada, H.H.: Measurement of finger posture and three-axis fingertip touch force using fingernail sensors. IEEE Trans. Rob. Autom. **20**(1), 26–35 (2004)
8. Yoshimoto, S., Kuroda, Y., Oshiro, O.: Estimation of object elasticity by capturing fingernail images during haptic palpation. IEEE Trans. Haptics **11**(2), 204–211 (2018)
9. Zolfaghari, M., Singh, K., Brox, T.: Eco: efficient convolutional network for online video understanding. In: Proceedings of the European Conference on Computer Vision (ECCV), pp. 695–712 (2018)

Spatial Resolution of Mesoscopic Shapes Presented by Airborne Ultrasound

Zen Somei[✉][iD], Tao Morisaki[✉][iD], Yutaro Toide[✉][iD],
Masahiro Fujiwara[✉][iD], Yasutoshi Makino[✉][iD], and Hiroyuki Shinoda[✉][iD]

The University of Tokyo, 5-1-5 Kashiwanoha, Kashiwa-shi,
Chiba-ken 277-8561, Japan
{somei,morisaki,toide}@hapis.k.u-tokyo.ac.jp,
Masahiro_Fujiwara@ipc.i.u-tokyo.ac.jp,
{yasutoshi_makino,hiroyuki_shinoda}@k.u-tokyo.ac.jp

Abstract. This study determined the spatial resolution of the virtual surface profile created by ultrasound haptic stimulation. We assumed the case where a finger moves along the surface of a virtual object. This object was produced by low-frequency lateral modulation that creates a pseudo static force. We defined the spatial resolution as the minimum distance required to discriminate between two virtual bumps. Several sensory channels are combined when a human feels the geometric features of a surface. This paper focuses on mesoscopic shapes, whose representative length ranges from a few millimeters to fingertip size. We considered two strategies to present mesoscopic shapes: changing either the contact position or the force strength. We measured the spatial resolutions in mesoscopic shapes created by each method, and discussed which factor is more effective to perceive mesoscopic features.

Keywords: Spatial resolution · Mesoscopic shapes · Ultrasound. · Midair haptics

1 Introduction

Among noncontact tactile presentation techniques using air jets [1], vortex rings [2], lasers [3], and ultrasound [4–6], the technique using ultrasound can reproduce several tactile sensations, including geometric features of surfaces. In this study, we assume the case where a finger moves along the surface of a virtual object and discuss the spatial resolution as the basic parameter. The base of this research is the pseudo-pressure presentation [7] recently demonstrated by Morisaki et al. Since static pressure sensation is generated by low-frequency lateral modulation (LM) [8], reproducing spatial unevenness by changing the presented pressure in synchronization with the finger movement is possible.

Multiple sensory signals are used for a human to perceive geometric features of a surface haptically. For example, when capturing a *macroscopic* shape whose

---

Supported in part by JST CREST JPMJCR18A2.

H. Seifi et al. (Eds.): EuroHaptics 2022, LNCS 13235, pp. 243–251, 2022.
https://doi.org/10.1007/978-3-031-06249-0_28

curvature radius is larger than the finger size, the curvature is determined by synthesizing the trajectory and surface angle of the perceived finger using proprioceptors and cutaneous receptors, respectively, during the finger motion [9]. Meanwhile, high-frequency vibration induced by the hand's stroking motion is the main factor when sensing the *microscopic* features of the surface. A fine surface structure whose representative length is smaller than the depth of the mechanical receptors in the skin or the spacing between the adjacent fingerprint ridges can induce overall vibration perceived using deep mechanoreceptors [10,11].

This paper focuses on *mesoscopic* shapes, whose representative length ranges from a few millimeters to fingertip size. These shapes can be perceived using superficial mechanoreceptors as the spatial pattern. Here, we distinguish between these mesoscopic and microscopic sensations though the word *roughness* encompasses both. In noncontact tactile displays, it remains unclear to what extent ultrasound can present detailed mesoscopic shapes, since it has a relatively large wavelength of 8.5 mm in 40 kHz in the air, which is the typical frequency of the current midair haptics.

In related studies, Matsubayashi et al. [12,13] proposed a method for displaying object shapes in a finger pad. However, these studies were limited to cases where the virtual objects vertically contact the skin. The surface roughness has already been considered in midair haptics studies [14,15]. Nevertheless, since static pressure sensation cannot be reproduced, presenting static shapes is difficult except for some special situations [16].

This study considers two basic strategies for presenting mesoscopic shapes. One is the contact position change (CPC) method, in which we change the center of the contact area between the surface shape and fingertip synchronously with the fingertip position. The other is the contact strength change (CSC) method, in which we change the sum of the forces using the fingertip position. This CSC method is similar to the approach proposed by Howard et al. [17], however, both strategies focus on different scales. The CSC method focuses on mesoscopic shapes, whereas Howard et al.'s research focused on macroscopic shapes.

In real contact, both the position and intensity of the contact change, nonetheless, we intentionally present them separately to clarify the spatial resolution in both strategies.

2   Methods

2.1   Overview of the Methods

Figure 1 shows the overview of the CPC and CSC methods. In the following experiments, the trajectory of the finger is on the horizontal plane, with no vertical motion. However, we assume that the finger moves on the uneven rigid body surface and estimate the contact point and contact force. In the calculation, we assume that the finger is a circular rigid body.

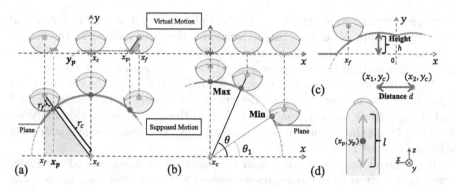

Fig. 1. Overview of the shape presentation. (a): Contact position change (CPC) method and (b): Contact strength change (CSC) method. (c): Method to present two adjacent convex surfaces. (d): Illustration of the LM linear movement modulation.

The CPC method, shown in Fig. 1(a), reproduces the surface irregularity by presenting the contact point estimated from the finger position, keeping the contact force constant. It is unclear whether the tactile stimuli reproduced using the CPC method expresses unevenness. However, since the tactile stimulus reproduced by the CPC method was perceived as "soft and easily deformable convex objects" in the preliminary experiment, it was considered a stimulus to express unevenness. When the skin traces the surface of a rigid object, the contact pressure changes according to the unevenness. Additionally, the height of the finger changes up and down. However, when the stimulus is presented by the CPC method, the presenting pressure is assumed to be constant and the trajectory of the finger is along the horizontal plane. In such a situation, the presented target surface is perceived as a very soft and easily deformable object.

Meanwhile, Fig. 1(b) shows a method for calculating the contact pressure using the CSC method. Here, the contact pressure is given proportional to the height of the contact point. In the case of contact with a real object, both the contact position and pressure change. In other words, by using CPC and CSC together appropriately, can reproduce slight irregularities on the surface of a rigid body. In this research, our scope excludes tactile reproduction, but we will examine how fine the unevenness pattern can be expressed by changing the position and intensity of the 40 kHz ultrasonic focus.

2.2   Contact Position Change Method

The focus position in the CPC method is given as follows. Assume the finger center is at $x = x_f$ and the ultrasound focus position is expressed as $(x, y) = (x_p, y_p)$. As shown in Fig. 1(a), (x_p, y_p) is obtained

$$x_p = \frac{r_c}{r_c + r_f} \cdot (x_f - x_c) + x_c, \quad y_p = r_f - \sqrt{r_f^2 - (x_f - x_p)^2}$$

from geometric considerations, where the radii of curvature of the finger and virtual object are r_f and r_c, respectively, and the x-axis center of the virtual convex surface is x_c. Here, we assume $r_f = 7$ mm. Moreover, we set r_c at 5, 15, 25, and 35 mm, respectively, to investigate the virtual convex surfaces.

Two adjacent convex surfaces are presented in Fig. 1(c). The contact point is calculated as follows: First, we determine the centers of the circles (x_1, y_c) and (x_2, y_c), circle radius r_c, and height h from the base plane. In this study, we fixed $h = 5$ mm throughout the experiments. Also, two peaks are symmetrically arranged around $x = 0$.

If the finger position x_f is negative or positive, it is considered the finger contacts the left or right circles, respectively. The contact point is only one and calculated as explained above, although, in real time, the finger might touch multiple points on the surface. The tactile stimuli reproduced using the CPC method in the preliminary experiment were perceived as two bumps.

2.3 Contact Strength Change Method

The CSC method is shown in Fig. 1(b). We assumed that the total contact force P is proportional to the height of the contact point, which is given as

$$P = \begin{cases} \frac{\theta'}{\pi/2} \cdot P_M & (\theta' \leq \pi/2) \\ -\frac{\theta'}{\pi/2} \cdot P_M & (\theta' > \pi/2) \end{cases},$$

where P_M is the Max Pressure, the maximum force strength of the device, and

$$\theta' = (\theta - \theta_1) \cdot \frac{\pi/2}{\pi/2 - \theta_1}.$$

where θ is the angle indicating the contact point between the finger and virtual object, while θ_1 corresponds to the minimal force strength when the finger touches the plane.

In the CSC method, the calculated pressure is applied to the finger pad center. Similar to the CPC method, the same parameters were used in the CSC-method experiments as $r_f = 7$ mm, $h = 5$ mm, and r_c at 5, 15, 25, and 35 mm, respectively.

2.4 Ultrasound Focus Point Presentation Method

In previous explanations, we simply used the term "contact pressure." In midair ultrasound tactile presentation, however, one problem is that the stimulus is too weak to perceive when a time-constant radiation pressure is applied to a fixed point on the skin. To avoid this problem, we used a low-frequency LM so that the stimulus is perceived even when the finger is stationary.

Recent studies have shown that LM modulation at low frequencies 10 Hz produces pseudo-pressure sensation [7]. Thus, the finger can feel pressure sensations with distinct spatial localization. In this study, we used the linear movement

LM as shown in Fig. 1(d). The focus is oscillated sinusoidally at f [Hz] along the finger at $z = l \sin(2\pi f t)$, where z is the focus position along the finger. We set $l = 1.2$ cm and $f = 15$ [Hz]. The number of points per cycle in the LM vibration was set to 50 to avoid unnecessary vibration.

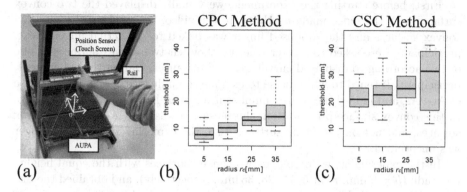

Fig. 2. Overview of the experiment. (a) is the photograph of the system. (b) and (c) are the results of the experiments by (b): CPC method and (c): CSC method.

3   Experiment

To clarify the threshold for discriminating convex surfaces, we developed a system and conducted experiments as follows.

3.1   System

Figure 2(a) shows the presentation system consisting of an infrared touch screen (ITS), GreenTouch, GT-IRTK156-1, 15.6 in., and an airborne ultrasound phased array (AUPA) [18].

ITS is a position sensor device that locates the two-dimensional position coordinates of a finger (or something else) using infrared transmitting and receiving tubes densely placed on the four sides of the screen. For this system, it was used to obtain the position coordinate in the moving direction of the finger (i.e., the x-axis direction in Fig. 2(a)). Meanwhile, AUPA is a device that creates an ultrasound focal point at an arbitrary three-dimensional position by controlling the amplitude and phase of each ultrasound transducer arranged in an array. Here, six AUPAs were used to form the focal point using the proposed methods described in Sect. 2.

The AUPA and ITS were synchronized by using virtual reality development environment (Unity 2020.3.13f1, produced by Unity Technologies) to put the ultrasound focus on the finger using the finger coordinates captured by the ITS. In addition, users were required to put their fingers on a linear rail to maintain a constant finger height.

3.2   Procedure

The following procedure was performed in the CPC and CSC methods. The procedure follows the declaration of Helsinki (2013), and participants provided written informed consent before the study.

First, before starting the experiment, we visually displayed the two convex surfaces with the video made by Unity. In the video, we prepared the assumed convex surfaces and the supposed finger was linked to the actual finger. Next, participants were asked to answer whether they felt two convex surfaces as one or two according to displayed curvatures and distances. During the experiment, participants listened to white noise to exclude the influence of the driving noise of AUPAs, and judged the convex number using their haptics without the video. In this convex surface presentation, the sensation was presented on the convex surfaces, and not on the plane surface. Also, the exploring speed of the finger was unspecified.

In the experiment, we prepared two convex surfaces with the equal heights and radii ($h = 5$ mm, r_c at 5, 15, 25, 35 mm, respectively), and obtained the discrimination threshold of the two convex surfaces using the 1-up, 1-down staircase method for each radius. In the beginning, the distance between the centers of both convex surfaces (henceforth, distance d, see Fig. 1(c)) was set to 0 mm. We asked participants to answer whether they felt one convex surface or two convex surfaces within 3-laps exploration. We changed the distance d by 2 mm according to their answers. If the participants felt one convex surface, d was increased until they could feel two convex surfaces. In contrast, if the participants felt two convex surfaces, d was decreased until they could feel one convex surface. We repeated this process seven times. The first reversal was ignored in subsequent analyses and the remaining six were averaged to obtain an estimate of the discrimination threshold. In this experiment, we had one measure per trial block. The presentation of the different radii was ordered randomly. The boundary of the two convex surfaces remained at the center of the rail (see Fig. 2(a)).

Participants consisted of ten males and two females, with a mean age ($\pm$ standard deviation) of 25.3 ± 1.2 yrs. All participants had touched the ultrasound in the past, but this was the first time using their fingertips. Before the experiment, subjects were asked to touch the surfaces with the video for them to understand the procedure. Since there were no videos during the experiment, this preparation did not develop the ability to discriminate thresholds well.

3.3   Results

Figure 2(b) and (c) show the experimental results. The discrimination thresholds of each participant obtained by the staircase method were summarized in a box plot for each radius. Among the participants, one participant always felt one convex surface even when the distance was over 50 mm at all radii in the CSC method, and another participant felt both convex surfaces even though the distance was 0 mm $r_c = 35$ mm of the CSC method. These discrimination thresholds were excluded from the box plots.

Comparing the CPC and CSC methods, we find that the discrimination threshold in the CPC method was generally smaller. In both methods, the smaller the radius of the convex surface, the smaller the discrimination threshold. The minimum threshold throughout the experiment was obtained at $r_c = 5$ mm in the CPC method, with a mean threshold of 8.0 mm and a standard deviation of 2.7 mm. Participants' comments included: "The CSC method was more difficult to feel," "Sometimes the number of convex surfaces was judged symbolically," and "It was more difficult to understand the change from two convex surfaces to one convex surface than from one convex surface to two convex surfaces." Although we did not specify the speed of the fingers, the measured speed ranged from about 20 to 100 [mm/s].

3.4 Discussion

Comparing the thresholds for both the CPC and CSC methods in each radius, the CPC method had a smaller discrimination threshold than the CSC method at all radii. Therefore, participants were identified the number of convex surfaces more easily using the CPC method. As the authors' subjective comment, the CPC method provided a clearer sensation of the soft and fine bump profile especially when the height h was lower than about 5 mm (see Fig. 1(c)).

A reason for the lower effectiveness of the CSC method is the weakness of the ultrasonic force (1.6 gf/cm^2 with 324 transducers [5]). Due to its small maximum strength, it was difficult to understand the difference when the intensity varied. This was also evident in the comments. In particular, when the radius was small in the CSC method, the stimulus occurred at a short horizontal distance, making it more difficult to judge. Therefore, some participants made logical judgments such as whether there was a stimulus difference, rather than report how they felt the curved surface, especially at $r_c = 5$ and 15 mm. Upon improving the presentation power in the future, the CSC method will become more effective and the threshold can be lowered.

However, the CPC method was effective as a method for transmitting mesoscopic features. In particular, the minimum threshold of 8.0 ± 2.7 mm obtained at $r_c = 5$ mm showed that we can discriminate the spatial pattern on a surface even if the distance d (see Fig. 1(c)) was within the diameter of a finger (about 14 mm) using the CPC method. As the authors' subjective comment, "the fine profile was naturally perceived though it feels very soft."

4   Conclusion

In this study, we discussed and evaluated the spatial resolution of mesoscopic shapes created using ultrasonic LM stimulation, in the situation where a finger moves along the surface of a virtual object. As an indicator of the spatial resolution, we used the minimum distance between two bumps by which we can distinguish them.

Both CPC and CSC methods were used to display the surface pattern. In each procedure, the contact position and the force strength were, respectively, controlled according to the finger position. The experimental results show that the CPC method more effectively conveyed mesoscopic features than the CSC method. In addition, it obtained the spatial resolution (minimum discriminable bump distance) of 8.0 mm with a standard deviation of 2.7 mm.

References

1. Suzuki, Y., Kobayashi, M.: Air jet driven force feedback in virtual reality. IEEE Comput. Graph. Appl. **25**(1), 44–47 (2005)
2. Wang, G.-Z., Huang, Y.-P., et al.: Bare finger 3D air-touch system using an embedded optical sensor array for mobile displays. J. Disp. Technol. **10**(1), 13–18 (2014)
3. Kim, H.-S., Kim, J.-S., et al.: Evaluation of the possibility and response characteristics of laser-induced tactile sensation. Neurosci. Lett. **602**, 68–72 (2015)
4. Iwamoto, T., Tatezono, M., Shinoda, H.: Non-contact method for producing tactile sensation using airborne ultrasound. In: Ferre, M. (ed.) EuroHaptics 2008. LNCS, vol. 5024, pp. 504–513. Springer, Heidelberg (2008). https://doi.org/10.1007/978-3-540-69057-3_64
5. Hoshi, T., Takahashi, M., et al.: Noncontact tactile display based on radiation pressure of airborne ultrasound. IEEE Trans. Haptics **3**(3), 155–165 (2010)
6. Carter, T., Seah, S.A., et al.: UltraHaptics: multi-point midair haptic feedback for touch surfaces. In: Proceedings of 26th Annual ACM Symposium User Interface Software Technology, pp. 505–514 (2013)
7. Morisaki, T., Fujiwara, M., et al.: Non-vibratory pressure sensation produced by ultrasound focus moving laterally and repetitively with fine spatial step width. IEEE Trans. Haptics (2021)
8. Takahashi, R., Hasegawa, K., et al.: Tactile stimulation by repetitive lateral movement of midair ultrasound focus. IEEE Trans. Haptics **13**(2), 334–342 (2020)
9. Wijntjes, M.W.A., Sato, A., et al.: Local surface orientation dominates haptic curvature discrimination. IEEE Trans. Haptics **2**(2), 94–102 (2009)
10. Lederman, S.J.: The perception of surface roughness by active and passive touch. Bull. Psychon. Soc. **18**(5), 253–255 (1981). https://doi.org/10.3758/BF03333619
11. Howe, R.D., Cutkosky, M.R.: Sensing skin acceleration for slip and texture perception. In: Proceedings of 1989 ICRA, pp. 145–150 (1989)
12. Matsubayashi, A., Oikawa, H., et al.: Display of haptic shape using ultrasound pressure distribution forming cross-sectional shape. In: Proceedings of IEEE WHC, pp. 419–424 (2019)
13. Matsubayashi, A., Makino, Y., Shinoda, H.: Rendering ultrasound pressure distribution on hand surface in real-time. In: Nisky, I., Hartcher-O'Brien, J., Wiertlewski, M., Smeets, J. (eds.) EuroHaptics 2020. LNCS, vol. 12272, pp. 407–415. Springer, Cham (2020). https://doi.org/10.1007/978-3-030-58147-3_45
14. Freeman, E., Anderson, R., et al.: Textured surfaces for ultrasound haptic displays. In: Proceedings of ICMI 2017, pp. 491–492 (2017)
15. Beattie, D., Frier, W., et al.: Incorporating the perception of visual roughness into the design of mid-air haptic textures. In: Proceedings ACM SAP 2020, Article No. 4, pp. 1–10 (2020)
16. Inoue, S., Makino, Y., et al.: Active touch perception produced by airborne ultrasonic haptic hologram. In: Proceedings of the 2015 IEEE WHC, pp. 362–367 (2015)

17. Howard, T., Gallagher, G., et al.: Investigating the recognition of local shapes using mid-air ultrasound haptics. In: 2019 IEEE WHC, pp. 503–508 (2019)
18. Suzuki, S., Inoue, S., et al.: AUTD3: scalable airborne ultrasound tactile display. IEEE Trans. Haptics **14**, 740–749 (2021)

Haptic Applications

Vibrotactile Similarity Perception in Crowdsourced and Lab Studies

Ramzi Abou Chahine[1], Dongjae Kwon[2], Chungman Lim[2], Gunhyuk Park[2], and Hasti Seifi[3]($\boxtimes$)

[1] San Francisco State University, San Francisco, CA 94132, USA
rabouchahine@mail.sfsu.edu
[2] Gwangju Institute of Science and Technology, Gwangju 61005, South Korea
{snoopy9502,chungman.lim}@gm.gist.ac.kr, maharaga@gist.ac.kr
[3] University of Copenhagen, 2100 Kobenhavn, Denmark
hs@di.ku.dk

Abstract. Crowdsourcing can enable rapid data collection for haptics research, yet little is known about its validity in comparison to controlled lab experiments. Furthermore, no data exists on how different smartphone platforms impact the crowdsourcing results. To answer these questions, we conducted four vibrotactile (VT) similarity perception studies on iOS and Android smartphones in the lab and through Amazon Mechanical Turk (MTurk). Participants rated the pairwise similarities of 14 rhythmic VT patterns on their smartphones or a lab device. The similarity ratings from the lab and MTurk experiments suggested a very strong correlation for iOS devices ($r_s = 0.9$) and a lower but still strong correlation for Android phones ($r_s = 0.68$). In addition, we found a stronger correlation between the crowdsourced iOS and Android ratings ($r_s = 0.78$) compared to the correlation between the iOS and Android data in the lab ($r_s = 0.65$). We provide further insights into these correlations using the perceptual spaces obtained from the four datasets. Our results provide preliminary evidence for the validity of crowdsourced VT similarity studies, especially on iOS devices.

Keywords: Crowdsourcing · Haptics · Similarity perception

1 Introduction

Designing vibrotactile (VT) icons has been the subject of much research in the last decades. VT patterns can help convey abstract information (e.g., message urgency) or emotions, provide cues for navigation, or correct user movement in physical rehabilitation and skill-training scenarios [5]. To identify salient VT parameters or design a set of distinct vibrations for a use case, haptics researchers often ask users to rate or group VT patterns according to their

This work was supported by the National Research Foundation of Korea (NRF) grant funded by the Korea government (MSIT) (NRF-2021R1C1C1008147).

similarity [3,6,7,9]. For example, Park and Choi evaluated pairwise similarities among one carrier and seven amplitude-modulated sinusoidal vibrations and showed that the envelope waveform is a salient parameter for VT design [7]. Ternes and MacLean assessed the perceptual structure of 84 rhythmic VT patterns by asking 6 experienced users to group the patterns according to their similarity [9]. They found that the length of VT pulses, their evenness, and frequency defined the pattern similarity. Because in-lab perceptual similarity studies are time consuming and expensive, researchers often use a small stimuli set and/or few users.

Online crowdsourcing platforms such as Amazon Mechanical Turk (MTurk) allow running large-scale perception studies at a fraction of time and cost [4], but little is known about how the results differ from in-lab experiments in haptics. Haptic studies often need specialized and/or calibrated hardware and focused attention which are lacking in crowdsourced settings. While smartphones increasingly have vibration actuators with improved rendering fidelity, the diversity of phone actuators [1], materials, internal make-up, and software can lead to differences in VT perception [8]. Despite these challenges, a few crowdsourced VT studies have been reported in recent years. The first study by Schneider *et al.* compared lab and crowdsourced ratings for 10 VT signals rendered on a voice-coil actuator and Android phones [8]. Android ratings for duration, roughness, and pleasantness were statistically equivalent to the voice-coil data for the majority of the patterns, but energy, speed, and roughness ratings showed mixed results. Demers *et al.* recently proposed a probabilistic method for efficient sampling of the VT engineering space by using the similarity data collected from Android users on MTurk [2]. These studies highlight the need for crowdsourcing in haptics, yet they do not assess the validity of crowdsourced data for perceptual similarity assessment tasks. Also, both studies focus on Android devices.

In this paper, we investigate two questions: 1) How comparable are VT similarity ratings obtained from crowdsourced and lab studies? and 2) How comparable are results from Android and iOS smartphones given their distinct hardware and software for producing VT stimuli? In contrast to typical VT similarity studies, we do not aim to identify the underlying VT perceptual parameters. Instead, we focus on testing the consistency of the data obtained from different device platforms and experiment settings.

We conducted online and lab experiments with iOS and Android devices to answer the above questions. Forty-eight participants rated the pairwise similarities of 14 VT patterns varying in rhythm and amplitude in four experiments (12 participants each). Pairwise similarities collected from MTurk and lab studies were highly correlated (RQ1), but iOS data showed stronger correlation ($r_s = 0.9$, $p < 0.001$) than Android results ($r_s = 0.68$, $p < 0.001$). Furthermore, iOS and Android data were highly correlated with each other (RQ2). To provide qualitative insights into these correlations, we obtained the perceptual spaces from the four experiments using dimensionality reduction techniques. Our results suggest that VT amplitude and even rhythms were perceived consistently across the four experiments, but perception of uneven VT rhythms varied in Android devices. Recruiting iOS users was faster than Android users which suggests that large-scale studies on iOS devices are viable. We discuss our findings and present future directions for haptic crowdsourcing.

2   Stimuli and Apparatus

We designed 14 VT patterns and iOS and Android applications for our studies.

VT Patterns. We used a subset of the VT patterns proposed by Ternes and MacLean [9]. They designed 21 rhythms consisted of 62.5 ms, 125 ms, and 375 ms pulses. They also modulated the VT frequency (200, 300 Hz) and amplitude (full, half) to obtain 84 patterns. Each rhythm was 500 ms long and repeated four times in a 2-second stimulus. They divide the rhythms into even and uneven groups depending on whether each part of the pattern feels regularly repeating (e.g., R1) or not (e.g., R21).

From the aforementioned patterns, we selected seven perceptually different rhythms with different pulse lengths and evenness (Fig. 1) in their reported perceptual space [9]. We modulated the amplitude for each rhythm to be the devices' upper amplitude limit or half of it. Because Android vibration API does not offer any frequency modulation options, we measured an Android phone's average vibration frequency for the patterns ($\mu = 159.34$ Hz; Samsung Galaxy S10) and matched a similar frequency on the iOS patterns ($\mu = 149.87$ Hz).

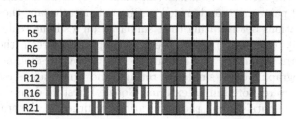

Fig. 1. Each row represents a 2-second VT pattern in our study. Gray and white denote vibration and silence. The red lines mark the 500-ms rhythms.

Android and iOS Applications. We designed functionally identical applications on iOS and Android with four sequential screens of initial, training, VT comparison, and submission. The initial screens provided the consent information and collected the participant's age, gender, nationality, and use of phone cover. The user also rated their haptics expertise on a 5-point scale from no experience to expert. Next, the applications instructed the user to remove their phone case, hold the phone with their left hand, and use their right index finger to interact with it. The training screen showed 14 buttons randomly assigned to

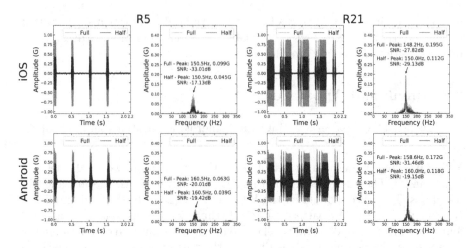

Fig. 2. Acceleration plots of exemplary VT patterns (R5 and R21) played by iPhone 11 Pro Max and Samsung Galaxy S10, and their power spectrums.

the 14 VT patterns. The user could press the VT buttons to feel the assigned vibrations, and the applications enabled proceeding to the VT comparison screen only when all the buttons were pressed at least once. The VT comparison screen allowed users to evaluate the similarity of 92 unique pairwise VT patterns. 91 pairs were selected from 14 patterns each compared once ($_{14}C_2$), and we added one pair that compared two identical vibrations (R1) as an attention test. The order of the 92 pairs were randomized per user. For each pair, the user had to play both vibrations before rating their similarity on a sliding scale of 0 (*totally different*) to 100 (*the same*). After the VT comparison screen, the user could provide free-form comments and submit the results by pressing a submit button. The demographic data, similarity ratings, and time stamps collected from the experiment were stored in an external Google Firestore database.

To reduce the data's noise, the applications returned to the training screen if the user spent more than 40 min in the training and main screens. Also, the applications did not proceed from the initial screens if the user's phone did not support the vibration API specifications in our study.

3   Similarity Rating Data Collection

We ran four between-subject experiments with a total of 42 participants, using two platforms (*iOS, Android*) and two experimental settings (*crowdsourced,lab*).

Crowdsourced Experiment. We recruited 27 users (14 iOS, 13 Android) on Amazon Mechanical Turk. To be eligible, MTurk users must have completed 5000 or more tasks on the platform with a success rate of 98% or greater. The experiment took 10–20 min and we offered 3 US\$s in compensation. Data from

3 participants (2 iOS, 1 Android) who provided a rating of less than 70 on the attention test were excluded, resulting in 12 participants per platform. The included participants were 24–70 years old (35.4 ± 12 years, 6F/6M) on iOS and 26–43 years old (33.4 ± 4.5 years, 3F/9M). The participants were from US (10 iOS, 6 Android), India (5 Android), and Brazil (2 iOS, 1 Android). All the recorded device models were unique, except for Galaxy S10+, iPhone XR, and iPhone 11 Pro Max that were used twice.

Lab Experiment. We recruited 25 participants (13 iOS, 12 Android) at Gwangju Institute of Science and Technology through an e-mail and flyers posted on digital and physical bulletins. No participants reported tactile disorder, and each participant read an instruction document, filled out a consent form, and used the same applications to complete the task as the MTurk participants did. The experiment took 30 min on average, and we offered 10 USD as compensation. We excluded data from one iOS participant who failed the attention test, resulting in 12 participants per platform. The included users were 23–31 years old (26.2 ± 2.73 years, 6F/6M) on iPhone 11 Pro Max and 20–27 years old (24.6 ± 1.75 years, 5F/7M) on Galaxy S10 platform. No phone case was used in the study. The participants listened to pink noise on headphones to block any sounds generated from the VT patterns.

4   Results

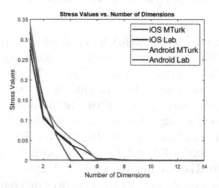

Fig. 3. Stress values of non-metric MDS over 14 dimensions.

We converted the similarity ratings into dissimilarity values and averaged over participants in each of the four conditions. Because the data is ordinal, we applied Spearman's Rank Correlation Coefficient (Spearman's ρ) to compare the platforms and experimental settings. We use non-metric multidimensional scaling (nMDS) to visualize the perceptual spaces. These methods have been used to compare perceptual spaces from pairwise similarity rating tasks [10].

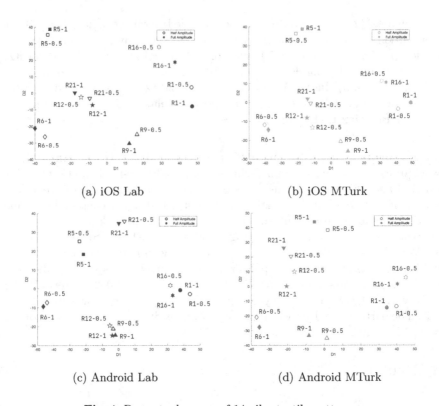

(a) iOS Lab

(b) iOS MTurk

(c) Android Lab

(d) Android MTurk

Fig. 4. Perceptual spaces of 14 vibrotactile patterns.

In the nMDS analysis, 3–4 dimensions adequately capture the perceptual dissimilarities of the VT stimuli as in the stress plots (Fig. 3), but we plot the perceptual spaces in 2D for better visibility (Fig. 4). The perceptual spaces from the four experiments show similar trends. One clear trend is that VT rhythm is more salient than the amplitude; the two amplitudes of the same rhythm are always close to each other. Also, the four perceptual spaces show similar VT clusters and configurations. Specifically, R1 and R16 consistently appear close to each other, whereas R5 and R6 are mapped far apart. Similarly, R9 appears equidistant to R1-R16 and R6 while R21 appears in between R5 and R6. Only R12 tends to notably vary its location over different conditions. Below, we analyze the VT similarities in relation to our research questions (RQ1, RQ2).

4.1   RQ1. How Comparable Are VT Similarity Ratings Obtained from Crowdsourced and Lab Studies?

For the iOS platform, the Spearman ρ correlation between the averaged pairwise dissimilarities in the crowdsourced and lab datasets show a very strong correspondence ($r_s = 0.90, p < 0.0001$). Also, the two perceptual spaces are nearly identical with minor variations in the proximity of the VT stimuli. The only notable difference is that the distance between R9 and R12-R21 is higher in the lab experiment. For the Android platform, the correlation of the lab and crowdsourced dissimilarities is strong ($r_s = 0.68, p < 0.0001$). The two perceptual spaces show overall alignment with some inconsistencies; R9, R12, and R21's relationship vary across the two spaces. The Android MTurk participants perceived R12 and R21 to be similar and different from R9, whereas the lab participants perceived R9 and R12 to be very similar and R21 to be dissimilar to the pair. The strong correlations between the lab and online settings and the conformity of the perceptual spaces, especially with iOS devices, suggest that MTurk similarity ratings can yield comparable results to the lab data.

4.2 RQ2. How Comparable Are Results from Android and iOS Smartphones Given Their Distinct VT Hardware and Software?

To answer this question, we compare Android and iOS data for each experimental setting (MTurk, lab). For the MTurk setting, the Spearman correlation is strong ($r_s = 0.78, p < 0.0001$) between iOS and Android data. We observe comparable configurations between the two perceptual spaces. The main difference is that Android participants found R12 and R21 to be closer to R5 and farther from R9 compared to the iOS participants. Also, the iOS ratings for R12 and R21 are closer to R6 than the Android ratings. For the lab setting, the Spearman correlation is strong ($r_s = 0.65, p < 0.0001$) for the two datasets. The perceptual spaces of iOS and Android show some differences. The Android users rated R9 and R12 as highly similar and regarded R21 dissimilar to the pair. In contrast, the iOS users perceived R12 and R21 as notably similar and R9 as different from the two stimuli. Also, the Android lab data shows a higher similarity between R1 and R16 than iOS lab data.

5 Discussion and Conclusion

We investigate the correspondence between data obtained from lab and crowdsourced VT studies. Thus, instead of analyzing the underlying perceptual parameters of vibrations, we discuss what vibration parameters are robust under variations in device platform and experimental settings. In all the four studies, rhythm dominates over amplitude in VT similarity evaluations, with participants rating patterns with the same rhythm as very similar despite the modulation of their amplitudes. The lengths of VT pulses influence the configuration of the perceptual space, with patterns with short pulses (e.g., R5) appearing distant from longer pulses (e.g., R6, R9). Also, the patterns with mixed pulse lengths (R12, R21) tend to appear between those with long and short pulses. Finally, even vibrations (e.g., R1, R16) cluster together in all perceptual spaces, whereas uneven vibrations are away from the pair.

iOS data from crowdsourcing and lab studies were better aligned than Android data. The similarity of iOS results could be attributed to the software, hardware, and build parity in iPhones. In contrast, the diversity of Android devices may have contributed to the lower correspondence in the Android's perceptual spaces. Upon comparing the stimuli generated by our iOS and Android lab devices, we noticed that the Android device's VT fall-off time varied for different rhythm patterns and it was nearly twice as long as that of the iOS device for some patterns (e.g., R21). This variation may have caused the disparities in the Android lab's perceptual space. These disparities were lower in the crowd-sourced Android data, possibly due to the washout effect from various devices. Surprisingly, the recruitment time for the iOS crowdsourced study was almost half the time of the crowdsourced Android experiment. This difference could be due to the Android version requirements in our experiment. Nevertheless, our data suggests that crowdsourcing on iOS devices is viable on MTurk.

Our research has a number of limitations. First, we did not modulate the frequency of the stimuli due to Android's inability to render this parameter. Second, our studies have a small sample size due to difficulties of recruiting lab participants during the COVID-19 pandemic. Third, we chose a subset of the stimuli from Ternes and MacLean's work [9] as the number of pairwise comparisons would increase rapidly with additional patterns. Lastly, the age range of participants across the experiments varied, which may have affected the results.

Our work shows promising results for crowdsourcing perceptual similarity studies on iOS and Android devices. We hope that future work further establishes the validity of data obtained from haptic crowdsourcing with different sets of VT stimuli and perceptual tasks.

References

1. Choi, S., Kuchenbecker, K.J.: Vibrotactile display: perception, technology, and applications. Proc. IEEE **101**(9), 2093–2104 (2012)
2. Demers, M., Fortin, P.E., Weill, A., Yoo, Y., Cooperstock, J.R., et al.: Active sampling for efficient subjective evaluation of tactons at scale. In: Proceedings of IEEE World Haptics Conference, pp. 1–6. IEEE (2021)
3. van Erp, J.B., Spapé, M.M., et al.: Distilling the underlying dimensions of tactile melodies. In: Proceedings of Eurohaptics, pp. 111–120. Springer, Heidelberg (2003)
4. Heer, J., Bostock, M.: Crowdsourcing graphical perception: using mechanical turk to assess visualization design. In: Proceedings of ACM SIGCHI Conference on Human Factors in Computing Systems, pp. 203–212 (2010)
5. Jones, L.A., Sarter, N.B.: Tactile displays: guidance for their design and application. Human Fact. **50**(1), 90–111 (2008)
6. Jones, L.A., Singhal, A.: Perceptual dimensions of vibrotactile actuators. In: Proceedings of IEEE Haptics Symposium, pp. 307–312 (2018)
7. Park, G., Choi, S.: Perceptual space of amplitude-modulated vibrotactile stimuli. In: Proceedings of IEEE World Haptics Conference, pp. 59–64. IEEE (2011)
8. Schneider, O.S., Seifi, H., Kashani, S., Chun, M., MacLean, K.E.: Hapturk: crowdsourcing affective ratings of vibrotactile icons. In: Proceedings of ACM SIGCHI Conference on Human Factors in Computing Systems, pp. 3248–3260. ACM (2016)

9. Ternes, D., MacLean, K.E.: Designing large sets of haptic icons with rhythm. In: Ferre, M. (ed.) EuroHaptics 2008. LNCS, vol. 5024, pp. 199–208. Springer, Heidelberg (2008). https://doi.org/10.1007/978-3-540-69057-3_24

10. Vardar, Y., Wallraven, C., Kuchenbecker, K.J.: Fingertip interaction metrics correlate with visual and haptic perception of real surfaces. In: Proceedings of IEEE World Haptics Conference, pp. 395–400. IEEE (2019)

Perception of Spatialized Vibrotactile Impacts in a Hand-Held Tangible for Virtual Reality

Pierre-Antoine Cabaret[1]([✉]), Thomas Howard[1], Claudio Pacchierotti[2], Marie Babel[1], and Maud Marchal[1,3]

[1] Univ Rennes, INSA Rennes, IRISA, Inria, CNRS, Rennes, France
{pierre-antoine.cabaret,thomas.howard,marie.babel,
maud.marchal}@irisa.fr
[2] CNRS, Univ Rennes, Inria, IRISA, Rennes, France
claudio.pacchierotti@irisa.fr
[3] Institut Universitaire de France (IUF), Paris, France

Abstract. Informative and realistic haptic feedback significantly enhances virtual reality (VR) manipulation. In particular, vibrotactile feedback (VF) can deliver diverse haptic sensations while remaining relatively simple. This has made it a go-to solution for haptics within hand-held controllers and tangible props for VR. However, VF in hand-helds has solely focused on monolithic vibration of the entire hand-held device. Thus, it is not clear to what extent such hand-held devices could support the delivery of spatialized information within the hand. In this paper, we consider a tangible cylindrical handle that allows interaction with virtual objects extending beyond it. This handle is fitted with a pair of vibrotactile actuators with the objective of providing in-hand spatialized cues indicating direction and distance of impacts. We evaluated its capability for rendering spatialized impacts with external virtual objects. Results show that it performs very well for conveying an impact's direction and moderately well for conveying an impact's distance to the user.

1 Introduction and Related Work

Vibrotactile feedback (VF) is a popular haptic feedback modality for virtual reality (VR) interaction because it combines relatively low technological complexity in its implementation with a wide variety of achievable haptic effects [5]. There is evidence for a positive impact of VF on many success metrics for interactions with virtual environments (VEs), such as improved task performances [2], improved user immersion [2], increased perceived realism [17] and increased presence [6,12]. VF can be used to communicate both physical cues relating to the VE (e.g. vibrating objects [16], contacts [4], impacts [8], interaction forces [3], texture roughness [7]) as well as abstract cues (e.g. for guidance, notification

This project has received funding from the EU Horizon 2020 program, grant agreement No 801413, project "H-Reality"; and from the Inria Défi project "DORNELL".

H. Seifi et al. (Eds.): EuroHaptics 2022, LNCS 13235, pp. 264–273, 2022.
https://doi.org/10.1007/978-3-031-06249-0_30

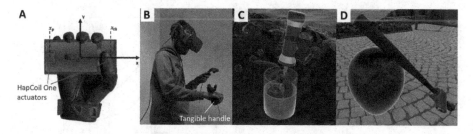

Fig. 1. (A) Close up CAD of the tangible handle being held by the user's avatar; (B) User manipulating the handle in VR; (C, D) VR interactions causing impacts at different distances and in different directions from the hand.

or communication [5]). Many technologies can deliver VF in VR, such as wearable [5], grounded [19], hand-held [1], and even mid-air haptic devices [10].

Despite the wide use of VF delivered through ungrounded hand-held devices in VR, the approach is mostly restricted to monolithic VF [5]. This has the advantage of simplicity as it requires only a single actuator. However it remains inadequate for providing spatial information. That is, cues originating from different directions relative to the user are identical and thus indiscriminable for the user, unless a mapping is created between direction and waveform parameters.

Conversely, localized VF through multiple actuators is widely used to convey spatial information to the user, in particular in wearables (e.g. [5,11,12]) and surface haptics (e.g. [15]). In this paper, we begin to explore the possibilities offered by localized VF within handheld tangible objects. In particular, we focus on rendering spatialized impacts happening on a virtual hand-held object larger than the tangible held by the user. To render the impacts, the tangible houses two vibrotactile motors at its extremities. We hypothesize that by using two actuators, we can provide localized vibrotactile feedback which can inform the user about where the impact occurred on the larger virtual object they are manipulating (see Fig. 1).

In early work on impact rendering in interactions within VEs, Wellman et al. [21] used a data-driven approach to play back recorded impact vibrations during virtual contacts on a voice-coil actuator embedded into a grounded force-feedback device handle. Okamura et al. expanded on this, compiling a vibration waveform library for impacts generated by fitting a simplified vibration model based on an exponentially decaying sinusoid to recorded impact data [14]. Because this model (see Sect. 2.2) provided an interesting compromise between perceived realism, impact property discrimination, and computing requirements, it has since been widely adopted in interactions with VEs [13,19]. Some work on spatialization in VR was performed by Gongora et al. [9]. They studied vibrotactile impacts delivered in a bimanual task using a pair of monolithic handheld vibrotactile devices, with the aim of rendering localized vibrotactile impacts along a virtual bar connecting both hands. There have also been a few research attempts at systems spatializing vibrotactile cues inside hand-held devices using multiple vibrotactors [18] or asymmetric vibrations [20] but to our knowledge none have been leveraged in VR interactions.

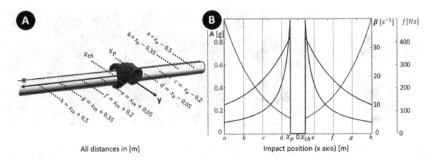

Fig. 2. (A) Virtual rod manipulated in the experiment with 4 possible impact distances extending symmetrically around the virtual hand. x_{th} and x_p respectively denote the thumb and pinkie side actuator positions. (B) Possible evolution of vibration amplitude A, decay β and frequency $f = 2\pi\omega$ as a function of impact distance for both actuators. Values were determined based on literature and a pilot study. We do not consider any impact occurring within the hand, hence the null values between x_{th} and x_p.

2   Experimental Design

2.1   Research Questions and Hypotheses

We seek to provide VF to render impacts between one manipulated virtual object and other virtual objects in a VE. Our question concerns the extent to which spatializing impact cues by distributing them between two actuators embedded in a cylindrical tangible handle (see Fig. 1-B) is effective in providing users with information on impact direction. We also seek to understand how this approach affects perceived realism and impact properties, and whether it is compatible with existing approaches to rendering impact distance in a setup using a single actuator (e.g. [9,19]). To investigate this, we compare distance and direction discrimination performances, as well as perceived realism and virtual object material properties in VR, using different impact vibration models (see Table 1). We formulate the following hypotheses:

H1 Spatialization of impacts in hand by assigning impact waveforms to distinct vibrotactors will allow discrimination of impact direction, regardless of the chosen impact vibration model.
H2 Impact models coding distance with more redundant parameters (see Sect. 2.2 for the details of the models) will yield better distance discrimination performance.

2.2   Rendering Impacts Distance and Direction

We use the simplified impact vibration model introduced by Okamura et al. [14], where $\alpha(x,t)$ denotes the waveform amplitude at the instant t for an impact at a distance x from the hand (see Fig. 2): $\alpha(x,t) = A(x)e^{-\beta(x)t}sin(\omega(x)t)$. In realistic impacts, the peak amplitude A, decay β and angular frequency ω would all

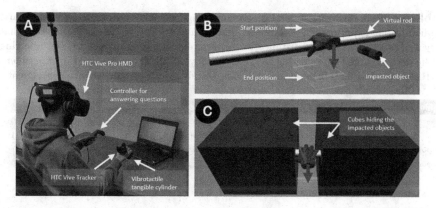

Fig. 3. (A) A subject performing the experiment. (B) VR view of the familiarization task, where the impacted objects are visible. (C) VR view of the actual task, where the impacted objects are hidden and only haptic feedback of impacts is provided.

be functions of impact distance as well as impact dynamics and properties of the materials involved. However, such impact models can sometimes be less effective at communicating usable information on impact distance [19]. An alternative is to select a subset of model parameters (A, β, ω) to encode impact distance, possibly leaving the remainder free for encoding other impact properties. Given these three parameters, there are seven different possibilities (see Table 1) for encoding impact distance (see Fig. 2).

2.3   Materials and Methods

To investigate the formulated hypotheses, we designed a pair of experiments assessing impact direction and distance perception in VR.

Hardware. Subjects sat at a table, wearing an HTC Vive Pro head-mounted display (HMD). They held the vibrotactile handle in their dominant hand which was tracked using an HTC Vive Tracker attached using an adhesive fixture to keep the palm and inside of the fingers unobstructed. They used an HTC Vive Controller held in their non-dominant hand to answer experimental questions (see Fig. 3-A). The handle was equipped with a pair of symmetrically mounted Actronika HapCoil One voice-coil actuators (see Fig. 1-A).

Experimental Task. The common experimental task for both experiments was inspired from Sreng et al. [19]. Subjects were asked to hold the tangible handle in their dominant hand. They observed the VE showing their virtual hand holding a virtual rod with the same diameter as the tangible handle but extending symmetrically outward 0.5m beyond the edges of the tangible handle. By moving this virtual rod up and down, it could impact a lightweight and unconstrained object at one of four distances $d_i = 0.05, 0.2, 0.35, 0.50$ m from

Table 1. Impact vibration models for encoding impact distance x from the hand studied in our experiments. The model names indicate the vibration parameters that vary as a function of impact distance, with *Amp* referring to amplitude A, *Dec* referring to the decay β and *Freq* referring to the frequency ω. (Left) Models used in experiment 1; (Right) Models used in experiment 2; *AmpDecFreq* was common to both experiments.

Model	No. of parameters encoding distance	Equation	Model	No. of parameters encoding distance	Equation
Amp	1	$a(x,t) = A(x)\, e^{-25\,t} \sin(2\pi\,300\,t)$	Freq	1	$a(x,t) = 0{,}6\, e^{-25\,t} \sin(\omega(x)\,t)$
Dec	1	$a(x,t) = 0{,}6\, e^{-\beta(x)\,t} \sin(2\pi\,300\,t)$	AmpFreq	2	$a(x,t) = A(x)\, e^{-25\,t} \sin(\omega(x)\,t)$
AmpDec	2	$a(x,t) = A(x)\, e^{-\beta(x)\,t} \sin(2\pi\,300\,t)$	DecFreq	2	$a(x,t) = 0{,}6\, e^{-\beta(x)\,t} \sin(\omega(x)\,t)$
AmpDecFreq	3			$a(x,t) = A(x)\, e^{-\beta(x)\,t} \sin(\omega(x)\,t)$	

either the thumb or the pinkie side of the hand (see Fig. 2-B,C). These impacts were rendered according to one of the impact models summarized in Table 1. During the experiment, the impacted object was obstructed from view so as to provide no visual feedback of the impact location (see Fig. 3-C). Subjects placed the rod at the starting location, then were prompted to move it downward. On the way down, the stick impacted a first virtual object which appeared randomly on the left or right at one of the distances d_i. Upon reaching the target location, subjects were prompted to return the stick to the start location and repeat the process. A second object appeared on the same side as the first, at one of the four possible distances, generating a second impact, after which subjects answered a pair of experimental questions:

Q1 Which side did the impacts occur on? (Left/Right)
Q2 Was the second impact further away from the hand than the first? (Y/N)

Experimental Design and Protocol. For achieving a shorter experiment, we split our investigation into two identical experiments containing 4 blocks each. Impacts were rendered respectively using *Amp, Dec, AmpDec, AmpDecFreq* for experiment 1 and *Freq, AmpFreq, DecFreq, AmpDecFreq* for experiment 2 (see Table 1). 24 subjects (19 m., 5 f., ages 21–30 (Mean:24.9 y), 20 right-handed) took part in the study after providing written informed consent. Subjects were randomly assigned to one of the two experiments.

In each experiment, subjects first performed a familiarisation task where the VE was fully visible, showing the hand-held virtual stick and the impacted virtual objects (see Fig. 3-B). During this task we ensured that subjects moved at a similar speed, though the vibration did not depend on it. They were informed that the rod and impacted object properties might vary during the course of the subsequent experiment. Subjects filled out an initial questionnaire indicating personal data and prior experience with haptics, VR and perception studies.

Each experiment contained one block per impact model, whose order was counterbalanced between subjects. Within each block, subjects performed 3 repetitions of the task for each of the 16 combinations of impact distances occurring on either side, totalling 96 trials presented in a fully random order. Post-block

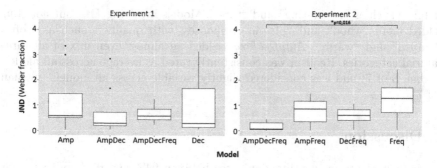

Fig. 4. JNDs for impact distance, expressed as Weber fractions, for both experiments.

questionnaires assessed perception of the stick and impacted object's material and geometric properties, their variability, and perceived impact realism.

3   Results

Impact directions were consistently correctly identified between 94% and 97% of the time across all impact models. Most errors occurred for pairs of low amplitude and duration stimuli.

To test for H2, subjects were assigned, for each impact model, to one of two groups based on whether they interpreted the impact model as intended (increased impact distance perceived as an increased impact distance) or in an inverted manner (increased impact distance perceived as a decreased impact distance). Inversion rates (percentage of subjects interpreting an increase in impact distance as a decrease) were around 50% for all models not involving *Freq*, and varied between 92% and 100% for all models involving *Freq*.

We then computed the 75%-just-noticeable-difference (JND) for distance discrimination as a Weber fraction for each subject by fitting cumulative Gaussians to the data. Finally, we compared the distribution of JNDs across impact vibration models (see Fig. 4). Data from the experiment 1 (Fig. 4-B) were not normally distributed, and a Friedman test showed no significant differences between conditions. Data from the experiment 2 were normally distributed, and a 2-way ANOVA showed a significant effect of impact model ($F(3) = 4.132$, $p = 0.021$) but no significant differences between participants. A post-hoc Tukey HSD test revealed the only significant difference to lie between the JNDs for the *Freq* and *AmpDecFreq* conditions ($p = 0.016$).

Rod properties were rated most consistent (median 2 of 7) in all conditions but *Amp* (median 3 of 7) and *AmpDecFreq* (median 4 of 7), however none of these differences were significant. The properties reported as changing between trials were rod material (*Freq, AmpDec, AmpDecFreq*), stiffness (all models except *Dec*), length (*Dec, AmpDec, AmpFreq*), fill (*Dec, AmpDec*), weight (*Freq, AmpDecFreq*). Subjectively reported rod materials were dominated by "metal" and "plastic" for all models involving *Amp*, as well as the *Dec* model, with qualifiers such as "resonating" and "tube". Models involving *Freq* but not *Amp* yielded more "wood" and "plastic" responses, with qualifiers such as "soft", "damped" and "warm". *AmpDecFreq* yielded an almost even mix of all three material categories. Realism was consistently rated as average across all models (median 3 of 7) and was considered slightly variable across all models (median 3 of 7).

4   Discussion

The impact direction identification rates between 94% and 97% indicate that regardless of the chosen impact model, spatializing the impacts between two actuators allowed subjects to correctly and intuitively identify the side on which the impact occurred with a high degree of accuracy. Hypothesis H1 is therefore verified. Looking at inversion rates, it is interesting to note that all models involving *Freq* tended to be systematically inverted (92% to 100% of subjects perceived an increase in distance as a decrease) which would indicate the evolution of ω may be the cause for this.

Weber fractions for distance discrimination were consistently high across all impact models except *AmpDecFreq* (m = 0.17), ranging from 0.6 (*DecFreq*, experiment 2) to 1.32 (*Amp*, experiment 1). This indicates that while distance discrimination was mostly possible, it was far from an easy task. The only statistically significant difference observed (*Freq-AmpDecFreq*, experiment 2) is in favor of hypothesis H2, and the mean JNDs seem to also support the hypothesis. However, given the poor performance of *AmpDecFreq* in experiment 1 and the fact that none but one of the differences are statistically significant, we cannot conclude that H2 is supported. This may be due to H2 being wrong, or to flaws in the stimulus or experimental task design. If H2 is not verified, there may be a lot of headroom to encode various impact properties by distributing them across different parameters without adversely impacting performance.

The high inversion rates due to using ω as a parameter led us to hypothesize that models combining ω with A, β or both may have been confusing to half the subjects that did not invert their interpretation of *Amp* and *Dec*. This hypothesis cannot be easily tested because subjects that performed *Amp* and *Dec* did not perform *AmpFreq* and *DecFreq*. Yet, analysing the results from experiment 1 revealed that 6 out of 12 subjects had inverted both *Amp* and *Dec* while 5 of 12 had not (the remaining subject inverted only one of both models). By looking at the JNDs for each of these groups of subjects in the *AmpDecFreq* condition, we note that the group that inverted both *Amp* and *Dec* performed better at *AmpDecFreq* (JNDs: 0.07 to 1.96, mean = 0.83) than the group that did not invert *Amp* and *Dec* (JNDs: 2.15 to 9.33, mean = 4.63). This would tend to support our interpretation and argue for the need to redesign the function $\omega(x)$ in our rendering approach. Furthermore, it may be necessary to consider the frequency dependence of vibration amplitude perception in such a redesign. However, given the very small sample size, this conclusion must be seen as tentative.

The spread in JNDs between subjects indicates a large inter-subject variability in the ability to perform the task. During the experiment, several subjects noted that the task was really difficult until they "chose" a way to understand the mapping of the stimuli to impact distance. Thus, we believe this variability shows that subjects displayed different capacities for adapting to the difficulty of the experimental tasks and choosing an effective response strategy. This means that the haptic representation of impact distance is far from intuitive or natural with the chosen models, although *AmpDecFreq* shows some promise in experiment 2. This may indicate poor model design, or the fact distance discrimination is really hard without any context such as e.g. visual feedback of impacts.

All models were perceived as equally (un)realistic, indicating that either the impact model used is unrealistic, that spatialization impacted realism, or both.

5   Conclusion and Perspectives

We presented an investigation into the use of spatialized in-hand vibrotactile feedback for VR interactions. Our study focused on the ability of a handle equipped with two vibrotactors to deliver realistic, discriminable and understandable sensations of impacts through which users could determine the location (direction and distance) of impacts on a virtual manipulated object.

We determined direction identification scores, JNDs for impact distance, and perceived impact realism for 7 impact vibration models. Results showed excellent direction identification performances, but distance discrimination performances were mediocre. Impacts were perceived as only moderately realistic, which may be due to the impact models studied as well as our spatialization technique.

In future work, we plan to investigate whether differently distributing the vibrations between both actuators can improve perceived realism and consistency between impacts while preserving distance and direction discrimination performance. This study also highlighted certain avenues for improving the perception of vibration impact models which we intend to investigate. Finally, we plan to extend the approach to 2D and 3D spatialization using more actuators.

Because of the good direction discrimination performance observed in this initial study, we believe there is potential for using multiple actuators in manipulated tangible objects or controllers. This seems particularly promising for VR applications which could benefit from the use of in-hand directional cues.

References

1. Adilkhanov, A., et al.: Vibero: vibrotactile stiffness perception interface for virtual reality. IEEE RAL **5**(2), 2785–2792 (2020)
2. Brasen, P.W., Christoffersen, M., Kraus, M.: Effects of vibrotactile feedback in commercial virtual reality systems. In: Brooks, A.L., Brooks, E., Sylla, C. (eds.) ArtsIT/DLI -2018. LNICST, vol. 265, pp. 219–224. Springer, Cham (2019). https://doi.org/10.1007/978-3-030-06134-0_25
3. Cheng, L.T., et al.: Vibrotactile feedback in delicate virtual reality operations. In: Proceeding ACM ICM, pp. 243–251 (1997)

4. Chinello, F., et al.: A three revolute-revolute-spherical wearable fingertip cutaneous device for stiffness rendering. IEEE Trans. Haptics **11**(1), 39–50 (2017)

5. Choi, S., et al.: Vibrotactile display: perception, technology, and applications. Proc. IEEE **101**(9), 2093–2104 (2012)

6. Cooper, N., et al.: The effects of substitute multisensory feedback on task performance and the sense of presence in a virtual reality environment. PloS One **13**(2), e0191846 (2018)

7. Culbertson, H., et al.: The penn haptic texture toolkit for modeling, rendering, and evaluating haptic virtual textures. Tech. Rep. (2014)

8. García-Valle, G., et al.: Evaluation of presence in virtual environments: haptic vest and user's haptic skills. IEEE Access **6**, 7224–7233 (2017)

9. Gongora, D., Nagano, H., Konyo, M., Tadokoro, S.: Experiments on two-handed localization of impact vibrations. In: Hasegawa, S., Konyo, M., Kyung, K.-U., Nojima, T., Kajimoto, H. (eds.) AsiaHaptics 2016. LNEE, vol. 432, pp. 33–39. Springer, Singapore (2018). https://doi.org/10.1007/978-981-10-4157-0_6

10. Howard, T., et al.: Pumah: pan-tilt ultrasound mid-air haptics for larger interaction workspace in virtual reality. IEEE Trans. Haptics **13**(1), 38–44 (2019)

11. de Jesus Oliveira, V., et al.: Designing a vibrotactile head-mounted display for spatial awareness in 3d spaces. IEEE TVCG **23**(4), 1409–1417 (2017)

12. Kaul, O.B., Meier, K., Rohs, M.: Increasing presence in virtual reality with a vibrotactile grid around the head. In: Bernhaupt, R., Dalvi, G., Joshi, A., K. Balkrishan, D., O'Neill, J., Winckler, M. (eds.) INTERACT 2017. LNCS, vol. 10516, pp. 289–298. Springer, Cham (2017). https://doi.org/10.1007/978-3-319-68059-0_19

13. Kuchenbecker, K., et al.: Improving contact realism through event-based haptic feedback. IEEE TVCG **12**(2), 219–230 (2006)

14. Okamura, A., et al.: Vibration feedback models for virtual environments. In: Proceeding IEEE ICRA, vol. 1, pp. 674–679 (1998)

15. Pantera, L., et al.: Multitouch vibrotactile feedback on a tactile screen by the inverse filter technique: vibration amplitude and spatial resolution. IEEE Trans. Haptics **13**(3), 493–503 (2020)

16. Passalenti, A., et al.: No strings attached: force and vibrotactile feedback in a virtual guitar simulation. In: Proceeding IEEE VR, pp. 1116–1117 (2019)

17. Peng, Y., et al.: Walkingvibe: reducing virtual reality sickness and improving realism while walking in vr using unobtrusive head-mounted vibrotactile feedback. In: Proceeding CHI, pp. 1–12 (2020)

18. Ryu, D., et al.: T-hive: vibrotactile interface presenting spatial information on handle surface. In: Proceeding IEEE ICRA, pp. 683–688 (2009)

19. Sreng, J., Lécuyer, A., Andriot, C.: Using vibration patterns to provide impact position information in haptic manipulation of virtual objects. In: Ferre, M. (ed.) EuroHaptics 2008. LNCS, vol. 5024, pp. 589–598. Springer, Heidelberg (2008). https://doi.org/10.1007/978-3-540-69057-3_76

20. Tappeiner, H., et al.: Good vibrations: asymmetric vibrations for directional haptic cues. In: Proceeding IEEE WHC, pp. 285–289 (2009)

21. Wellman, P., et al.: Towards realistic vibrotactile display in virtual environments. In: Proceeding ASME Dynamic Systems & Control Division, vol. 57, pp. 713–718 (1995)

Wearable Haptics in a Modern VR Rehabilitation System: Design Comparison for Usability and Engagement

Cristian Camardella[✉][iD], Massimiliano Gabardi[iD], Antonio Frisoli[iD], and Daniele Leonardis[iD]

Scuola Superiore Sant'Anna - Institute of Mechanical Intelligence, Pisa, Italy
{cristian.camardella,massimiliano.gabardi,antonio.frisoli,
daniele.leonardis}@santannapisa.it
https://www.santannapisa.it/en/institute-mechanical-intelligence

Abstract. Modern immersive virtual reality (VR) systems include embedded hand tracking, stand-alone and wireless operation, fast donning and calibration: these features are precious for usability of rehabilitation serious games in the clinical practice, envisaging also home-care applications. Can wearable haptics well integrate with the above features? Different designs result in a trade-off between wearability and richness of feedback. Yet, engagement of the user is also one of the key-features for rehabilitation serious games. We developed two novel fingertip devices aiming the first at lightweight and wearability, the second at rich and powerful cutaneous feedback. We compared the two designs in terms of usability and users' engagement within a modern rehabilitation system in immersive VR.

Keywords: Wearable · Haptics · Immersive · Virtual reality · Rehabilitation · Serious games

1 Introduction

Immersive Virtual Reality (VR) systems, and in particular Head Mounted Displays (HMD), have seen in the past decade an impressive progress in terms of quality of the visual feedback, involving precise head tracking, wider field of view and higher resolution of the rendered image. Even more, in the last couple of years certain off-the-shelf products (i.e. Oculus Quest 2) offer advanced characteristics that considerably improve usability and comfort of the user: they include wireless stand alone operation, simple donning of the system and fast calibration of the workspace, and embedded hand tracking. The latter is relevant

This work was supported by Project "TELOS - Tailored neurorehabilitation thErapy via multi-domain data anaLytics and adapative seriOus games for children with cerebral palSy", funded by Tuscany Region, Italy (CUP J52F20001040002).

H. Seifi et al. (Eds.): EuroHaptics 2022, LNCS 13235, pp. 274–282, 2022.
https://doi.org/10.1007/978-3-031-06249-0_31

for VR applications that include manipulation: the HMD alone becomes now enabled for immersive manipulation task without the need of any additional device. Together with the above features improving overall usability, such recent technology advances might significantly change the use of VR in certain critical applications, such as rehabilitation and serious games for educational and training in general [2,12,15]. In neurorehabilitation, VR systems have been proposed and experimented in the shape of serious-games, combining engagement of the user with task oriented motor exercises [5]. Virtualization of the exercise brings the advantage of intrinsic parametrization, adaptability and repeatability of the exercises.

Fig. 1. The 'Potions' virtual experience, played through the Oculus Quest 2 VR headset with embedded hands tracking and light wearable haptics

In a recent study, cutaneous feedback devices have been used into a prolonged pilot clinical study, integrated into a VR serious game for children with cerebral palsy [1]. Richness and intensity of the feedback versus lightweight and compactness of the device represents a common trade-off in the design of wearable haptic devices, especially in the design of fingertip haptic devices. In the last decade a lot of innovative fingertip haptic or multimodal devices have been proposed in the scientific literature [11,14]. Each device is though to render one or more specific feature of the interaction between object and fingerpad such as contact orientation [3,4,13], friction action [6,10], thermal transients [7] or high frequency vibrations due to texturized surface exploration [8,9]. In most of the cases wearability and user's comfort have been considered as secondary objectives. In this paper novel fingertip haptic devices, able to provide the user with tactile stimuli in a wide frequency range and specifically designed focusing on the usability, are in terms of comfort, wearability, usability in a modern VR rehabilitation system designed for both high immersivity and involvement of the patient (Fig. 1). The proposed highly wearable thimbles named "LightThimbles" are compared with more powerful fingertip haptic devices named "HapticThimbles" and designed for the rendering of contact no-contact transitions and tactile stimuli in the range of 0–350 Hz. The novelty of this work is given by the app-roach of understanding the trade-off between comfort and usability, and the richness of haptic feedbacks in a VR application. Specifically this trade-off is crucial when dealing with rehabilitation contexts, in which patients often strug-gle to achieve even simple hand or fingers movements and the comfort become essential for delivering the therapy. As a preliminary study, devices evaluation was performed with healthy subjects only.

2   Materials and Methods

In this section we describe design of the two haptic devices used in the experi-ment, the developed serious game in immersvie VR, and the experimental pro-cedure.

2.1   Device 1: The Light Haptic Thimble

A light and thin haptic thimble has been developed to enhance lightweight, compact size and wearability. We focused on rendering high frequency compo-nents and fast transients of the haptic feedback, while eliminating static and slow force components. This allows to reduce power requirements and size of the actuators, at the cost of limited rendering capabilities (i.e. no out of contact transition and no static grasping force). A miniaturized custom-made electro-magnetic voice coil (outer diameter 12 mm, output force 0.4 N) was implemented due to the high quality and wide frequency response of this type of actuators. The above design choices allowed other advantages improving wearability: the moving part of the actuator (1° of freedom) is designed to be kept always in contact with the fingerpad through a soft and compliant structure of the whole thimble (no calibration phase is needed). The thimble is fabricated in soft resin (Photocentric UV DLP Flexible) making it adaptable to different finger sizes. Stereo-lithography 3D printing method allows for a high degree of customiza-tion. The thimble shape increases flexibility and compliance of the frontal part, supporting the embedded voice coil, and of the two lateral brackets adapting to different finger size. Total mass of the device is just 7 g (Fig. 2).

2.2   Device 2: The Haptic Thimble

The HapticThimble has been designed with focus on both the quality and rich-ness of the feedback provided to the user. The device has the same actuated DoF of the LightThimble but the HapticThimble allows a 4 mm stroke of the contact plate performing also the rendering of contact no-contact transitions between fingerpad and virtual object. The moving part is thought to be lightweight and it moves on low friction plastic bushings in order to increase the maximum feed-back frequency allowed (i.e. 350 Hz). Moreover, the compact custom designed voice coil placed on top of the device, is able to provide the user also with con-stant forces. To increase the users comfort the device does not constraints the

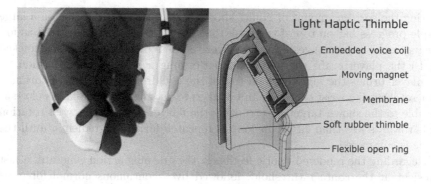

Fig. 2. The novel Light Thimble haptic device with design focused at lightweight and compactness.

finger distal joint, in fact it is thought to be fastened on the middle phalanx of the finger, whereas the actuated part, placed on the distal phalanx, is connected to the fastened part by means of a revolute flexure hinge coaxial with the finger joint axis. A small contact area between the actuated part of the device and the fingertip guarantee that the two moves together. Finally, the part of the device covering the user's finger has been white painted in order to better perform with the vision based markerless tracking system (Fig. 3).

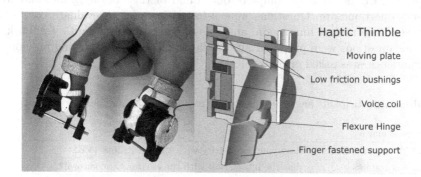

Fig. 3. The novel HapticThimble with design aimed at richness and high quality contact feedback.

2.3 The Virtual Reality Serious Game: Potions

The virtual reality experience is part of a serious game developed in our laboratory for neurorehabilitation of children with cerebral palsy. It takes place into an alchemist laboratory where the player is asked to prepare a magic potion, mixing several ingredients in a cauldron (Fig. 1). The game requires the player to perform parametrized pick-and-place and prono-supination motor tasks, in order to pour the right ingredients into the cauldron. The game asks for precision in pouring the correct amount of liquid, shown by a green reference line on a large purple progress bar on top of the cauldron. The game loop is structured as follows: a) a red arrow suggests to the player the ingredient that has to be picked up, b) the player grabs the ingredient performing a hand-pinching motion, c) the player brings the ingredient on top of the cauldron, d) starts to pouring it after rotating the wrist over 80°, and e) tries to get the progress bar as close as possible to the shown target, interrupting the flow decreasing the wrist rotation angle towards less than 70°. This loop is repeated for five ingredients, until the recipe is completed.

Regarding the rendered haptic feedback, the grasping action generates a fast transient at the contact threshold, followed by a continuous normal force, as long as the object is grasped. The pouring phase adds a continuous vibration, synthesized from the audio signal of a pouring liquid.

2.4   Experimental Setup and Protocol

Nine subjects (3 females), age 29.40 ± 3.59 years, height 1.74 ± 10.43 m have been enrolled in this study and signed an informed consent before joining the experiment. The study was approved by the Joint Ethical Committee of the Scuola Superiore Sant'Anna and Scuola Normale Superiore of Pisa, Italy. Each subject experienced three gaming sessions in random order, one for each experimental condition: hands-free (HF), wearing the Light Thimble (HS), and wearing the Haptic Thimble (HB). Each session required the subject to wear the Oculus Quest 2 VR headset, to complete one round of the Potions game, and to answer a questionnaire. Questions were selceted in order to evaluate comfort and ease of interaction, immersion, and role of the visual and haptic feedback. The same questions were proposed after each condition. Q1, regarding previous VR experience, was asked once (Table 1).

Table 1. Participants answered questions from Q2 to Q9 after each condition.

Questionnaire	
Q1 VR experience	Rate your familiarity with Virtual Reality (VR)
Q2 Naturalness	Was the interaction with the environment natural?
Q3 Hand coherency	Was the pose of the real and virtual hands coherent?
Q4 Ease of the task	Was the task easy to accomplish ?
Q5 Comfort	Rate the comfort of the wearable equipment
Q6 Immersivity	Did you feel immersed in the virtual experience?
Q7 Virtual environment	How much the 3D environment enhanced the experience?
Q8 Haptic feedback	How much the haptic feedback enhanced the experience?
Q9 Overall experience	How would rate the overall experience?

Regarding objective results, time duration and the pouring precision were chosen as performance metrics. The time duration included the overall time spent to accomplish each session. The pouring precision is the difference between the reference and the poured amount, averaged over all the five potions used in the session. Data were analyzed using the Kruskal-Wallis non-parametric statistical method with Dunn's post-hoc test.

3   Results

Results showed noticeable differences among the three conditions, reflecting the expected trends in both in-game recorded data and questionnaires, although not all the differences were statistically significant. For the former case, Time and Pouring Precision performance metrics were analyzed and Kruskal-Wallis test found a statistical difference among the three groups for both metrics (see Fig. 4). For Time, Dunn's test reported a significant difference between HF and HS groups (p = .0040) and HF and HB groups (p = .0004) but not among HS and HB groups (p = .7984). For Pouring Precision, a significant difference was found among HF-HB groups (p = .0369) but not among HF-HS (p = .1755) and HS-HB groups (p = .7763).

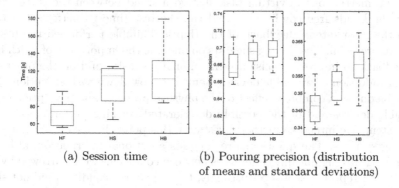

(a) Session time   (b) Pouring precision (distribution
                              of means and standard deviations)

Fig. 4. In-game performance metrics graphs across the three experimental conditions

For what concerns questionnaire results, Kruskal-Wallis test did not found any significant difference among conditions. Answwers to the questionnaire are reported in Fig. 5 for each experimental condition. The mean value of the Q1 (VR experience) for the participants was indicating prevalence of naive subjects.

4   Discussion and Conclusions

With respect to the past, modern VR systems noticeably improve wearability and dexterity to the user, adding wireless connectivity and embedded hand tracking. These features are of great advantage in certain applications, such as virtual serious games for neurorehabilitation. In this work we evaluated two different haptic thimble designs (plus the hands-free condition) in terms of both

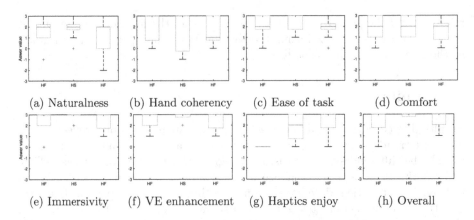

(a) Naturalness (b) Hand coherency (c) Ease of task (d) Comfort

(e) Immersivity (f) VE enhancement (g) Haptics enjoy (h) Overall

Fig. 5. Questionnaire answers distribution among all subjects

effectiveness of the haptic feedback and perception of the immersive VR equipment and environment. In terms of quantitative measurements, the Time and Precision metrics of the virtual task showed a trend between the three conditions: the hands free condition scored the shortest time to perform the task, while the most intense feedback of the Haptic Thimble performed the best in terms of precision. The finding suggests on one side the importance of a rich haptic feedback to improve precision in manipulation tasks, and on the other side how the addition of wearable devices seemed to slow down task execution. This might be due to a combined effect of the physical presence and perception of the wearable devices and of the possibly deteriorated tracking performance of the VR system. We intend to further investigate this point, by means of redundant tracking systems. The questionnaire suggests additional information, although not statistically significant. Differences between conditions were narrowed by the overall high reported scores Surprisingly, the hands-free condition did not show noticeably higher score in questions regarding comfort, naturalness and ease of the task. However, the virtual experience was relatively short and possibly not sufficient to emphatize different levels of fatigue. The LightThimble reported in general the highest scores related to the overall evaluation of the experience, naturalness of the interaction, and in particular immersivity. The HapticThimble condition was rated the highest score in terms of haptic effectiveness, with lower scores than the LightThimble for coherency of hand tracking, natural interaction and comfort. Concluding, the experimental experience investigated how different designs of wearable devices influence interaction and perception into an immersive virtual experience. Quantitative metrics of task execution showed a clear trend at increasing size of devices and richness of the haptic feedback, resulting in a trade-off between speed and precision. Participants preferences suggested that a highly wearable but limited intensity haptic device was the most effective for the given environment. Other design choices (or hands-free) can be more suitable for different scenarios involving precise manipulative tasks, or, at the opposite, faster movements and dynamic interaction.

References

1. Bortone, I., et al.: Immersive virtual environments and wearable haptic devices in rehabilitation of children with neuromotor impairments: a single-blind randomized controlled crossover pilot study. J. NeuroEngineering Rehabil. **17**(1), 1–14 (2020)
2. Bortone, I.: Wearable haptics and immersive virtual reality rehabilitation training in children with neuromotor impairments. IEEE Trans. Neural Syst. Rehabil. Eng. **26**(7), 1469–1478 (2018)
3. Chinello, F., Malvezzi, M., Prattichizzo, D., Pacchierotti, C.: A modular wearable finger interface for cutaneous and kinesthetic interaction: control and evaluation. IEEE Trans. Ind. Electron. **67**(1), 706–716 (2019)
4. Chinello, F., Pacchierotti, C., Malvezzi, M., Prattichizzo, D.: A three revolute-revolute-spherical wearable fingertip cutaneous device for stiffness rendering. IEEE Trans. Haptics **11**(1), 39–50 (2017)
5. Deutsch, J., McCoy, S.W.: Virtual reality and serious games in neurorehabilitation of children and adults: prevention, plasticity and participation. In: Pediatric Physical Therapy: The Official Publication of the Section on Pediatrics of the American Physical Therapy Association 29(Suppl 3 IV STEP 2016 CONFERENCE PROCEEDINGS), S23 (2017)
6. Fani, S., Ciotti, S., Battaglia, E., Moscatelli, A., Bianchi, M.: W-fyd: a wearable fabric-based display for haptic multi-cue delivery and tactile augmented reality. IEEE Trans. Haptics **11**(2), 304–316 (2017)
7. Gabardi, M., Chiaradia, D., Leonardis, D., Solazzi, M., Frisoli, A.: A high performance thermal control for simulation of different materials in a fingertip haptic device. In: Prattichizzo, D., Shinoda, H., Tan, H.Z., Ruffaldi, E., Frisoli, A. (eds.) EuroHaptics 2018. LNCS, vol. 10894, pp. 313–325. Springer, Cham (2018). https://doi.org/10.1007/978-3-319-93399-3_28
8. Gabardi, M., Solazzi, M., Leonardis, D., Frisoli, A.: a new wearable fingertip haptic interface for the rendering of virtual shapes and surface features. In: 2016 IEEE Haptics Symposium (HAPTICS), pp. 140–146. IEEE (2016)
9. Leonardis, D., Gabardi, M., Solazzi, M., Frisoli, A.: A parallel elastic haptic thimble for wide bandwidth cutaneous feedback. In: Nisky, I., Hartcher-O'Brien, J., Wiertlewski, M., Smeets, J. (eds.) EuroHaptics 2020. LNCS, vol. 12272, pp. 389–397. Springer, Cham (2020). https://doi.org/10.1007/978-3-030-58147-3_43
10. Leonardis, D., Solazzi, M., Bortone, I., Frisoli, A.: A 3-rsr haptic wearable device for rendering fingertip contact forces. IEEE Trans. Haptics **10**(3), 305–316 (2016)
11. Pacchierotti, C., Sinclair, S., Solazzi, M., Frisoli, A., Hayward, V., Prattichizzo, D.: Wearable haptic systems for the fingertip and the hand: taxonomy, review, and perspectives. IEEE Trans. Haptics **10**(4), 580–600 (2017)
12. Simões, M., Bernardes, M., Barros, F., Castelo-Branco, M.: Virtual travel training for autism spectrum disorder: proof-of-concept interventional study. JMIR Serious Game. **6**(1), e8428 (2018)
13. Solazzi, M., Frisoli, A., Bergamasco, M.: Design of a novel finger haptic interface for contact and orientation display. In: 2010 IEEE Haptics Symposium, pp. 129–132. IEEE (2010)
14. Wang, D., Ohnishi, K., Xu, W.: Multimodal haptic display for virtual reality: a survey. IEEE Trans. Ind. Electronics **67**(1), 610–623 (2019)
15. Wiemeyer, J.: Serious games in neurorehabilitation: a systematic review of recent evidence. In: Proceedings of the 2014 ACM International Workshop on Serious Games, pp. 33–38 (2014)

Perceiving Sequences and Layouts Through Touch

Richa Gupta(✉) 🄳

Indraprastha Institute of Information Technology Delhi, Okhla Phase 3, New Delhi, India
richa.gupta@iiitd.ac.in

Abstract. Accessing graphical information is a challenge for persons with individuals with blindness and visual impairment (BVI). The primary method for making graphical information more accessible to BVI is to translate visual graphics into tactile graphics (TGs), sometimes called "raised line" graphics. Effective design of tactile graphics demands an in-depth investigation of perceptual foundations of exploration through touch. This work investigates primitives in tactile perception of spatial arrangements (i.e. sequences and layouts). Two experiments using tiles with different tactile shapes were arranged in a row on tabletop or within a 5×5 grid board. The goal of the experiments was to determine whether certain positions offered perceptual salience. The results indicate that positional primitives exist (e.g. corners, field edges and first and last positions in sequences), and these reinforce memory of spatial relationships. These inferences can influence effective tactile graphic design as well as design of inclusive and multi-modal interfaces/experiences.

Keywords: Tactile graphics · Accessible graphics · Design for disability · Persons with visual impairment · Blindness · Tactile perception · Spatial perception

1 Introduction

Tactile graphics (TGs) are embossed outline drawings which provide a means for acquiring pictorial information through touch particularly by persons with visual impairment and blindness (BVI) [1]. Researchers have discussed the advantages of use of tactile graphics in education over verbal or textual descriptions of graphics as they allow users to actively explore the diagrams [2]. Since, a graphic is a layout of symbols on a two-dimensional (2D) surface, the study of tactile perception of 2D spaces (both exploration and retention strategies) by persons with blindness is necessary for effective TG design. Additionally, with the rise in need for advanced computer interfaces, new opportunities to develop more complex multi-sensory displays have emerged. When designing such displays it is necessary to consider human perceptual capabilities and understand how people find patterns, recognize forms and organize individual elements into structures and groups [3]. The inferences of this work can contribute to the development of design guidelines for designing intuitive multi-modal interfaces as well as TGs.

H. Seifi et al. (Eds.): EuroHaptics 2022, LNCS 13235, pp. 283–291, 2022.
https://doi.org/10.1007/978-3-031-06249-0_32

Despite the fact that every TG is essential an arrangement of tactile information organised in a 2D field, there has been very little research into the "compositional semantics", i.e. the perceptual salience and potential semantic significance of various positions. For example, it is unknown if BVI students perceive TG shapes positioned in the centre of a framed field as semantically central to the overall meaning of the theme of that field. One study [4] found that BVI subjects using vibro-tactile touchscreens remember information better when it was associated with a specific position in document layout. Auditory feedback cued positions on a touch screen can also help BVI subjects interpret graphics [5]. It is interesting to note that many blind individuals play chess (personal observation, school of the National Association of the Blind, New Delhi) which requires spatial awareness and spatial learning strategies such as chunking [6].

This current work investigates if participants with blindness can explore and remember sequential or framed arrangements of tactile shapes and asks whether the recreation of such arrangements is facilitated by semantic memory (remembering where a particular symbol was positioned); and/or by positional memory (remembering specific positions as occupied by specific shape). A set of experiments were conducted to determine if certain positions in an ordered arrangement or framed field are perceived and remembered better than other positions and thus may be considered "positional primitives". It is also noteworthy that a majority of earlier work in this area have been conducted with sighted (sometimes blindfolded) participants and the conclusions cannot be extrapolated for blindness. Therefore, the studies in this work were conducted with BVI subjects.

2 Related Work

Limited works investigate the retention of tactually acquired information from tactile-stimuli (in contrast to the large number of studies that have addressed retention of visual stimuli or visuo-tactile stimuli). Even fewer have addressed the exploration and retention of meaningful tactile representations (i.e. tactile graphics with some information) and have rather used pictures of common day to day objects or abstract shapes in experiments. Some works study the "compositional semantics", i.e. the perceptual salience and potential semantic significance of various positions.

Significant work by Berla & Butterfield (1977) demonstrated that BVI children learn and remember tactile geometric maps better if they use certain learning strategies: a) systematic tactual/haptic exploration within a framed field; b) methodical tracing of each contour; and c) "distinctive feature analysis", i.e. mentally noting when a contour has unique graphic features that other shapes lack. Two other beneficial strategies for remembering positions of multiple TG shapes: d) noting shape-to-shape relationship, e.g. this shape is *next to* or *on top of* another; and b) noting a shape's position relative to the top, bottom and sides of its field [8]. The subject's orientation to the frame becomes another relevant factor in tactuo-spatial learning. In experiments reported by Newell et al. blindfolded subjects learned an arrangement of seven distinct haptic objects in a frame, then were challenged to recognize which two objects had been repositioned by researchers [9]. Performance decreased significantly when the frame was rotated 60° between learning and testing. This finding resonates with another work [10].

Although definite conclusions may be elusive, these issues suggest an opportunity for designing innovative strategies for tactile compositions in framed fields. The objective of this work is to investigate the tactile exploration and retention strategies for learning sequences and layouts of small tactile shapes by persons with BVI and identify if there are any positional primitives (or positions of salience that facilitate information acquisition and encoding). The aim of this work is to explore information organization strategies for tactile perception and utilize that knowledge into making effective tactile graphics and better information organization for BVI.

3 Experiments

Two experiments were conducted to test memorability of sequences and layouts of tiles with abstract TG shapes (see Fig. 1). The goal was to identify positional primitives i.e. sequence or grid positions with greater perceptual salience that facilitate information acquisition and its retention.

To achieve the said goal, participants were asked to explore and recreate sequences and layouts. Students from Indiana School for the Blind and Visually Impaired, Indianapolis, USA (ISBVI) and National Association for the Blind, New Delhi, India (NAB) participated in the experiments. All participants had blindness with no cognitive impairments. All participants had prior experience with TGs as part of their school education. Ethical clearance was obtained from ethical committee at IIT Delhi. Informed consent was obtained from participants and school authorities.

A set of 30 abstract TG shapes (see Fig. 1) were prepared on 5 cm- × -5 cm plastic tiles with a single symbol on each tile. Tiles were placed in a sequence of 4 to 8 tiles on a table-top for Exp.1 and in a 5 × 5 gridded board for Exp.2 (see Fig. 2).

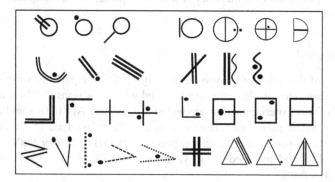

Fig. 1. Set of abstract tactile shapes used in experiments 1 & 2

Fig. 2. (left) Sequence of tiles on table-top for Exp.1; (right) Tactile tiles in a grid for Exp. 2

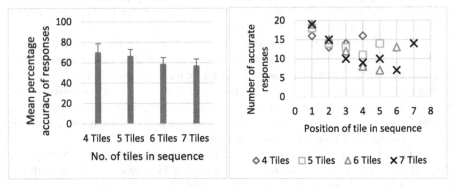

Fig. 3. (left) A bar chart showing mean (±SEM) percentage accuracy of responses for Exp. 1, (right) A scatter plot showing the frequency of accurate responses for each position for 4, 5, 6 and 7 tile sequences

3.1 Experiment 1

The objective of this experiment was to investigate if participants can explore, learn, and recreate sequences of tiles from memory and which positions facilitate acquisition and retention. This experiment was conducted at ISBVI in April 2018 and at NAB in October 2018. Seven participants (four males, three females) aged 15–18 years (M = 15.57, SD = 1.13) participated in this experiment at ISBVI in April 2018 and 14 (9 males, 5 females) students aged 12–17 years (M = 14.78, SD = 1.25) participated at NAB in October 2018.

Sequences of randomly selected tiles were presented to participants who were asked to freely explore. Then a jumbled set of the same tiles was presented to them and they were asked to recreate the sequence from memory. The experiment was repeated with four different sequences of 4, 5, 6 and 7 random tiles. The responses of the participants were visually observed and noted by the research team and number of accurate placement (correct tile in correct position) was counted for each recreation.

Results. Figure 3 shows mean percentage accuracy of correct placement of tiles in a sequence (mean includes all participants from both locations). The results show a degradation in performance with the increase in number of sequenced tiles. However, a one-way ANOVA indicated that the difference between the accuracy for sequences with different number of tiles was not significant (F(3,80) = 0.788, p = 0.504).

The "primacy effect" was observed: Subjects remembered the correct placement of tiles that were at the beginning of the sequence better than tiles in middle, no matter how long the sequence. The same can be seen in Fig. 3. However, further empirical and statistical validation is needed to strengthen this observation.

3.2 Experiment 2

This experiment aimed to evaluate subjects' memory of different layouts of tiles in gridded board, to determine if: (1) some positions in the grid are remembered better than others (e.g. corners vs center); (2) a type of layout is remembered better than others (e.g. structured vs randomized); (3) some structures are remembered better than other (e.g. row vs column, chunks vs line, cross vs chunks or lines). This used 6 structured layouts and 1 random layout (See Fig. 5). The experiment tested 21 participants from NAB and ISBVI during Oct-Nov 2018. Seven students (four females, three males) aged 15–18 years (M = 16.29, SD = 0.951) participated at ISBVI and 14 students (5 females, 9 males) aged 12–17 years (M = 14.78, SD = 1.25) participated at NAB. The participants were presented with test layouts, were asked to explore it for 5 min and then recreate the layout on an empty grid from memory (called 'response layouts' in following text).

Results. Figure 4 shows the mean percentage accuracy of the subjects' (including data from both test locations) ability to recall: (1) 'only positions' from test layouts (i.e. placing incorrect tile in correct position); and (2) the correct target tile and its correct positions (referred to as 'identity+position' in chart). A one-way ANOVA for 'only position' data indicated a significant difference between the performance for these layouts (F(6,140) = 13.520, p < 0.001). A subsequent Tukey comparison showed that performance for a 4-tile layout was significantly better than that for 9-tile layout (p < 0.05). Additionally,

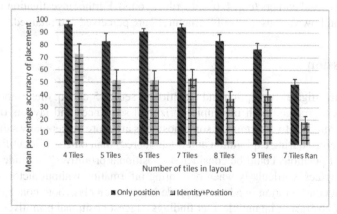

Fig. 4. A bar chart showing mean (±SEM) percentage accuracy of responses for tiles in correct positions and correct tiles in correct in correct positions for the various layouts in Exp. 2

performance for a random 7-tile layout was found to be significantly worse than other structured layouts (p < 0.001 for all pairs).

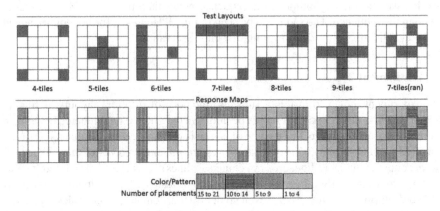

Fig. 5. Test layouts and heat maps of responses in Exp. 2

As can be seen in Fig. 4, the participants remembered the position of tiles better that the identity of the tiles in those positions. A one-way ANOVA for 'identity+position' data indicated a significant difference in accurate tiles placement in layouts with different numbers of tiles (F(6,140) = 5.773, p < 0.001). The analysis was followed by a Tukey comparison of means which indicated that performance for 4-tile layouts was significantly better than for 8-tile (p < 0.01) and 9-tile layouts (p < 0.05). Additionally, the performance for randomized 7-tile layout was significantly worse than for structured layouts with 4, 5, 6 and 7 tiles (p < 0.05 in all cases). A heat diagram presented in Fig. 5 shows the test layouts versus the cumulative placements of tiles by participants in response. This indicates overlap in test and response particularly for layouts with adjacency to edges and corners (e.g. 4, 6 and 7 tile layouts). Further investigation is required to understand how various kinds of "structures" are perceived and retained.

4 Discussion

This work investigates perception and retention of sets of 5-cm-sq. tiles embossed with tactile shapes, arranged both in simple horizontal sequences and layouts on gridded boards. The aim was to determine if some positions are more perceptually salient (more readily perceived during exploration and remembered) than others. If identified, particularly salient positions could be considered "positional primitives" to guide designers of tactile graphics, particularly when presenting information without pictorial sources, an important consideration in advanced grade levels when classroom concepts become increasingly abstract or intangible. Our findings suggest positional primitives exist, and that their use as semantic drivers needs a deeper understanding for effective classroom applications and pedagogical strategies.

4.1 Salience of Boundary and Structure

In Exp. 1, it was observed that the tiles in the beginning and end of the sequence (mostly the participants started from left although they were not directed to do so) was remembered better than the ones in the middle. This demonstrates the presence of primacy and recency effects in serial memory of tactile stimuli, a well-known phenomenon called "serial position effect" [11] (recency effect was seen as well, though not very strongly). This was observed in participants from both US and India.

In Experiment 2, tiles placed in corners or along board edges of a tactile layout (composition of tiles on a 5 × 5 gridded board) were more easily remembered for 'identity+position' (i.e. provides salience for both positional and semantic memory). This suggests that boundaries of any perceptual field offer a high level of positional/identity salience for BVI subjects and subject can most easily locate a tactile landmark positioned in a corner or along the edge of a framed field. It is noteworthy that this observation contrasts with common understanding visual perception, which suggests that centre of a visual composition has higher salience and hence attracts more attention [12]. Additional empirical validation is required to confirm this phenomenon. This salience persists in semantic memory, i.e. subjects can identify a tile removed from the grid and easily remember its former position (however, this was not formally tested).

It was seen that "structured" layouts were easier to remember than randomized layouts (focussed empirical investigation is needed to understand the role of different kinds of structures in aiding memory). This finding resonates with the concept of finding of 'figural goodness' being favorable to memory [13]. This insight resonates with the concept that coherently configured tile sets offer "chunks" of positional information. Chunking is a well-studied mental technique for remembering series of numbers or other data sets [14]. Interestingly, chunking also functions in playing chess [6] where practiced players cognitively chunk piece positions and plan moves. However, not surprisingly, even with chunking and corner/edge adjacency, overall retention of tile position/identity degrades as the number of distinct tiles in the layout increases. This suggests that compositional and juxta positional simplicity is the most effective design strategy. We speculate there may be an optimum number of positioned elements that subjects can easily learn in a framed field using the very simple target layouts (as seen in Experiment 2). Notably, this number matches well-known estimates of working memory capacity as 7 elements, plus/minus 2 [15]. These inferences pose interesting research questions and need further empirical validation.

4.2 Exploration and Retention Strategies

In post-experiment discussions, subjects revealed self-generated tactual-semantic strategies to encode the position/identity of tiles. These strategies included: (1) chunking, i.e. dividing the presented tiles into self-defined sets of 3–4 items even when tiles were not adjacent; (2) noting adjacency of tiles to board corners/edges; (3) using basic internal, directional self-instructions such as, "this tile was on the left side of the board"; (4) naming, i.e. picking a word to associate with the tactile symbol in order to remember it easily; (5) self-narrating a simple story to provide positional context for the tiles, e.g.

one participant imagined the upper half of the grid board as sky and the lower half as ground, then imagined *this* tile "flying in air" and *that* tile "lying on the ground".

We also observed that participants used different techniques of manual exploration. Some began their exploration in the centre of the grid while others started at the top left corner (like reading). These differences were not factored into our analysis; it is unknown how this observation impacted retention and requires further investigation.

5 Conclusion

This work demonstrates that 1) compositions of abstract tactile tiles in a framed field can be learned and remembered by BVI subjects using tactual exploration; 2) tactile elements adjacent to field corners and edges were retained better; 3) coherent or structured configurations of adjacent tiles; 4) in linear sequences of tiles, the serial position effect was observed. These compositional principles comprise "positional primitives" for TG design. Additionally, retention quality degrades as the number of tiles increases in both sequences and layouts. Though, these insights offer an understanding into tactile perception of sequences and layouts, there are several parameters whose role remains to be understood such as complexity of the tactile images used, discriminability of these images and similarity or categorization of the tile images. The role of factors such as orientation, movement and experiment surrounding and alignment with the experimental setup pose additional intriguing research questions.

These findings can inform the design of more effective tactile learning resources for persons with BVI. Further investigation is needed to understand how tactile compositions facilitate semantic relationships and its pedagogical validation. This work can also find application in the design of tactile/multi-modal interfaces.

Acknowledgement. This research is supported by the Center for Design and New Media (A TCS Foundation Initiative supported by Tata Consultancy Services) at IIIT-Delhi. Special thanks to Prof. S. Mannheimer, Prof. P.V.M. Rao and Prof. M. Balakrishnan for their guidance and support.

References

1. Brady, E., Morris, M.R., Zhong, Y., White, S., Bigham, J.P.: Visual challenges in the everyday lives of blind people. In: Conference on Human Factors in Computing System, no. i, pp. 2117–2126 (2013). https://doi.org/10.1145/2470654.2481291
2. Gupta, R., Balakrishnan, M., Rao, P.V.M.: Tactile diagrams for the visually impaired. IEEE Potentials 36(1), 14–18 (2017)
3. Goldstein, E.B., Brockmole, J.L.: Sensation and Perception. Cengage Learning, Boston (2016)
4. Maurel, F., et al.: Haptic perception of document structure for visually impaired people on handled devices. Procedia Comput. Sci. 14, 319–329 (2012)
5. Li, J., Kim, S., Miele, J.A., Agrawala, M., Follmer, S.: Editing spatial layouts through tactile templates for people with visual impairments. In: Conference on Human Factors in Computing System, pp. 1–11 (2019). https://doi.org/10.1145/3290605.3300436

6. Waters, J., Gobet, F., Leyden, G.: Visuospatial abilities of chess players. Br. J. Psychol. **93**(4), 557–565 (2002)
7. Berla, E.P., Butterfield, L.H.: Tactual distinctive features analysis : training blind students in shape recognition and in locating shapes on a map. J. Spec. Educ. **11**(3), 335–346 (1977)
8. Klatzky, R.L., Lederman, S.J.: Touch. In: Weiner, I.B. (ed.) Handbook of Psychology, pp. 147–236. Wiley (2013)
9. Newell, F.N., Woods, A.T., Mernagh, M., Bülthoff, H.H.: Visual, haptic and crossmodal recognition of scenes. Exp. Brain Res. **161**(2), 233–242 (2005). https://doi.org/10.1007/s00 221-004-2067-y
10. Gentaz, E., Baud-Bovy, G., Luyat, M.: The haptic perception of spatial orientations. Exp. Brain Res. **187**(3), 331 (2008). https://doi.org/10.1007/s00221-008-1382-0
11. Murdock, B.B., Jr.: The serial position effect of free recall. J. Exp. Psychol. **64**(5), 482 (1962)
12. Tatler, B.W.: The central fixation bias in scene viewing: selecting an optimal viewing position independently of motor biases and image feature distributions. J. Vis. **7**(14), 1–17 (2007). https://doi.org/10.1167/7.14.4
13. Wagemans, J., et al.: A century of Gestalt psychology in visual perception: II. Conceptual and theoretical foundations. Psychol. Bull. **138**(6), 1218 (2012)
14. Gobet, F., et al.: Chunking mechanisms in human learning. Trends Cogn. Sci. **5**(6), 236–243 (2001). https://doi.org/10.1016/S1364-6613(00)01662-4
15. Jensen, O., Lisman, J.E.: Novel lists of 7+/−2 known items can be reliably stored in an oscillatory short-term memory network: interaction with long-term memory. Learn. Mem. **3**(2–3), 257–263 (1996)

Whole-Hand Haptics for Mid-air Buttons

Martin Maunsbach[✉][iD], Kasper Hornbæk[iD], and Hasti Seifi[iD]

University of Copenhagen, Universitetsparken 1, 2100 Copenhagen, Denmark
{mama,kash,hs}@di.ku.dk

Abstract. Mid-air buttons are currently slow and error-prone. One reason is that their haptic feedback are attempts at replicating physical button feedback instead of being designed specifically for interaction in mid air. We present an approach to haptics for mid-air buttons that extends the feedback beyond the fingertip. Our approach is inspired by recent findings that show how skin vibrations from fingertip presses extend to the whole hand. We apply the haptic feedback across the whole hand to simulate the pull-up effect that triggers users to withdraw their finger upon button activation. We conduct a user study with two tasks to evaluate the whole-hand feedback and compare it with prior work. Our results show that the whole-hand haptic feedback reduces the overall button press duration and allows for more successful button activations compared to the localized haptic feedback. We discuss the reasons behind the improved performance and further steps to improve mid-air presses.

Keywords: Ultrasound haptics · Virtual reality · User interface

1 Introduction

Mid-air buttons are commonly used in extended reality or holographic interactions. In a pandemic-touched and increasingly germophobic world, contactless interactions cause less hygiene concerns for multi-user buttons. While current technology allow their visual and audio feedback to resemble physical buttons, mid-air buttons have poor performance due to their lack of physicality [1,8,9]. The fingertip can rest on the surface of physical buttons, but mid-air buttons have no surface to rest on. Where physical buttons reach a hard barrier upon being fully pressed, mid-air buttons can be pressed far beyond their activation point as seen in Fig. 1. The physical barrier creates a natural pull-up sensation with physical buttons that is not present in mid-air buttons, leading to an increase in the press duration or failed button activations.

Researchers have proposed various solutions to add haptic feedback in mid-air [4] including wearable devices like haptic gloves, encounter-type devices that can move to make contact with users, or airborne feedback. The latter approach is power efficient and does not require users to wear a device. Ultrasound feedback has been shown to improve user interactions with mid-air widgets. Adding ultrasound haptics to the gestural control of an automotive dashboard reduced

© The Author(s) 2022
H. Seifi et al. (Eds.): EuroHaptics 2022, LNCS 13235, pp. 292–300, 2022.
https://doi.org/10.1007/978-3-031-06249-0_33

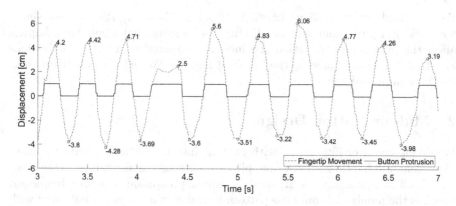

Fig. 1. A sample timeline from a participant performing a rapid tapping task with a 3D mid-air button. The timeline shows the finger and button movements. The local maximums show the fingertip displacements between button presses (inter-press displacement) while the local minimums show the displacement beyond the button activation point (press-through displacement). All measurements were captured using the Oculus Quest 2 hand tracking.

the eyes-off-the-road time and the driver's mental workload [12]. Martinez et al. showed that ultrasound haptics can increase users' sense of agency over mid-air buttons [2]. Sand et al. studied text entry in virtual reality (VR) by adding an array of ultrasound transducers to a VR headset. They found that ultrasound feedback significantly reduced user report of temporal demand compared to no haptic feedback, but there was no significant difference in user performance [11]. Ito et al. presented a dual-layer button with varied ultrasound intensities and quality for the button press and activation phases, but they only showed that people could locate the two haptic layers [5]. The current solutions only apply ultrasound feedback to the surface of the hand that is in contact with the button. They report little performance improvements, possibly due to the limited intensity of ultrasound haptics.

We propose whole-hand haptic feedback for mid-air button presses and report user performance with it. When tapping a physical surface with the fingertip, the skin vibrations propagate down the hand [13]. As the intensity of ultrasound haptics is low, we induce the vibrations to the whole hand by moving the ultrasound's focal point over the hand instead of relying on the strength of the fingertip feedback to propagate down the hand. The objective of the whole-hand haptics is to mitigate the lack of pull-up effect in mid-air. To achieve this, we consider the three following measurements as the main indicators of an improved pull-up effect: a decrease in the pull-up time after an activation, a decrease in the finger displacement beyond the activation point, and an increase in the amount of presses in a time frame. We ran a user study to evaluate user performance with mid-air buttons using our whole-hand haptic feedback approach. Participants interacted with 2D and 3D mid-air buttons in a VR environment while they received either localized haptic feedback at their fingertip or whole-hand feedback over their dominant hand. Our results show that whole-hand haptics affect the button press duration and increase the number of button activations in a time frame compared to the localized haptics. We discuss possible reasons for these improved results.

2   Mid-air Button Design

We divide the user interaction with buttons into four phases where the finger movement and haptic feedback from physical buttons are distinct [6]. 3D buttons include all four phases, while 2D buttons have no protrusion. The draw frequency, which is the number of times the pattern is drawn in a second, is indicated with Hz. The sensations are rendered using Spatiotemporal Modulation [3]. Figure 2 shows how the haptics are spread across the palmar side of the hand. We design whole-hand haptics for user interaction in these phases based on a pilot study:

1. Proximity: When the user's finger is near the button without touching it. We set this value to 10 mm above the button and present a haptic sensation over the whole finger by moving the focal point between the distal, middle and proximal phalanx (70 Hz).
2. Protrusion: When the fingertip is pressing the button. We set the button depth to 10 mm and present a haptic effect that is a 10 mm in radius circle (Hz = 70). The sensation is focused on the fingertip to match the visual contact area of the button.
3. Activation: A transient phase starting from when the button is successfully pressed. The sensation is a 80 mm wide line that traverses from the fingertip to the root of the hand (90 Hz) in 100 ms.
4. Release: From the end of the activation phase until the button is unpressed. We provide no sensation in this phase.

We kept the visual feedback minimal to avoid interference with the haptic effect. Based on a pilot study, we opted for 10 mm press distance and used a 40 by 40 mm button to ensure participants could accurately hit it. The 2D button had

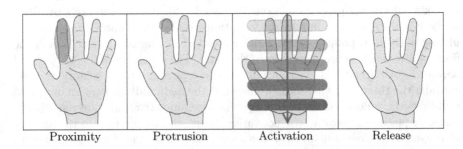

| Proximity | Protrusion | Activation | Release |

Fig. 2. The whole-hand sensation set showing the spread of the haptic sensation in each of the phases. Picture of the palm is adapted for non-commercial use [10].

no visual feedback for any phases. The 3D button's visual protrusion followed the fingertip, indicating to users that it was activated or released when it no longer followed the fingertip.

3   User Study

To evaluate the impact of whole-hand haptic feedback on user performance with 2D and 3D buttons, we conducted a user study with two user tasks.

Participants. We recruited 20 participants (4F/16M) 23–57 years old (M = 31, SD = 9.93). No one reported any sensory impairment in their dominant hand. Seven users had prior experience with mid-air haptics[1]. The experiment took around 30 min and participants received a gift worth around 100 DKK.

Tasks. The first task was a rapid tapping task, where participants pressed the button as quickly as possible. This task was chosen to capture whether more presses are possible with whole-hand feedback and the effect it has on the press duration and finger movement. Each trial of the task lasted 20 s. The second task was to double-click the mid-air button as quickly as possible after hearing a randomly timed sound cue. This reaction task is a modified version of the moving target selection task [7] without the visual element. We require two presses (i.e., a double click) to capture measures on the inter-press duration.

Study Design. The experiment used a within-subjects design with the three independent variables *dimension* and *haptic* and *trial*. The button dimension was either 2D or 3D to study the effect of the protrusion phase on user performance. The haptic sensation was either the whole-hand haptics or a localized sensation used as a baseline. The localized haptics was based on the activation sensation by Martinez *et al.*, that showed improvement to the sense of agency for mid-air buttons [2]. Instead of the five focal points they used to cover $1\,cm^2$, we used a 200 ms transient version similar to the protrusion sensation in Fig. 2. The combination of button dimension and haptics yielded four conditions. Each condition was repeated three times for a total of 12 trials in each task.

Apparatus. We used the ultrasound device STRATOS Explore to induce haptic feedback on the palmar side of the hand. The participants wore an Oculus Quest 2 VR head-mounted display with a refresh rate 90 Hz to interact with the mid-air button. The button was calibrated to be approximately 20 cm above the ultrasound device's surface, where the ultrasound focal point is the strongest. We used Oculus Quest 2's hand tracking (60 Hz refresh rate) to measure the movements of all limbs in the hand as well as the visual representation of the hand. The displacement of the button followed the fingertip while being constrained to one dimension with an upper and lower limit. The Leap Motion's hand tracking (120 Hz refresh rate) was used to position the haptic feedback only. Measurements were recorded 200 Hz.

---

[1] Recruitment used the same mailing list as a previous study involving mid-air haptics.

Procedure. After signing a consent form, the participants were introduced to the mid-air technology by feeling common 2D shapes like circles and lines. They were informed of the two tasks and that they needed to press a variation of 2D or 3D virtual buttons with haptic sensations. They were not told how the sensations would differ. The participants were instructed to complete the press interaction as quickly as possible. They were told that the goal of the rapid tapping was to press as many times as possible and that the time from the sound cue to the release of the second click was important for the reaction task. Before the first trial of each condition, the participants could practice the button press for up to a minute to get conditioned to the button dimension and haptic feedback. During this training, the participants were given visual feedback after each interaction on their performance. After the training, each trial was started. There was a five to ten second break between each trial of the same condition and a two minute break between the two tasks. The condition order was counterbalanced. Pink noise was played to mask the sound of the device.

4 Results

We recorded the movements of the button and the index finger for both tasks. With the button movements we captured the number of times the button was successfully activated (i.e., the *press count*). We obtained the *press duration* as a sum of the *down-press duration* (from button contact until the button is activated) and the *pull-up duration* (from activation until the button is released). The *press-through displacement* quantifies how far the finger moves beyond the button activation point. The *inter-press displacement* is how far the finger moves away from its released state between each press. Additionally, the reaction task included the *reaction time* from the sound cue until contact with the button, the duration between the two presses (*inter-press duration*), and the *success rate*.

Rapid Tapping Task. A total of 16,609 successful presses were recorded. Outliers in each participant's trials were removed using the IQR method, where outliers are values more than 1.5 times above or below the trial's inter-quartile range.

We ran three-way repeated measures ANOVAs with the six measurements as the dependent variables and dimension, haptics, and trial as the within-subjects factors (Table 1). The results showed main effects of haptics for *press count, press duration,* and *pull-up duration* without any significant interaction effects with dimension nor trial. The participants pressed significantly quicker (9.52%) and had more successful button presses (4.76%) with whole-hand haptics. The *pull-up duration* was significantly decreased (10.87%). The *press-through displacement* showed an improvement of 12.19%, but this difference was not significant due to the high standard errors. The *inter-press displacement* was lower with localized haptics but not significantly.

The repeated measures ANOVA also showed a main effect of dimension for the *press count, press duration* and the *inter-press displacement* (Table 1). These

Table 1. The differences in press count, durations and displacements (Disp.) between the localized and whole-hand haptics (top) and 2D and 3D buttons (bottom) in the rapid tapping task.

Measurement	Localized	Whole-hand	p-value	η_p^2
Press Count [per 20 s]	67.517 ± 2.690	70.892 ± 2.678	**0.010**	0.298
Press Duration [s]	0.147 ± 0.008	0.133 ± 0.006	**0.023**	0.243
Pull-Up Duration [s]	0.139 ± 0.008	0.124 ± 0.006	**0.016**	0.268
Down-Press Duration [s]	0.009 ± 0.001	0.009 ± 0.001	0.370	0.042
Press-Through Disp. [cm]	2.912 ± 0.289	2.557 ± 0.219	0.141	0.111
Inter-Press Disp. [cm]	3.169 ± 0.235	3.261 ± 0.233	0.451	0.030
Measurement	3D	2D	p-value	η_p^2
Press Count [per 20 s]	66.783 ± 3.203	71.265 ± 2.205	**0.010**	0.303
Press Duration [s]	0.155 ± 0.008	0.126 ± 0.007	**<0.001**	0.538
Pull-Up Duration [s]	0.137 ± 0.008	0.126 ± 0.007	0.094	0.141
Down-Press Duration [s]	0.018 ± 0.001	N/A	N/A	N/A
Press-Through Disp. [cm]	2.635 ± 0.267	2.834 ± 0.265	0.470	0.028
Inter-Press Disp. [cm]	3.746 ± 0.254	2.685 ± 0.230	**<0.001**	0.603

values are significantly different for 2D buttons due to no down-press time in the dimension. Importantly, we found no interaction among dimension and haptics for these dependent variables ($p > 0.713$). Figure 3 shows the results for the six measurements under the four conditions.

Reaction Task. We ran three-way repeated measures ANOVAs with the nine measurements as the dependent variables and dimension, haptics, and trial as the within-subjects factors. There were no significant effects of the haptic conditions on any of the dependent variables ($p \geq 0.064$ for all the measures). The 2D condition showed a significant improvement in *press-through displacement* ($p = 0.024$) and *inter-press displacement* ($p < 0.001$).

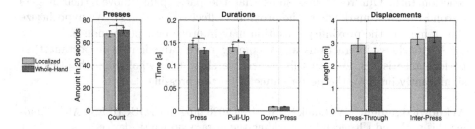

Fig. 3. The mean and standard error of the localized and whole-hand conditions in the rapid tapping task. The number of presses in 20 s increased significantly with the whole-hand haptics. A significant decrease is evident in the press and pull-up duration, while the differences for the displacements were not significant.

5   Discussion and Conclusion

This paper set out to improve haptics for mid-air buttons, exploring the idea that whole–hand haptics could improve the interaction due to an enhanced pull-up effect. To support this idea we examined the pull-up duration, press depth, and press count as main indicators. The results suggest that it is possible to design haptic feedback that improves the performance for rapid presses. In the rapid tapping task, whole-hand haptics significantly improves the pull-up duration. Since the pull-up duration accounts for 94% of the overall press duration according to our measurements, it is a major factor in the improved performance. We believe the decrease in the press duration is the main factor behind the increase in the press count. Improving the press count shows that whole-hand haptics is an improvement to the overall press interaction, not just the release phase. While the reduction in press-through displacement was not significant, it did trend towards improvement. The standard error of the whole-hand condition (0.219 cm) is lower than the localized (0.289 cm), suggesting it is easier to control the finger movement with the whole-hand haptics.

There are multiple reasons for improved performance with the whole-hand haptics. One is the overall intensity felt on the hand. Some participants mentioned they perceived the whole-hand haptics as stronger compared to the localized, even though their focal point intensity is equal. The design of the haptic feedback upon activation can also be a reason. The sweep-back sensation from the fingertip to the root of the hand can signal users to follow that direction. An instant whole-hand sensation, like a clap, may not have the same effect on user performance.

The lack of significant improvements in the reaction task can be due to the high frequency of repetitions in the rapid tapping task compared to the controlled double clicks. As participants continuously press in the rapid tapping task, their motor system likely takes over. With each press, it tunes the motor command to infer the moment of activation [8]. The intensity of the rapid tapping task forces participants to rely more on the haptic feedback, whereas the reaction task allows them to rely on the visual feedback in the short period between each double click. The proximity sensation also did not seem to improve user performance in the reaction task. Our recordings show that the participants moved their fingers beyond the 2-cm mark where the proximity effect stops (Fig. 1). We hypothesize that removing the proximity sensation may improve user performance.

As touchless interactions are becoming more valued, it is important that performance is not lost. In this paper, we show how whole-hand haptics can significantly improve the performance of button pressing.

Acknowledgement. We thank EU Horizon 2020 program TOUCHLESS AI for funding this work and Ultraleap for providing the ultrasound haptic device.

References

1. Bermejo, C., Lee, L.H., Chojecki, P., Przewozny, D., Hui, P.: Exploring button designs for mid-air interaction in virtual reality: a hexa-metric evaluation of key representations and multi-modal cues. Proc. ACM Hum. Comput. Interact. 5(EICS), 1–26 (2021). https://doi.org/10.1145/3457141
2. Cornelio Martinez, P.I., De Pirro, S., Vi, C.T., Subramanian, S.: Agency in mid-air interfaces. In: Proceedings of the 2017 CHI Conference on Human Factors in Computing Systems, Denver, Colorado, pp. 2426–2439. ACM (2017). https://doi.org/10.1145/3025453.3025457
3. Frier, W., et al.: Using spatiotemporal modulation to draw tactile patterns in mid-air. In: Prattichizzo, D., Shinoda, H., Tan, H.Z., Ruffaldi, E., Frisoli, A. (eds.) EuroHaptics 2018. LNCS, vol. 10893, pp. 270–281. Springer, Cham (2018). https://doi.org/10.1007/978-3-319-93445-7_24
4. Hoshi, T., Abe, D., Shinoda, H.: Adding tactile reaction to hologram. In: RO-MAN 2009 - The 18th IEEE International Symposium on Robot and Human Interactive Communication, pp. 7–11 (2009). https://doi.org/10.1109/ROMAN.2009.5326299
5. Ito, M., Kokumai, Y., Shinoda, H.: Midair click of dual-layer haptic button. In: 2019 IEEE World Haptics Conference (WHC), Tokyo, pp. 349–352. IEEE (2019). https://doi.org/10.1109/WHC.2019.8816101
6. Kim, S., Lee, G.: Haptic feedback design for a virtual button along force-displacement curves. In: Proceedings of the 26th Annual ACM Symposium on User Interface Software and Technology, UIST 2013, pp. 91–96. Association for Computing Machinery (2013). https://doi.org/10.1145/2501988.2502041
7. Lee, B., Kim, S., Oulasvirta, A., Lee, J.I., Park, E.: Moving target selection: a cue integration model. In: Proceedings of the 2018 CHI Conference on Human Factors in Computing Systems, Montreal, pp. 1–12. ACM (2018). https://doi.org/10.1145/3173574.3173804
8. Oulasvirta, A., Kim, S., Lee, B.: Neuromechanics of a button press. In: Proceedings of the 2018 CHI Conference on Human Factors in Computing Systems, Montreal, pp. 1–13. ACM (2018). https://doi.org/10.1145/3173574.3174082
9. Ozkul, C., Geerts, D., Rutten, I.: Combining auditory and mid-air haptic feedback for a light switch button. In: Proceedings of the 2020 International Conference on Multimodal Interaction, Virtual Event, Netherlands, pp. 60–69. ACM (2020). https://doi.org/10.1145/3382507.3418823
10. PNGWING: Hand finger palm. https://www.pngwing.com/en/free-png-mubka. Accessed 9 Dec 2021
11. Sand, A., Rakkolainen, I., Isokoski, P., Kangas, J., Raisamo, R., Palovuori, K.: Head-mounted display with mid-air tactile feedback. In: Proceedings of the 21st ACM Symposium on Virtual Reality Software and Technology, Beijing, pp. 51–58. ACM (2015). https://doi.org/10.1145/2821592.2821593
12. Shakeri, G., Williamson, J.H., Brewster, S.: May the force be with you: ultrasound haptic feedback for mid-air gesture interaction in cars. In: Proceedings of the 10th International Conference on Automotive User Interfaces and Interactive Vehicular Applications AutomotiveUI 2018, Toronto, pp. 1–10. Association for Computing Machinery (2018). https://doi.org/10.1145/3239060.3239081
13. Shao, Y., Hayward, V., Visell, Y.: Spatial patterns of cutaneous vibration during whole-hand haptic interactions. Proc. Nat. Acad. Sci. U.S.A 113(15), 4188–4193 (2016). https://doi.org/10.1073/pnas.1520866113

Proximity-Based Haptic Feedback for Collaborative Robotic Needle Insertion

Robin Mieling[1]([✉])(iD), Carolin Stapper[1]([✉])(iD), Stefan Gerlach[1],
Maximilian Neidhardt[1], Sarah Latus[1], Martin Gromniak[1],
Philipp Breitfeld[2], and Alexander Schlaefer[1]

[1] Institute of Medical Technology and Intelligent Systems, Hamburg University
of Technology, Hamburg, Germany
{robin.mieling,carolin.stapper}@tuhh.com
[2] Department of Anesthesiology, University Medical Center Hamburg-Eppendorf,
Hamburg, Germany
https://mtec.et8.tuhh.de/

Abstract. Collaborative robotic needle insertions have the potential to improve placement accuracy and safety, e.g., during epidural anesthesia. Epidural anesthesia provides effective regional pain management but can lead to serious complications, such as nerve injury or cerebrospinal fluid leakage. Robotic assistance might prevent inadvertent puncture by providing haptic feedback to the physician. Haptic feedback can be realized on the basis of force measurements at the needle. However, contact should be avoided for delicate structures. We propose a proximity-based method to provide feedback prior to contact. We measure the distance to boundary layers, visualize the proximity for the operator and further feedback it as a haptic resistance. We compare our approach to haptic feedback based on needle forces and visual feedback without haptics. Participants are asked to realize needle insertions with each of the three feedback modes. We use phantoms that mimic the structures punctured during epidural anesthesia. We show that visual feedback improves needle placement, but only proximity-based haptic feedback reduces accidental puncture. The puncture rate is 62% for force-based haptic feedback, 60% for visual feedback and 6% for proximity-based haptic feedback. Final needle placement inside the epidural space is achieved in 38%, 70% and 96% for force-based haptic, visual and proximity-based haptic feedback, respectively. Our results suggest that proximity-based haptic feedback could improve needle placement safety in the context of epidural anesthesia.

Keywords: Force feedback · Collaboration · Human-robot interaction · Epidural anesthesia · Optical coherence tomography

---

R. Mieling and C. Stapper—Contributed equally.

H. Seifi et al. (Eds.): EuroHaptics 2022, LNCS 13235, pp. 301–309, 2022.
https://doi.org/10.1007/978-3-031-06249-0_34

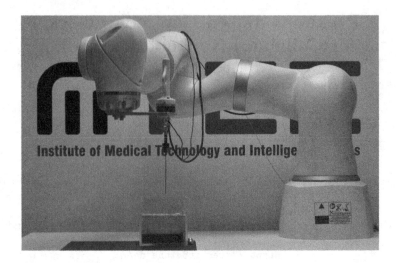

Fig. 1. Robotic system with proximity-based haptic feedback for collaborative needle insertions into tissue mimicking phantoms.

1   Introduction

Epidural anesthesia plays an important role for perioperative pain management, e.g. during orthopedic, urologic and general surgery. The procedure requires placing a needle that guides a catheter into the epidural space (ES). The ES is located directly behind the ligamentum flavum (LF) and surrounds the dura that protects the spinal cord and the cerebrospinal fluid. Major complications during epidural anesthesia are not common but potentially serious, including hematoma, post-operative neurologic deficits, infections and even death [1]. Nerve injury and long-term headache result from accidental dural perforation [2,3]. Placement within the ES without dura injury can be challenging given the small size of only 2 mm to 6 mm [4].

The most common technique for identifying the correct needle placement is loss-of-resistance (LOR). It is based on the different tissue densities in LF and ES. The entry into ES is visually or haptically perceived by the performing surgeon [5]. However, LOR requires frequent training, false-positives are possible and dura punctures still occur [5,6].

We consider robot-assisted needle insertions in the context of epidural anesthesia. CT-guided robotic needle insertions have shown promising results for soft tissue biopsy sampling [7]. In epidural anesthesia, external image guidance is challenging and tissue deformation as well as patient movement make a fully automated needle placement difficult. We consider a collaborative approach where the trajectory is guided by a robot but additional feedback is required to enable the correct axial placement by the operator. Besides LOR, experienced physicians rely on the haptic impression at the needle shaft to navigate. Consequently, haptic feedback based on force measurements has been intuitively considered for robotic needle insertions. Multiple force sensors [8] or force modeling [9] can provide enhanced feedback on needle tip forces that are otherwise superimposed with friction forces. However, force measurements always require physical contact first, potentially damaging delicate structures.

Instead, we propose a method that can detect structures before physical contact occurs. We have recently shown that an optical coherence tomography (OCT) fiber embedded into an epidural needle can enable the detection of rupture events during needle insertions [10]. We now employ high resolution OCT needles to measure the distance to structures. During collaborative needle insertions, the distance is converted to a resistive force and employed as haptic feedback. We perform experiments on tissue mimicking phantoms simulating the epidural cavity. We compare haptic feedback based on force measurements as well as the visual representation of the proximity with and without additional haptic feedback. We evaluate our methods in a user study with ten participants that each conduct needle insertions with the three different feedback modes.

2 Methods

Our system setup contains an optical needle probe, a 7-degree-of-freedom (DOF) medical robot with a custom handle for collaborative robot manipulation and a tissue mimicking phantom (see Fig. 1).

2.1 Sensor and Phantom Setup

The proximity sensor is based on OCT imaging. An optical fiber is fitted into a standard Tuohy needle with a diameter of 1.4 mm for forward-facing, common-path OCT imaging. Axial depth scans (A-scans) are acquired with a spectral domain OCT system (Telesto I, Thorlabs) with a axial resolution of 6.5 μm in air. The maximum imaging depth in tissue is approximately 1.77 mm, assuming a constant refractive index of 1.45 for tissue. Our proximity sensor output is the distance d to structures, which are positioned along the insertion trajectory in front of the needle tip. It is obtained from detecting the closest intensity peak within the processed A-scan (see Fig. 2, top right).

To evaluate our system, we employ phantoms from tissue mimicking gelatin gels. We replicate the ES within the gelatin gels with two successive layers of cellulose (see Fig. 2, bottom right). The layers are spaced 3 mm apart and represent the LF and the dura respectively. The synthetic dura is supported by polyethylene foam that represents the area of the subdural space.

2.2 Robotic System

The robotic system consists of a 7-DOF light-weight robot (LBRMed 14, KUKA) and a specially manufactured handle for the collaborative control by the surgeon. The handle (see Fig. 2, left) includes a 6-DOF force-torque sensor (M3703, Sunrise Instruments) that measures the forces and torques applied to the handle

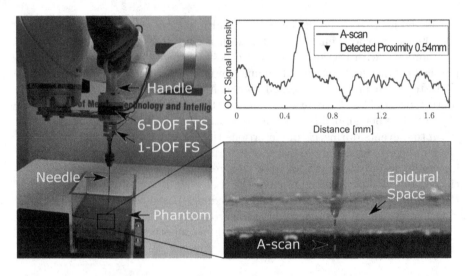

Fig. 2. The custom handle for collaborative needle insertions is shown on the left. A 6-DOF force-torque sensor (FTS) detects the users input and the 1-DOF force sensor (FS) registers forces acting on the needle. A close-up of the needle tip within the ES mimicking phantom and the visualization of approaching structures via the A-scan are shown on the right.

by the surgeon. Additionally, a 1-DOF force sensor (KD24s, ME-measurement systems GmbH) measures the forces acting on the needle shaft.

The outer control loop is designed with an admittance controller. Prior to insertion, the 6-DOF force-torque sensor allows the operator to freely position the needle axis along a desired trajectory. During collaborative insertion, the task space is restricted to the needle axis. The operator controls the 1-D movements by the forces exerted on the handle. Haptic feedback is implemented by an opposing resistive force. We employ a feedback control loop as illustrated in Fig. 3. The input value of the control loop is the control error e. It is defined as the difference between the feedback force F_{Fb} and the handle force applied in the insertion axis F_{Handle}. Negative values of e are not considered in order to prevent stability problems caused by oscillations. The control error is converted to the desired movement $\dot{x}_{i,d}$ by the PID controller. Execution by the robot results in the new insertion depth x of the needle tip. For $F_{\text{Handle}} < 0$ the handle force is directly mapped to the corresponding motion without any feedback to allow retraction.

The handle enables three different feedback modes. Switching between the three modes changes the source of the applied feedback force F_{Fb}. In mode 1, haptic feedback consists of the measured needle force. In mode 2, the user is provided no haptic feedback, but a visual representation of the OCT signal (see Fig. 2, top right). In mode 3, this visual feedback is supplemented by the haptic feedback of the computed distance. During mode 1, we measure the force acting on the needle, which is then used as direct feedback $F_{\text{Fb}} = F_{\text{Needle}}$. As mode 2 contains only visual feedback, the feedback loop is not closed with $F_{\text{Fb}} = 0$ at

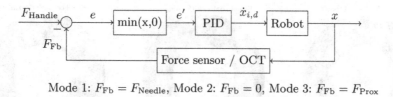

Mode 1: $F_{\text{Fb}} = F_{\text{Needle}}$, Mode 2: $F_{\text{Fb}} = 0$, Mode 3: $F_{\text{Fb}} = F_{\text{Prox}}$

Fig. 3. Control Scheme during Insertion. Three modes are considered for the feedback loop. In mode 1, the needle shaft force is used as feedback. In mode 2, no feedback is used and in mode 3, a force is generated based on the proximity according to Eq. 1.

all times. For the proximity-based haptic feedback (mode 3), the sensor output d computed from the OCT signal is mapped to a corresponding resistive force $F_{\text{Fb}} = F_{\text{Prox}}$, for which

$$F_{\text{Prox}}(d) = \tanh(2 * (1 - \frac{d}{d_{\text{Max}}})), d_{\text{Max}} = 1.77 \text{ mm} \tag{1}$$

is used. Robot communication and control is realized with the Robot Operating System (ROS). The controller runs with a frequency of 1 kHz, both force sensors update with 200 Hz and the computed distance is updated with a frequency of 100 Hz. The latency of the system is 30 ms.

2.3 Experiments

Ten participants with limited experience in needle insertions are asked to position the needle tip within the ES. We conduct five insertions per participant per feedback mode. Pullback is permitted and the insertion is stopped once the participant releases the handle. The participants are granted one test run in each mode to familiarize with the system behaviour. The order in which the three modes are employed is randomized for each participant.

For evaluation, we determine the position of our mimicked dura relative to our robot coordinate system. Based on the end-effector robot poses, we consider the distance to the target height in mm. We report mean and standard deviation at maximum extension d_{Max} and for the final position d_{End}. Insertions with correct needle placement refer to all insertions where $0 \text{ mm} < d_{\text{End}} < 3 \text{ mm}$. Insertions with $d_{\text{Max}} < 0 \text{ mm}$ correspond to dura puncture. The insertion is aborted if the operator inserts the needle more than 15 mm beyond the dura.

3   Results

In total, we conduct 150 insertions. A single insertion is aborted as the participant fails to identify the ES in mode 2. Figure 4 shows the dura puncture rate and the successful needle placement separated by the three feedback modes. For force-based haptic feedback (mode 1), placement within the ES is successful in 38% of cases. 62% of insertions result in dura puncture and no insertions are stopped before the LF. For purely visual feedback (mode 2), one insertion is

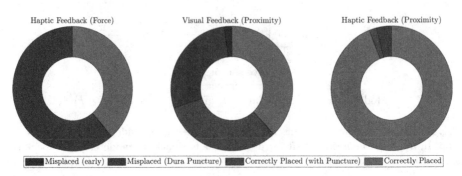

Fig. 4. Combined results for the dura puncture rate (including overshoot) and the placement success rate based on the maximum and final insertion position respectively.

Table 1. Needle placement accuracy for each feedback mode in comparison of the first and last insertion.

Feedback mode		1	2	3
d_{Max} [mm]	First	-0.35 ± 1.86	-1.96 ± 4.37	0.62 ± 0.38
	Last	-0.75 ± 2.07	-0.15 ± 1.25	0.67 ± 0.53
d_{End} [mm]	First	-0.08 ± 1.66	-1.37 ± 4.63	0.76 ± 04.08
	Last	-0.67 ± 2.06	0.39 ± 1.38	0.80 ± 0.66

stopped prematurely and 38% of insertions are correctly stopped in front of the dura. Dura puncture occurs in 60% of cases, but is detected and corrected by a subsequent pullback in 32% of all insertions. The proximity-based haptic feedback (mode 3) increases the correctly placed insertions without dura puncture to 94%. The dura is punctured during three insertions (6%), one position is successfully corrected.

We further report the distance to the dura for the three feedback modes (see Fig. 5). Considering d_{Max} (red boxes), needle insertions with both mode 1 and 2 are extended below the dura, with a mean of (-0.33 ± 2.01) mm and (-0.65 ± 2.37) mm, respectively. On average, insertions with proximity-based haptic feedback (mode 3) are sto pped (0.58 ± 0.68)mm before the dura. Mean distances to the dura at the final needle position d_{End} are (-0.19 ± 1.98) mm, (-0.19 ± 1.98) mm and (0.21 ± 0.62) mm for mode 1 to 3, respectively.

In order to evaluate the learning effect of the participants, the positions from the first and the last attempt of each mode are displayed in Fig. 5. The mean accuracy and standard deviation (SD) is also given in Table 1.

4   Discussion and Outlook

The results from our phantom study show that the needle shaft forces (mode 1) provide insufficient haptic feedback for the operator to accurately place an epidural needle within the ES. The large variation in final needle placements

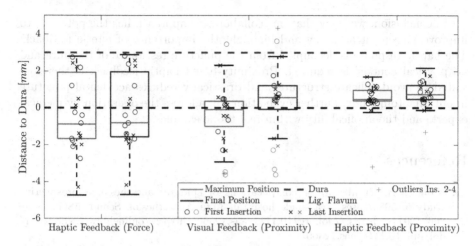

Fig. 5. Distance to the dura for the maximum and final extension for each of the three feedback modes. Additionally, the first and the last attempt from each participant is displayed. Outliers are marked in red, the aborted insertion is not shown.

implies that the participants have difficulty distinguishing the two boundary layers. Multiple insertions are stopped after only partially rupturing the mimicked LF signified by points directly under the 3 mm line in Fig. 5. With visual feedback (mode 2), the puncture rate is nearly identical to mode 1. The higher transparency compared to mode 1 and 3 also results in the only aborted attempt posing a significant safety issue. However, the participants are able to detect and correct the overshoot in half of the cases. This implies that they are able to detect the dura but fail to react in time. Higher damping for finer movements and a more intuitive visual representation could help mitigate the overshoot. In mode 3, the number of needle placements where no puncture occurs and the needle is stopped inside the ES increases to 94% from 38% in mode 1 and 2. This indicates that users have less difficulty distinguishing the two boundary layers and can more intuitively insert the needle within the ES.

Regarding the learning effect, the comparison between the first and last attempt for each mode results in relatively small differences compared to their standard deviations. This makes definitive conclusions difficult, considering the small sample size. Nevertheless, it indicates that the lack of adequate feedback in mode 1 and 2 is not intuitively compensated within five attempts.

Previously, proximity-based haptic feedback has been proposed for ultrasound imaging [11] and endovascular catheterization [12]. However, these approaches provide an insufficient resolution, are not designed to work in-vivo or rely on external imaging that is not typically available in epidural anesthesia. Our proximity sensor with μm resolution can resolve small scale structures like the ES and can be easily integrated into medical needles. Our system does not decouple the needle tip from shaft forces which has been shown to improve needle placement in [8]. However, this requires a more complex needle tip sensor.

In conclusion, we show that our collaborative approach has the potential to improve placement accuracy and highlight the importance of haptic feedback. The haptic response is decoupled from the puncture resistance of the dura and no physical contact is required. Proximity-based haptic feedback is therefore suited to avoid delicate structures and drastically reduces accidental puncture in our phantom study. Further evaluation will address the operation by medical experts and the applicability within real tissue samples.

References

1. Kang, X.H., et al.: Major complications of epidural anesthesia: a prospective study of 5083 cases at a single hospital. Acta Anaesthesiol. Scand. **58**(7), 858–866 (2014). https://doi.org/10.1111/aas.12360. https://onlinelibrary.wiley.com/doi/full/10.1111/aas.12360
2. Bezov, D., Lipton, R.B., Ashina, S.: Post-dural puncture headache: Part I diagnosis, epidemiology, etiology, and pathophysiology. Headache J. Head Face Pain 50(7), 1144–1152 (2010). https://doi.org/10.1111/j.1526-4610.2010.01699.x. https://headachejournal.onlinelibrary.wiley.com/doi/full/10.1111/j.1526-4610.2010.01699.x
3. Webb, C.A.J., et al.: Unintentional dural puncture with a tuohy needle increases risk of chronic headache. Anesth. Analg. **115**(1), 15–25 (2012). https://doi.org/10.1213/ANE.0b013e3182501c06. https://journals.lww.com/anesthesia-analgesia/Fulltext/2012/07000/Unintentional_Dural_Puncture_with_a_Tuohy_Needle.22.aspx
4. Manchikanti, L., Atluri, S.: Chapter 152 - lumbar epidural nerve block. In: Waldman, S.D., Bloch, J.I. (eds.) Pain Management, pp. 1281–1293. W.B. Saunders, Philadelphia (2007). https://doi.org/10.1016/B978-0-7216-0334-6.50156-4. https://www.sciencedirect.com/science/article/pii/B9780721603346501564
5. Dhansura, T., Shaikh, T., Maadoo, M., Chittalwala, F.: Identification of the epidural space-loss of resistance to saline: an inexpensive modification. Indian J. Anaesth. **59**(10), 677–679 (2015). https://doi.org/10.4103/0019-5049.167483
6. Yang, J., et al.: The development of a novel device based on loss of guidewire resistance to identify epidural space in a porcine model. J. Healthc. Eng. **2020**, 8899628 (2020). https://doi.org/10.1155/2020/8899628
7. Neidhardt, M., et al.: Robotic tissue sampling for safe post-mortem biopsy in infectious corpses. IEEE Trans. Med. Robot. Bionics **4**, 94–105 (2022)
8. de Lorenzo, D., Koseki, Y., de Momi, E., Chinzei, K., Okamura, A.M.: Coaxial needle insertion assistant with enhanced force feedback. IEEE Trans. Biomed. Eng. **60**(2), 379–389 (2013). https://doi.org/10.1109/TBME.2012.2227316
9. Okamura, A.M., Simone, C., O'Leary, M.D.: Force modeling for needle insertion into soft tissue. IEEE Trans. Bio-med. Eng. **51**(10), 1707–1716 (2004). https://doi.org/10.1109/TBME.2004.831542
10. Latus, S., et al.: Rupture detection during needle insertion using complex OCT data and CNNS. IEEE Trans. Biomed. Eng. (2021). https://doi.org/10.1109/TBME.2021.3063069

11. Antonello, R., Oboe, R.: Force controller tuning for a master-slave system with proximity based haptic feedback. In: IECON 2014–40th Annual Conference of the IEEE Industrial Electronics Society, pp. 2774–2779 (2014). https://doi.org/10.1109/IECON.2014.7048900
12. Dagnino, G., Liu, J., Abdelaziz, M.E.M.K., Chi, W., Riga, C., Yang, G.Z.: Haptic feedback and dynamic active constraints for robot-assisted endovascular catheterization. In: 2018 IEEE/RSJ International Conference on Intelligent Robots and Systems (IROS), pp. 1770–1775 (2018). https://doi.org/10.1109/IROS.2018.8593628

Furekit: Wearable Tactile Music Toolkit for Children with ASD

Di Qi[1]([✉]), Mina Shibasaki[1], Youichi Kamiyama[1], Sakiko Tanaka[2],
Bunsuke Kawasaki[2], Chisa Mitsuhashi[2], Yun Suen Pai[1],
and Kouta Minamizawa[1]

[1] Keio University Graduate School of Media Design, Yokohama, Japan
`teki.sai@kmd.keio.ac.jp`
[2] Miraikan - The National Museum of Emerging Science and Innovation,
Tokyo, Japan

Abstract. Children with autism spectrum disorder (ASD) face the challenge of social interaction and communication, leading to them often requiring significant support from others in their daily lives. This includes challenges like basic communication to convey their emotions to comprehension in early education. To aid with their early development, we propose Furekit, a wearable toolkit that encourages physical interaction via audio and tactile stimuli. Furekit can be attached to various parts of the body, can be operated wirelessly, and is equipped with both a speaker and a vibrotactile actuator. The audio and tactile stimuli are triggered when touched via a conductive pad on the surface, aiming to aid these children's learning and social experience. From our conducted workshop with children with ASD, we found that Furekit was well-received and was able to encourage their spontaneous physical movement. In the workshop, Furekit shows its potential as an educational and communication tool for children with ASD.

Keywords: Autism Spectrum Disorder · Haptic interaction · Assistive learning

1 Introduction

Symptoms of Autism Spectrum Disorder (ASD) are commonly found in children because they tend to appear at a very young age [1]. As a spectrum disorder, each children with ASD has their own unique challenges to overcome. Generally though, autism usually leads to impaired posture and motion, poor explicit (verbal, etc.) and implicit (eye contact, etc.) communication, lack of willingness to interact with others and social awkwardness [1]. In recent years, researchers and clinicians are beginning to focus on the deficits in sensory-motor abilities, which is responsible for the rhythm and synchrony of social interaction [2]. If left without intervention, this will lead to the children eventually facing a lot of challenges living independently [3]. Though the cause of autism is still unclear until this point, early intervention can help children with ASD. However, research findings about the conventional solutions so far mostly rely on improving methods

H. Seifi et al. (Eds.): EuroHaptics 2022, LNCS 13235, pp. 310–318, 2022.
https://doi.org/10.1007/978-3-031-06249-0_35

for the caregiver or teacher to interact with the children, such as having better communication or physical interaction.

We hypothesize that a tool specifically designed to aid with physical interaction can further assist these children with their social development and learning experience. Therefore, we propose Furekit, a tactile music toolkit that is modular, wireless, easily deployable and easy to use. The hardware comprises of a wide touch-enabled 3D-printed surface with integrated vibrotactile actuators, speaker, and wireless communication modules. Snap bottoms are used to easily attach the modules onto any parts of the body allowing it to facilitate various physical interactions. We conducted a workshop where Furekit was integrated into the lessons for the children in a special need school. The workshop required them to move their bodies and dance with Furekit by following the music. In summary, our contributions are therefore threefold:

1. We propose a wearable tactile music toolkit interface that can provide a sensory response to each goal-directed movement to augment the learning and social experience of children with ASD.
2. We conducted a workshop at a special needs education school to determine the effectiveness of our prototype.
3. We found that Furekit was well-received, enjoyable, and encourages spontaneous physical movement from the children.

2   Related Works

2.1   Interventions for Children with ASD

There has been several research in the past that looked into understanding the development of Children with ASD [4–6]. When interacting or educating children with ASD, it is important to be aware of their non-verbal behavior, create personal interactions among them, and always provide other forms of feedback modality for easier comprehension [6]. This led to the research on interventions for children with ASD. Most interventions should be deployed at an early age due to their increased neuroplasticity [5,7] and signs of autism typically start emerging between the child's birth to the third birthday [4].

Past interventions have looked into modifying the interaction with children with ASD as a form of rehabilitation. For example, a free operant procedure via delayed reinforcement was used, and it was found that there was a need for physical contact among them [8]. This was also reinforced by Dohsa-hou, a well-known Japanese rehabilitation method for children with ASD where an intended movement is trained to correlate with the physical movement [9]. This method has been widely used in research and has been proven to support mutual interaction [10,11]. The results from these findings form the basis of our system which is meant to facilitate spontaneous physical movement and interaction from the children. Fraunberger et al. [12] found that for a tool be designed to assist children with ASD, it needs to be highly engaging, provide visual support, have a structured interaction, can provide means to express themselves, and cannot be complex. We draw from these findings to design both our prototype and the conducted workshop.

2.2 Effects of Tactile and Audio Feedback

Tangible user interfaces (TUIs) have been shown to bring social benefits to children with ASD due to its tactile nature [13,14] and also suited for therapy [15]. Lego is a common tool due to its playful appearance, yet having infinite possibilities for interaction [16]. However, TUIs simply mean the use of physical interfaces; in addition to that, having an object that reacts to touch with tactility creates new interventions. For example, vibrotactile feedback has been shown to be able to enhance their learning skill [17] as it provides an "act of meaning" to an interaction for children with ASD [18].

On the other hand, music therapy has also been used to support the development of children with ASD. For example, the Observation of Social Motor Synchrony with an Interactive System (OSMOsSIS) is an interactive musical system that transforms body movements into sound [19]. Qi et al. [20] also worked on a prototype that provides audio feedback via an attachable wireless speaker on the children's body. Though both works showed an overall increase in engagement, OSMOsSIS only allowed the interaction between the researcher/facilitator and the child, whereas both works did not provide any form of tactile feedback. Another similar tool was proposed by Alessandrini et al. [21] that uses an audio-augmented paper as a tool for the therapist instead to engage the child in storytelling. However, the prototype was not designed for children to interact with each other. Le et al. [22] implemented a haptic and audio-based interactive painting activity which showed positive engagement. Yet, to our knowledge, there is little work on an intervention toolkit that is able to provide both tactile and audio feedback to children with ASD, as well as being easily integrated into their educational and social activities.

3 Furekit Development

The goal of Furekit is to design, develop and test a toolkit that can aid the communication and interaction with children with ASD. It will be attached to the children's body and used as an interactive modality for socializing, learning and basic interaction. To achieve this, we focus on the following features: 1) modular design, 2) usable wirelessly, 3) easily wearable on the body and 4) allows for simple interactions that promotes physical contact. For the design process and workshop integration, we collaborated with the Tokyo Metropolitan Rinkai Aomi School for Special Needs Education[1] (the following is abbreviated as Rinkai Aomi School).

3.1 Hardware Prototyping

When designing the appearance of the toolkit, we emphasize the need for it to look fun, simple and playable so as to encourage the children to interact with them. Additionally, the size and shape of each toolkit will depend on which

---

[1] http://www.rinkai-aomi-sh.metro.tokyo.jp/site/zen/.

area of the body it is attached to. We adopted an iterative design process, with feedback gathered from Rinkai Aomi School for each design. The TouchMusic form factor was initially used, comprising of a separate black touch surface. However, it was found that a separated touch surface make it hard to understand where to touch. Additionally, different sizes are preferable so that the toolkit can be attached to body parts that are harder to reach, yet can still be interacted with easily, such as the back. We finally opted for two variations; a general rounded design without any sharp edges, and a wider version so as to require lesser precision of touch. The designs are shown in Fig. 1. (left) as Version 2 and 2+. The interior of the toolkit was initially fitted with the ESP32-WROOM-32D[2] module. However, we found that it tends to overheat inside the enclosure, it lacks an SD card for data logging, as well as a power switch for ease of use. Therefore, we opted to design our own printed circuit board as shown in Fig. 1. (right), based on the aforementioned ESP32 board but in a circular form factor of 64 mm × 64 mm to fit the enclosure well and with additional features like a micro SD card slot, touch sensor interface, two amplifiers, an LED light and a switch.

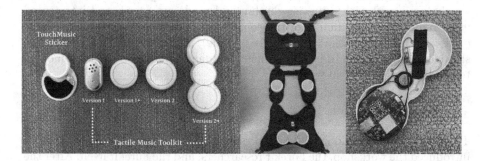

Fig. 1. (left) Various Furekit modules; (middle) attached to a wearable vest; (right) custom board fitted into a Furekit

3.2   System Interface Design

We developed a system interface in Max[3] that acts as the server machine that controls each toolkit independently and simultaneously. By interfacing with the Arduino[4] program on the board, we can control the system to toggle the sound wirelessly and in real-time, modify the sound effects, record time and behavioral data of each toolkit. When any of the modules are touched, a haptic and audio feedback is generated from them. Additionally, the program also supports parallel feedback; when a module is touched, haptic and audio feedback can also be activated on other modules. This allows for solo as well as collaborative use.

---

[2]   https://www.sparkfun.com/products/15663.
[3]   https://cycling74.com/products/max.
[4]   https://www.arduino.cc/.

4   Workshop in Special Needs Education School

We conducted a workshop for students of Rinkai Aomi School. The workshop obtained ethical approval from Keio University Graduate School of Media Design ethics committee and its contents were notified to the children's parents in advance. The school grouped the children according to their impairment status. The workshop participants were nine elementary school males in the second grade who are receiving a curriculum corresponding to autism spectrum disorders. Due to the limited number of Furekit, we divided group into team a and team b. There were four children in team a and five in team b.

According to the teachers, the more severe the children's intellectual disabilities are, the more limited they can express themselves with verbal language. Additionally, children with ASD have their own communication style and culture. In this workshop, we aimed to examine if Furekit is an effective tool for their social expression through physical interaction.

Following the teacher's suggestion, we designed a session in which participants were allowed to follow the music and touch the Furekit freely without instruction. To make it easier for children with ASD to get on with the activity, we chose a song[5] with a precise rhythm and moderate speed, and intercepted 90 s from the very beginning of the song for the session. We attached Furekit on a wearable vest as shown in Fig. 1. (middle), and made the modules on the shoulders and stomach to produce a hand-clapping sound, and the back module to produce a ride cymbal sound.

4.1   Procedure

This Workshop was held in a 30-min format. In the first five minutes, the teacher introduced the workshop, the Furekit researchers, and the Furekit itself. After the introductory session, children were asked to dance by following the video of a simplified version "Haptic Exercise", a touch-based gymnastic that aims to rediscover the sense of touch in body movement and environment, which is produced by Haptic Design Project[6]. Then, the children of team a were asked to stand up, put on the vests with Furekit modules and dance freely with the song (shown in Fig. 2). Next, the team a and b exchanged the position, the team b children stood up, and completed the same activities as the team a. In this session, each team had 3 min to prepare and 90 s to dance. Before putting the vests on the children in team b, the staff disinfected each Furekit thoroughly. Lastly, the teacher ended the workshop with a 5-min summary of the today's content.

4.2   Results and Discussion

Observations. Three of the nine children started out without touching, but their proactive actions increased in the middle of the song. Six of them tapped

---

[5] https://www.youtube.com/watch?v=278TJLCJ8RQ.
[6] http://hapticdesign.org/.

Fig. 2. The children are dancing with Furekit

the Furekit to the rhythm, and all of them touched the Furekit modules on their shoulders, stomach, and back. Participant 1(abbreviated as P1), P6, and P8 tapped Furekit following the order of shoulders, stomach, and back. P2 and P9 mainly looked at the next child and touched Furekit less frequently; however, when it was the other team's turn, P2 swayed his body from side to side with the music and tapped his body to the rhythm. P3 and P4 were swaying left and right with the music while tapping Furekit. P5 sat on the floor with his hands waving along to the rhythm when the other team danced. When it came to his turn, he first put the shoulders and stomach modules against his ears to hear the sound; when the song was halfway through, he began to tap Furekit frequently as well as the floor. P7 walked around the room while tapping Furekit on his body.

Feedback from School Teachers. Overall, the children performed actively with Furekit. The interactive experience of Furekit was very entertaining for them. Several children who are sensitive to things on their bodies did not show antipathy to wearing the Furekit, and they even tried hard to interact with the Furekit. Though some children were confused at the beginning of the song, most of them became active by imitating the children next to them. The school rarely gives free expression classes to children with ASD, and this workshop allowed the teachers to observe each child during this free activity and gain a deeper understanding of the child's unique communication style. Moreover, teachers believe that Furekit can also be used in other forms of teaching, such as placing Furekit modules throughout the classroom and allowing children to walk around and move their bodies to explore the environment.

Discussion. Through the workshop, we found that Furekit, which integrates tactile and sound stimuli, can serve as an effective tool to encourage touch behavior in children with ASD. It acts as a reward-based system giving sensory responses to each goal-directed movement, which shows its potential in developing the intentional movement of autism. By wearing the Furekit, children with

ASD can be motivated to express themselves through physical movement. We also expect Furekit to be used as an educational toolkit to accommodate assistive learning for children with ASD, as well as a tool that can increase empathy between the teacher and children. However, in this workshop, we attached the Furekit to the vest to make it easier to put on and take off, which limited the behavior pattern of the children. Furthermore, due to the COVID-19, the school does not allow physical contact between children, which affected the observation of the effectiveness of Furekit in promoting interaction among children.

5   Conclusion

We propose Furekit, a wearable tactile music toolkit to assist children with ASD in social development and learning. We found that overall, Furekit was well-received and was able to encourage spontaneous physical movement in children with ASD. In the future, we plan to design more wearable forms of Furekit and continue to explore the applications as an educational and communication tool with its collaborative use. We will further discuss how Furekit can be more integrated into their social lives, as well as how it can also potentially be used for other forms of neurodevelopmental disorders.

Acknowledgement. This work was supported by teachers from Tokyo Metropolitan Rinkai Aomi School for Special Needs Education, and was funded by JST Moonshot R&D Program "Cybernetic being" Project (Grant number JPMJMS2013).

References

1. Johnson, C.P., Myers, S.M.: Identification and evaluation of children with autism spectrum disorders. Pediatrics **120**(5), 1183–1215 (2007)
2. Trevarthen, C., Delafield-Butt, J.T.: Autism as a developmental disorder in intentional movement and affective engagement. Front. Integr. Neurosci **7**, 49 (2013)
3. Howlin, P., Goode, S., Hutton, J., Rutter, M.: Adult outcome for children with autism. J. Child Psychol. Psychiatry **45**(2), 212–229 (2004)
4. Landa, R., Garrett-Mayer, E.: Development in infants with autism spectrum disorders: a prospective study. J. Child Psychol. Psychiatry **47**(6), 629–638 (2006)
5. Johnson, M.H., Munakata, Y.: Processes of change in brain and cognitive development. Trends Cogn. Sci. **9**(3), 152–158 (2005)
6. National Research Council: Educating Children with Autism. The National Academies Press, Washington (2001)
7. Kuhl, P.K., Tsao, F.M., Liu, H.M.: Foreign-language experience in infancy: effects of short-term exposure and social interaction on phonetic learning. Proc. Natl. Acad. Sci. **100**(15), 9096–9101 (2003)
8. Takahashi, M., Ohno, H.: Early intervention with a child suspected to have autism: a case study using free operant procedures. Jpn. J. Special Educ. **42**(5), 329–340 (2005)
9. Imura, O., et al.: Introduction to Dohsa-hou: an integrated Japanese body-mind therapy (2015)

10. Sasagawa, E., Oda, H., Fujita, T.: Effectiveness of the Dohsa-hou on mother-child interactions: children with Down syndrome and autism. Jpn. J. Special Educ. **38**(1), 13–22 (2000)
11. Kawano, J., Fujino, H.: Dohsa-hou intervention for reciprocal interpersonal interaction for a girl with Kabuki syndrome and autism spectrum disorder. Clin. Case Rep. **9**(6), e04296 (2021)
12. Frauenberger, C., Good, J., Alcorn, A., Pain, H.: Supporting the design contributions of children with autism spectrum conditions. In: Proceedings of the 11th International Conference on Interaction Design and Children, pp. 134–143 (2012)
13. Farr, W., Yuill, N., Raffle, H.: Social benefits of a tangible user interface for children with autistic spectrum conditions. Autism **14**(3), 237–252 (2010)
14. Nakano, T., Kato, N., Kitazawa, S.: Superior haptic-to-visual shape matching in autism spectrum disorders. Neuropsychologia **50**(5), 696–703 (2012)
15. Vaucelle, C., Bonanni, L., Ishii, H.: Design of haptic interfaces for therapy. In: Proceedings of the SIGCHI Conference on Human Factors in Computing Systems, pp. 467–470 (2009)
16. LeGoff, D.B.: Use of LEGO© a therapeutic medium for improving social competence. J. Autism Dev. Disord. **34**(5), 557–571 (2004)
17. Mustafa, M., Arshad, H., Zaman, H.B.: Framework methodology of the autism children–vibratory haptic interface (AC-VHI). In: 2013 International Conference on Advanced Computer Science Applications and Technologies, pp. 201–206. IEEE (2013)
18. Trimingham, M.: Touched by meaning: haptic effect in autism. Affect. Perform. Cogn. Sci.: Body, Brain Being, 229–240 (2013)
19. Ragone, G.: Designing embodied musical interaction for children with autism. In: The 22nd International ACM SIGACCESS Conference on Computers and Accessibility, pp. 1–4 (2020)
20. Qi, D., Hynds, D., Shibasaki, M., Pai, Y.S., Minamizawa, K.: Tactile music toolkit: supporting communication for autistic children with audio feedback. In: 2021 IEEE World Haptics Conference (WHC), p. 1156. IEEE (2021)
21. Alessandrini, A., Cappelletti, A., Zancanaro, M.: Audio-augmented paper for therapy and educational intervention for children with autistic spectrum disorder. Int. J. Hum Comput Stud. **72**(4), 422–430 (2014)
22. Le, H.H., Loureiro, R.C., Dussopt, F., Phillips, N., Zivanovic, A., Loomes, M.J.: Soundscape and haptic cues in an interactive painting: a study with autistic children. In: 5th IEEE RAS/EMBS International Conference on Biomedical Robotics and Biomechatronics, pp. 375–380. IEEE (2014)

318   D. Qi et al.

A Database of Vibratory Signals from Free Haptic Exploration of Natural Material Textures and Perceptual Judgments (ViPer): Analysis of Spectral Statistics

Matteo Toscani[1,2]([✉]) and Anna Metzger[1,2]

[1] Justus-Liebig University, Giessen, Germany
[2] Bournemouth University, Poole, UK
{mtoscani,ametzger}@bournemouth.ac.uk

Abstract. We recorded vibratory patterns elicited by free haptic exploration of a large set of natural textures with a steel tool tip. Vision and audio signals during the exploration were excluded. After the exploration of each sample, participants provided judgments about its perceptual attributes and material category. We found that vibratory signals can be approximated by a single parameter in the temporal frequency domain, in a similar way as we can describe the spatial frequency spectrum of natural images. This parameter varies systematically between material categories and correlates with human perceptual judgements. It provides an estimate of the spectral composition of the power spectra which is highly correlated with the differential activity of the Rapidly Adapting (RA) and Pacinian Corpuscle (PC) afferents.

Keywords: Statistics of natural textures · Materials · Perception · Touch

1 Introduction

When we touch natural surfaces, we are extremely good at distinguishing different materials (e.g. silk from satin) despite the complexity of the patterns of stimulation elicited by tactile exploration [1].

Touching a surface causes patterns of vibrations on our skin which are sensed by the mechanoreceptors embedded in the skin. Softness and temperature information is also available. The vibratory signals play an important role for perceiving natural textures [2–4]. Perceptual representations based only on vibratory signals acquired indirectly with a tool are remarkably similar to representations obtained with bare hand exploration [4]. These vibratory signals are highly

This work was supported by Deutsche Forschungsgemeinschaf (DFG, German Research Foundation) – project number 222641018 – SFB/TRR 135.

H. Seifi et al. (Eds.): EuroHaptics 2022, LNCS 13235, pp. 319–327, 2022.
https://doi.org/10.1007/978-3-031-06249-0_36

dependent on exploration movements (e.g. speed) and on the local properties of textures.

In order to understand the relationship between the vibratory signals elicited by free exploration of natural textures and how we perceive these textures while we touch them, we built a database of vibratory patterns recorded while human participants freely explored a large set of natural textures with a steel tool tip. After the exploration of each signal, participants provided judgments about their perceptual attributes (e.g. roughness or friction) and rated how much the explored material feels like each of the seven material categories used in the experiment (wood, plastic, fabric, paper, metal, stone and animal).

Here we describe the properties of the vibratory signals we recorded in relation to the perceptual judgements provided by human participants. The spectral power P relates to the temporal frequency f according to a power law ($P = \frac{1}{f^s}$). We found that vibratory signals can be approximated by a single parameter s in the temporal frequency domain. The same relationship characterizes the spatial frequency spectrum of natural visual textures (e.g. [5]). Crucially, this parameter s varies systematically between material categories such that it can be used to predict physical category labels (e.g. metal vs. plastic) better than chance. Classification performance improves when classifying perceptually assigned labels, suggesting a relationship between perception of material categories and the spectral statistic described by s. In fact, s correlates with human judgements of some of the perceptual attributes participants rated.

In our previous research, we showed that haptic perceptual representations emerge by efficient encoding of vibratory signals [4]. These representations resemble the responses of the RA and PC afferents, tuned to lower temporal frequencies (peak 50 Hz) and higher (250 Hz), respectively [6]. Here we show that s correlates with the ratio of the PC to the RA responses; i.e. the higher the contribution of high temporal frequencies as compared to lower frequencies, the higher s. Thus, s provides a concise measure of the temporal frequency composition of the vibratory signals elicited by the exploration of natural images, provides information about material categories and perceptual attributes, and can be computed by comparing the responses of mechanoreceptors.

2 Methods

2.1 Participants

Eleven students volunteered to participate in the experiment; all were naïve to the purpose of the experiment and were reimbursed for their participation. The study was approved by the local ethics committee LEK FB06 at Giessen University and was in line with the declaration of Helsinki from 2008. Written informed consent was obtained from each participant.

2.2 Stimuli

Our natural textures consisted of 81 different material samples which were glued on wooden pieces (14×14 cm, Fig. 1). These materials samples are the same used by [7]. They belong to seven material categories: plastic, paper, fabric, fur and leather, stone, metal and wood, and were chosen to represent the large variety of materials we encounter in everyday life.

Fig. 1. Photographs of all material samples. Rows indicate different material categories: plastic, paper, fabric, fur and leather, stone, metal and wood, from top to bottom.

2.3 Procedure and Apparatus

Participants set at a table looking at a computer monitor elevated by a support. Material samples were positioned by the experimenter in front of them through a hole in the support so that they could easily be touched but no visual information was available. The sound from the exploration of materials was covered by earplugs and white noise presented via headphones.

Participants freely explored the 81 surfaces with a 3D printed pen containing a steel tip at its end and a mounted accelerometer (ADXL345). This way we could record the vibrations elicited by the interaction between the steel tip and material samples. Each participant explored each material once.

The onset of the white noise signalled that they could begin the exploration. They were instructed to slide the pen over the material's surface. After the exploration of each material, participants rated how much the explored material felt like each of the seven material categories (paper, fabric, animal, stone, plastic, wood, metal) on a scale from "very different" to "very similar". Then, they rated how much the material could be described by each of seven opposing adjective pairs. We used a subset of the descriptors used by [7]: rough vs. smooth, hard vs. soft, orderly vs. chaotic, warm vs. cold, elastic vs. not elastic, high friction vs. slippery, textured/patterned vs. homogeneous/uniform. The experimental software was written with Psychopy [8].

2.4 Vibratory Signals

For each material, 10 s of recording were acquired at 3200 Hz temporal resolution. We started the recording after 2 s of exploration and stopped 2 s before the exploration was terminated (i.e. participants explored 14 s each sample), to prevent that signals are affected by contact onset and offset. We filtered out frequencies 10 Hz as they may be ascribed to exploratory hand movements [9–12]. We cleaned the signals by removing frequencies 800 Hz, which are not relevant for perception of texture properties of materials [1–3, 13, 14] and may be caused by measuring noise.

2.5 Analysis

We approximated the relationship between temporal frequency f and the amplitude power P with the following function $P = \frac{1}{f^s}$ In a log-log space, this equals to the following linear relationship $P = -sf$. We determined s by fitting a line in log-log space. From now on we refer to s as the *slope* (of the line). We assessed how well this linear relationship can approximate the power spectrum by computing R^2, i.e. the proportion of variance explained by the linear fit. *Slope* was computed for each material and each participant separately, i.e. for each exploration trial. Since signals only included power at frequencies 10 Hz and 800 Hz, the linear fits were performed within this interval.

We used *slope* to classify the material categories by means of a linear classifier. For classification, we averaged *slope* across participants yielding one average *slope* per material. To prevent over-fitting, we iteratively left out one material and trained the classifier on the remaining slopes, then computed performance on the left-out material. We used bootstrap analysis to test whether classification performance was higher than chance: we repeated the classification analysis 5000 times shuffling the category labels every time. Thus we computed the distribution of classification accuracy under the null hypothesis of chance-level classification. The 95% confidence interval was computed by reading out the 2.5^{th} and the 97.5^{th} percentiles of the distribution. The empirical chance level corresponded to the mean of that distribution. We repeated the classification analysis based on perceptual labels. To determine perceptual category labels, first we assigned to each material the highest rated category by each participant, then we chose the most frequent category, i.e. the one chosen by the majority of participants.

To relate *slope* to the responses of the mechanoreceptors, we estimated the responses of the RA and PC afferents using the TouchSim toolbox [15].

3   Results

Figure 2 shows one example of vibratory signal per material category. For all categories the linear model seems to provide a good approximation of the spectral profile.

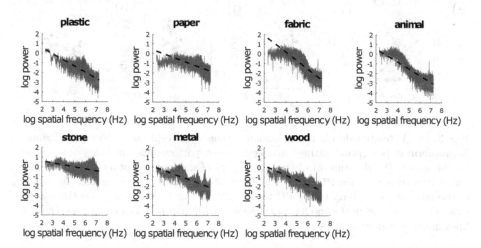

Fig. 2. Spectral profiles of example materials for each of the seven categories. Natural logarithm of the amplitude spectrum on the y-axis, as a function of the natural logarithm of temporal frequency (x-axis). Each diagram represents the spectral profile of one example material, explored by one participant in one experimental trial.

On average, the linear fits could explain 43% of the variance of the spectral profiles, indicating that almost half of the variability in such high dimensional signals (32000 dimensions) can be approximated by two parameters (slope and intercept). We used a linear classification analysis to determine whether *slope* variations across different vibratory patterns provide information about material properties. Based on *slope* we could classify material categories better than chance (classification performance = 33.3%; empirical chance level = 14.8 %, with [6.17 23.46] 95% Confidence interval). Performance increased when we repeated the classification analysis based on perceptually assigned categorical labels (classification performance = 38.27%; empirical chance level = 16.67%, with [6.17 27.16] 95% Confidence interval). This means that human misclassifications could be explained by differences in *slope*, suggesting that perception is at least in part based on *slope*. To explore this possibility, we investigated the relationship between *slope* and human ratings of perceptual attributes (Fig. 3).

Correlation analyses show that shallower *slope* is significantly associated with low roughness, elasticity, fiction and with high hardness judgments ($r = -0.31$, $r = -0.45$, $r = -0.31$, $r = 0.35$; all p-values $< \alpha$, with ($\alpha = 0.00714$), according to Bonferroni correction for seven post-hoc comparisons).

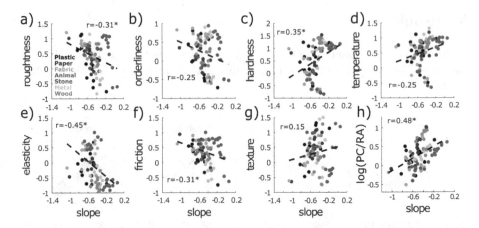

Fig. 3. (a–g) Relationship between human ratings of perceptual attributes and *slope*. Z-transformed perceptual ratings averaged across participants, on the y-axis. *Slope* on the x-axis. (h) Relationship between activity of mechanoreceptors and *slope*. The differential activity of the PC and RA afferents is represented by the natural logarithm of the ratio of the activity of the PC to the RA afferent (y-axis). *Slope* on the x-axis. Colors in the legend in the top left panel indicate different material categories. Dashed lines indicate regression lines.

RA and PC afferents seem crucial for perception of natural textures [3,4]. RA respond to relative low temporal frequencies, whereas PC to higher frequencies. Therefore, the differential activity of these afferents provides information about the composition of the power spectrum of the tactile input. We quantified this information as the natural logarithm of the ratio between the activity of the RA and the PC afferents (PC/RA). For each vibratory signal we computed the RA and PC responses using the TouchSim toolbox [15], and related this quantity to *slope* (Fig. 3).

The ratio PC/RA correlates with *slope* ($r = 0.48$, $p < 0.00001$), indicating that they employ a similar way to summarize the power spectrum of vibratory signals elicited by exploration of natural materials.

4   Discussion

We recorded the vibratory signals elicited by free exploration of a large number of natural materials. Each exploration is associated to perceptual judgments. Other databases of vibratory signals have been published. Vibratory signals were measured with the same system we used at controlled speed and free exploration [9,10,12]. To our knowledge, our database is the only one to present perceptual judgments, effectively ruling out the influence of visual and audio information. Furthermore, we focused on everyday life materials rather than following industrial conventions for sampling and naming materials. Our database with vibratory signals and perceptual judgements (ViPer) is publicly available (https://github.com/matteo-toscani-24-01-1985/ViPer). Other attempts have been made to relate perceptual judgments to vibratory signals elicited by natural materials. However, the vibration signals were elicited by exploratory motion by robots [16,17].

We showed that vibratory signals can be approximated by a single parameter representing the *slope* of the linear relationship between spectral power and temporal frequency in log-log space in the frequency domain. This parameter can be used to classify material categories. Classification performance improves when classifying perceptual labels, suggesting that *slope* can account for human misclassifications. The bond between *slope* and perception is strengthened by the correlations with human judgements of perceptual attributes like roughness, hardness, elasticity and friction.

The same linear relationship characterizes the spatial frequency spectrum of natural visual textures (e.g. [5]). This may indicate scale invariance of the visual world. Spatially rescaling an image by a factor of α implies rescaling the corresponding frequency domain axes by a factor of $\frac{1}{\alpha}$. A power spectrum that falls as a power law will retain its shape under this transformation, i.e. would show scale invariance. We speculate that the *slope* we used to describe vibratory signals may exhibit time invariance, i.e. would change minimally for explorations at different speed. Hence, *slope* could be estimated based on the differential activity of the RA and PC afferents (as suggested by Fig. 3h) and used to perceive material properties despite different velocities of exploratory movements, i.e. to achieve speed invariance [18]. However, we did not measure the movement speed, therefore we cannot test how much it affects *slope*.

We previously showed that haptic perceptual representations emerge by efficient encoding of vibratory signals [4]. Such representations can be described within a space whose dimensions resemble the activity of RA and PC afferents. In this space, the representations of different materials tend to lay on a line along which it is possible to distinguish between different material categories, i.e. materials can be distinguished based on the differential activity of the RA and PC afferents. As *slope* is able to capture the differential activity of the RA and PC afferents, we speculate that *slope* may be a prominent feature of the compressed representation we previously discovered.

Our analyses showed that a simple statistical property of the Fourier spectrum is able to capture nearly half of the variability within the vibratory signals elicited by a large number of natural textures. This property systematically differs between material categories and correlates with perceptual judgments. Our results may represent a significant step ahead for tactile rendering, just like in vision pink noise is used for synthesizing naturalistic textures (e.g. [19]).

References

1. Manfredi, L.R., et al.: Natural scenes in tactile texture. J. Neurophysiol. **111**, 1792–802 (2014)
2. Bensmaïa, S., Hollins, M.: Pacinian representations of fine surface texture. Percept. Psychophys. **67**(5), 842–854 (2005). https://doi.org/10.3758/BF03193537
3. Weber, A.I., et al.: Spatial and temporal codes mediate the tactile perception of natural textures. Proc. Natl. Acad. Sci. **110**(42), 17107–17112 (2013)
4. Metzger, A., Toscani, M.: Unsupervised learning of haptic material properties. eLife **11**, e64876 (2022)
5. Simoncelli, E.P., Olshausen, B.A.: Natural image statistics and neural representation. Annu. Rev. Neurosci. **24**(1), 1193–1216 (2001)
6. Mountcastle, V.B., LaMotte, R.H., Carli, G.: Detection thresholds for stimuli in humans and monkeys: comparison with threshold events in mechanoreceptive afferent nerve fibers innervating the monkey hand. J. Neurophysiol. **35**(1), 122–136 (1972)
7. Baumgartner, E., Wiebel, C.B., Gegenfurtner, K.R.: Visual and haptic representations of material properties. Multisensory Res. **26**(5), 429–455 (2013)
8. Peirce, J., et al.: Psychopy2: experiments in behavior made easy. Behav. Res. Methods **51**(1), 195–203 (2019). https://doi.org/10.3758/s13428-018-01193-y
9. Strese, M., Lee, J.Y., Schuwerk, C., Han, Q., Kim, H.G., Steinbach, E.: A haptic texture database for tool-mediated texture recognition and classification. In: 2014 IEEE International Symposium on Haptic, Audio and Visual Environments and Games (HAVE) Proceedings, pp. 118–123 (2014)
10. Strese, M., Boeck, Y., Steinbach, E.: Content-based surface material retrieval. In: 2017 IEEE World Haptics Conference (WHC), pp. 352–357. IEEE (2017)
11. Strese, M., Boeck, Y., Steinbach, E.: Content-based surface material retrieval. In: 2017 IEEE World Haptics Conference (WHC), pp. 352–357. IEEE (2017)
12. Culbertson, H., Lopez Delgado, J.J., Kuchenbecker, K.J.: The Penn haptic texture toolkit for modeling, rendering, and evaluating haptic virtual textures (2014)
13. Hollins, M., Bensmaïa, S.J., Washburn, S.: Vibrotactile adaptation impairs discrimination of fine, but not coarse, textures. Somatosens. Mot. Res. **18**(4), 253–262 (2001)
14. BensmaIa, S.J., Hollins, M.: The vibrations of texture. Somatosens. Mot. Res. **20**(1), 33–43 (2003)
15. Saal, H.P., Delhaye, B.P., Rayhaun, B.C., Bensmaia, S.J.: Simulating tactile signals from the whole hand with millisecond precision. Proc. Natl. Acad. Sci. **114**(28), E5693–E5702 (2017)
16. Chu, V., et al.: Robotic learning of haptic adjectives through physical interaction. Robot. Auton. Syst. **63**, 279–292 (2015)
17. Chu, V., et al.: Using robotic exploratory procedures to learn the meaning of haptic adjectives. In: 2013 IEEE International Conference on Robotics and Automation, pp. 3048–3055. IEEE (2013)
18. Boundy-Singer, Z.M., Saal, H.P., Bensmaia, S.J.: Speed invariance of tactile texture perception. J. Neurophysiol. **118**(4), 2371–2377 (2017)
19. Ebert, D.S., Musgrave, F.K., Peachey, D., Perlin, K., Worley, S.: Texturing & modeling: a procedural approach. Morgan Kaufmann, Massachusetts (2003)

Appendix

Developing a VR Training Environment for Fingers Rehabilitation with Haptic Feedback

Alireza Abbasimoshaei(✉)📧 ⓘ, Ahmed Aly Ibrahim Alyⓘ, and T. A. Kernⓘ

Hamburg University of Technology, Hamburg, Germany
al.abbasimoshaei@tuhh.de
https://www.tuhh.de/

Abstract. There are many people worldwide who suffer from hand disability due to stroke. The inability to extend the thumb and fingers is a hindrance for these patients, as they cannot successfully grasp and release in daily life. Gamification of the VR -based rehabilitation system encourages patients to actively participate in rehabilitation. This could lead to better recovery of their fingers and ideally complement conventional therapy. This research focuses on the combination of the VIVE headset, the haptic feedback Sense Glove, and the software development of a game that provides an engaging VR training environment for finger rehabilitation. The game is aimed at patients with hemiparesis who can train to grasp and place virtual objects of various shapes and sizes in a virtual environment. The performance of the patients is analyzed by the system as follows. The main evaluation parameters are the speed and acceleration of the hands. After each level, a report is displayed to evaluate the performance.

Keywords: Hemiperasis · Rehabilitation · Gamification · Virtual reality · Sense glove

1 Introduction

Stroke is a major cause of long-term disability in most countries, especially in the elderly population, where the stroke incidence is highest. Of the 795,000 new stroke patients, 26% remain disabled in basic activities of daily living and 50% have some mobility problem due to hemiparesis [1]. Among these stroke survivors, the most common loss is extension of the fingers. Due to the high cost associated with conventional therapy, the duration of sessions is always quite limited. Therefore, safe-to-use automated rehabilitation systems can be used independently at home and ideally complement traditional therapy, as patients can perform rehabilitation exercises performed by physiotherapists and therapists [2]. VR application combined with haptics would increase the efficiency of rehabilitation of stroke patients. The use of haptic-based therapy contributes significantly to reduce of transfer cost. Gamification of VR -based rehabilitation can improve patient motivation, resulting in a better overall experience. This can be achieved through a mini VR designed for hemiparesis patients that provides simple and inexpensive rehabilitation exercises. These exercises should be

© The Author(s) 2022
H. Seifi et al. (Eds.): EuroHaptics 2022, LNCS 13235, pp. 331–333, 2022.
https://doi.org/10.1007/978-3-031-06249-0

repetitive, consistent, and provide a high level of engagement [3]. Compared with conventional therapy, VR rehabilitation can monitor and store patient performance data for evaluation and assessment [2]. The game immerses the patient in tasks that guide the same rehabilitation exercises and movements that would be provided in a physical therapy session. Cues, challenges, pacing, and feedback are critical to keep the patient engaged [4]. The patient should complete the task efficiently by overcoming challenges in each level. As he gains this confidence and becomes familiar with the various actions, it will be easier for him to perform the tasks and focus on the game [5]. Many forms of home-based technology targeting stroke rehabilitation and a number of human factors are provided in a comprehensive review [6].

2   Game Development

The game should be viable in virtual reality and it should support the execution of the rehabilitation movement. Unity 3D is one of the most popular platforms for creating virtual environments. Because the target patient is people who suffer from hemiparesis, the room environment game was chosen to practice repetitive grasping and placement exercises in a training environment. Grasping performance is evaluated throughout the game and during interaction with virtual objects in the virtual environment. The ability to extend all fingers and the thumb can be tested using VIVE and the Sense Glove System. The VR -based rehabilitation game consists of three levels, allowing patients to progress from one level to the next based on their performance. Patients are asked to take a certain number of objects with different radii from a table and place them in a different location in the environment. If the patients succeed in placing the objects with different sizes and dimensions at the target location within a certain period of time, they advance to the next stage. Objects and material properties can be designed to facilitate rehabilitation. Therefore, it is important to select the objects based on size, shape, and resistance that are most appropriate. Depending on the patient's ability level, objects of different sizes will be used. Through haptic force feedback, a different sensation can be felt depending on the virtual object and its material. To motivate patients, there is also a point system and audio feedback in the environment.

3   Experimental Results

The game consists of three different difficulty levels. At level 1, patients select large objects with workspace range of 0 to 30°; at level 2, medium sized objects with a working range of 30 to 70°; and at level 3, small objects with workspace range of 0 to 90°. In addition, the game provides the ability to display all joints angles for each finger individually, allowing the therapist to analyze the patient's performance during the game. A survey with six healthy participants was conducted to evaluate the game and the responses are collected. The questionnaire showed that most of the participants felt comfortable in using the system and were comfortable with the environment.

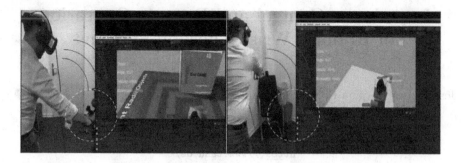

Fig. 1. Hand movement in the game environment while the object is grasped (from white circle to hand place)

4   Conclusion

This work focused on the design and implementation of a responsive VR training environment game in the Unity3D game engine that can be used for hemiparesis patients for hand rehabilitation. This system extracts the total flexion angle, angular velocity and angular acceleration of finger movement with below than one percent precision and can also estimate the range of motion of the hand. To achieve this, communication is established between the Sense Glove, the VR headset, and the tracker, which was challenging and required many configurations in Unity. Patient performance is evaluated after each level to assess them and move to the next level of difficulty, which is based on grasping and releasing objects within a certain time. Angular velocity and angular acceleration are recorded throughout the duration of the game to provide a continuous report of the patient's progress. Data analysis has been implemented to record the fingers flexion in real time when it touches the object and when it grasps the object.

References

1. Katan, M., Luft, A.: Global burden of stroke. In: Seminars in neurology. 38(2), pp. 208–211. Thieme Medical Publishers (2018)
2. Sadihov, D., et al.: Prototype of a VR upper-limb rehabilitation system enhanced with motion-based tactile feedback. In: 2013 World Haptics Conference (WHC), pp. 449–454. IEEE (2013)
3. Connelly, L., et al.: Use of a pneumatic glove for hand rehabilitation following stroke. In: 2009 Annual International Conference of the IEEE Engineering in Medicine and Biology Society, pp. 2434–2437. IEEE (2009)
4. Burke, J.W., et al.: Optimising engagement for stroke rehabilitation using serious games. In: The Visual Computer. 25(12), pp. 1085–1099 (2009)
5. Ma, M., Bechkoum, K.: Serious games for movement therapy after stroke. In: 2008 IEEE International Conference on Systems, man and Cybernetics, pp. 1872–1877. IEEE (2008)
6. Yu, M., et al.: Home-based technologies for stroke rehabilitation: a systematic review. Int. J. Med. Inf. **123**, 11–22 (2019)

Impedance Control and Remote Operation of Robotic Wrist Therapy Systems

Jose Pedro Kitajima Borges$^{(\boxtimes)}$, Alireza Abbasimoshaei , and T. A. Kern

Hamburg University of Technology, Hamburg, Germany
al.abbasimoshaei@tuhh.de
https://www.tuhh.de/

Abstract. In this paper, a concept for tele-rehabilitation between the patient and the physical therapist is presented and in order to provide haptic feedback between these two parties, an overview of an impedance controller for a two-part wrist rehabilitation is presented. For checking the output, the implemented control algorithm simulated an arbitrarily designed spring-damper system. In the practical part, the remote implementation of the impedance control algorithm to connect the two robots with a virtual spring and damper was also tested. To verify the haptic feedback, the position, velocity, and torque data from both robots were collected and analyzed and remote communication of the two-part robot is enabled by developing a control algorithm.

Keywords: Impedance control · Telemanipulation · Haptic feedback

1 Introduction

Certain conditions such as hemiparesis and hemiplegia result in weakness or complete loss of strength in the upper extremities of the body. Physical therapy is required to treat these conditions to prevent muscle spasticity and joint stiffness [1]. However, this process is costly and repetitive, which has led to an increasing demand for remote physical therapy. A cost-effective alternative to physical therapy is the use of robotic systems that enable remote interaction between physical therapist and patient by providing haptic feedback. Telerehabilitation is defined as the delivery of rehabilitation services via information and communication technologies [2].

The iMEK Institute has proposed a wrist and forearm therapy system consisting of two identical robots. Each robot provides wrist abduction/adduction, flexion/extension, and pronation/suppination. In this paper, an impedance controller was designed and implemented to connect a virtual spring and damper between the two sides to create the sensation of haptic feedback.

2 Impedance Controller

The goal of impedance control is to control the system so that it behaves like a spring-damper system given the desired stiffness and damping constants [3, 4].

© The Author(s) 2022
H. Seifi et al. (Eds.): EuroHaptics 2022, LNCS 13235, pp. 334–336, 2022.
https://doi.org/10.1007/978-3-031-06249-0

To create an impedance controller that connects the end effector to a reference via a spring and damper while reducing the effects of inertia and bearing friction, the control Algorithm 1 described by [5] was used, where $\dot{\theta}_d$ and θ_d are the desired velocity and displacement profiles. B is the desired damping of the system and K is the desired stiffness of the system, K_f is a gain that the designer can set arbitrarily to reduce the influence of inertial and damping forces, T_a is the driving torque, T_e is the external torque, b is the rotational damping, and θ represents the angular displacement.

$$T_a = K(\theta_d - \theta) + B(\dot{\theta}_d - \dot{\theta}) + K_f[T_e + K(\theta_d - \theta) + B(\dot{\theta}_d - \dot{\theta})] \qquad (1)$$

3   Haptic Feedback

Once the system was modeled and the control algorithm for each side was determined, haptic feedback could be generated by setting the reference position and velocity of each side to its remote counterpart. In this way, both robots are connected with a virtual spring and damper, as shown in Fig. 1, where a linear model is shown for illustration. To communicate the position and velocity between the two sides, the Google Cloud Pub/Sub service was used.

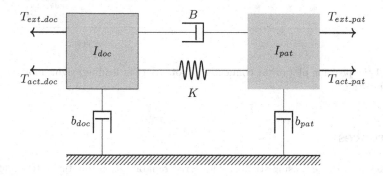

Fig. 1. Mechanical model of the 2 DOF robotic system

4   Experiments and Conclusion

In order to verify whether the haptic feedback was achieved, some experiments were conducted. The first experiment consisted of moving the patient's end effector and observing whether the physician's end effector followed the movement. The experiments were performed with a stiffness of $0.5\,\mathrm{N\,m^{-1}}$ and a damping constant of $0.1\,\mathrm{N\,s\,m^{-1}}$. The right figure in Fig. 2 shows that the physician's end effector successfully tracked the patient's position. The average value of the position error and delay were calculated to be 0.055 rad and 0.24 s which are small and hardly noticed.

The second experiment tested whether haptic feedback was achieved by analyzing the torque sent to the motors on each side based on the position difference between the two robots. The left figure in Fig. 2 shows the positions of the robots and their following situations. The right figure in Fig. 2 shows that when the two robots are in different positions, the torque commanded to the motors has similar magnitudes but different directions. This is explained by the effect of the virtual spring that was modeled, and it creates the haptic force and a sense of haptic feedback as both robots exert torques so that the positions are again aligned. Future work includes determining the optimal spring-damper parameters for better haptic feedback and increasing the stiffness of the system.

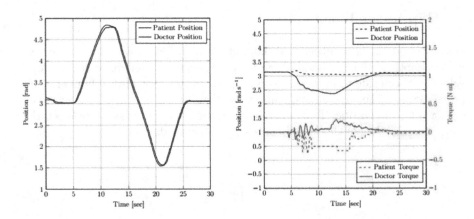

Fig. 2. Left: position plot of the connected robots (patient leading) right: motor torque versus position

References

1. Oelkers, P.: Force sensing strategies for wrist rehabilitation robotic system and use in teleoperation. Institute for Mechatronics in Mechanics (2021)
2. Brennan, D.M., et al.: A blueprint for telerehabilitation guidelines-October 2010. Telemedicine e-Health. **17**(8), 662–665 (2011)
3. Gmerek, A., Jezierski, E.: Admittance control of a 1-DoF robotic arm actuated by BLDC motor. In: 17th International Conference on Methods and Models in Automation and Robotics, pp. 633–638. IEEE (2012)
4. Atlihan, M., Akdoğan, E., Arslan, M.S.: Development of a therapeutic exercise robot for wrist and forearm rehabilitation. In: 2014 19th International Conference on Methods and Models in Automation and Robotics (MMAR), pp. 52–57. IEEE (2014)
5. Hogan, N., Buerger, S.: Impedance and interaction control. In: Robotics and Automation, pp. 375–398 (2004). https://doi.org/10.1201/9781420039733.ch19

Feasibility of Smartphone Vibrations as a Sensory Diagnostic Tool

Rachel A. G. Adenekan[1]([✉]), Alexis J. Lowber[1], Bryce N. Huerta[1], Allison M. Okamura[1], Kyle T. Yoshida[1], and Cara M. Nunez[1,2,3]

[1] Stanford University, Stanford, CA 94305, USA
adenekan@stanford.edu
[2] Harvard University, Cambridge, MA 02138, USA
[3] Cornell University, Ithaca, NY 14850, USA

Abstract. Traditionally, clinicians use tuning forks as a binary measure to assess vibrotactile sensory perception. This approach has low measurement resolution, and the vibrations are highly variable. Therefore, we propose using vibrations from a smartphone to deliver a consistent and precise sensory test. First, we demonstrate that a smartphone has more consistent vibrations compared to a tuning fork. Then we develop an app and conduct a validation study to show that the smartphone can precisely measure a user's absolute threshold. This finding motivates future work to use smartphones to assess vibrotactile perception, allowing for increased monitoring and widespread accessibility.

Keywords: Clinical diagnostics · Smartphone · Vibrotactile perception

1 Introduction

Clinical tuning forks are commonly used to diagnose diminished vibrotactile sensory function and monitor changes over time. This method requires a clinician to manually conduct a vibration sensitivity test (VST) during which the clinician strikes a tuning fork, places it on the patient's skin, then asks the patient to verbally indicate if vibrations are perceived. The highly variable tuning fork vibrations and the use of only binary responses to a single vibration stimulus leads to an imprecise VST.

Due to the ubiquity of haptic actuators in mobile phones, vibrotactile perception can be tested outside of the lab or clinic [1]. Smartphones have been used for sensory diagnostics [3, 4], but prior studies suffer from lack of characterization of the vibration stimulus, confounding factors, and use of only a binary measurement (similar to the tuning fork VST). We intend to mitigate these issues and

This work was supported in part by the Stanford Precision Health and Integrated Diagnostics Center, National Science Foundation Graduate Research Fellowships, Stanford Graduate Fellowships, and National Science Foundation grant 1830163.
Kyle T. Yoshida, Cara M. Nunez—These authors contributed equally to this work.

H. Seifi et al. (Eds.): EuroHaptics 2022, LNCS 13235, pp. 337–339, 2022.
https://doi.org/10.1007/978-3-031-06249-0

assess the reliability of mobile phones as a research and diagnostic tool. Specifically, we aim to develop a non-binary smartphone-based VST that enhances the precision and accessibility of diagnostic exams for tactile deficits.

2   Instrument Vibration Characterization

To measure tuning fork vibrations present during a VST, we attached an accelerometer (Analog Devices, EVAL-ADXL354CZ, 3-axis $\pm 2g$) to the base of 128 Hz tuning fork (CynaMed). Then, we measured acceleration using a DAQ (National Instruments, NI9220), interfaced with MATLAB (Mathworks) (Fig 1A). The resulting vibration amplitudes varied between trials (Fig 1B). This aligns with a previous finding that tuning fork vibration waveforms are sensitive to the strength of the blow used to generate the vibrations [5]. We also found that 128 Hz tuning fork resonated 178 Hz instead 128 Hz (Fig 1C).

We also measured vibrations on the front-center of an Apple iPhone 12 Pro Max with the same accelerometer (Fig 1D). Using Apple's Core Haptics framework, continuous vibrations were delivered with a "hapticSharpness" value of 1.0 and "hapticIntensity" value of 0.25 ($n = 3$). Vibration acceleration waveforms (Fig 1E) and FFTs (Fig 1F) indicate that the smartphone can relay more consistent vibration amplitudes than the tuning fork. The iPhone had a peak frequency 230 Hz, which is slightly higher than the tuning fork, but still in the range that stimulates the Pacinian corpuscles which respond to vibration [2].

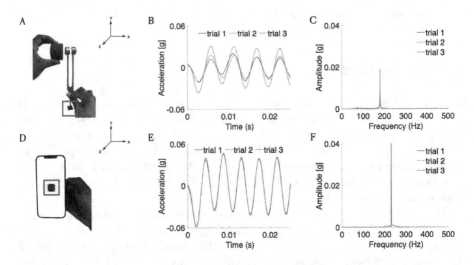

Fig. 1. Vibration measurement setup and data. A: 128 Hz tuning fork accelerometer placement (green box). B: Tuning fork filtered vibration waveforms. C: Tuning fork FFT for each trial (peak frequency = 178 Hz with varying amplitudes). D: iPhone 12 Pro Max accelerometer placement (green box). E: iPhone filtered vibration waveforms. F: iPhone FFT for each trial (peak frequency = 230 Hz with similar amplitudes).

3   Preliminary Tool Validation

We developed an iOS app that controls Apple Core Haptics variables and implements a staircase method (reversals = 8) to determine absolute intensity threshold. The "hapticIntensity" varies with a step size of 0.05, while the "duration" (0.1 s) and "hapticSharpness" (1.0) are held constant. Time intervals between vibrations are randomized to reduce bias, and response times greater than 1.5 s are counted as false positives. Absolute intensity threshold is calculated by averaging the vibration intensity readings at the reversal indices.

To test the precision of our tool, we conducted ten trials with the app on one healthy participant in a single day. For consistency between trials, we instructed the participant to hold the phone such that all four fingertips are in contact with the back of the phone and to use the thumb to provide responses via a button within the app. Physiologically, we expect absolute intensity threshold to remain stable during this time period. For this participant, we calculated an absolute intensity threshold of 0.348 ($\pm$0.040). Since the standard deviation is less than the "hapticIntensity" step size of 0.05, we conclude that our approach can consistently and precisely measure a participant's absolute intensity threshold.

4   Conclusions and Future Work

We demonstrate that smartphone-based vibrations can be used for a reliable, mobile VST. Next, we will quantify the clinical significance of our app's absolute intensity threshold by correlating it to clinical benchmarks, including 128 Hz tuning fork and the Semmes-Weinstein monofilament exam, for both healthy users and patients with sensory neuropathy ($n > 20$). Lastly, we will complete a comprehensive mechanical characterization of smartphone vibrations and refine our app so that it can be widely distributed as an in-home diagnostic tool.

References

1. Blum, J.R., et al.: Getting your hands dirty outside the lab: a practical primer for conducting wearable vibrotactile haptics research. IEEE Trans. Haptics 12(3), 232–246 (2019)
2. Brisben, A.J., Hsiao, S.S., Johnson, K.O.: Detection of vibration transmitted through an object grasped in the hand. J. Neurophysiol. 81(4), 1548–1558 (1999)
3. Jasmin, M., Yusuf, S., Syahrul, S., Abrar, E.A.: Validity and reliability of a vibration-based cell phone in detecting peripheral neuropathy among patients with a risk of diabetic foot ulcer. Int. J. Lower Extremity Wounds, 1–8 (2021)
4. May, J.D., Morris, M.W.J.: Mobile phone generated vibrations used to detect diabetic peripheral neuropathy. Foot Ankle Surg. 23(4), 281–284 (2017)
5. Rossing, T.D., Russell, D.A., Brown, D.E.: On the acoustics of tuning forks. Am. J. Phys. 60(7), 620–626 (1992)

Deixis Without Sight: Use of Spatial Demonstratives in the Absence of Vision

Gozdem Arikan[✉] and Kenny R. Coventry

University of East Anglia, England, UK
g.arikan@uea.ac.uk

Abstract. Spatial demonstratives (e.g. *this/that* in English) in combination with deictic gestures (e.g., pointing) are universal communicative tools, commonly used when establishing joint attention of interlocutors and drawing attention to a specific point of reference. There is a large body of evidence suggesting that the use of demonstratives is related to perceived proximity/reachability of the referent, mapping onto the distinction between peripersonal vs. extrapersonal space. However, most of the research investigating the relationship between spatial perception and demonstratives has focused on vision as the main sensory domain. In a series of experiments the spatial demonstrative choices of individuals was recorded, when they rely on either haptic, visual or a combination of these two sensory modalities for spatial information. Sighted (SI) and blind participants (BI) were asked to memorize and indicate the positions of various 3-D items placed on the sagittal plane.

Keywords: Spatial demonstratives · Haptic perception · Spatial perception

1 Introduction

Spatial demonstratives (e.g. *this* and *that* in English) in combination with pointing gestures (deictic communication) are universal, and demonstratives are among the first words children learn and are used frequently [1]. In English it has been shown that there is a mapping between choice of demonstrative (*this* versus *that*) and object reachability; when referring to an item which is perceptually proximal/reachable (peripersonal space), people tend to word *this* more often, and when the referent is distal (extrapersonal space), people tend to use the word *that* more often [2, 3]. However, while spatial perception had been argued as a strong factor in demonstrative choice, most of the research has focused on vision as the main sensory domain for perception. We do not know much about multisensory properties of peripersonal-extrapersonal space, especially regarding its mapping to language use.

In a study investigating blind and sighted individuals' auditory spatial perception, Kolarik, Pradhan, Cirstea and Moore (2017) suggest that blind individuals tend to perceive the proximal sound sources as more distal, and distal as more proximal than their original positions. Previous research looking at differences between visual and haptic modalities suggests some degree of difference between two in perception of orientation [5], location [6] and scenery [7]. Therefore, one might also expect that

© The Author(s) 2022
H. Seifi et al. (Eds.): EuroHaptics 2022, LNCS 13235, pp. 340–342, 2022.
https://doi.org/10.1007/978-3-031-06249-0

language used to describe objects presented at different distances may be affected by tactile experience of the world and the presence/absence of visual cues.

In four conditions (SI-haptic only, SI-visual only, SI-multisensory and BI), we manipulated the sensory modalities participants can use when perceiving object positions and recorded demonstrative choice of participants for objects varying in distance from the participant in the sagittal plane. The main aim of this project is to bridge the gap in literature concerning language use and multisensory spatial perception. We predicted that the items within reach will be labelled with word *this* more often that the word *that*. Moreover, we expected that the exact position of "within reach" will differ for sighted and blind individuals, with corresponding differences in demonstrative choice.

2 Methodology

We adapted Memory Game Paradigm [8] to haptic perception (active touch to explore setup and grasp objects). During the experiment, sighted and blind participants memorized and indicated the positions of 3-D [9] geometrical objects (different shape and surface texture but same size). Participants interacted with the stimuli either only haptically, only visually or with combination of these two sensory modalities (Fig. 1).

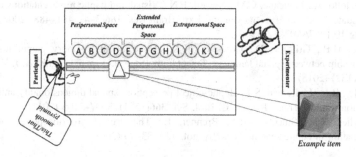

Fig. 1. Schematic representation of experimental set-up. There were 12 possible object positions (A to L). The positions varied across participants depending on reaching capacities. The objects were placed on a 150 cm rail, one object at a time. Items within peripersonal space are reachable with simple arm stretch. Items outside peripersonal space will require stretching from torso. Items A-D are within peripersonal space, E-F within extended peripersonal space and I-L in extrapersonal space.

In first condition, using both of their hands [10], blindfolded-sighted participants were asked to haptically explore the stimuli to memorize object location. After the haptic exploration, participants are asked to label/indicate the object position with a sentence such as "*this/that* rough triangle" and pointing gesture. Once participants labelled items three times, they perform the memory task in which they are asked to move the object on the glider to match the previous object locations. In second condition, sighted participants were asked to observe the items only visually. In the third condition, sighted participants are asked to touch and look at the items. In final condition, blind individuals are asked to perform the same task.

Results: Data collection is still in progress and will be completed by May 2022.

References

1. Diessel, H.: Demonstratives, joint attention, and the emergence of grammar. Cogn. Linguist. **17**(4), 463–489 (2006)
2. Coventry, K.R., Griffiths, D., Hamilton, C.J.: Spatial demonstratives and perceptual space: describing and remembering object location. Cogn. Psychol. **69**, 46–70 (2014)
3. Coventry, K.R., Valdés, B., Castillo, A., Guijarro-Fuentes, P.: Language within your reach: near-far perceptual space and spatial demonstratives. Cognition **108**(3), 889–895 (2008)
4. Kolarik, A.J., Pardhan, S., Cirstea, S., Moore, B.C.J.: Auditory spatial representations of the world are compressed in blind humans. Exp. Brain Res. **235**(2), 597–606 (2016). https://doi.org/10.1007/s00221-016-4823-1
5. Postma, A., Zuidhoek, S., Noordzij, M.L., Kappers, A.M.L.: Haptic orientation perception benefits from visual experience: evidence from early-blind, late-blind, and sighted people. Percept. Psychophys. **70**(7), 1197–1206 (2008)
6. Feng, G., Hu, Q., Shao, Y.: Blindfolded adults' use of geometric cues in haptic-based relocation. Psychon. Bull. Rev. **29**, 88–96 (2021). https://doi.org/10.3758/s13423-021-01994-x
7. Pasqualotto, A., Finucane, C.M., Newell, F.N.: Visual and haptic representations of scenes are updated with observer movement. Exp. Brain Res. **166**(3–4), 481–488 (2005). https://doi.org/10.1007/s00221-005-2388-5
8. Gudde, H.B., Griffiths, D., Coventry, K.R.: The (Spatial) memory game: testing the relationship between spatial language, object knowledge, and spatial cognition. J. Visualized Exp. (132) (2018)
9. Klatzky, R.L., Lederman, S.J.: Haptic object perception: spatial dimensionality and relation to vision. Philos. Trans. R. Soc. B: Biol. Sci. **366**(1581), 3097–3105 (2011)
10. Overvliet, K.E., Smeets, J.B.J., Brenner, E.: The use of proprioception and tactile information in haptic search. Acta Psychol. **129**, 83–90 (2008)

A Scalable Haptic Circuit for Multi-digit Grasps

Cristian-Tiberius Axinte[1,2] , Ciprian Stamate[3] ,
Robert Gabriel Lupu[1] , Alexandru Bârleanu[1] ,
and Georgiana Juravle[2(✉)]

[1] Faculty of Automatic Control and Computer Engineering,
Department of Computer Science and Engineering, Gheorghe Asachi Technical
University of Iasi, 700050 Iasi, Romania
[2] Sensorimotor Dynamics Laboratory, Faculty of Psychology and Education
Sciences, Alexandru Ioan Cuza University, 700554 Iasi, Romania
georgiana.juravle@uaic.ro
[3] Faculty of Mechanics, Gheorghe Asachi Technical University of Iasi,
700050 Iasi, Romania

Abstract. This paper summarizes the design, build, and tests of a 4-sensor haptic circuit designed to assess grip force in various bimanual weight assessment tasks. Experiment 1 details the haptic circuit, together with the initial tests on force and corresponding voltage. Results underline that the system is reliable at assessing normal grip force and can assess forces corresponding to a range of weights between 150g and 1000g. In Experiment 2, we present hand-specific differences in grip normal force data collected from 16 participants who used both of their hands to hold a cylindrical jar given by an experimenter. These results are discussed in the framework of bimanual weight perception research.

Keywords: Haptic · Multi-digit · Bimanual · Weight perception

Reaching for and grasping various objects in our surroundings requires the integration of specific sensory information with the descending motor commands. While sensorimotor integration for goal-directed grasping has received extensive attention [1], the specific contribution of grip force to naturalistic grasping actions is still under debate. This paper proposes a robust, low-cost and scalable haptic solution that measures the fingertip normal applied force. Potential applications include, amongst others, typical weight assessments [2] and measurements of naturalistic arbitrary grasps [3].

1 Experiment 1

The haptic circuit consisted of 4 FSR transducers (FSR03CE, Ohmite Mfg. Co, Warrenville IL, USA). Each FSR was part of a Wheatstone bridge inputting one of four channels of an operational amplifier (TLC274CN, Texas Instruments Inc., Dallas TX, USA). Fixed resistors values were chosen to ensure an output voltage from 0 to 10 V, for a FSR resistance value of up to 1100 Ω. Filtering capacitors and transient-voltage-suppression diodes were also added, see Fig. 1(a). To determine voltage-force

H. Seifi et al. (Eds.): EuroHaptics 2022, LNCS 13235, pp. 343–346, 2022.
https://doi.org/10.1007/978-3-031-06249-0

correspondence, Experiment 1 assessed the circuit by pairing the FSR with a USB6009 data acquisition device with 1 kHz sample rate and average of 100 samples (NI, Austin TX, USA). A tribometer (CETR UMT-2, Center for Tribology, Campbell, CA, USA) was used for calibrating 3 different FSRs, by applying 5 sequences of force through a silicone fingertip; see Fig. 1(b). Each sequence comprised 40 values of force, from 0.5 to 20 N, in steps of 0.5 N. The tribometer held each given force constant for 4.5 s, paused for a second, then transitioned to the next force [4]; see Experiment 1 results in Figs. 1(e), 1(f).

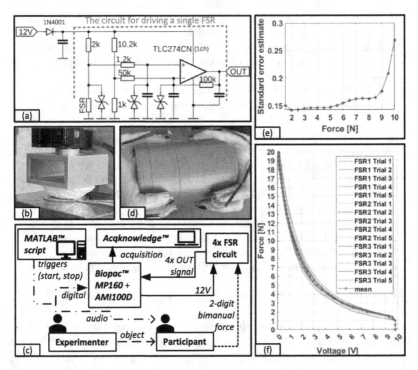

Fig. 1. (a) Schematic of the circuit driving a single FSR. (b) Tribometer applying force on a single FSR in Experiment 1. (c) Data acquisition setup in Experiment 2. (d) Example of participant holding experimental object in Experiment 2. (e) Standard error estimate of the least-squares 3rd degree polynomial used to fit the mean sensor in the range of 1.5 to 10 N. (f) Raw trial data together with mean for voltage to force correspondence in Experiment 1.

2   Experiment 2

16 participants (M_{age} = 22.75 years, SD = 3.89 years; 2 male; 2 left-handed) received course credit for taking part. 4 FSRs were attached with tape to the index and thumb fingers of both hands throughout the 32 trials of the experiment (15 min). Each trial participants used both their hands to hold, for 3.5 s, one of four cylindrical jars of equal weight (425 g, 12 cm height, 25 cm circumference); see Fig. 1(c) for experimental set-

up and 1(d) for the object. Normal force data was collected at 2 kHz and resampled at 250 Hz, by linear interpolation. For each finger, force was derived from voltage, and for each hand we calculated the resultant normal force by subtracting the index data from the thumb data [3]; see Fig. 2(a) for single trial data.

A 2 HANDS x 4 JARS ANOVA was conducted on the averaged force data. Results indicated a significant main effect of HAND, with the left hand gripping the target object significantly more forcefully (M = 2.23; SE = .28), as compared to the right hand (M = .47; SE = .28; $F(1,15) = 24.47$, $p = .001$, $\eta_p^2 = .620$). As expected, no significant differences in normal force were detected between the jars tested ($F(3,45) = 2.19$; $p = .102$; $\eta_p^2 = .127$). An interaction between the two factors was found ($F(3,45) = 3.66$; $p = .019$; $\eta_p^2 = .196$. Post-hoc tests indicated however that this was driven by the grip force applied by the left hand during the bimanual grip of the object, with the left hand applying significantly more force for all of the four jars (all $ps < .006$) and no difference in normal grip force between the jars for the right hand (all $ps > .062$); see Fig. 2(b).

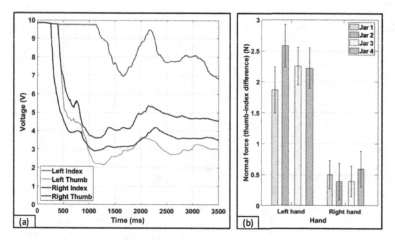

Fig. 2. (a) Raw trial data collected from an exemplary participant in Experiment 2. (b) Means (± SEs) for average normal force collected in Experiment 2.

3 Discussion

The present haptic circuit performs satisfactorily in measuring the normal applied force in a multi-digit grasp scenario. Hand dominance is known not to affect the control of the fingertip forces [2], and objects appear of lighter weight in bimanual assessments [5]. The present results indicating significant differences in normal forces applied by the left and the right hands highlight the important contribution of the non-dominant hand to bimanual naturalistic multi-digit grasping.

Acknowledgement. This work is funded by a grant from the Romanian Ministry of Education and Research, CNCS-UEFISCDI, project number PN-III-P1–1.1-TE-2019–1699.

References

1. Juravle, G., Binsted, G., Spence, C.: Tactile suppression in goal-directed movement. Psychon. Bull. Rev. **24**(4), 1060–1076 (2016). https://doi.org/10.3758/s13423-016-1203-6
2. Buckingham, G., Ranger, N.S., Goodale, M.A.: Handedness, laterality and the size-weight illusion. Cortex **48**, 1342–1350 (2012)
3. Zatsiorsky, V.M., Latash, M.L.: Digit forces in multi-digit grasps. In: Sensorimotor Control of Grasping: Physiology and Pathophysiology, pp. 33–51 (2009)
4. Barnea, A., Oprisan, C., Olaru, D.: Force sensitive resistors calibration for the usage in gripping devices. In: Diagnosis and Prediction in Mechanical Engineering Systems, pp. 1–7. Galati, Romania (2012)
5. Giachritsis, C., Wing, A.: Unimanual and bimanual weight discrimination in a desktop setup. In: Ferre, M. (ed.) EuroHaptics 2008. LNCS, vol. 5024, pp. 378–382. Springer, Heidelberg (2008). https://doi.org/10.1007/978-3-540-69057-3_49

Instrumented Force and Optical Point Tracking to Measure Manual Brushing in Pleasant Touch

J. Michael Bertsch[✉], Zackary Landsman, and Gregory Gerling

University of Virginia, Charlottesville, VA 22903, USA
{xyp5sj, ztl4bm, gg7h}@virginia.edu

Abstract. Several groups studying pleasant touch use either controlled robots or human proctors to deliver brush strokes to the skin, showing that optimal pleasantness is experienced near 3–10 cm/s. However, there has been little physical evaluation of the forces, displacements, and velocities produced as a brush contacts and strokes the skin surface. Using a custom-instrumented brush and point tracking method, we describe a potential means to evaluate the physics of brush strokes and the variability present in human-administered brushing. Preliminary results show that human delivery of brush strokes generates consistent force and velocity within a session involving several, sequential brush strokes, but greater variance between sessions. The work begins to frame considerations for the naturalistic, manual delivery of brush strokes used to study pleasant touch.

Keywords: Pleasant touch · Brush · Haptic technologies

1 Introduction

Touch can be therapeutic and help trigger a sense of calm in those who receive it. Prior works describe gentle brush strokes at speeds near 3–10 cm/s as optimal for evoking pleasant touch [1]. However, when participants have difficulty remaining stationary (e.g., young children, subjects with disabilities), those therapies are difficult to administer and control. To substitute human-delivered therapies, researchers have implemented highly-controlled, brushing robots [2]. Though more consistent, robotic stimulation is typically less adaptable to participant positioning and natural than human touch. We aim to understand the delivery of pleasant touch with tools to measure the force, velocity, and variability of human-administered brush strokes.

2 Methods

To further understand the contact behavior of gentle brush strokes, we developed a custom-instrumented brush and used 2D point tracking to measure forces, displacements, and velocities upon manual contact and stroking of the skin, Fig. 1.

© The Author(s) 2022
H. Seifi et al. (Eds.): EuroHaptics 2022, LNCS 13235, pp. 347–350, 2022.
https://doi.org/10.1007/978-3-031-06249-0

The brush was outfitted with a 6-axis load cell (Nano 17, ATI Industrial Automation, Apex, NC). The voltage output from the load cell was recorded at 40 Hz using an analog-to-digital converter (NI-USB6210, National Instruments, Austin, TX) and used to generate force and torque using a calibration matrix. Force (Fy) measured brush strokes into and down the forearm, while torque (Tx) measured bristle bend.

Fig. 1. Expanded view of brush instrumentation; Free-body diagram of forces; Point tracking of brush tip during 3 cm/s stroking of a participant's forearm over a 25 s session.

Videos of brush strokes were recorded with a manual focus camera (PA150S 720P, Papalook, China). Publicly-available software DeepLabCut (version 2.2 [3]) was used to track 2D points at the tip of the bristles and the metal ferrule. A network was trained using video of a subject's forearm being brushed at various stroke speeds, lengths, and depths. Forty training frames were labeled, and a MobileNetV2–1.0 based neural network utilized 230,000 training iterations. 2D point locations on the brush were defined via a coordinate plane along the arm to analyze motion between contact and release.

3   Experiments and Results

We measured brush stroking of the forearm skin. Two sessions were conducted for two brush depths, a first causing minimal bend of the bristles, a second maximizing bristle bend. The results indicated that force delivery was consistent within a session, but differed between sessions of a single depth, Fig. 2. Between depths 1 and 2, force delivered was distinguishable. The R-squared value between force and torque was greater than 0.92 each session, indicating a strong positive correlation.

We analyzed the pixel locations in the video data to calculate the vertical displacement of the brush strokes towards the skin, as well as their velocity and length, Fig. 3. Though variance is natural in manual, human delivery, we found a higher consistency of displacement towards the skin with the more forceful strokes in the depth 2 trials.

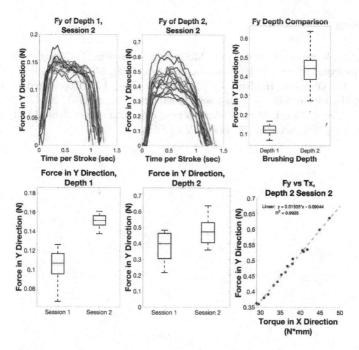

Fig. 2. (Top row) Fy traces of each brush stroke in session 2 of depths 1 and 2; Comparison of maximum Fy per stroke between depths. (Bottom row) Comparison of maximum Fy per stroke between sessions at depths 1 and 2; Correlation of force and torque.

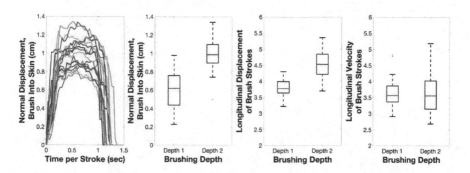

Fig. 3. Normal displacement traces of each brush stroke in session 2 at depth 2; Aggregated across sessions: normal displacement, longitudinal displacement, and stroke velocity.

We believe these methods can be used to measure human delivery of pleasant touch. With force and velocity measurements, real-time feedback can be used to reduce the intra-trial variability.

Acknowledgements. This work was supported by the National Science Foundation (Grants IIS-1908115 and NRT-1829004) and National Institutes of Health (Grant NINDS R01NS105241).

References

1. Löken, L.S., Wessberg, J., Morrison, I., McGlone, F., Olausson, H.: Coding of pleasant touch by unmyelinated afferents in humans. Nat. Neurosci. **12**(5), 547–548 (2009)
2. Triscoli, C., Olausson, H., Sailer, U., Ignell, H., Croy, I.: CT-optimized skin stroking delivered by hand or robot is comparable. Front. Behav. Neurosci. **7**, 208 (2013)
3. Mathis, A., et al.: DeepLabCut: Markerless pose estimation of user-defined body parts with deep learning. Nat. Neurosci. **21**, 1281–1289 (2018)

Instrumented Object for Finger Pad Imaging During Active Manipulation

David Córdova Bulens[1]([⊠]) [iD], Benoit Delhaye[2] [iD], Philippe Lefèvre[2] [iD], and Stephen Redmond[1] [iD]

[1] University College Dublin, Dublin, Ireland
davidcordovabulens@gmail.com
[2] Université Catholique de Louvain, Ottignies-Louvain-la-Neuve, Belgium

Abstract. We introduce a device capable of imaging the finger pad during active manipulation tasks while also recording the forces applied by the participant. This device was created using 3D-printed and low-cost off-the-shelf components.

Keywords: Haptics · Manipulation · Finger

1 Introduction

Tactile feedback is known to influence our ability to dexterously manipulate objects without dropping them. Indeed, we continuously adapt the forces we apply with our fingers in response to the mechanics of the task (friction, weight, etc.) [1]. However, which tactile feedback is actually used during manipulation has only recently started to be explored [2, 3], through the development of specialised equipment allowing skin imaging during manipulation. Currently, this specialized equipment is cumbersome and requires counter-weighting as it is too heavy. Here, we present an instrumented object weighing only 300g capable of imaging the finger pad while simultaneously measuring the forces applied by the participant during natural manipulation.

2 Design and Testing

The instrumented object is designed to be manipulated in a precision grip; i.e., pinched between the thumb and the index finger. The main design goal was to image skin deformations at the finger pad of either the thumb or the index finger while the object is being manipulated, while simultaneously measuring the applied forces and torques with an ATI Mini 40 six-axis sensor (ATI-IA, USA), and orientation and accelerations with a 9°-of-freedom IMU (Adafruit, BNO055). The images are captured using a half-mirror (Edmund optics, 50 ×

This work was funded by a Science Foundation Ireland Future Research Leaders Award (17/FRL/4832).

H. Seifi et al. (Eds.): EuroHaptics 2022, LNCS 13235, pp. 351–353, 2022.
https://doi.org/10.1007/978-3-031-06249-0

50 mm, 50R/50T, Plate Beamsplitter), light source (Neewer SL-12), and camera (Raspberry pi camera v2, 720 × 1280p at 60 FPS) equipped with a lens (T angxi, CWBL2.8-12-3MP-C) (Fig. 1A and B). The total mass of the object is 300 g and the dimensions are W × L × H = 75 × 77 × 225 mm. All other parts of the device are 3D printed. A Raspberry Pi 4B is used to acquire all data, using Python 3.7 code.

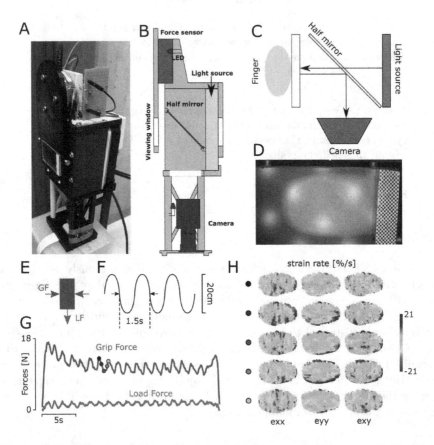

Fig. 1. A) Photograph and B) cross-section of the instrumented object showing force sensor, light source, mirror, and camera. C) The half-mirror is used to align the light source and the axis of the camera. Arrows show the trajectory of light reflected on the glass and the half mirror. D) Example of an image captured by the camera. E) Grip force and load force applied on the object; the object is presented as a gray box. F) Illustration of the amplitude and frequency of the movement performed by participants. G) Grip force and load force evolution across a trial of 20 s. H) Strain rates extracted from the images of the finger pad in contact with the object.

The instrumented object is equipped with an optical system allowing the imaging of the fingertip-plate contact at one of the fingers. A direct illumination system was selected as it allows for a more contrasted image despite leading to a less compact structure [4] (Fig. 1C). Light from the light source goes through the half-mirror, reflects on the plate in contact with the finger, goes back to the half mirror before being reflected into the camera lens. The image obtained has a high contrast due to Differences in refractive index between glass-skin and glass-air interfaces. This technique has been used to image the fingerprint during static loading [5, 6] and dynamic manipulation [2].

To test the instrumented object, we performed an experiment where participants lifted the object in a precision grip, with the index finger being imaged, and performed vertical oscillations of 20 cm amplitude at a rate of 0.75 Hz for 30 s (Fig. 1E and F). This experiment followed the same protocol as Delhaye et al. (2021) [2]. Participants performed 15 trials of 20 s each.

As the object is moved up and down, the grip force applied correlates with the load force (Fig. 1G). We can analyze how the skin deforms during movement from the captured video. The contact area was segmented using a semi-automatic machine learning algorithm. The shear displacement field was obtained at feature points. The contact area was divided into small elements using Delaunay triangulation, taking features as vertices. Strain rates were computed for each triangular element (Fig. 1H). All of those procedures are described in detail in [7].

3 Discussion

Here we present an apparatus capable of imaging the finger pad of the index or thumb during free active object manipulation. This instrumented object is the first such apparatus that is light enough to be lifted unassisted and allow for more natural manipulation. This will enable a large range of manipulation experiments to be performed. Early results show its potential to synchronously measure the forces applied by participants on the object and finger pad deformation during dexterous manipulation. We aim to release all the necessary information to fabricate this object as open source once testing is complete.

References

1. Westling, G., Johansson, R.S.: Factors influencing the force control during precision grip. Exp. Brain Res. **53**(2), 277–284 (1984). https://doi.org/10.1007/BF00238156
2. Delhaye, B., et al.: Measuring fingerpad deformation during active object manipulation. J. Neurophysiol. **126**(4), 1455–1464 (2021)
3. Schiltz, F., et al.: Grip force is adjusted at a level that maintains an upper bound on partial slip across friction conditions during object manipulation. IEEE Trans. Haptics, 1 (2021)
4. Bochereau, S., et al.: Characterizing and imaging gross and real finger contacts under dynamic loading. IEEE Trans. Haptics **10**(4), 456–465 (2017)
5. Delhaye, B., et al.: High-resolution imaging of skin deformation shows that afferents from human fingertips signal slip onset. eLife, **10**, e64679 (2021)
6. Delhaye, B., et al.: Dynamics of fingertip contact during the onset of tangential slip. J. R. Soc. Interface **11**(100), 20140698 (2014)
7. Delhaye, B., et al.: Surface strain measurements of fingertip skin under shearing. J. R. Soc. Interface **13**(115), 20150874 (2016)

Vibrotactile Stimuli are Perceived More Intense at the Front than at the Back of the Torso

Bora Celebi[✉], Müge Cavdan, and Knut Drewing

Giessen University, 35390 Giessen, Germany
bora.celebi@psychol.uni-giessen.de

Abstract. Vibrations effectively transmit information from objects, surfaces or events to the human skin through the cutaneous sense. However, due to the diverse densities of receptive fields and mechanoreceptor populations vibrotactile sensitivity differs across body parts. Hardware that utilizes vibrotactile information should consider such differences. Here, we examined perceived intensity of vibrotactile stimuli applied to the front and back of the human torso. Participants wore a vibrotactile vest. They had to judge if a vibration from the back side of the vest was larger or smaller than a fixed vibration given from the front side; the intensity of the stimulus at the back was adapted using staircase methods. We found that, stimuli at the back had to be physically more intense by 12.3% than stimuli at the front to be perceived equally intense: Presentation of vibrotactile information through wearables could equalize for differential sensitivity, e.g., to equalize attention-capturing effects.

Keywords: Vibrotactile perception · Human torso

1 Introduction

The human cutaneous sense receives information on object properties and events in the environment through skin contact. Vibrations provide one important way to transmit information–being elicited, e.g., through movement across textures, fast contacts with objects or artificial sources. Haptic displays such as vests use this sensory channel by transmitting vibrations to the skin [1]. Specialized mechanoreceptors in the skin gather information: Pacinian Corpuscles (PC) are highly responsive to vibrations, in particular to frequencies above 40 Hz. Their sensitivity achieves its peak around 200–300 Hz. Also, rapidly adapting (RA) Meissner receptors contribute to the perception of vibrotactile stimulation in lower frequency ranges below 100 Hz [2]. Vibration sensitivity can differ between body parts, e.g., between upper and lower leg, indicating

Research was supported by the EU FET-OPEN Project "ChronoPilot" (H2020 – Grant Agreement: 964464 and Deutsche Forschungsgemeinschaft (DFG, German Research Foundation) – project number 222641018 – SFB/TRR 135, A5. Experimental procedures were approved by the local ethics committee of Giessen University, LEK FB 06, in accordance with the Declaration of Helsinki without preregistration.

H. Seifi et al. (Eds.): EuroHaptics 2022, LNCS 13235, pp. 354–357, 2022.
https://doi.org/10.1007/978-3-031-06249-0

differences in perceived intensity [3, 4]. Knowledge on perceived intensity can improve the design and use of vibrotactile garments by guiding actuator choices and allowing to equate perceptual effects across the body. Here, we studied perceived intensities of vibrations at front versus back of the torso in the context of haptic vest design.

2   Methods

15 participants (8 female; age range: 21–25 years; $M = 23.1$; $SD = 1.3$) participated in the experiment (none reported sensory impairments). Participants provided written informed consent prior to the experiments. An already available vibrotactile vest, bHaptics TactSuit X40 (40 Eccentric Rotating Mass actuators with vibration frequency $\sim$ 90 Hz), was used. The vest was tightly fit to the body equally both at the front and the back with straps. Using one actuator at each time point (four in total), we gave vibrotactile stimulation for 300 ms on the front or back of the torso at height of the upper middle chest (Fig. 1B). Active noise cancelling headphones (Sennheiser Momentum 3) plus white noise masked actuator sounds. We used the adaptive staircase method to estimate points of subjectively equal vibration intensity (PSEs) at the back side of the torso as compared to the front side. In a within-participant design, fixed *front intensity* levels of vibration (acceleration) amplitude in different blocks were 11.3, 12.9, and 14.5 m/s^2 root mean square [5]. Stimuli were given either at the left or the right *body location* (two different sessions, order balanced). Each trial started with a vibration at the front, followed by a 100 ms interstimulus interval, and then a vibration at the back. Participants indicated whether the intensity at the back was larger than at the front. In each staircase, back intensity was reduced by 1.4% after a 'yes'-response in the previous trial and increased after 'no'. If a participant responded oppositely in 2 consecutive trials, it was considered a reversal. Each staircase stopped after 6 reversals or after 100 trials (average staircase length: 55 trials). PSEs were calculated as the average back intensity at the last 3 reversals. Each block comprised randomly interleaved trials from 6 staircases (starting at front intensity +44% or +24%).

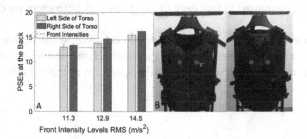

Fig. 1. A. PSEs at the back as a function of *front intensity* and *body location*. Error bars represent standard errors. **1B.** Vibrotactile vest and the actuator locations used in the experiment.

3 Results

We calculated condition-wise individual average PSEs (Fig. 1A). t-tests between PSEs at the back and the physical front intensities (2 body locations × 3 front intensities) were significant, each $t(14) > 15.7$, $p < .001$ (Bonferroni-corrected), showing that physical intensities at the back need to be higher than at the front to be perceived equal. Further, a repeated measures ANOVA compared PSE values between different conditions. As expected, there was no significant main effect of *body location*, $F(1, 14) = 2.7$, $p = .122$, nor a significant interaction with *front intensity*, $F(2, 28) = .9$, $p = .40$, but a significant main effect of *front intensity*, $F(2, 28) = 184.9$, $p < .001$. Follow-up t-tests for pair-wise comparisons (Bonferroni-corrected) were significant, indicating that PSEs at the back were larger for higher front intensities.

4 Discussion

Here, we investigated the perceived intensity for vibrotactile stimuli applied at the front of the human torso as compared to the back. The front torso turned out to be clearly more sensitive: Physical intensities at the back had to be higher by 12.3% than at the front to be perceived as being equally intense. The difference was observed both in the left and right side of the body. Similar sensitivity differences between front and back were found for spatial acuities for point stimuli [6]. Such sensitivity differences may indicate differences in mechanoreceptor distributions between front and back.

Knowledge on differences in perceived intensity across the body can be beneficial in vibrotactile vest design and use. Finding suggest, e.g., that less intense actuators are needed in the front of the torso as compared to back. Also, attention-capturing effects of vibrotactile stimuli are known to increase with intensity [7]. Considering differences in perceived intensity could be used to capture attention at different body sites in equalized and precise ways. Here, however, we only investigated the PSEs of two locations at the back. Moreover, data may have slight bias because we had a fixed order of front and back stimulation and initial back stimuli were higher than in the front. Further studies are required to precisely model perceived differences on different body locations.

We conclude that stronger vibrations need to be given to the back of the torso in order to match vibrotactile sensations at the front. Haptic rendering displays could benefit from equalizing the perceived intensities, e.g., in attention capturing.

References

1. Plaisier, M.A., Sap, L.I.N., Kappers, A.M.L.: Perception of vibrotactile distance on the back. Sci. Rep. **10**(1), 1–7 (2020)
2. Corniani, G., Saal, H.P.: Tactile innervation densities across the whole body. J. Neurophysiol. **124**, 1229–1240 (2020)

3. Shah, V.A., Casadio, M., Scheidt, R.A., Mrotek, L.A.: Spatial and temporal influences on discrimination of vibrotactile stimuli on the arm. Exp. Brain Res. **237**(8), 2075–2086 (2019). https://doi.org/10.1007/s00221-019-05564-5
4. Wentink, E.C., Mulder, A., Rietman, J.S., Veltink, P.H.: Vibrotactile stimulation of the upper leg: effects of location, stimulation method and habituation. In: Proceedings of the Annual International Conference of the IEEE EMBS, pp. 1668–1671 (2011)
5. Morioka, M., Griffin, M.J.: Thresholds for the perception of hand-transmitted vibration: dependence on contact area and contact location. Somatosens. Mot. Res. **22**(4), 281–297 (2009)
6. Weinstein, S.: Intensive and extensive aspects of tactile sensitivity as a function of body part, sex and laterality. First Int' l Symp. Skin Sens. (1968)
7. Zheng, Y., Morrell, J.B.: Haptic actuator design parameters that influence affect and attention. In: Haptics Symposium 2012, HAPTICS 2012 – Proceedings, pp. 463–470 (2012)

Imaging Sub-surface Skin Strain Patterns During Fingertip Sliding

Giulia Corniani[1,2]([✉]) [iD], Zing Lee[3], Matt J. Carré[3], Roger Lewis[3],
Benoit P. Delhaye[4,5], and Hannes P. Saal[1,2]

[1] Active Touch Lab, Department of Psychology, University of Sheffield, Sheffield, UK
g.corniani@sheffield.ac.uk
[2] Sheffield Robotics, University of Sheffield, Sheffield, UK
[3] Department of Mechanical Engineering, University of Sheffield, Sheffield, UK
[4] Institute of Information and Communication Technologies, Electronics and Applied Mathematics, Université Catholique de Louvain, Louvain-la-Neuve, Belgium
[5] Institute of Neuroscience, Université Catholique de Louvain, Brussels, Belgium

Abstract. Mechanoreceptors in the human fingertip are separated from external tactile stimuli by several skin layers with different mechanical properties and complex morphology. Understanding the biomechanics of sub-surface skin layers is therefore fundamental for understanding the human tactile sensory system. Here, we employ Optical Coherence Tomography (OCT) to image the skin's internal morphology during natural sliding interaction with objects. We demonstrate that this approach can be used to reconstruct strain rates within different skin layers during dynamic object contact.

Keywords: Optical coherence tomography · Skin mechanics

1 Introduction

The fingertip is a highly effective tactile sensing organ populated by thousands of mechanoreceptors that translate different aspects of skin deformations into neural responses. The receptors themselves are located within the skin and separated from the surface by several layers of skin tissue with different mechanical properties and complex morphology. Those receptors closest to the surface are situated at the dermis-epidermis junction, and mechanical stimuli will have to traverse the stratum corneum and viable epidermis to reach them (see illustration in Fig. 2A). Understanding how a tactile stimulus applied to the skin surface is translated into neural responses requires knowledge of how the stimulus will affect the local strain patterns at the location of the receptors. However, little is known about the biomechanics of the skin beneath its immediate surface, especially in dynamic conditions due to the technical difficulties of measuring

This work was supported by the EU Horizon 2020 research and innovation programme under grant agreement 813713 (NeuTouch).

H. Seifi et al. (Eds.): EuroHaptics 2022, LNCS 13235, pp. 358–361, 2022.
https://doi.org/10.1007/978-3-031-06249-0

these aspects in-vivo. Here, building on a previous study [2], we employed Optical Coherence Tomography (OCT), a non-invasive imaging technique, to capture the skin's internal morphology as it is undergoing deformation due to contact with external objects. Specifically, we acquired OCT images of the fingertip during sliding interactions with flat plates embossed with tactile features to investigate dynamic strain rates associated with different skin layers. We demonstrate the feasibility of our experimental pipeline and present initial results that demonstrate differences in the mechanical behaviour of different skin layers.

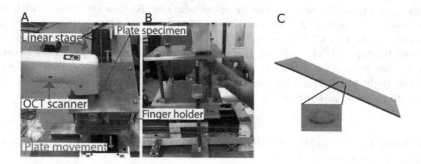

Fig. 1. Experimental setup and plate specimen model A) View of the motorized stage, OCT scanner, and finger holder. B) Side view showing finger in holder and pressed against the transparent plate. C) Model of the transparent plate with embossed dot.

2   Materials and Methods

The experimental protocol was approved by the ethical review board of the Department of Psychology, University of Sheffield (Ethics Number 039144). The experiment was performed on the left index fingertip of two participants. A transparent plate of poly(methyl methacrylate) (PMMA) embossed with a round-dot of 1 mm base radius and 0.4 mm height was custom made for the experiment (see Fig. 1C). The plate was held in place using a support rig fixed on a linear stage, moving distally or proximally, sliding the plate against the subjects' fingertip, which was fixed to a finger holder (see Fig. 1A and B). The normal force and sliding speed were kept constant at 2 N and 2 mm/s, respectively, throughout the experiment. The total displacement of the plate was 10 mm, repeated four times in each direction. The clinically approved Vivosight OCT system (Michelson Diagnostics, Kent) was used to acquire images of the fingertip surface and subsurface through the transparent plate at an image capture rate of 20 frames per second with 7.5 μm lateral and 5 μm axial resolution.

After the acquisition, images were preprocessed using the ImageJ software and stabilized using auto-correlation across subsequent frames. The top and valley of each visible fingertip ridge were manually tracked at three skin layer borders (see Fig. 2C). Strain rates were computed using the Green-Lagrange

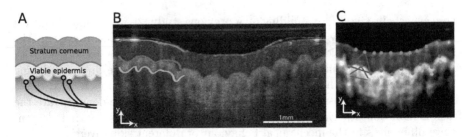

Fig. 2. OCT images and feature tracking. A) Illustration of skin layers in the fingertip and typical type-1 mechanoreceptor locations. B) Raw OCT image with skin layer borders in colour. A faint outline of the transparent plate with embossed dot is visible at the top. C) Processed OCT image with tracked features. Coloured dots indicate the top and valley of each ridge at the surface (green), at the border between the stratum corneum and the viable epidermis (purple), and at the dermis-epidermis junction (light blue). Triangles (blue and red lines) were constructed from the tracked points to calculate strain rates.

strain equations as in [1], resulting in ε_{xx} and ε_{yy} aligned with the x and y axes, and a shear component ε_{xy}.

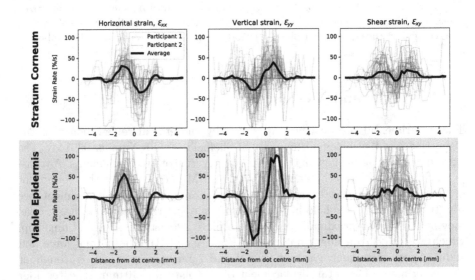

Fig. 3. Strain rates as a function of dot position. Horizontal, vertical and shear strain rates measured in the stratum corneum (top) and in the viable epidermis (bottom). Negative values on the horizontal axis denote the dot approaching, positive values the dot leaving, at 0 it is overhead the tracked ridge. The variability of individual trials can be explained by noise in measurement and features tracking, but might also be partly due to different mechanical behaviour of individual ridges.

3   Preliminary Results and Future Work

We reconstructed strain rate patterns within the stratum corneum and the viable epidermis during the movement of the plate. We found repeatable, stereotypical spatiotemporal patterns in all three strain components (Fig. 3). Specifically, the skin stretches horizontally and compresses vertically as the dot approaches and then reverts to its initial state as it is leaving. Interestingly, our preliminary analysis suggests that strains rates are larger in the viable epidermis than closer to the surface in the stratum corneum. These results demonstrate that the methodology and analysis pipeline developed is able to reliably track sub-surface features and reconstruct strain rates during natural contact of the fingerpad with external objects. Future analysis will focus on how mechanical and geometrical features of the object and the skin influence the resulting strain patterns and include more complex tactile features. The findings will help understand how surface strains propagate into deeper layers towards mechanoreceptor locations and shed light on the function of skin morphology in encoding tactile stimuli.

References

1. Delhaye, B., Barrea, A., Edin, B.B., Lefèvre, P., Thonnard, J.-L.: Surface strain measurements of fingertip skin under shearing. J. R. Soc. Interface **13**(115), 20150874 (2016)
2. Lee, Z.S., Maiti, R., Carré, M.J., Lewis, R.: Morphology of a human finger pad during sliding against a grooved plate: a pilot study. Biotribology **21**, 100114 (2020)

Sense of Agency Over Hands-free Gestural Control is Modulated by the Timing of Haptic Feedback

Calvin Deans-Browne[1], Antonio Cataldo[1,2(✉)], William Frier[3], Hannah Limerick[3], David Beattie[3], and Patrick Haggard[1]

[1] Institute of Cognitive Neuroscience - University College London, London WC1N 3AZ, UK
a.cataldo@ucl.ac.uk
[2] Institute of Philosophy – School of Advanced Study, London WC1E 7HU, UK
[3] UltraLeap, Bristol BS2 0EL, UK

Abstract. Several studies reported that implicit feelings of control over a device (commonly known as 'sense of agency' or SoA) is enhanced by haptic feedback. These findings extend to haptics generated by touchless technology. Little research has investigated SoA when haptic feedback accompanies abstract gestures that characterize hands-free control. We report a study in which participants made a forward movement of the hand and arm, that caused a tone after a short, variable delay. Participants gave an absolute numerical estimate of this delay. According to the "intentional binding" concept, shorter estimates, indicating temporal compression between action and effect, indicate greater SoA. When haptic feedback to the palm from an UltraLeap STRATOS device accompanied the tone, people showed reduced interval estimates, and thus a stronger SoA, compared to the same feedback at the time of the action. This suggests that when haptics are used in hands-free gestural control, they may be best placed after the action as a form of feedback, as opposed to simultaneous with the action, as occurs in natural touch. "Haptic echos" of an action, at the time of the action's outcome, serve to enhance the binding between action and outcome, and thus the Sense of Agency.

Keywords: Touchless haptics · Sense of agency · Hands-free gestural control

1 Introduction

Several studies suggest that implicit perceptions of control (commonly known as 'sense of agency' or SoA) over an interface increases as a function of its haptic feedback [1, 2]. Thus, haptic feedback boosts SoA when pushing a virtual button [3]. Few studies have added haptics to more abstract hand gestures, such those used for pushing or swiping actions. This is important for considering how haptics can be applied to the hand for gesture confirmation during hands-free control. Our study aims to investigate whether applying haptic-feedback in conjunction with such movements increases users' SoA over touchless interfaces.

© The Author(s) 2022
H. Seifi et al. (Eds.): EuroHaptics 2022, LNCS 13235, pp. 362–365, 2022.
https://doi.org/10.1007/978-3-031-06249-0

2   Method

2.1   Participants

Thirty-six participants from an opportunity sample volunteered for the experiment. Twelve participants were excluded for failing to perceive differences in time intervals (see Fig. 2B), a necessary precondition for our temporal measure of SoA. The final sample size was n = 24 (15 males, mean age ± SD: 29.8 ± 9.6).

2.2   Experimental Design and Procedure

In this study, hands-free manual actions (a forward movement of the arm and hand) caused a tone after a random delay of 300, 500 or 700 ms (Fig. 1A-B). Participants performed 4 blocks (passive movements applied to the arm by the experimenter, active movements without haptic feedback, active movements with haptic feedback at the time of action, active movements with haptic feedback at the time of the tone; Fig. 1C). The haptic stimulus was a line travelled distal-proximal along the palm for 250 ms. Participants sat with their dominant arm on a mobile armrest, palm facing a STRATOS Explore stimulator (Fig. 1A). They were instructed to make a forward movement with their arm after hearing a cue. In a passive block, the participants' arm was moved by the experimenter. Either movement (active/passive) was tracked by the STRATOS camera, triggering a tone after a short, unpredictable delay. Participants made an absolute numerical estimate of the time between the arm movement and tone onsets (Fig. 1B). Block order was randomized. Each of the three action-tone intervals was presented 12 times per block.

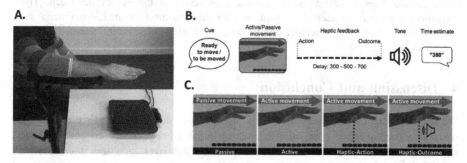

Fig. 1. A. Experimental setup. **B.** Trial structure. **C.** Experimental conditions.

3   Results

A time estimate error (TEE) was calculated in each trial by subtracting participants' estimate of the interval from the actual interval. Using the intentional binding concept, perceptual compression of the time between action and outcome can be taken as a proxy measure of SoA [4], with a smaller or more negative TEE indicating greater SoA. Figure 2A shows the mean TEE in each block. A paired t-test showed that the mean TEE in the active block was significantly lower than in the passive block (t(23) = 2.69, p = .013, d = 0.42). This replicates the classic finding that participants have SoA for their own volitional actions, but not for passive movements. Omnibus ANOVA showed a significant main effect of block (F(2,46) = 7.00, p = .002, MSE = 0.68, $\widehat{\eta}_p^2$ = .22). The TEE was significantly higher when the haptic feedback was paired with the action, than when no feedback was present (Dunnett's test, z = 2.29, p = .041, d = 0.35). In contrast, pairing haptic feedback with the tone resulted in a lower TEE than when no feedback was present, although this difference was not significant (Dunnett's test, z = −1.42, p = .27, d = 0.21). This suggests that participants felt less SoA when haptics accompanied their action compared to when it was absent or paired with the tone.

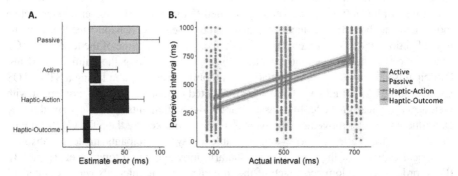

Fig. 2. A. Mean time estimate error for each condition. Error bars represent the SD of the mean. **B.** Perceived time interval as a function of the actual interval duration. Dots represent individual trials; colored lines represent means; shaded lines represent the SD of the mean. These regressions were run for each participant, and for them to meet the inclusion criteria, each regression line had to be significantly positive (α = 0.05, Bonferroni adjusted for four comparisons).

4   Discussion and Conclusion

We speculate haptic re-presentation of an action alongside that action's outcome may serve as a "haptic echo" promoting subjective binding between actions and outcomes, and therefore SoA. We conclude that maximizing SoA over touchless interfaces may require placing haptics that accompany a user's gestures *after* the gesture as a form of echo feedback, rather than simultaneously with the gesture as 'feed-forward'.

Acknowledgements. Supported by a European Union Horizon 2020 research and innovation program (TOUCHLESS, project no. 101017746). CD-B's research was further supported by UltraLeap through UCL's BiX programme.

References

1. Coyle, D., Moore, J., Kristensson, P.O., Fletcher, P., Blackwell, A.: I did that! Measuring users' experience of agency in their own actions. In: Proceedings of the SIGCHI Conference on Human Factors in Computing Systems, pp. 2025–2034 (2012)
2. Limerick, H., Moore, J.W., Coyle, D.: Empirical evidence for a diminished sense of agency in speech interfaces. In: Proceedings of the 33rd Annual ACM Conference on Human Factors in Computing Systems, pp. 3967–3970 (2015)
3. Martinez, P.I.C., De Pirro, S., Vi, C.T., Subramanian, S.: Agency in mid-air interfaces. In: Proceedings of the 2017 CHI Conference on Human Factors in Computing Systems, pp. 2426–2439 (2017)
4. Haggard, P., Clark, S., Kalogeras, J.: Voluntary action and conscious awareness. Nat. Neurosci. 5(4), 382–385 (2002). https://doi.org/10.1038/nn827

Exploiting a Wearable Extra-Finger for Haptic Applications

Mihai Dragusanu[1], Zubair Iqbal[1], Domenico Prattichizzo[1,2],
and Monica Malvezzi[1(✉)]

[1] Department of Information Engineering and Mathematics, University of Siena,
53100 Siena, Italy
malvezzi@dii.unisi.it

[2] Humanoids & Human Centered Mechatronics Research Line, Istituto Italiano
di Tecnologia, Genoa, Italy

Abstract. This extended abstract presents the design of a wearable device for haptic stimulation of hand palms and phalanges. Most of the wearable haptic devices for hand palms are based on a parallel structure, that guarantees good precision and stiffness but presents workspace limitations and encumbrance problems. In this work, we improve the design of a wearable extra-finger, previously designed to augment human hands and to provide assistance for people affected by hand and upper-limb diseases to apply as a haptic device. To employ this device for haptics applications, we provided it an additional adduction/abduction degree of freedom and we modified the fingertip/end-effector to include a micro force sensor.

1 Introduction

Human augmentation by means of supernumerary robotic limbs (SRLs) represents a recent and lively research topic [1]. SRLs are developed to allow humans to perform complex actions with increased strength and precision, and also to enlarge the human workspace. Our research group has been involved in particular in the development of supernumerary extra-fingers [2, 3]. Adding a single or double extra robotic finger can improve the grasping capabilities of the human hand and its dexterity even in complex actions.

In this work we propose to exploit supernumerary extra fingers as wearable haptic devices for cutaneous stimuli of hand palm and phalanges. Although the hand is one of the primary interfaces between humans and the surrounding environment, most of the haptic devices that return tactile stimuli are focused on the fingers [4], while fewer of them are developed specifically for the palm. Most of the wearable haptics devices developed for hand palm are based on parallel mechanisms, i.e. mechanical systems using multiple serial chains (typically varying from 3 to 6) to support a single platform, or end-effector [5–7]. One of the main drawbacks of parallel mechanisms is that their workspace, i.e. the set of possible configurations that can be reached by the end effector is rather limited,

H. Seifi et al. (Eds.): EuroHaptics 2022, LNCS 13235, pp. 366–368, 2022.
https://doi.org/10.1007/978-3-031-06249-0

and in haptics context this means that a limited part of the hand can be stimulated. Serial mechanisms, on the other hand, typically are more flexible and less precise than parallel ones, but their open structure allows to reach a wider workspace.

2   The Improvement of the Sixth Finger

As mentioned in the previous section, in this work we improved the wearable extra finger developed in [3, 8] to exploit it in haptic applications.

The robotic extra finger [8] has a modular structure, each module is composed of a rigid part realized in ABS (Acrylonitrile Butadiene Styrene, ABSPlus, Stratasys, USA) and flexible part, made of TPU (thermoplastic polyurethane, Lulzbot, USA). We selected TPU in particular because the high elongation of this material allows for repeated movement and impact without wear or cracking proving. Seven modules were employed in order to achieve a length of the finger similar to the average size of human hand. The robotic extra finger is tendon-driven and is actuated by only one motor, providing the flexion movement.

To exploit the extra finger in haptic applications, we increased its workspace by providing it the adduction/abduction motion, by connecting the proximal module to a platform orientable by means of a gear system. To reduce the encumbrance on the wrist, the motor was substitured by a linear actuator. Furthermore the distal module was sensorised with a microforce sensor, necessary to

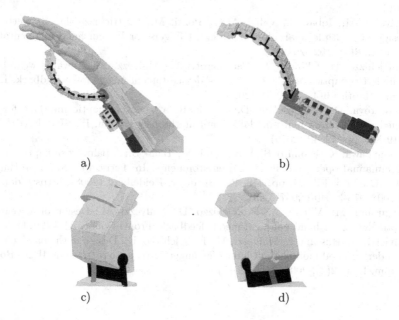

Fig. 1. (a) CAD model of the wearable extra finger exploited for haptic applications, worn by the user. (b) CAD model of the wearable extra finger, showing the actuation. (c) Fingertip module. (d) Sensing elements in the fingertip module.

control haptic interaction with human hand. CAD model of the proposed device is shown in Fig. 1. Furthermore, the stiffness of the flexible elements connecting finger phalanges has been calculated using the procedure proposed in [8] so that when the actuator pull the tendon, the extra-finger flexes by keeping the distal phalanx approximately perpendicular to hand palm.

3   Future Works

The device is currently in the prototyping phase. A first set of experimental tests will be aimed at verifying its functional characteristics, in particular, we will identify the part of the hand that can be stimulated with the device and the corresponding force that can be applied. Then in a second experimental phase we will compare it with the device presented in [7]. Then, users studies will be carried out to assess device usability in VR and AR contexts.

References

1. Prattichizzo, D., et al.: Human augmentation by wearable supernumerary robotic limbs: review and perspectives. Progress Biomed. Eng. **3**(4), 042005 (2021)
2. Prattichizzo, D., Salvietti, G., Chinello, F., Malvezzi, M.: An object-based mapping algorithm to control wearable robotic extra-fingers. In: 2014 IEEE/ASME International Conference on Advanced Intelligent Mechatronics, pp. 1563–1568. IEEE (2014)
3. Malvezzi, M., Iqbal, Z., Valigi, M.C., Pozzi, M., Prattichizzo, D., Salvietti, G.: Design of multiple wearable robotic extra fingers for human hand augmentation. Robotics **8**(4), 102 (2019)
4. Prattichizzo, D., Chinello, F., Pacchierotti, C., Malvezzi, M.: Towards wearability in fingertip haptics: a 3-DoF wearable device for cutaneous force feedback. IEEE Trans. Haptics **6**(4), 506–516 (2013)
5. Trinitatova, D., Tsetserukou, D.: TouchVR: a wearable haptic interface for VR aimed at delivering multi-modal stimuli at the user's palm. In: SIGGRAPH Asia 2019 XR, pp. 42–43 (2019)
6. Minamizawa, K., Kamuro, S., Kawakami, N., Tachi, S.: A palm-worn haptic display for bimanual operations in virtual environments. In: Ferre, M. (ed.) EuroHaptics 2008. LNCS, vol. 5024, pp. 458–463. Springer, Heidelberg (2008). https://doi.org/10.1007/978-3-540-69057-3_59
7. Dragusanu, M., Villani, A., Prattichizzo, D., Malvezzi, M.: Design of a wearable haptic device for hand palm cutaneous feedback. Front. Robot. AI 8 (2021)
8. Salvietti, G., Hussain, I., Malvezzi, M., Prattichizzo, D.: Design of the passive joints of underactuated modular soft hands for fingertip trajectory tracking. IEEE Robot. Autom. Lett. **2**(4), 2008–2015 (2017)

Finger Pad Deformation Under Torsion

Sophie du Bois de Dunilac[1]([⊠]) [iD], David Córdova Bulens[1] [iD],
Philippe Lefèvre[2] [iD], Stephen J. Redmond[1] [iD], and Benoit P. Delhaye[2] [iD]

[1] University College Dublin, Dublin, Ireland
sophie.dubois@ucdconnect.ie
[2] Université catholique de Louvain, Louvain-la-Neuve, Belgium

Abstract. The deformation of finger pad skin in contact with a glass
plate is measured during the onset of slip created by a rotating motion.
Similar to the case of a translating stimulus, slip starts at the contact
periphery before propagating inwards, giving rise to local strains. We
characterized the evolution of the contact and the propagating slip front.

Keywords: Biomechanics · Partial slip · Incipient slip

1 Introduction

When we touch an object with our finger, the resulting deformation of our skin is
transduced into nerve activity by sensory afferents, which is the basis of sensation
and perception. Local strains at the skin surface, which can be correlated with
the imminence of slippage, have notably been linked to the responses of sensory
afferents [6]. However, previously-used stimuli, being limited to a tangentially
translating plate, do not capture the richness of real-world interactions; notably
absent is tangential torque, which arises when the lifting force applied by the
fingers to the object does not pass through the object's centre of mass.

For translating stimuli, slippage propagates as an annulus from the periphery
of the contact area towards its centre, giving rise to local strain patterns which
are signalled by specific sensory afferents [6]. As contact theory predicts that
a slip annulus will also arise under tangential torque loading [1], local strain
patterns are expected, impacting afferents uniquely in function of the location
of their receptive field within the contact area. Indeed, recording nerve activity
in response to a stimulus rotating in the tangential plane highlighted complex
responses to torque magnitude and direction [2]. Some of the observed inter-
afferent variability might be explained by specific local strains present in their
receptive fields.

This work was partly funded by a Science Foundation Ireland Future Research Leaders
Award (17/FRL/4832), and by a grant from the European Space Agency, Prodex
(BELSPO, Belgian Federal Government). BPD is supported by a grant from the Fonds
de la Recherche Scientifique – FNRS (Belgium).

H. Seifi et al. (Eds.): EuroHaptics 2022, LNCS 13235, pp. 369–371, 2022.
https://doi.org/10.1007/978-3-031-06249-0

2   Methods

A platform, previously used by Delhaye *et al.* ([4, 5]) to apply translations, was used here to rotate a glass plate against the finger pad of seven healthy human volunteers (aged 23–35, 4 males). The plate was rotated, in its own plane, at a constant angular velocity (5-100°/s), up to 80°, followed by a 1 s pause. Each condition was repeated 5 times per rotation direction. See Fig. 1A for details.

Computations of contact area segmentation, skin displacement field, and surface strains are described in [5]. A triangular skin element (from Delaunay triangulation) was considered to have slipped if its rate of rotation differed from that of the rigid plate by more than 0.6°/frame). Rate of rotation was defined as the average angular displacement of the triangle vertices around its centroid. The stick ratio was then defined as the area stuck over the contact area at each video frame. Full slip was defined as a stick ratio smaller than 0.03.

3   Results

As expected, at the start of rotation, slippage started at the contact area periphery before propagating inwards until full slip occurred (Fig. 1B,C). At small normal force levels (0.5 to 2 N), a larger normal force correlated with a larger required rotation angle to reach full slip. Above a normal force of 2 N, the rotation angle at which full slip occurred remained relatively constant (Fig. 1D). Increasing the angular velocity had no impact on the rotation angle at full slip (Fig. 1E); hence, time available to react decreases linearly with angular velocity.

The contact area reduced as the stimulus plate was rotated, due to two distinct effects: peeling and deformation (Fig. 1B). Normal force did not affect area reduction (Fig. 1F), but angular velocity did increase the reduction, mainly driven by an increase in peeling (Fig. 1G). This dependency on velocity highlights a viscoelastic influence.

4   Discussion

An important variation in the biomechanical response of the skin was observed across subjects. A typical example of such variation concerns the angle needed to reach full slip: At 2 N and 20°/s, subject averages ranged from 15.81° to 58.93°, almost a four-fold difference.

As partial slips could signal an unsecured grip, with associated local skin deformations being encoded in afferent firings [6], these differences in the evolution highlight the need to pay attention to biomechanical variables such as local strains when searching for what causes sensory afferents to fire and grip force to be adjusted. Furthermore, skin losing contact at the periphery of the contact area (called peeling here) might also trigger sensory afferents to signal the slip; slow adapting types might be sensitive to this effect due to their known sensitivity to pressure change.

Further analysis of the strains generated by torsional loading will be performed in future work to highlight patterns specific to torsional loading.

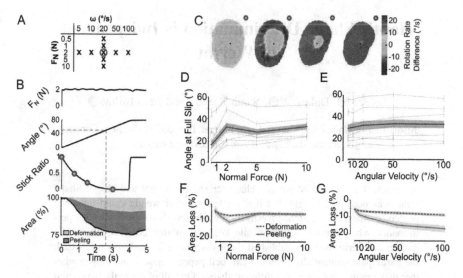

Fig. 1. (A) Experimental conditions. (B) Normal force, stimulus angle, stick ratio and contact area relative to the beginning of the rotation for the example condition circled in panel A (2 N, 20°/s, clockwise). Lines displayed are averages of the 5 repeats under this condition for a single participant. Green dashed line represents the time and angle when full slip occurs. (C) Difference in rotation rate between skin and stimulus during an example trial. A non-zero difference indicates slippage. The centre of rotation is marked with a black dot. (D,E) Averages across participants of the angle to reach full slip as a function of normal force (D), and of angular velocity (E). Participant averages indicated as grey lines. (F,G) Averages across participants of the area reduction after completion of the 80° rotation as a function of normal force (F) and of angular velocity (G). Error bars and shaded areas represent 95% confidence intervals.

References

1. Johnson, K.L.: Contact Mechanics. Cambridge University Press, Cambridge (1987)
2. Birznieks, I., et al.: Encoding of tangential torque in responses of tactile afferent fibres innervating the fingerpad of the monkey. J. Physiol. **588**(7), 1057–1072 (2010)
3. André, T., et al.: Effect of skin hydration on the dynamics of fingertip gripping contact. J. R. Soc. Interface **8**(64), 1574–1583 (2011)
4. Delhaye, B., et al.: Dynamics of fingertip contact during the onset of tangential slip. J. R. Soc. Interface **11**(100), 20140698 (2014)
5. Delhaye, B., et al.: Surface strain measurements of fingertip skin under shearing. J. R. Soc. Interface **13**(115), 20150874 (2016)
6. Delhaye, B.P., et al.: High-resolution imaging of skin deformation shows that afferents from human fingertips signal slip onset. eLife, 10:1–21 (2021)

Haptic Shape Discrimination is Independent of Weight

Jessica M. Dukes[(⊠)][iD], Sarah Kiendl, and Jutta Billino[iD]

Justus-Liebig University, Otto-Behaghel-Street 10F, 35394 Giessen, Germany
jessica.dukes@psychol.uni-giessen.de

Abstract. Objects are commonly characterized by weight and shape. Shape influences perceived weight, but it is unclear whether weight could influence perceived shape. The current investigation evaluated shape discrimination performance when varying weight, while keeping volume constant. Thirty-one participants performed an established shape discrimination task. For every trial, the participant sequentially explored two bell pepper replicas and judged whether they possessed the same or different shapes. On half of the trials, the weight of the two replicas were equivalent (i.e., "same weight" trials) while the other half of trials had replicas with different weights (i.e., "different weight" trials). Results reveal that discrimination performance was equivalent for both trial sets. We conclude that weight has no effect on shape discrimination performance, and is therefore not a significant factor when perceiving haptic shape.

Keywords: 3D shape perception · Weight perception · Natural objects

1 Introduction

Haptic perception is heavily dependent on our interactions with objects. Depending on how we interact with an object, we can extract information about various object properties [1]. Contour following, unsupported holding, and enclosure are types of hand movements that result in sufficient judgments (judgments with 50–65% accuracy) for weight and global shape of an object [1]. Due to shared exploratory strategies, it is possible that weight could affect shape perception, or vice versa.

Shape types affect weight perception [2]. A cube is consistently perceived as heavier than a tetrahedron, but there are individual differences when comparing weights of cubes and spheres, or tetrahedrons and spheres [2]. However, it is unknown whether weight affects shape perception. The current investigation will require participants to make shape judgments while ignoring potential weight differences. If weight does not influence shape judgments, performance will be unaffected by a presented weight difference. If weight does influence shape judgments, shape judgments will be significantly worse when a weight difference is presented.

H. Seifi et al. (Eds.): EuroHaptics 2022, LNCS 13235, pp. 372–375, 2022.
https://doi.org/10.1007/978-3-031-06249-0

2 Methods

2.1 Materials

The stimulus objects were 3D bell pepper replicas (printed on an Ultimaker S5 Fused Deposition Modeling-3D printer using Polylactic Acid plastic). The life-sized replicas used for the current experiment are part of a subset used in previous research (objects 1, 2, 3, 5, 7, 8, 11, 12) [3, 4]. There were two sets printed for the current experiment: a "light set" and a "heavy set". The light set has an average mass of 115 g and infill density 15%, while the heavy set has an average mass of 200 g and infill density of 25–28%. The volume of all objects are equivalent to the original replicas [4].

2.2 Procedure

On any given trial, participants were handed two objects sequentially behind an occluding aperture (each presented for 3s, ISI 3s). They haptically explored, including lifting, both objects and were then required to indicate whether both have had the same shapes or not. For the same shape trials, one of the 8 stimulus objects was paired with itself (e.g., object 1 presented twice). For the different shape trials, two different stimulus objects would be presented (e.g., object 1 and object 3). The different object pairings are: objects 1 and 3, objects 1 and 7, objects 2 and 11, objects 3 and 7, objects 3 and 8, and objects 5 and 12. Please note that the object volumes within each pairing were equivalent. The chosen pairings have been demonstrated to be "easily confusable" [4] and therefore would be challenging stimulus objects for participants (Fig. 1).

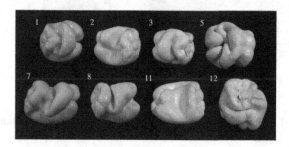

Fig. 1. Stimulus objects, numbered according to the original stimulus

Twenty-four of the 48 same shape trials contain objects from one of the two weight sets ("same weight"), and the remaining half contain objects from both weight sets ("different weight"). The different shape trials also had 24 same weight trials and 24 different weight trials. Therefore, there are 96 total trials for the shape discrimination task. The performance for the same weight and different weight trials were quantified as a measure of d', a measure of perceptual sensitivity [5].

2.3   Participants

Thirty-three younger adults participated in the current experiment. Two were excluded from analyses due to d' values that were 0 or negative. The remaining participants (n = 31, M = 23.3 years, SD = 3.94) were included in the analysis below. All participants gave written informed consent according to the Declaration of Helsinki.

3   Results

A paired-sample t-test revealed that shape discrimination performance between the same weight and different weight trials were not significantly different ($t(30)$ = −0.411, p = 0.684). The same weight and different weight performance have mean d' of 1.061 (SE = 0.099) and 1.106 (SE = 0.082) respectively. Since the absence of an effect would support the null hypothesis that differences in weight do not reduce the haptic sensitivity to shape, we calculated the corresponding Bayes Factors (BF) to back up this. Analyses of BF support that haptic shape discrimination is robust to weight differences, BF = 0.21. The results are plotted in Fig. 2 below.

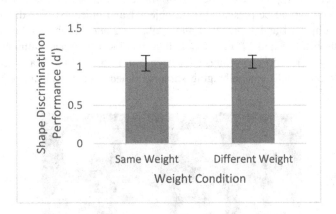

Fig. 2. Shape discrimination performance for both weight conditions. Error bars indicate ± 1 SE.

4   Discussion

The results of the current investigation indicate that natural shape can be discriminated independent of weight difference between objects (i.e., weight does not significantly influence shape judgments). During shape judgments, participants tend to use the exploratory procedures of enclosure and contour following, while weight judgments tend to require "hefting" motions [1]. Participants employ exploratory procedures based on the requirement of the task. Current task demands required enclosure and contour following. Therefore, the participants employed exploratory procedures that are typically not associated with weight (hefting). The current results confirm that weight does not affect enclosure or contour following; otherwise, there would have been a significant effect between conditions.

References

1. Lederman, S.J., Klatzky, R.L.: Hand movements: a window into haptic object recognition. Cogn. Psychol. **19**, 342–368 (1987)
2. Kahrimanovic, M., Bergmann Tiest, W.M., Kappers, A.M.L.: The shape-weight illusion. In: Kappers, A.M.L., van Erp, J.B.F., Bergmann Tiest, W.M., van der Helm, F.C.T. (eds.) EuroHaptics 2010. LNCS, vol. 6191, pp. 17–22. Springer, Heidelberg (2010). https://doi.org/10.1007/978-3-642-14064-8_3
3. Norman, J.F., Dukes, J.M., Palmore, T.N.: Aging and haptic shape discrimination: the effects of variations in size. Sci. Rep. **10**, 14690 (2020)
4. Norman, J.F., Norman, H.F., Clayton, A.M., Lianekhammy, J., Zielke, G.: Visual and haptic perception of natural object shape. Percept. Psychophys. **66**(2), 342–351 (2004). https://doi.org/10.3758/BF03194883
5. Macmillan, N.A., Creelman, C.D.: Detection Theory: A User's Guide. Cambridge University Press, Cambridge (1991)

Human-to-Human Strokes Recordings for Tactile Apparent Motion

Basil Duvernoy[(✉)] [iD] and Sarah McIntyre[(✉)] [iD]

Center for Social and Affective Neuroscience, Department of Biomedical and Clinical Sciences, Linköping University, Linköping 58183, Sweden
{basil.duvernoy,sarah.mcintyre}@liu.se
https://liu.se/en/research/csan

Abstract. The main objective of this study is to investigate whether one can use recordings of human-to-human touch, such as a caress, to improve tactile apparent motion interfaces to make them feel more natural. We report here preliminary recordings of natural and continuous human-to-human caresses. To do this, six accelerometers were positioned on the receiving hand next to the stimulated area while a finger gently stroked the skin. The results suggest that we are able to capture signals from real human caresses that can be compared to signals produced by apparent motion stimuli. This is encouraging for our plan to continue the study in the second stage, which consists of tuning vibrotactile actuators to reproduce a similar pattern of vibrational responses in the accelerometers. In this way, the actuators mimic human behavior.

Keywords: Naturalistic touch · Haptics interface · Tactile apparent motion

1 Introduction

At the beginning of the century, vision researchers showed interesting results using a natural approach to study movement that can reveal more information than more classic methods [1, 2]. In touch, most studies on apparent tactile motion investigate the perceptual impact of this technique by varying the values of predefined parameters [3–5]. Even though this approach is classic, it is likely that the methods used so far to create apparent motion are not optimal. We expect that it should be possible to achieve more natural-feeling stimuli for the tactile apparent motion if they are tuned with recordings of natural tactile motions. Similar to previous apparent motion studies, we plan in the near future to use a set of vibrotactile actuators activated in sequence to produce the apparent motion illusion. With in mind to tune the actuators' behavior in a more natural approach, we recorded the vibrations generated by human-generated caresses, which constitute our case study. To do so, we used six accelerometers placed on the receiving hand. We were primarily interested in the speed and intensity of the overall motion, as well as the content of the frequency domain.

© The Author(s) 2022
H. Seifi et al. (Eds.): EuroHaptics 2022, LNCS 13235, pp. 376–378, 2022.
https://doi.org/10.1007/978-3-031-06249-0

2   Materials and Method

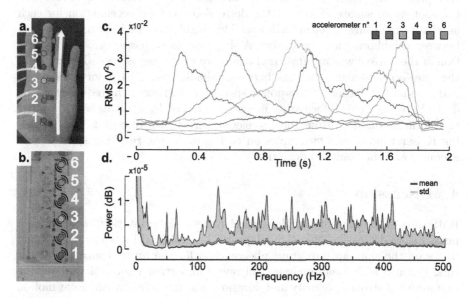

Fig. 1. Natural observation: **a.** Accelerometers' locations and path of the stimuli; **b.** Apparent motion display for future work; **c.** Example of root mean square for each accelerometer during a single stroke from a finger; **d.** Overall mean and standard deviation of the power spectrum from the full data set. Color figure online

2.1   Apparatus

Partly shown in Fig. 1a, the recording apparatus is composed of six 3-axis accelerometers (ADXL335 rev.B), three ADC (MCP3008), and a microcontroller (Teensy 4.1). The overall system is optimised to collect data up 500 Hz at a sampling rate of 2 kHz.

2.2   Procedure

The tactile interaction involved the application of gentle stroking motions to a "receiver" hand using a single "transmitter" finger. The transmitter was required to stroke using her/his fingertip the receiving left hand from the bottom of the palm to the fingertip of the fourth finger (or vice versa). This location was chosen as the surface of the skin is relatively flat, which is required for the planned apparent motion device. The glabrous skin is also a sensitive area, where tactile cues can be more easily discriminated than other body places, such as the forearm. Therefore, the accelerometers were arranged linearly along the stimulated area, spaced at regular intervals of approximately 20 mm.

3   Results

Our preliminary result is based on recordings of 40 caresses. Figure 1c shows the root mean square (RMS) of the detrended recorded acceleration for each accelerometer using a 200 ms window. This highlights the information shared between neighboring accelerometers. A single caress is shown here, as the duration of the strokes was unsupervised and more processing is needed to compare the inter-accelerometers timings between stimuli. Figure 1d reports the mean and standard deviation of the power spectrum (PS) for each frequency in the 0–500 Hz range. This information has been obtained by applying a fast Fourier transform to the data from each accelerometer of each recording using a sliding Kaiser window of 128-samples on each sample; the resulting spectra were averaged and the standard deviation was extracted.

4   Discussion

RMS shows a similar pattern along accelerometer data: intensity increases to a peak, before decreasing. Signal overlaps can be explained by the propagation of waves on the skin, an important phenomenon for the future control of actuators. Planned additional analyses will reveal important information about the uniformity of stroking velocity and duration, which will inform apparent motion stimulus parameter optimisation. PS analysis shows that using one frequency to emulate apparent tactile motion may not be the best option. As previous studies of tactile apparent motion often used only one frequency, these findings may indicate a good way to provide more convincing tactile apparent motion. To achieve this, it is planned to use a bespoke vibrotactile display (see Fig. 1b) capable of delivering vibrations up 500 Hz (based on this actuator design [6]).

References

1. Felsen, G., Dan, Y.: A natural approach to studying vision. Nat. Neurosci. 8, 1643–1646. Nature Publishing Group (2005)
2. Maallo, A.M.S., Duvernoy, B., Olausson, H., McIntyre, S.: Naturalistic stimuli in touch research. In: Current Opinion in Neurobiology (in review) (2022)
3. Israr, A. Poupyrev, I.: Control space of apparent haptic motion. In: 2011 IEEE World Haptics Conference, pp. 457–462. IEEE (2011)
4. Hachisu, T., Suzuki, K.: Tactile apparent motion through human-human physical touch. In: Prattichizzo, D., Shinoda, H., Tan, H.Z., Ruffaldi, E., Frisoli, A. (eds.) EuroHaptics 2018. LNCS, vol. 10893, pp. 163–174. Springer, Cham (2018). https://doi.org/10.1007/978-3-319-93445-7_15
5. Kwon, J., Park, S., Sakamoto, M., Mito, K.: The effects of vibratory frequency and temporal interval on tactile apparent motion. IEEE Trans. Haptics 14(3), 675–679 (2021)
6. Duvernoy, B., Farkhatdinov, I., Topp, S., Hayward, V.: Electromagnetic actuator for tactile communication. In: Prattichizzo, D., Shinoda, H., Tan, H.Z., Ruffaldi, E., Frisoli, A. (eds.) EuroHaptics 2018. LNCS, vol. 10894, pp. 14–24. Springer, Cham (2018). https://doi.org/10.1007/978-3-319-93399-3_2

Between-Tactor Display Using Dynamic Tactile Stimuli

Ryo Eguchi[1]([⊠]) , David Vacek[1], Cole Godzinski[2], Silvia Curry[2], Max Evans[2], and Allison M. Okamura[1]

[1] Stanford University, Stanford, CA 94305, USA
{eguchir,dvacek,aokamura}@stanford.edu
[2] Triton Systems, Inc., Chelmsford, MA 01824, USA
{cgodzinski,scurry,mevans}@tritonsys.com

Abstract. Display of illusory vibration locations between physical vibrotactile motors (tactors) placed on the skin has the potential to reduce the number of tactors in distributed tactile displays. This paper presents a between-tactor display method that uses dynamic tactile stimuli to generate illusory vibration locations. A belt with only 6 vibration motors displays 24 targets consisting of on-tactor and between-tactor locations. On-tactor locations are represented by simply vibrating the relevant single tactor. Between-tactor locations are displayed by adjusting the relative vibration amplitudes of two adjacent motors, with either (1) constant vibration amplitudes or (2) perturbed vibration amplitudes (creating local illusory motion). User testing showed that perturbations improve recognition accuracy for in-between tactor localization.

Keywords: Vibrotactile feedback · Wearable devices · Haptic illusions

1 Introduction

Torso-worn vibrotactile displays providing spatial cues can be used for directional navigation. In these displays, a vibrotactile motor (tactor) vibrates at a point on the human transverse plane to indicate a subjective direction defined by a vector between the torso center and the tactor location. The torso is an attractive location because it has a large skin area and wearable devices can be easily attached, typically in the form of a belt. Applying tactile feedback on the torso also leaves more functional body parts (e.g., hands, fingers, and feet) free for interaction with the environment. Previous studies have examined localization

This research was funded by a Small Business Innovation and Research (SBIR) grant W81XWH21C0051 through the Defense Health Agency and technical point of contact, Dr. Christopher Brill, through the Air Force Research Laboratory. The views expressed in this paper are those of the authors and do not reflect the official views or policy of the Department of Defense or its Components. Mention of any specific commercial products, process, or service does not constitute or imply its endorsement, recommendation, or favoring by the United States Government, Department of Defense, or Department of Air Force.

H. Seifi et al. (Eds.): EuroHaptics 2022, LNCS 13235, pp. 379–381, 2022.
https://doi.org/10.1007/978-3-031-06249-0

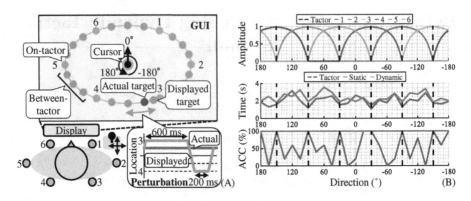

Fig. 1. (A) Experimental setup. Only the orange and gray targets are visible to the user. (B) Amplitude of each tactor and demonstration results (time from stimulus onset to target acquisition and recognition accuracy).

using a single vibrotactile stimulus (e.g., [1, 3]) or recognition of a direction using two successive stimuli (e.g., [4, 5]). Representation of locations or movements between physical tactors would increase the resolution of such a display without increasing the number of tactors [2]. In this paper, we propose a novel between-tactor display using dynamic tactile stimuli to achieve this representation.

2   Methods

Vibrotactile Patterns: We used a vibrotactile belt consisting of an elastic strap and six eccentric rotating mass motors (VZ7AL2B1692082, Vybronics), which we refer to in this paper as tactors. The tactors are evenly spaced at 12 cm and attached to the belt using hook-and-loop fasteners. A microcontroller (Nano 33 BLE, Arduino) and motor driver sends commands to the tactors, and the microcontroller communicates with a laptop PC via a USB cable. A customized MATLAB (2021b, MathWorks) graphical user interface (GUI) commands the tactor voltage (which corresponds to vibration amplitude), displays candidate targets, and records a user's response via joystick movement 100 Hz.

As shown in Fig. 1(A), 24 directions are presented using the 6 tactors. An on-tactor location is represented by vibrating a single tactor. A between-tactor target location is represented by adjusting relative vibration amplitudes of two adjacent tactors according to their distances to the target. Target locations are defined as an angle indicating the direction of the target from the center of the torso. The amplitude y ($\in [0, 1]$) with respect to the target angle x is:

$$y(x) = \begin{cases} \max\left(1 - \exp\left(-\frac{x - d + 60°}{T}\right), 0\right) & x \leq d \\ \max\left(1 - \exp\left(-\frac{d - x + 60°}{T}\right), 0\right) & x > d \end{cases} \tag{1}$$

where d is the angle describing the location of a tactor and T is a time constant and set to 15. The exponential function maintains sufficient vibration amplitudes of adjacent tactors such that the midpoint between two tactors can be perceived.

During testing, we found that identification of between-tactor locations was poor with constant vibration amplitudes. To address this, we perturbed the vibration amplitudes for between-tactor locations, creating illusory motion. The perturbation is a trapezoidal function with a period of 1 s. The between-tactor target takes 100 ms to move between the two adjacent tactors and stays on each tactor for a different duration calculated as the inverse of the ratio of distances from that tactor to the between-tactor target (e.g., 600 ms on the nearest tactor and 200 ms on the other for describing the quantile point).

Demonstration: We demonstrated the method with a single healthy user. The user wore the belt so that two tactors on the front were equally distant from the navel and then learned to identify both on-tactor and between-tactor locations via simultaneous display of a tactile stimulus from the belt and visual indication of its respective location from the GUI. Twenty-four directions were presented five times each using either static (not perturbed) or dynamic (perturbed) stimuli for the between-tactor locations. On-tactor stimuli were always static. The participant was asked to use the tactile stimuli to identify target locations, and indicate the perceived location by moving a cursor from the center of the ellipse to targets on the GUI using a joystick (Hotas Warthog Flight Stick, Thrustmaster).

3   Discussion

The relative amplitudes of tactors, recognition accuracy, and time from stimulus onset to target acquisition (response time) for each direction are shown in Fig. 1(B). Although perturbations slightly prolonged the response time, they significantly enhanced the recognition accuracy for between-tactor locations. This indicates that the user can quickly and easily identify directional cues from spatiotemporal information, consisting of the illusory motion and differences in tactor vibration duration. Thus, the proposed between-tactor display can present directional cues with higher resolution than the number of physical tactors. In future work, we will perform a full user study measuring accuracy and response time for various between-tactor resolutions and measure effects of cognitive load.

References

1. Cholewiak, R.W., Brill, J.C., Schwab, A.: Vibrotactile localization on the abdomen: effects of place and space. Percept. Psychophys. **66**(6), 970–987 (2004)
2. Israr, A., Poupyrev, I.: Tactile brush: drawing on skin with a tactile grid display. In: Proceedings of the SIGCHI Conference on Human Factors in Computing Systems, pp. 2019–2028 (2011)
3. Jones, L.A., Sarter, N.B.: Tactile displays: guidance for their design and application. Hum. Factors **50**(1), 90–111 (2008)
4. Van Erp, J.B.: Presenting directions with a vibrotactile torso display. Ergonomics **48**(3), 302–313 (2005)
5. Van Erp, J.B.: Vibrotactile spatial acuity on the torso: effects of location and timing parameters. In: IEEE World Haptics Conference, pp. 80–85 (2005)

Measuring Oddball Responses to Vibrotactile Textures

Giulia Esposito[1]([✉]), Sylvie Nozaradan[1], Olivier Collignon[1,2,3],
and André Mouraux[1]

[1] Institute of Neuroscience (IoNS), UCLouvain, 1200 Woluwe-Saint-Lambert,
Belgium
giulia.esposito@uclouvain.be
[2] Institute for Psychological Research (IPSY), UCLouvain,
1348 Louvain-la-Neuve, Belgium
[3] School of Health Sciences, HES-SO Valais-Wallis, 1950 Sion, Switzerland

Abstract. Using a fast-periodic oddball paradigm, together with EEG frequency-tagging, we aimed to assess the possibility of measuring periodic responses to rapid changes in a vibrotactile texture. Sequences consisting of standard (A) and oddball stimuli (B) were presented in an AAAAB pattern, with a base and oddball presentation rate of 8 Hz and 1.6 Hz (8/5 Hz), respectively. A and B stimuli either differed in frequency and intensity, or in terms of their complex spectrotemporal composition. Preliminary results suggest that the protocol can be successfully used to record EEG correlates of the cortical processing of haptic textures.

Keywords: Somatosensory · Vibrotactile · EEG

1 Introduction

Oddball paradigms, together with EEG frequency-tagging, can be employed to investigate responses to changes occurring within fast, continuous sequences of stimuli, and represent a useful tool to investigate somatosensory processing of haptic textures, with an extensive body of literature employing such methods in the auditory [1] and visual [2] domains. In recent years, it was shown that EEG frequency-tagging can be used to study haptic processing in conditions of passive dynamic touch using a textured surface varying periodically in intensity while it slides against the participants' fingers [3]. Whether the discrimination of complex textures varying in terms of their spectrotemporal pattern involves the primary somatosensory cortex (S1) is debated. Our aim was to determine whether it is possible to record tactile oddball responses to changes in the spectrotemporal content of a complex vibrotactile texture, and to assess whether these responses follow the somatotopical organization of S1.

© The Author(s) 2022
H. Seifi et al. (Eds.): EuroHaptics 2022, LNCS 13235, pp. 382–385, 2022.
https://doi.org/10.1007/978-3-031-06249-0

2 Methods

Vibrotactile stimuli were designed in MATLAB (version R2020b) and consisted of sequences of 125 ms vibrations presented according to an established (AAAAB) oddball fashion. In a first condition, we aimed to achieve a frequency/intensity contrast, where As and Bs had a frequency of 300 Hz and 200 Hz, respectively. In a second condition, vibrations were composed of bandpass-filtered (40–200 Hz) white noise sequences, where As and Bs were matched in terms of intensity and average frequency content but differed in their spectrotemporal distribution. Each AAAAB sequence lasted 625 ms and was repeated 64 times per trial, for a total of 40 s. Stimuli were delivered to the participants' right hand or foot using a vibrating device (Minishaker Type 4810, Bruel & Kjaer) connected to a power amplifier (Type 2718, Bruel & Kjaer). A 64 Ag-AgCl electrode cap (ANT) was used for EEG recordings and placed on the participants' scalp according to the International 10/10 system. Signals were recorded using an average reference. Sample rate was 2000 Hz and impedances were kept below 10 kΩ.

EEG data analysis was performed using the Letswave6 (http://letswave.org) Matlab toolbox. After preprocessing, EEG signals were averaged across trials in the time domain, and a fast Fourier Transform (FFT) was applied to obtain amplitude spectra with a resolution of 0.025 Hz. Finally, a baseline subtraction was performed to remove, at each frequency bin, the average of the 24 neighboring frequency bins, excluding immediately adjacent bins.

3 Preliminary Results and Conclusion

Preliminary results (N = 3) show that for both hand and foot stimulation, the control condition elicited responses at the base (8 Hz) and oddball frequencies (1.6 Hz and harmonics) for both hand and foot stimulation (Fig. 1).

Similarly, responses to spectrotemporal contrasts could be recorded for both stimulation sites at the base and oddball frequencies (Fig. 2).

The topographical distribution of EEG responses to spectrotemporal contrasts was localized over contralateral frontal and parietal electrodes following right hand stimulation, and over central electrodes following right foot stimulation, consistent with activity originating from the hand and foot representation of S1. In conclusion, EEG frequency tagging can be used to capture cortical activity specifically associated with a contrast in the spectrotemporal distribution of complex vibrotactile stimuli matched in terms of intensity and average frequency distribution, with implications for the design of virtual textures for haptic displays. The topographical distribution of the EEG responses suggests that this activity originates, at least in part, from the contralateral S1.

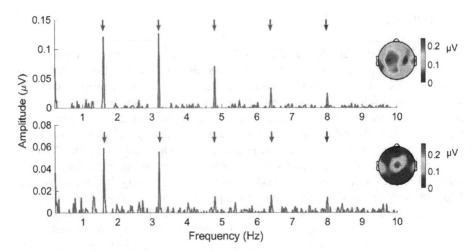

Fig. 1. Baseline-subtracted spectrum of the EEG responses to frequency/intensity contrasts delivered to the right hand (top) and foot (bottom) (average across all scalp channels). Oddball and base frequencies and their harmonics are shown in red and blue, respectively. The topographical maps shows the response at the oddball frequency (1.6 Hz).

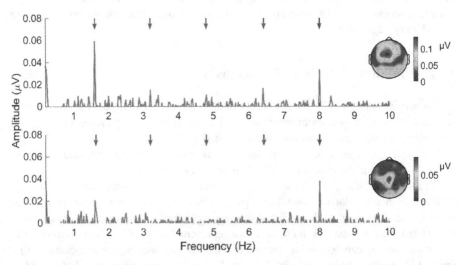

Fig. 2. Baseline-subtracted spectrum of the EEG responses to spectrotemporal contrasts delivered to the right hand (top) and foot (bottom) (average across all scalp channels). Oddball and base frequencies and their harmonics are shown in red and blue, respectively. The topographical maps shows the response at the oddball frequency (1.6 Hz).

Acknowledgements. This study has received funding from the European Union's Horizon 2020 Research and Innovation Program, under Grant Agreement No. 860114.

References

1. Nozaradan, S., Mouraux, A., Cousineau, M.: Frequency tagging to track the neural processing of contrast in fast, continuous sound sequences. J. Neurophysiol. **118**(1), 243–253 (2017)
2. Norcia, A.M., Appelbaum, L.G., Ales, J.M., Cottereau, B.R., Rossion, B.: The steady-state visual evoked potential in vision research: a review. J. Vis. **15**(6), 4 (2015)
3. Moungou, A., Thonnard, J.L., Mouraux, A.: EEG frequency tagging to explore the cortical activity related to the tactile exploration of natural textures. Sci. Rep. **6**, 20738 (2016)

Distinguishable Virtual Haptic Textures to Understand Multisensory Sensation

Jenna Fradin$^{(\boxtimes)}$ and David Gueorguiev

CNRS, Sorbonne Université, ISIR, Paris, France
`radin@isir.upmc.fr`

Abstract. The perception of natural surfaces involves several senses including vision and touch. Several studies have shown cross-modal interactions between the two modalities. However, the integrative mechanisms underlying this process remain unclear. Therefore, we aim to explore the mechanisms involved in the visuo-tactile perception of textures, namely how tactile and visual information are integrated and merged to generate the subjective perception of textures. But one of the difficulties encountered in haptic studies of real textures is to generate tactile feedback that faithfully reproduces the natural surfaces. To overcome this problem, we selected several texture whose vibrotactile signature were recorded. From these signature, we created a database that will be used in the coming electrophysiological and psychophysical studies. First, we performed a preliminary psychophysical task in which the participants were asked to distinguish pairs of textures based on haptic feedback only. The results showed a success rate of 0.82 ± 0.04 and thus demonstrated that discrimination of the rendered textures is possible without being trivial. The next step would be to perform a study about how people perceive visuo-tactile textures with co-localised visual and tactile dimensions.

Keywords: Haptic textures · Psychophysics · Multisensory interaction

1 Introduction

When we explore an object or touch a natural surface, both vision and touch provide information about the properties of the object or the surface. The information gathered by each individual sense have been widely investigated [1, 3]. In the past, the visual sense was considered dominant over touch. But later studies have shown that visuo-haptic integration can be achieved in a statistically optimal way [2]. However, it remains unclear how the information gathered by the two modalities are fused together to enable the perception of textures. To fill this gap, we propose an original approach that aims to create a new dataset of visuo-haptic textures recorded with the bare finger, which will be used to perform psychophysical experiments and electrophysiological recordings.

Supported by the ANR WAVY grant.

H. Seifi et al. (Eds.): EuroHaptics 2022, LNCS 13235, pp. 386–388, 2022.
https://doi.org/10.1007/978-3-031-06249-0

2   Methods

2.1   Texture Recording

We recorded tactile interaction with everyday fabrics to use these recordings in psychophysical tasks. We used five common surfaces: black corduroy, regular denim, stretch denim, black flag banner and a transparent plastic with 3mm-embossed dots. The selected textures were chosen to provide rather similar haptic and visual feedback. Each sample is a 3 x 9.5 cm rectangle glued to a piece of wood of the same dimension. During the recordings, each surface was placed on a top of a force sensor (Nano 17 Titanium, ATI Inc.). A lightweight accelerometer (PCB 352A21) was strapped around the right index fingertip of the experimenter to measure the vibrations that were generated during the texture exploration. The normal and tangential forces applied on the finger and the acceleration were measured simultaneously during the exploration of the different surfaces with a sampling rate of 10 kHz. We displayed a slider on a Touchscreen (DFRobot LCD screen, 183 x 100 mm) placed next to the recording setup, which moved at a constant speed of 4 cm/s. It served as a visual guide to keep the exploration speed constant. A total of three trials were recorded for each texture.

2.2   Psychophysical Task

In the second part, we used the recorded tactile signals to perform a psychophysical experiment on haptic feedback. Three subjects (3 men from 22 to 24 years old, all right-handed) were recruited to participate in this experiment. The possible combinations of textures within the previously collected set of fabrics were successively presented to the participants by the mean of an tactile actuator (MM3C, Tactile lab, 36 x 9.5 x 9.5 mm) held between the thumb and the index of their dominant hand. The participants could see the experimental setup and the tactile actuator. For each trial, they were asked to compare two successive vibrations. After the trial, they answered whether the vibration were the "same" or "different". To enhance the reliability, all pairs of textures were repeated 10 times in a pseudo-random order. During the entire experiment, participants listened to white noise delivered through headphones to avoid auditory cues.

3   Results

We conducted an experiment to test whether participants could discriminate the selected textures using only tactile feedback. The results of all the participants were used to compute a confusion matrix Fig. 1. The rows of the confusion matrix represent the textures presented first to the participants and the columns represent the one presented second. The average ratio of correct answers was 0.82 ± 0.04. These results show that people mostly felt the difference across the rendered textures when played by the vibrotactile actuator. The participants reported two pairs as quite similar. The texture 3 and 4 were considered different 23% and

30% of the time depending on the presentation order. When textures 2 and 4 are compared, they are only perceived in 57% and 42% of the cases as different. Interestingly, they correspond respectively to black flagbanner and stretch denim, which are rather distinct.

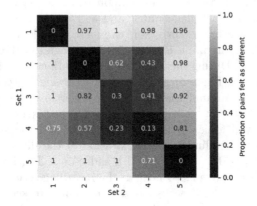

Fig. 1. Confusion matrix of the experiment's responses. The columns and rows represent the different haptic textures presented, namely: 1-plastic, 2-black flag banner, 3-regular denim, 4-stretch denim, 5-black corduroy. The matrix entries represent the frequencies for which the two vibrations were considered different.

4   Discussion

In this study, we attempted to select and record in active touch conditions a dataset of textures on which tactile confusion could occur despite rather clear overall dissimilarity. To test the success of the attempt, we conducted an experiment on the perception of these recorded textures. This experiment showed promising results since some level of confusion occurred. The next step is to update this dataset and explore how humans perceive visuo-tactile surfaces when the visual and tactile dimensions are collocated. We aim to investigate which parameter alters multimodal perception the most and if a subtle change in the tactile or visual dimension is noticed by humans.

References

1. Bergen, J.R., Adelson, E.H.: Early vision and texture perception. Nature **333**(6171), 363–364 (1988)
2. Ernst, M.O., Banks, M.S.: Humans integrate visual and haptic information in a statistically optimal fashion. Nature **415**(6870), 429–433 (2002)
3. Saal, H.P., Bensmaia, S.J.: Touch is a team effort: interplay of submodalities in cutaneous sensibility. Trends Neurosci. **37**(12), 689–697 (2014)

The Role of the Visuo-Tactile Congruence Between the Size of a Collider and Virtual Contents in the Interactions with Virtual Objects

Girondini Matteo[1,2(✉)] ⓘ, Clerici Monica[1,2], Montanaro Massimo[2], and Gallace Alberto[1,2] ⓘ

[1] Department of Psychology, Università Milano-Bicocca, Milan, Italy
m.girondini@campus.unimib.it
[2] Mind and Behavior Technological Center Department of Psychology, Università Milano-Bicocca, Milan, Italy

Abstract. Multisensory integration is a central aspect in respect to cognitive perception. Visuo-haptic incongruence implies distortions in terms of perceptual evaluation. Through Virtual Reality, the present experimental paradigm creates a mismatch between visual and haptic feedback during an object interaction with virtual tools. Two hypotheses were tested in this preliminary study: how visuohaptic mismatch influence task execution and if the manipulation effect influence size evaluation during a categorization task. The results shown difficulty to complete the task in case of small collider (point of interaction inside the virtual object). Moreover, a performance decrease (in term of correct response) was found after manipulation regarding comparison between target cube and slightly larger cube. Exploratory analysis revealed a relationship between number of interactions with small collider and performance decrease in the categorization task. Such distortion was discussed in relationship to other illusion due to visuo-haptic mismatch, and a standardize version of experimental set up is currently under-testing.

Keywords: Multisensory perception · Virtual reality · Cognitive perception

1 Introduction

Multisensory perception is fundamental in our daily life sensory-motor and cognitive processes [1]. As Gestalt theory suggests, perception is the result of the integration between multiple inputs in order to obtain a plausible reality impression [2]. This theory has been confirmed by several multisensory illusions, some of them concerning visual and haptic integration. One of them is the Uznadze haptic aftereffect, for which two identical spheres appear different in size due to previous haptic manipulation (holding a bigger or smaller sphere before the test [3, 4]. The size-weight illusion is another demonstration about how sensory inputs from haptic and visual channels influence cognitive or perceptual evaluations. Nowadays, Virtual Reality (VR) seems a promising tool to assess these illusions in a feasible way, dissociating different sensory

H. Seifi et al. (Eds.): EuroHaptics 2022, LNCS 13235, pp. 389–393, 2022.
https://doi.org/10.1007/978-3-031-06249-0

inputs through realistic and immersive experiences. In particular, VR allows us to study how multisensory incongruence during object interaction affects perceptual evaluations. In this work, we present preliminary results concerning the visuo-haptic mismatch effect on size evaluation through Virtual Reality. The experiment is composed by a two-size evaluation task (judgment of a comparison between two cubes size) alternated by a manipulation task. During the manipulation, participants can move a virtual cube using two virtual sticks from an initial (spawning) position to a random target point in the virtual environment (a simple room). Three different cubes, each characterized by a different interaction collider (i.e. point of interaction between the cube and the sticks), were presented in a random order. The invisible collider could be bigger, equal, or smaller than the cube, and controller-generated haptic feedback was provided when the stick hit the collider. Our first exploratory hypothesis focused on how the colliders can affect performance during the manipulation task, measured as the number of interactions between the virtual stick and collider used to complete each trial. In addition, we analyzed the number of correct responses in a size evaluation task (where cubes similar to those used in the manipulation task were presented) executed before and after the manipulation.

2 Methods

Participants. Through G*Power 3.1 (Faul, et al., 2009) software we calculated our sample size. A within subjects design (α = .05 power $(1-\beta)$ = .80, and small effect size (0.25) with an estimated sample size of 34 participants. A total amount of 40 participants were recruited by self-enrollment using the University recruiting platform. The VR equipment used for the experiment included an Oculus Rift Head mounted display (HMD), with a 1280 × 1440 pixel resolution per eye (refresh rate 80 Hz).

Categorization Task. During the categorization task, a sequence of three different cubes was presented to the participant (Fig. 1). The first cube had a fixed size (0.25 m) while the third cube changed its size in each trial with a range between .10 and .40 (.10; .15; .20; .25; .30; .35; .40 m). Between the first and third cube, a mask was presented (0.75 cm). The participant had to evaluate (in a fixed window of 4 s) if the third cube was bigger (or not) than the first cube, using the controller as guidance.

Fig. 1. Small, equal, and big colliders (from left) during interaction

Manipulation Task. The manipulation task had a duration of 16 min. During this period, participants interacted with the three different cubes (each one with a collider of different size) in a random order and had to move the cube with the two virtual sticks (controlled through the headset proprietary controllers) to reach a target point. During the interaction between cubes, collider and stick, vibration feedback was provided through the VR controller.

3 Results

Manipulation Task. A significant effect of the type of collider (F = 1642, p < .001) was found in respect of the number of interactions with the cube needed to complete each trial. Post-Hoc analysis revealed a significant difference between small and equal colliders, small and big colliders, and between small and big (all ps < .001) (Fig. 2).

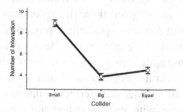

Fig. 2. Number of interaction with each collider

Categorization Task. A significant effect of size ratio (F = 38.02, p < .001) and of the interaction between order and size ratio (F = 3.65, p = .028) were found concerning correct response. The only trend shown in the plot (Fig. 3) suggested a decrease of performance between pre and post categorization in the case of big size ratio.

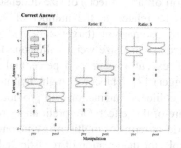

Fig. 3. Number of correct responses for each size ratio

A T-Test analysis has been used to compare differences in terms of number of correct responses between pre- and post-categorization task for each size ratio. No significant effects were found regarding correct responses, when comparing pre- vs post-categorization for small size ratio (t = − .998, p = .838) and equal size (t = − 1.09, p = .856). However, a significant effect was found in the pre- vs post-categorization for big size ratio (t = 1.72, p = .045; d = .395).

Exploratory Analyses. A significant relationship between interactions with the small collider and the number of correct responses was found (t = 1.31, 0 = .035), while other parameters did not reach significant value (for big collider t = 0.54, p = .054; for equal collider t = − 0.310 p = 432). Results clearly shows an inverse relationship between the number of interactions with the small collider and the correct responses in the post categorization.

4 Discussion

Our preliminary study investigated the effect of visuo-haptic input integration during an interaction with virtual objects. More specifically, we created an incongruence between the visual size of cubes and the size of the visuo-haptic interactive space surrounding it (i.e., the collider). We also investigated through exploratory analyses, the carry-over effect of such manipulation in respect of perceptual processes (size evaluation). Virtual Reality gives the opportunity to examine multisensory incongruence through an innovative approach, decomposing sensory inputs in ways otherwise not possible (and to understand the relative neural mechanisms of sensory/cognitive adaptation). The results showed that the small collider required more interactions to complete the task compared to the performance with the equal and the bigger ones.). Regarding the categorization task, a significant decrease in terms of correct responses was found (but only in the comparison between the target cube and the slightly larger cube). Given these results, an exploratory analysis revealed an inverse relationship between small collider interaction and correct responses. Premised that our results should be interpreted carefully, due to their preliminary nature, leading to a few methodological issues. Thus, just as shown elsewhere (similarly to what occurs in prism visuomotor adaptation procedures), interacting for a long time with small colliders would seem to affect the perception (in a sort of 'after effect') of size dimension [6]. In fact, a decrease in performance appears to be related to the number of interactions and seems to influence only one kind of comparison (target cube vs big size cubes). Another hypothesis could be related to the size-weight illusion, for which small collider impairment (regarding task execution) entails a perceptual size bias (more effort to move, so perceived as bigger) [7]. Thus, we might hypothesize that using a fixed amount of time during the colliders' interaction will provide confirmation and clearer answers regarding the nature of this effect. Our preliminary results suggest that virtual reality mirrors the real world in terms of multisensory incongruence, and it should be taken into account reasonably for the creation of realistic and reliable experiences [8]. As far as this aspect is concerned, it would be of interest to generate multisensory incongruences that cannot be (or are more difficult to be) reproduced in real interactions (such as varying independently the size of the tactile and the visual collider). While the visual component is well recreated inside the virtual environment, multisensory integration (especially involving the tactile modality; see [9; 10; 11] are poor and often missing (and so are the knowledge on the neural mechanisms of sensory motor adaptation to 'new realities'). Thus, further investigation in this field is required.

References

1. Ernst, M.O., Banks, M.S.: Humans integrate visual and haptic information in a statistically optimal fashion. Nature **415**(6870), 429–433 (2002)
2. Koffka, K.: Perception: an introduction to the Gestalt-Theorie. Psychol. Bull. **19**(10), 531 (1922)
3. Kappers, A.M., Tiest, W.M.B.: Haptic size aftereffects revisited. In: 2013 World Haptics Conference (WHC), pp. 335– 339. IEEE, April 2013
4. Daneyko, O., Maravita, A., Zavagno, D.: see what you feel: a crossmodal tool for measuring haptic size illusions. i-Perception **11**(4), 2041669520944425 (2020)
5. Flanagan, J.R., Beltzner, M.A.: Independence of perceptual and sensorimotor predictions in the size–weight illusion. Nat. Neurosci. **3**(7), 737–741 (2000)
6. Redding, G.M., Wallace, B.: Generalization of prism adaptation. J. Exp. Psychol. Hum. Percept. Perform. **32**(4), 1006 (2006)
7. Buckingham, G.: Examining the size–weight illusion with visuo-haptic conflict in immersive virtual reality. Q. J. Exp. Psychol. **72**(9), 2168–2175 (2019)
8. Maehigashi, A., Sasada, A., Matsumuro, M., Shibata, F., Kimura, A., Niida, S.: Virtual weight illusion: weight perception of virtual objects using weight illusions. In: Extended Abstracts of the 2021 CHI Conference on Human Factors in Computing Systems, pp. 1–6, May 2021
9. Gallace, A., Spence, C.: In Touch with the Future: the Sense of Touch from Cognitive Neuroscience to Virtual Reality. OUP Oxford (2014)
10. Gallace, A., Ngo, M.K., Sulaitis, J., Spence, C.: Multisensory presence in virtual reality: possibilities and limitations. In: Multiple Sensorial Media Advances and Applications: New Developments in MulSeMedia, pp. 1–38. IGI Global (2012)
11. Gallace, A., Girondini, M.: Social touch in virtual reality. Curr. Opin. Behav. Sci. **43**, 249–254 (2022)

Thermal Sensation Display Controlling the Thermal Conductivity of a Surface

Johan Gonzalez Sanjuan[1(✉)] and Norihisa Miki[2]

[1] Universidad Politecnica de Valencia (UPV), Cami de Vera,
46022 Valencia, Spain
johan10gonz@gmail.com
[2] Keio University, 3-14-1 Hiyoshi, Kohoku, Yokohama, Kanagawa, Japan

Abstract. In this work, a new approach to a thermal sensation display is being researched. The thermal sensation when touching an object depends on several factors like the temperature of the object and its thermal conductivity. Although thermal sensation using the temperature is well known, this is an attempt in controlling the thermal conductivity of a surface using a new device.

This device contains two cavities, one full of air with low thermal conductivity and another full of liquid metal with high thermal conductivity. When the liquid metal fills the air cavity, the thermal conductivity of the device increasing, thus changing the thermal perception A mechatronic system based on Arduino is controlling the volume of liquid metal of the device.

Based on this, some tests on volunteers are being made to test the usability of this technology as a mean to transmit information. It can also be used to test the thermal perception of humans.

Keywords: Tactile display · Thermal sensation · Thermal conductivity · Liquid metal · Mechatronics

1 Introduction

1.1 Human Touch Sense

Our skin has receptors to collect information from our surroundings. Every receptor has a specific trigger and there are receptors for cold and warm touch. This receptors trigger when they cool down or heat up. For this reason, it is impossible to distinguish an absolute temperature.

Nevertheless, it is possible to use this property by changing the rate at which our receptor changes its temperature. Two surfaces at 20ºC, one made from wood and other made from metal. When touching them, our skin will start to transfer heat to the surface, cooling down and triggering the cold receptors. However, the surface made of metal will absorb heat faster than the wood surface, leading to a feeling of "colder" surface. This is the property this screen is trying to use.

1.2 Previous Work

Previous prototypes using peltier cells to change the temperature of the surface showed that it is possible to use the thermal sensation to recognize easy patterns after some

© The Author(s) 2022
H. Seifi et al. (Eds.): EuroHaptics 2022, LNCS 13235, pp. 394–396, 2022.
https://doi.org/10.1007/978-3-031-06249-0

training [1]. Other research changing the thermal conductivity using a stream of water under the surface [2] also proved to be useful to transmit specific sensations [3]. Our group is developing the device [4] and this is the first implementation of an array of pixels and automated control. The mechanism is detailed in the following section.

1.3   Description of the Technology

The surface is divided in smaller units (2×2 cm) that can be controlled individually, called pixels. Each pixel is composed of 4 pieces. The tactile surface is a 0.02 mm thickness sheet of titanium. Under it, there is a piece of acrylic with holes. These holes are filled or emptied with liquid metal (Galinstan) to change the conductivity as seen in Fig. 1. At the bottom there is copper piece that behaves as a reservoir or liquid metal. Finally, the acrylic and the copper parts are separated by a 20 μm thickness layer of latex, that effectively encapsulates the liquid metal without altering the thermal conductivity of the system. A pixel without the titanium cover and the mechatronic system can be seen in Fig. 1 as well.

Fig. 1. Left: Working principle with real pictures. Center: pixel assembly. Notice the holes in the acrylic plate that will be filled with liquid metal. Right: complete system. A 2 × 2 screen can be seen on the right of the top layer.

2   Testing

2.1   Test with Thermal Camera

In this test, a thermal camera is being used to see if the state of the pixel can be deducted from the evolution of the temperature of the surface. This is a way of verifying that the device works as intended. In this test, a human touches the surface of the screen, holds the position for 30 s, and then releases the screen. Since the conductivity of the titanium layer is high and the thermal inertia of the system is very small, it is difficult to capture the feeling of the screen, however, some differences can be seen. Figure 2 shows the same screen with the top side with high conductivity. In these figures, the gradient from white to blue represent the temperature, going from higher to lower respectively.

Pixels with high thermal conductivity seem to absorb more heat from the hand and have a bigger thermal inertia, so after some time in contact with the hand, display a higher temperature than low pixel, as shows the red color of those pixels.

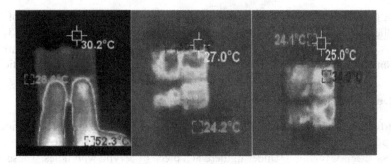

Fig. 2. Example of test and results after a 30 s long touch in a 2 × 2 screen. Red parts indicate a higher temperature and thermal conductivity.

It is important to note that even though the temperature of those pixels is higher, the volunteers felt a "colder" sensation when touching them, which means that the thermal conductivity plays in fact an important role in out thermal perception. This is being research further as discussed in 2.2.

2.2 Test Human Perception

Early tests on a group of four volunteers showed that it is possible to feel a difference when touching the pixel one by one with the tip of the fingers, specially if it is located exactly on one of the four cavities of the pixel. This feeling is described consistently as "cooling down" or "colder" pixels. Nevertheless, tests on the forearm or multiple pixels at the same time are still unconclusive. More research is being done to verify the replicability and trust of the thermal sensation.

References

1. Singal, A.: Development of thermal displays for haptic interfaces (2016). http://hdl.handle.net/1721.1/104139
2. Sakaguchi, M., Imai, K., Hayakawa, K.: Development of high-speed thermal display using water flow. In: Yamamoto, S. (ed.) HCI 2014. LNCS, vol. 8521, pp. 233–240. Springer, Cham (2014). https://doi.org/10.1007/978-3-319-07731-4_24
3. Hayakawa, K., Imai, K., Honaga, R., Sakaguchi, M.: High-Speed thermal display system that synchronized with the image using water flow. In: Kajimoto, H., Ando, H., Kyung, K.-U. (eds.) Haptic Interaction. LNEE, vol. 277, pp. 69–74. Springer, Tokyo (2015). https://doi.org/10.1007/978-4-431-55690-9_13
4. Hirai, S., Norihisa, M.: A thermal tactile sensation display with controllable thermal conductivity. Micromachines **10**(6), 359 (2019). https://doi.org/10.3390/mi10060359

Simulating the Influence of Surface Hardness on the Contact Area Formation for Dry and Moist Finger Skin Contacts

Michael Grießer, Thomas Ules[(⊠)], Christian Schipfer, Peter Fuchs, and Dieter P. Gruber

Polymer Competence Center Leoben GmbH, Roseggerstraße 12,
8700 Leoben, Austria
thomas.ules@pccl.at

Abstract. This contribution examines the relationships between surface hardness, mechanical properties of human skin and the implications on contact area formation. To explore the contact behavior and its dependence on these parameters finite element method simulations were conducted. The finger skin model is constituted by three different layers reflecting the Stratum Corneum, the Epidermis and the Dermis. Different mechanical properties were assigned to the layers to mimic human skin in dry, intermediate and moist condition. To test the influence of the hardness of the counter surface these finger skin models were brought in contact with a soft surface, a firmer surface and a rigid counter body. While the contact area mainly depends on the moisture level of the skin model, expressed via the mechanical properties of the skin layers, the hardness of the counter surface influences the contact area just for dry and stiff finger skin. In the case of moist and soft finger skin the contact area is determined by the mechanical properties of finger skin alone.

Keywords: Haptics · Contact area · Finite element method

1 Introduction

The study of contact mechanics and in particular the contact area, formed between human skin and counter bodies with various mechanical properties is crucial to understand the interaction between human skin and everyday life objects. Direct observation of the area in contact, in particular at the micrometer scale, is difficult and frictional properties may be used to study contact formation indirectly. Using a living biological material that is difficult to characterize and whose mechanical properties may change due to occlusion and sweat accumulation at the interface in the course of an experiment results may be insufficiently precise. Another approach is to simulate the contact mechanics via the finite element method [1]. This approach is pursued in this contribution. The aim was to illuminate the influence of skin moisture and surface hardness of the counter body on the contact area formation via finite element method simulations.

© The Author(s) 2022
H. Seifi et al. (Eds.): EuroHaptics 2022, LNCS 13235, pp. 397–400, 2022.
https://doi.org/10.1007/978-3-031-06249-0

2   Contact Modelling Methodology

The surface topography of the finger pad model used in the simulations are obtained from microscopy measurements of a human finger pad. The total surface area is 1.2×1.6 mm². The topography was measured with the Leica DCM8 in the focus variation mode and 10x-magnification. After lowpass filtering a CAD model was generated to further mesh the whole cutout. Using the Finite Element software Abaqus (R2017, Dassault Systemes, Simulia, Vélizy-Villacoublay, France), a structured mesh of C3D8RH Hex-Elements with a size of 10 μm was employed. In order to assign mechanical properties to the different skin layers, a set of appropriate stiffness values was adopted from Derler et al. [2]. The FEM finger model was built from three different layers, with a Neo-Hookean material model assigned to each layer (see Fig. 1). Three different stiffness variations were considered for the Stratum Corneum layer in order to represent dry, intermediate and moist conditions, as it is mainly this layer that is affected by hydration [2]. The simulation boundary conditions were defined as follows: The contact was enforced by the lowering the counter surface towards the Stratum Corneum while the bottom surface of the Dermis layer was fixated. Periodic boundary conditions were applied to the side areas as the cutout is assumed to be a representative unit cell of the finger skin. Furthermore, the influence of the counter surface was analysed using three different settings: rigid analytical surface, soft elastic solid (Young's modulus = 10 MPa) and firm elastic solid (Young's modulus = 2500 MPa).

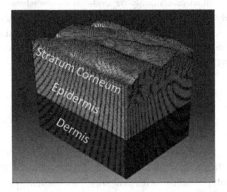

Fig. 1. FEM-model of the cutout with the three layers representing skin tissue.

3   Results and Discussion

Figure 2 shows the evolution of the calculated contact area for three types of skin with rising reaction force, against a rigid flat substrate. As expected, the respective contact areas increase significantly with decreasing hardness of the outermost skin layer. This behavior is also observed in friction measurements of human skin. In the case of adhesion dominated friction, where contact area determines the friction force, moist skin reveals substantially higher frictional forces compared to dry skin. This behavior is attributed to the softening of the stratum corneum and the enhanced ability to conform to the counter surface [2]. It is also observed that the contact area for dry skin depends linearly on the normal force F. For moist skin conditions a Hertzian contact formation with the contact area being proportional to $F^{1/3}$ is observed. To study the influence of the hardness of the counter surface on the contact area formation, the dry and moist finger skin was simulated in contact with the soft and the hard counter surface. For the chosen mechanical properties of the counter surfaces moist finger skin reveals a much larger contact area compared to dry skin on both surfaces as observed for the rigid counter surface. For moist skin the calculated contact area is solely determined by the properties of the compliant finger skin. This can be seen by equal contact area curves for the two counter surfaces (blue curves in Fig. 2). This contradicts dry skin, where the contact area is different for the two counter surfaces (red curves in Fig. 2). This is important for understanding tribological measurements with human skin, where the contact area is assumed to depend solely on the mechanical properties of a contacting surface when the skin is rather dry. This suggests that for adhesion dominated friction the friction forces of dry finger skin should be strongly determined by the hardness of the respective counter surface. In contrast, moist finger skin will reveal higher friction forces however with slight differences on surfaces with varying hardness since the contact area is expected to be determined by the soft and compliant skin tissue.

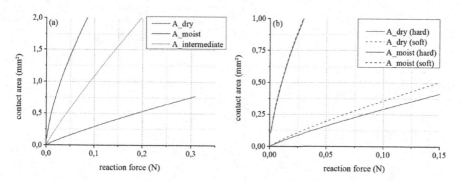

Fig. 2. Contact area versus reaction force for (a) three types of skin with different mechanical properties reflecting different hydration levels and a rigid counter body and (b) dry (red) and moist (blue) skin against a hard (continuous line) and soft (dotted line) counter surface.

4   Conclusion

In this contribution the dependence of contact area formation on the mechanical properties of human finger skin and hardness of the counter body was explored. While moist skin showed substantially larger contact areas compared to dry and stiff finger skin the implications of the hardness of the counter surface was restricted to dry finger skin. It may be expected that only dry finger skin will yield different frictional forces when contacting surfaces of different hardness.

References

1. Leyva-Mendivil, M.F., Lengiewicz, J., Page, A., Bressloff, N.W., Limbert, G.: Skin microstructure is a key contributor to its friction behaviour. Tribol. Lett. **65**(1), 1–17 (2016). https://doi.org/10.1007/s11249-016-0794-4
2. Derler, S., Gerhardt, L.-C.: Tribology of skin: review and analysis of experimental results for the friction coefficient of human skin. Tribol. Lett. **45**(1), 1–27 (2012). https://doi.org/10.1007/s11249-011-9854-y

Embedded 3D Printing: A Cost-effective Development Platform for Tactile Sensors

Sonja Groß[1,2](✉) ⓘ, Silija Breimann[1](✉) ⓘ, Sascha Schwarz[1] ⓘ,
Amartya Ganguly[1] ⓘ, and Sami Haddadin[1,2] ⓘ

[1] Munich Institute of Robotics and Machine Intelligence, Technical University
of Munich, Munich, Germany
`sonja.gross@tum.de`
[2] Centre for Tactile Internet with Human-in-the-Loop (CeTI), Dresden, Germany

Abstract. Soft tactile sensor design is commonly implemented with conductive, stretchable materials that are either soft themselves or embedded in an elastomeric matrix. It is well documented that embedded 3D printing (e-3D) overcomes the limitations of these processes, such as delicate hard-soft interfaces, poor shape fidelity, intricate manufacturing processes, and extensive multi-step casting. This paper proposes a precise and accessible e-3D setup by combining a cost-effective fused deposition modeling (FDM) printer with a high-precision extruder suitable for soft materials and liquids. We designed and fabricated four sensor patterns that were evaluated in stretch and normal force experiments. The results indicate, that the three sensors under normal force were able to detect qualitative contact changes, while the one exposed to stretch performed similar to previous work in the literature. These preliminary results prove the feasibility of the proposed setup, which will be used as an optimization and evaluation platform for future soft tactile sensor development.

Keywords: Embedded 3D printing · Soft tactile sensors · Biomimetic

1 Introduction

In recent years, e-3D has shown numerous promising applications for efficient fabrication of highly stretchable, functional systems including strain and pressure sensitive skins [1, 2] . This method allows for the design of complex 2D/3D structures in a single fabrication step, using the freedom-of-design-aspect of traditional additive manufacturing without the need for support structures or multi-step elastomer casting [3]. We demonstrate the feasibility of an accessible e-3d platform by combining a high-resolution extruder with a low-cost FDM

Funded by the German Research Foundation (DFG, Deutsche Forschungsgemeinschaft) as Part of Germany's Excellence Strategy - EXC 2050/1 - Project ID 390696704 - Cluster of Excellence "Centre for Tactile Internet with Human-in-the-Loop" (CeTI) of Technische Universität Dresden

H. Seifi et al. (Eds.): EuroHaptics 2022, LNCS 13235, pp. 401–403, 2022.
https://doi.org/10.1007/978-3-031-06249-0

printer. This system enables extensive and affordable material testing allowing for efficient sensor design and optimization. Preliminary experimental results with developed strain and pressure sensors underline the feasibility of the system and provide a realistic outlook on future experiments.

2   Materials and Methods

We adapted a commercial FDM printer (Ender 5 plus, Creality) by replacing the original extruder with a soft materials print head (vipro-HEAD 3, ViscoTek) as shown in Fig. 1(I). Pneumatic pressure drove the material through the 30 cc cartridge (502758, Vieweg), which is screwed to the print head via Luer-Lock. We mixed Ecoflex 00–30 silicone with different additives to synthesize two material layers [3]. The printing reservoir layer of high viscosity lay under a low-viscosity filler layer that fills the cavities created by the nozzle movement (all reagents from Smooth-On Inc.). Carbon grease (846-1P, MG Chemicals, resistivity $63\Omega cm$) served as conductive ink. Manually generated G-code commands controlled the path of the conductive ink in the reservoir material to achieve desired sensor patterns, see Fig. 1(II). Optimized parameters influenced the pattern quality – the movement rate (v mm/min), the extrusion rate (e mm^{-1}), and the extrusion offset (eo mm), as shown in Fig. 1(III). We set the extrusion rate and printing speeds at $e = 5mm^{-1}$ and $v = 100, 200$ mm/s for slow (change in direction) and fast (straight line printing), respectively. The experimental setup consisted of a 3-axis force Sensor (K3D40-50N, ME-Messsysteme) mounted on the end-effector of a light-weight robot (Panda, Franka Emika). During stretch experiments, we clamped the sensor probe between the force sensor and the ground plate and stretched it with periodic increments (F_Z, 0-1N) while recording the relative resistance change ($R_{Rel} = \Delta R/R_0$). During force experiments, the robot applied normal force on the sensor probe in periodic increments (F_N, 0-10N).

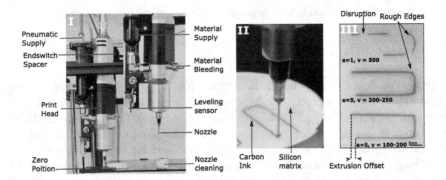

Fig. 1. (I) The adapted printer setup combines an FDM printer with a soft materials print head. (II) The printer deposits conductive ink inside the reservoir material. (III) Influence of printing parameters.

3   Results and Discussion

In Fig. 2, the middle panel depicts the applied stretch force, with $R_{rel_A} = 0.5$ at $F_Z = 0.3N$ and $R_{rel_{A_{MAX}}} = 1.1$ at $F_Z = 1N$. The right panel shows three sensors subjected to normal force with maximum $R_{rel_{B_{MAX}}} = 0.058$; $R_{rel_{C_{MAX}}} = 0.047$ and $R_{rel_{D_{MAX}}} = 0.044$. All patterns show relaxation and drift which is well known for resistive elastomer sensor materials [4]. The data of the stretch experiment supports the results of [1]. The normal force experiments with three different patterns show that the material is less sensitive to pressure application which leads to a higher influence of the non-linear behavior on the signal evaluation. While for the stretch sensor, the resistance change reflects the increasing stretch force, the normal force experiments show that the tested patterns are not feasible for quantitative force measurement, but rather qualitative contact event monitoring.

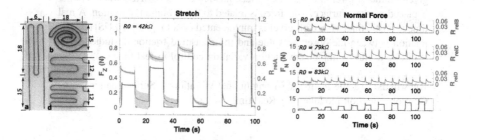

Fig. 2. Relative change in electrical resistance of different sensor patterns (a-d, left) subjected to cyclic stretch (a, middle) and cyclic normal force (b, c, d, right).

In summary, the successful implementation of an accessible and high-resolution e-3D system was shown. Although further research is needed, this platform enables efficient investigation and evaluation of soft tactile sensing applications.

References

1. Muth, J.T., et al.: Embedded 3D printing of strain sensors within highly stretchable elastomers. Adv. Mater. **26**(36), 6307–6312 (2014)
2. Senthil Kumar, K., Chen, P.Y., Ren, H.: A review of printable flexible and stretchable tactile sensors. Research **2019**, 1–32 (2019)
3. Truby, R.: Embedded three-dimensional printing of autonomous and somatosensitive soft robots. Doctoral Dissertation, Harvard University, Graduate School of Arts & Sciences (2018)
4. Georgopoulou, A., Sebastian, T., Clemens, F.: Thermoplastic elastomer composite filaments for strain sensing applications extruded with a fused deposition modelling 3D printer. Flexible and Printed Electronics 5 (2020)

Acoustic Finger Force Observation in a Longitudinal Ultrasonic Surface Haptic Device and Its Relation to Friction

Diana Angelica Torres[(✉)], Anis Kaci, Betty Lemaire-Semail,
Christophe Giraud-Audine, Frederic Giraud, and Michel Amberg

University of Lille, Arts Et Metiers Institute of Technology, Centrale Lille, Junia,
ULR 2697 - L2EP, 59000 Lille, France
diana.torres-guzman@univ-lille.fr

Abstract. Acoustic finger force is the additional force needed to maintain a given vibration level in ultrasonic surface haptic devices (USHD) when a finger is placed on their surface. This force is related to the evolution of friction. A real time acoustic finger force observer is implemented on a longitudinal USHD. The relation between the observed value and the measured friction is studied. This could be useful to implement friction variation control in real time.

Keywords: Tactile feedback · Longitudinal vibrations · Surface haptics · Control · Friction modulation · Real time

1 Introduction

Friction reduction due to longitudinal vibration may be obtained using transverse as well as longitudinal ultrasonic vibration [1]. A common issue present on both modes, which raises the question of viability and robustness of ultrasonic devices, concerns the problem of tactile perception standardization. Creating a more uniform perception of friction modulation throughout the population is desirable, since this would help to improve the construction and rendering of texture models for USHDs. In previous work [2], a closed loop amplitude control was proposed as a means of compensating for the impact of the finger on the vibration amplitude. However, even with the vibration amplitude control, it is still difficult to predict the value of friction for each user. An alternative solution could therefore be to relate this friction with a real time observation of the acoustic finger force.

In this paper we use the amplitude-controlled longitudinal USHD explained in [3], to implement a real time observation of the acoustic finger force and evaluate its relation with friction.

H. Seifi et al. (Eds.): EuroHaptics 2022, LNCS 13235, pp. 404–407, 2022.
https://doi.org/10.1007/978-3-031-06249-0

2 Friction vs. Acoustic Finger Force Experiments

The setup for the experiments carried out in this study is shown in Fig. 1. The longitudinal USHD tested is the one described in [3]. The device is placed over a three-axial force sensor (GSV-4USB from ME-Meßsysteme). The USHD and force sensor are adapted over the moving section of a tribometer. The observer is programmed in the DSP together with the controller. A computer is connected to the DSP and force sensor.

Three experiments are performed, provisionally with a single participant, to evaluate the behaviour of the observed acoustic finger force and the friction.

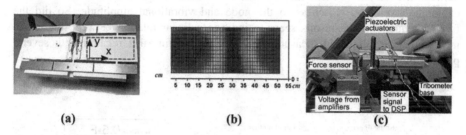

Fig. 1. (a) Longitudinal USHD. The x axis is defined in the horizontal direction and the y axis in the vertical direction. (b) Cartography of one facet of the device, where the stationary longitudinal ultrasonic wave is represented. The maximum longitudinal displacement (at the anti-node) is depicted in red and the minimum (at the node), in blue. (c) Setup for lateral acoustic finger force and friction measurements.

For the first and second experiments, the tribometer base remains static. The objective of the first experiment was to evaluate the influence of the motion and speed of the finger in the acoustic finger force and friction measurements. To answer this question, a closed-loop amplitude reference of $1\mu m_{p-p}$ was then given to the USHD. A finger was then placed on the device. First it remained static and after about 1–2 s, it initiated rapid motion (about 150 mm/s) along the y axis over the anti-node, in the radial-ulnar direction. The friction and acoustic finger force observation were recorded.

As explained in [3], the wavelength of the longitudinal mode is very large. For this reason, the friction reduction and haptic feedback are not uniform along the x axis. The second experiment is therefore designed to explore the effect of this issue on friction and acoustic finger force. For this, a closed-loop amplitude reference of $1\mu m_{p-p}$ was then given to the USHD. A finger was then slowly (about 20 mm/s) slid over the surface of the device, along the x axis in the radial-ulnar direction. Due to this trajectory, the finger explored a vibration anti-node, followed by a node and another antinode, as can be seen in Fig. 1(b). The friction and acoustic finger force observation were recorded.

The third experiment was aimed at exploring the effect of vibration amplitude in friction and acoustic finger force measurements. In this case, the tribometer base was programmed to perform reciprocating motions at a constant speed of 30 mm/s. A finger was placed on the device statically, while the device slid below the finger over a vibrational anti-node along the y axis. A visual aid helped maintain the normal pressure force at about 0.5 N. Friction and acoustic finger force measurements were acquired simultaneously for a closed loop amplitude reference of 0.5 μm_{p-p} for a number of periods. The measurements are then repeated with a reference of 1.5 μm_{p-p}.

3   Results and Conclusions

The friction and acoustic finger force measurements for the first, second and third experiments are plotted in Fig. 2 (a)–(c). The preliminary results show a correlation between friction and acoustic finger force. The acoustic finger force was decreased by exploration velocity, closeness to the node and vibrational amplitude. So did the friction reduction. The results seem to indicate a linear relation between these two measurements. Additional measurements will help define this relation with several participants.

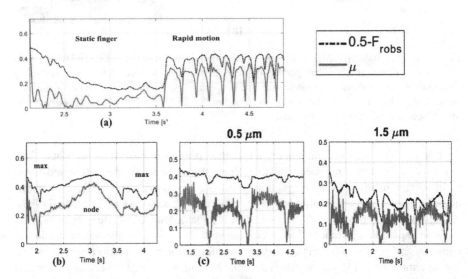

Fig. 2. Acoustic finger force vs. Friction. (a) Static finger vs. rapid motion. (b) Exploration through the x axis. (c) Measurements at 0.5 μm_{p-p} and 1.5 μm_{p-p}.

Acknowledgment. This work has been carried out within the framework of the Mint Project of IRCICA (CNRS Service and Research Unit 3380).

References

1. Guzman, D.A.T., Lemaire-Semail, B., Kaci, A., Giraud, F., Amberg, M.: Comparison between normal and lateral vibration on surface haptic devices. In: 2019 IEEE World Haptics Conference (WHC), Tokyo, Japan, July 2019, pp. 199–204 (2019)
2. Ghenna, S., Giraud, F., Giraud-Audine, C., Amberg, M.: Vector control of piezoelectric transducers and ultrasonic actuators. IEEE Trans. Industr. Electron. **65**(6), 4880–4888 (2018)
3. Torres, D.A., Lemaire-Semail, B., Giraud-Audine, C., Giraud, F., Amberg, M.: Design and control of an ultrasonic surface haptic device for longitudinal and transverse mode comparison. Sens. Actuators, A **331**, 113019 (2021)

Design and Implementation of a Cartesian Impedance Control in a Bilateral Telemanipulation System Using UR10e Robots

Ali Hammoud$^{(\boxtimes)}$ ⓘ, Fady Youssef ⓘ, and Thorsten A. Kern ⓘ

Institute for Mechatronics in Mechanics, Hamburg University of Technology, Hamburg, Germany
ali.hammoud@tuhh.de
https://www.tuhh.de/imek/

Abstract. This paper presents an overview of a work-in-progress project, where a telemanipulation system is designed and its stability, delay and transparency are discussed. In this telemanipulation system a cartesian impedance control system was implemented, tested and evaluated.

Keywords: Impedance control · Telemanipulation · Robot-environment interaction · UR10e

1 Introduction

The basic objective of impedance control is to achieve a desired dynamic relationship between the robot motion and the external forces acting on it [1]. This relationship plays a major role in the field of telemanipulation, where interaction with the robot, and transparency of feedback forces are essential. The main concept of this controller can be derived from the physical behavior of a mass-spring-damper system.

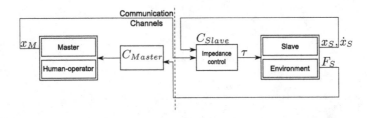

Fig. 1. Block-diagram of impedance control within a telemanipulation system

Fig. 1 shows the basic elements of a telemanipulation system (Human-Operator, Master, Communication, Slave and Environment) [3]. C_{Slave} is the

© The Author(s) 2022
H. Seifi et al. (Eds.): EuroHaptics 2022, LNCS 13235, pp. 408–411, 2022.
https://doi.org/10.1007/978-3-031-06249-0

controller on the slave, where x_m, x_s, and $\dot{x}_s$ represent the desired position, current position and velocity of the slave robot, respectively. The output τ stands for the motors' torques applied to the robot motors. The two main factors in controlling telemanipulation system are stability and transparency. If a force occurs on the slave, measured by force sensors, then reflected back to the master, this is referred to as bilateral control of the teleoperator [2, 3]. Reflection of forces allows the human operator to rely on both his/her haptic and visual senses. A delay in the communication channels can lead to instability of the system, which can be solved using many methods e.g. wave variables [6].

2   Basic Concepts

2.1   Cartesian Impedance Control

Given two homogeneous matrices H_t^0 and H_v^0 describing the current and the desired poses of a robot's Tool center point (TCP), the cartesian impedance controller for a 6-DOF robot is physically described by a spatial multidimensional spring with symmetric stiffness matrix $K \in R^{6\times6}$ connected between both poses and attempts to match them along with a spatial multidimensional damper. The wrench W_s exerted on the robot caused by the spatial spring is given as [4]:

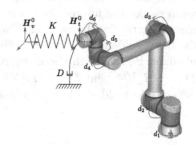

Fig. 2. Cartesian impedance control: compliance is added via a spatial multidimensional spring $K \in R^{6\times6}$ and damping is introduced either via a spatial multidimensional damper $D \in R^{6\times6}$ or space dampers d_i connected to each joint.

$$W_s = \begin{pmatrix} m_s \\ f_s \end{pmatrix} = \begin{pmatrix} K_o & K_c \\ K_c^T & K_t \end{pmatrix} \cdot \begin{pmatrix} \delta\theta_t^v \\ \delta p_t^v \end{pmatrix}$$

Where $\delta\theta_t^v$ and δp_t^v are infinitesimal twists in vector form and K_t, K_o, and K_c, are the symmetric translational, rotational and coupling stiffness matrices, which represent the spatial compliance. Starting with W_s and using the related equations in [4], the wrench can be calculated and transformed in the base frame W_s^0. Adding a damper can be done using the same procedure or as follows:

$$W_{sd}^0 = \begin{pmatrix} m^0 \\ f^0 \end{pmatrix}_{sd} = W_s^0 - \begin{pmatrix} D_o & 0 \\ 0 & D_t \end{pmatrix} (v_o v_t) \tag{1}$$

where the indices s and sd represent spring and (spring + damper), D_t and D_o are the translational and the rotational damping matrices, with $D_i = d_i I_3$. Finally, v_t and v_o are the translational and rotational velocities of the TCP, respectively.

2.2   Telemanipulation System Using Cartesian Impedance Control

The controller presented in the previous section was implemented for both robots, master and slave, within the Kinesthetic Force Feedback (*KFF*) architecture to build a telemanipulation system. According to [5, 7], in this architecture the position of the master is transmitted to the slave, and the environment force acting on the slave is sent back to the master, that is passed to the operator as a feedback force. A block diagram of this architecture can be seen in Fig. 1.

3   Experiments

Two *UR10e* robots were used Fig. 2. The forces measured on the slave were filtered using a Kalman filter Fig. 3b before being reflected as feedback force to the master. The *forcemode* function provided by the robot was used, that accept wrench directly without the need to determine the motors' torques τ. On the master side, the filtered force was used as wrench for the cartesian impedance controller, that's why no wrench was calculated and the controller was used for the hand guidance and force feedback. In the evaluation a cartesian trajectory planning Fig. 3a, with the gains $K_t = 3000I_3[\frac{N}{m}]$, $K_o = 50I_3[\frac{N}{rad}]$, $K_c = 0_3$, $D_t = 50I_3[\frac{Ns}{m}]$ and $D_o = 5I_3[Ns/rad]$ are used. One can see that the delay is between both sides is less than 200 ms.

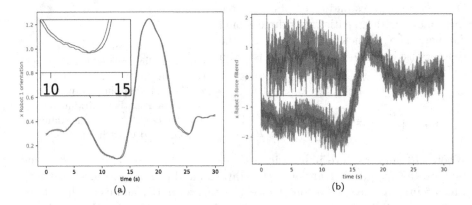

(a)   (b)

Fig. 3. (a): the Orientation of the Master (blue) and Slave (red) Robots on the X-axis. (b): force measured (blue) and filtered using a Kalman filter (red)

4   Summary

The performance of cartesian impedance control in telemanipulation systems was studied, where a delay of less than 200 ms, and stability in contact was achieved. Optimizating the performance can be made with, for example, adding Shared Compliance Control, that can improve the stability, dealing with time delays and perception. Also, the effect of using a higher accuracy external F/T sensor to be discussed instead of internal force sensor used.

References

1. Hogan, N.: Impedance control: an approach to manipulation: Part I-Theory (1985)
2. Hatzfeld, C., Kern, T.A. (eds.): Engineering Haptic Devices. SSTHS, Springer, London (2014). https://doi.org/10.1007/978-1-4471-6518-7
3. Hokayem, P.F., Spong, M.W.: Bilateral teleoperation: an historical survey. Automatica **42**(12), 2035–2057 (2006)
4. Raiola, G., Cardenas, C.A., Tadele, T.S., De Vries, T., Stramigioli, S.: Development of a safety-and energy-aware impedance controller for collaborative robots. IEEE Robot. Autom. Lett. **3**(2), 1237–1244 (2018)
5. Klomp, F.M.: Haptic control for dummies: an introduction and analysis. Eindhoven, August (2006)
6. Anderson, R.J., Spong, M.W.: Bilateral control of teleoperators with time delay. In: Proceedings of the 1988 IEEE International Conference on Systems, Man, and Cybernetics, vol. 1, pp. 131–138. IEEE (1988, August)
7. Melchiorri, C.: Robotic telemanipulation systems: an overview on control aspects. IFAC Proc. Vol. **36**(17), 21–30 (2003)

Mid-air Haptic Biosignal Transfer

Daria Hemmerling[1,2], Maciej Stroinski[1,2], Kamil Kwarciak[1,2],
Maciej Szymkowski[1,3], William Frier[4], Orestis Georgiou[4(✉)],
and Mykola Maksymenko[5]

[1] Softserve, Wrocław, Poland
dhemm@softserveinc.com
[2] AGH University of Science and Technology, Krakow, Poland
[3] Warsaw University of Technology, Warsaw, Poland
[4] Ultraleap, Bristol, UK
[5] Softserve, Lviv, Ukraine

Abstract. The widespread of virtual communications highlights the importance of enriching audio-visual communications with haptic content. Typically, this haptic content is directly related to the audio-visual media stream. Here, we focus on the detection and haptic enhancement of non-verbal biometric cues and signals that can add an emotional backdrop to virtual interactions. To that end, we present a 20-participant study that demonstrates a successful biosignal transfer in a fully contactless way, utilizing camera-based heartrate readings and ultrasonic mid-air haptic technology to affect the audience. Our analysis of biometric data and subjective responses hint towards a higher arousal and affect in the haptic condition.

Keywords: Haptics · Biosignals sharing · Communication engagement

1 Introduction

In expressive communications, our words and body-language are coupled to changes in our psychological state. Our heartrate goes up when we feel anxious or uneasy. Our skin momentarily becomes more conductive when we are aroused. On the receiving end, the audience's psychological state also changes, often reacting to or mirroring that of the source. In digital communications however (e.g., audio and video calls), our capacity to express ourselves is limited, thus dramatically reducing our ability to convey the peculiarities of emotional colour and personal experiences. For example, subtle non-verbal cues such as respiration patterns, sweating or heartrate biosignals which we are naturally capable of directly observing and interpreting in our face-to-face experiences, are often lost during online communications. Sharing such biosignals, e.g., during a video call can enhance empathy, the sense of co-presence and intimacy [1].

Motivated by this opportunity, different haptic interfaces have stepped in to enrich online and video content. For example, Frey et al. [2] show how sharing personalised biofeedback and tactile sensations impact emotion recognition intimacy. To that end, a number of studies have explored the use of contact haptic interfaces (wearable or

H. Seifi et al. (Eds.): EuroHaptics 2022, LNCS 13235, pp. 412–415, 2022.
https://doi.org/10.1007/978-3-031-06249-0

tangible objects) to communicate emotional affect, social touch, and biosignal transfer [1, 3], however very little has been done using non-contact haptic interfaces such as ultrasound mid-air haptics [4, 5]. Ablart et al. [4], observe a positive effect of haptic feedback during repeated viewings of short movie experiences, however the haptics were not related to any biosignal input. Romanus et al. [5] created an AR hologram of a beating heart organ that users could reach out and touch and feel using a mid-air haptic pattern that was temporally synced to a wearable heart rate reader device.

In this work-in-progress paper, we present an initial study that demonstrates a successful biosignal transfer in a fully contactless way. Twenty people were shown a short video clip1[1] of a sad narrative performed by an actor. We then utilized camera-based biosignal readings[2] to extract the actor's heart rate, and then ultrasonic mid-air haptic technology[3] to affect the audience participant with a temporal tactile pattern representing the actor's heartrate. Our analysis of objective biometric data and subjective responses hints towards increased values of arousal and affect due to the added haptics.

2 System and Methods

The study setup and conditions are shown in Fig. 1. A between-subject experiment was approved by ethics and conducted with two groups of 10 people (5 females and 5 males, mean age 20.2) each experiencing one of the two conditions (30 min per participant). The Biosense SDK[2] was used to extract the actor's hear rate (HR) from the video, and the participant's HR. The actor's HR was used to adjust the rate of the mid-

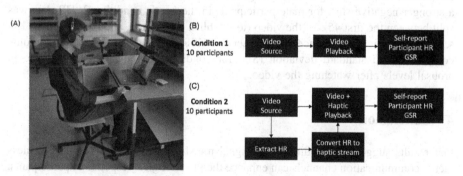

Fig. 1. (A) Study setup showing a participant with his left palm above the mid-air haptic device (enclosed in a box) and the right hand with the GSR sensors. (B)(C) The two conditions used in the experiments. In Condition 1, participants watched a 3-min video. In Condition 2 participants experienced the same video with added mid-air tactile feedback in the form of a heart rate pattern that was synchronized to the person in the video.

---

[1] https://youtu.be/C3hABRHmQoo/.

[2] https://demo.softserveinc.com/biosense/.

[3] https://www.ultraleap.com/datasheets/STRATOS_Explore_Development_Kit_datasheet.pdf.

air haptic pattern being displayed by an Ultraleap device[3] in Condition 2. The haptic pattern presented to the participant's palm was a circle with a dynamically changing radius oscillating between 2 and 5 cm, like that used in [5]. Before and after the experiment, each of the participants was asked to fill out a self-report measure of affect using a Positive and Negative Affect Scale on a 5-point scale. Finally, the participant's skin-conductance was collected using a GSR sensor during the experiment. For Condition 2, a mid-air haptic familiarization phase was conducted prior to the experiment.

3 Results

	Condition 1		Condition 2	
	Δ + ve Affect	Δ -ve Affect	Δ + ve Affect	Δ -ve Affect
All	1.91	−0.92	1.36	**−3.82**
Female	**4.00**	−1.67	1.33	−1.50
Males	−0.60	−0.17	1.40	**−6.60**
	HR mean	HR SD	HR mean	HR SD
All (0–45 s)	62	2.0	74	3.1
All (45–180 s)	72	15.3	71	15.3
	GSR tonic mean	GSR SD	GSR tonic mean	GSR SD
Female	1.00	0.80	1.21	0.93
Males	0.99	1.04	1.27	1.40

Comparing the before and after Δ-change in + ve and -ve affects between the two conditions, we observe significant changes in the PANAS self-report data, most notably a stronger negative shift for male participants in the haptic condition 2. HR data was split between the first 45 s of the video (to establish a baseline), and the remaining 135 s, but no significant differences were observed. GSR data showed a higher mean tonic component and standard deviation (SD) for condition 2, indicating an increase in arousal levels after watching the video.

4 Conclusion

Our results suggest that additional biosignal transfer via non-invasive and touchless tactile communication channels can enhance the physiological response of participants. Mid-air haptics can therefore potentially enhance experience transfer and communication engagement of video calls, however further research is needed to support such claims. In the future, we intend to expand our study to include videos with happy emotions, and also to include a third condition using a wearable haptic. We also intend to create a demonstrator that facilitates for a two-way biosignal feedback mid-air haptic enhanced video call.

This project has received funding from the EU Horizon 2020 research and innovation programme under grant agreement No 101017746, Touchless.

References

1. Feijt, M.A., et al.: Sharing biosignals: an analysis of the experiential and communication properties of interpersonal psychophysiology. Hum.–Comput. Interact. 1–30 (2021)
2. Frey, J., et al.: Breeze: sharing biofeedback through wearable technologies. In: Proceedings of ACM CHI, pp. 1–12 (2018)
3. Huisman, G.: Social touch technology: a survey of haptic technology for social touch. IEEE ToH **10**(3), 391–408 (2017)
4. Ablart, D., et al.: Integrating mid-air haptics into movie experiences. In: ACM TVX, pp. 77–84 (2017)
5. Romanus, T., et al.: Mid-air haptic bio-holograms in mixed reality. In: IEEE ISMAR, pp. 348–352 (2019)

Ultrasonic Friction Modulation Modifies the Area of Contact, Not the Shear Strength

Nicolas Huloux[1]([✉]), Laurence Willemet[2], and Michaël Wiertlewski[2]

[1] MIRA, Aflokkat, Sarrola-Carcopino, France
nicolas@aflokkat.com
[2] Delft University of Technology (TU Delft), Delft, Netherlands
{l.willemet,m.wiertlewski}@tudelft.nl

Abstract. Adhesive friction of skin is the consequence of two factors: the area of real contact and the shear strength of each microscopic junctions. The frictional force of skin can be modulated by ultrasonic vibration. The transverse waves levitate the skin away from the plate, reducing the real area of contact. However, it is not clear if the shear strength of the junction is also affected by the vibration. Here, we image the evolution of the real area of contact as a function of the lateral force and of the amplitude of vibration. We find the decrease of the area of contact before the transition from stick-to slip is consistent across all vibration amplitudes. The findings suggest that the shear strength of the contact is not affected by the ultrasonic vibrations.

Keywords: Surface haptics · Squeeze film · Fingertip friction · Shear strength

1   Introduction

Friction is fundamental to the perception of material properties and the motor control of grasping and manipulation [1]. The adhesive model of friction recognizes that the resistive force of friction emerges from the adhesion of a collection of microscale junctions. Each of these junctions support a maximum interfacial shear, called the shear strength τ_0 of the contact. The total frictional force is the integral of the shear strength over the real area of contact A^R, which is made by every junction, such that the friction force reads $f_t = \tau_0 \cdot A^R$ [3]. This adhesive law of friction has been experimentally verified for the contact of soft elastomers and of biological tissues such as skin [4].

Since friction perception is an essential dimension of tactile perception, several haptic devices modulate its amplitude to create vivid effects. Among them, ultrasonic friction modulation can reduce the friction force of skin on glass by more than 90%. Imaging of the contact showed that ultrasonic friction modulation has a profound effect on the area of contact A^R [5]. However, it is still unclear if ultrasonic vibration impacts the shear strength τ_0.

We studied the evolution of the area of real contact under an increasing lateral force at different amplitudes of ultrasonic vibration. Before sliding, the

© The Author(s) 2022
H. Seifi et al. (Eds.): EuroHaptics 2022, LNCS 13235, pp. 416–418, 2022.
https://doi.org/10.1007/978-3-031-06249-0

real area of contact decreases with increasing lateral force, and that its value at the transition between stick-to-slip relates directly to the shear strength [5].

We find that every subject has the same shear strength over all trials, providing evidence that ultrasonic levitation phenomenon is unlikely affecting the shear strength. Moreover, the findings reinforce the idea that acoustic levitation pressure has a similar effect than the pressure from the normal force.

2 Methods

Ten volunteers placed their index finger on the apparatus shown in Fig. 1 at a 20° angle. The apparatus could measure the real area of contact using frustrated total internal reflection and change the friction owing to the ultrasonic glass plate. The plate moved back and forth lateral, while the friction randomly changed at each stroke from over seven levels.

The normal force was controlled with a balance mechanism, and the lateral motion was speed-controlled with a DC-motor. Forces were captured from a 6-axis force sensor (ATI Nano 43). A high-speed camera imaged the contact made by the finger on the glass plate at 300 fps with a pixel resolution of 0.0535 mm.

The FTIR illumination highlights the asperities in contact, creating a high contrast between contacting and non-contacting asperities [5]. The raw image was first masked by applying Otsu thresholding followed by closing and then opening operation. This operation removed any leakage of light. From the masked image, we estimate the value of the area of real contact A^R using the light scattering measurement described in [2].

3 Results

Each measurement of the area of contact as a function of the shear force followed a parabola, see Fig. 2a and was fitted with a quadratic function, leading to a $R^2 \approx 0.65 \pm 0.09$. For a given participant, the transition between the stick and slip regimes is aligned for all frictional conditions, as shown in blue in Fig. 2a. We then regressed a linear function ($R^2 \approx 83 \pm 0.11$), unique for each subject. It can be shown that the slope of this line is the reciprocal of the shear strength τ_0.

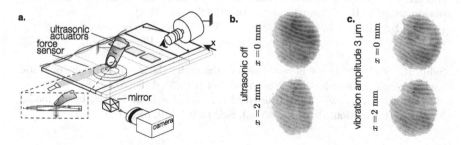

Fig. 1. (a) Apparatus for contact imaging at the onset of sliding. Contact images of a typical subject when the finger is resting (top row) and at the moment of sliding (bottom row) (b) without ultrasonic actuation, and (c) with a vibration of 3 μm.

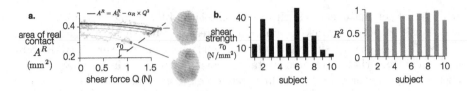

Fig. 2. (a) Evolution of the area of real contact of a typical subject, dots represent individual datum, blue ones represent the very moment of sliding, and line the quadratic fit, $A^R = A_0^R - \alpha_R \times Q^2$. (b) Shear strength of each subject (black) and the goodness of fit of the linear function (gray).

In Fig. 2a, we report the shear strength of the contact of each subject and the R^2 correlation coefficient. The linear fit led to an average goodness of fit R^2 of 83%, showing good agreement with a linear model.

4   Conclusion and Discussion

The results are consistent with an independence of the frictional strength of fingertip/glass contact against ultrasonic modulation. The found value of the shear strength exhibited large variance, further demonstrating the non-trivial nature of fingertip friction. Moreover, the quadratic trend of the area of real contact at the onset of sliding for different amplitudes of vibration follows the same trend as the effect of normal force shown in [4]. Ultrasonic vibration changes friction force by changing the area of contact, and seems to leave the shear strength of each remaining junction unchanged. These findings also hint that modulating the area of contact before sliding can be key to rendering virtual materials with varying static friction.

References

1. Cadoret, G., Smith, A.M.: Friction, not texture, dictates grip forces used during object manipulation. J. Neurophysiol. **75**(5), 1963–1969 (1996)
2. Huloux, N., Willemet, L., Wiertlewski, M.: How to measure the area of real contact of skin on glass. IEEE Trans. Haptics **14**(2), 235–241 (2021)
3. Persson, B., Albohr, O., Creton, C., Peveri, V.: Contact area between a viscoelastic solid and a hard, randomly rough, substrate. J. Chem. Phys. **120**(18), 8779–8793 (2004)
4. Sahli, R., et al.: Evolution of real contact area under shear and the value of static friction of soft materials. Proc. Nat. Acad. Sci. **115**(3), 471–476 (2018)
5. Wiertlewski, M., Friesen, R.F., Colgate, J.E.: Partial squeeze film levitation modulates fingertip friction. Proc. Nat. Acad. Sci. **113**(33), 9210–9215 (2016)

Fingertip Friction and Tactile Rating
of Wrapping Papers

Kim Michèle Jost[1,2] (iD), Knut Drewing[3] (iD),
and Roland Bennewitz[1,2(✉)] (iD)

[1] INM – Leibniz-Institute for New Materials, 66123 Saarbrücken, Germany
roland.bennewitz@leibniz-inm.de
[2] Physics Department, Saarland University, 66123 Saarbrücken, Germany
[3] Department of Psychology, Justus Liebig University, Giessen, Germany

Abstract. The tactile exploration and perception of wrapping papers is investigated in terms of fingertip friction and rating of sensory, affective, and evaluative adjectives. Friction coefficients, which vary significantly between samples, are correlated with factors such as valence which are identified in a principal component analysis of subjective ratings. We found that affective appraisals of valence and arousal as well as evaluations of novelty, but not of value, decreased with increasing friction.

Keywords: Human tactile perception · Fingertip friction · Paper

1 Introduction and Methods

Paper wrapping is a key element of making a gift or selling a valuable article. The affective and persuasive role of its touch has become subject of haptics research [1, 2]. Here, we study the role of fingertip friction for the tactile sensory, affective, and evaluative appraisal of wrapping papers. Participants were asked to rank adjectives which describe their perception while exploring the papers with their fingertip and we measured exploration forces (details in [3]). Every ranking can thus be related to the applied normal and friction forces.

The 5×5 cm^2 samples are 8 expensive and 2 cheap gift-wrapping papers as well as 5 quality papers from an artist supplier. Papers were mounted to stiff support plates by double-sided adhesive tape. Participants (24 volunteer students and scientists, age 18 to 31, naïve with respect to the goal of the study, native German speakers) explored the papers with circular movements of the fingertip of their preferred hand. They applied mean normal forces between 0.3 and 3 N while exploring the paper samples. Some participants kept the normal force between 0.2 and 0.4 N, others varied between 0.4 and 3.6 N. They did not see the samples and wore noise-reducing headphones. While exploring the samples, each participant ranked 10 papers according to each of the 26 adjectives in Table 1 on a 5-level Likert scale (adjective applies... not at all – not really – so/so – pretty much – fully). The orders of adjectives and papers were completely randomized.

H. Seifi et al. (Eds.): EuroHaptics 2022, LNCS 13235, pp. 419–421, 2022.
https://doi.org/10.1007/978-3-031-06249-0

2 Fingertip Friction Results

The mean friction coefficients (friction force divided by applied normal force) are between 0.3 and 0.65. To give equal weight to all participants, we normalize the friction coefficients by division with the average value for each participant. Friction on different samples is compared in Fig. 1a. There are significant differences between mean normalized friction coefficients of different samples (ANOVA, $F(14,225) = 69$, $p < 0.001$). Of the 105 pairwise comparisons between samples, 56 show significant differences in the normalized friction coefficient (105 t-tests, Bonferroni-corrected, overall $\alpha = 0.05$). Three samples (6,10,11) show similar high friction values, seven samples (1–4,7,13,14) similar low friction values, and one sample (15) a particularly low friction.

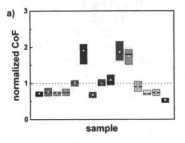 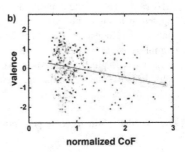

Fig. 1. a) Normalized friction coefficients for all 15 samples. Bars indicate median values, boxes 25%/75% quartiles, and squares mean values. b) Factor "valence" from PCA plotted versus the normalized friction coefficient for each sample and participant. The slope of the linear fit differs significantly from zero ($n = 240$, Pearson $r = -0.23$, $p < 0.001$).

The skin moisture at the fingertip was determined before and after tactile exploration of the paper samples by means of a Cornemeter (CM 825, Courage + Khazaka, Germany). The median friction coefficient of participants increased significantly with increasing moisture reading ($n = 24$, Pearson $r = 0.50$, $p = 0.013$). Three quarters of the participants exhibited higher fingertip skin moisture after exploring the paper samples.

3 Perception Results

We performed a principal component analysis (PCA) with varimax-rotation, where adjectives were attributed to factors based on the covariances in participants' ratings. The number of factors was determined with the Kaiser criterion. Adjectives were attributed to a factor when they explained more than 40% of the average variance of a single adjective. We found three sensory, two affective, and three evaluative factors which are named and listed in Table 1 together with the associated adjectives. "Sticky" was the only adjective with could not be attributed to a factor, and "naturalness" turned out as a single-adjective factor.

As example for the correlation of friction with tactile perception, we plot all values for the factor valence versus the individual friction coefficients in Fig. 1b. There is a significant negative correlation, in extension of previous reports on the negative correlation of pleasantness and friction [5]. We find significant negative correlations of normalized friction with all factors except smoothness, novelty, and value (overall $\alpha = 0.05$, Bonferroni-correction for 8 tests, Pearson r in Table 1) and conclude that fingertip friction on wrapping papers is important for their tactile affective and evaluative appraisal. The next step of this work-in-progress will invite participants back to a forced-choice study of the papers for wrapping a valuable gift.

Table 1. Adjectives for judgement by participants and factors resulting from the PCA, with the explained variance for each factor in brackets and the Pearson r for the correlation with the normalized friction coefficient. The German words used in the study can be found in Ref. [5], except for two sensory and all evaluative adjectives (samtig, unregelmäßig; natürlich, ungewöhnlich, hochwertig, schlicht, verspielt, innovativ, überladen, unbrauchbar).

sensory		affective		evaluative	
adjectives	factors	adjectives	factors	adjectives	factors
rough	smoothness (30%) $r=-0.01$ (NS)	relaxing	valence (36%) $r=-0.25$	innovative	novelty (32%) $r=-0.15$ (NS)
smooth		pleasant		unusual	
slippery		calming		playful	
bumpy	bumpiness (24%) $r=-0.30$	unsatisfying		simple	
irregular		uncomfortable		simple	value (18%) $r=-0.02$ (NS)
		irritating		high-quality	
velvety	silkiness (16%) $r=-0.32$	exciting	arousal (27%) $r=-0.19$	flamboyant	
silky		excitatory		useless	
		attention-grabbing			naturalness (17%) $r=-0.29$
sticky		boring		natural	

Acknowledgements. We acknowledge financial support by the Volkswagen Foundation.

References

1. Chen, X., Barnes, C.J., Childs, T.H.C., Henson, B., Shao, F.: Materials' tactile testing and characterisation for consumer products' affective packaging design. Mater. Des. **30**(10), 4299–4310 (2009)
2. Briand Decré, G., Cloonan, C.: A touch of gloss: haptic perception of packaging and consumers' reactions. J. Prod. Brand Manage. **28**(1), 117–132 (2019)
3. Sahli, R., et al.: Tactile perception of randomly rough surfaces. Sci. Rep. **10**(1), 15800 (2020)
4. Drewing, K., Weyel, C., Celebi, H., Kaya, D.: Systematic relations between affective and sensory material dimensions in touch. IEEE Trans. Haptics **11**(4), 611–622 (2018)
5. Klöcker, A., Wiertlewski, M., Theate, V., Hayward, V., Thonnard, J.L.: Physical factors influencing pleasant touch during tactile exploration. PLoS ONE **8**(11), e79085 (2013)

A Psychological and Behavioral Study on Remote Audio-Visual-Tactile Communication System with Pressure Stimulation

Mayuko Ito[1(✉)], Takumi Kuhara[1], Takuto Matsuhashi[2], Chihiro Hosoda[2], Keigo Inukai[3], Junji Watanabe[4], Kouta Minamizawa[5], and Yoshihiro Tanaka[1]

[1] Nagoya Institute of Technology, Nagoya, Japan
m.ito.748@nitech.jp, tanaka.yoshihiro@nitech.ac.jp
[2] Teikyo University, Tokyo, Japan
[3] Meiji Gakuin University, Tokyo, Japan
[4] NTT Communication Science Laboratory, Atsugi, Japan
[5] Keio University Graduate School of Media Design, Tokyo, Japan

Abstract. This study developed a tactile transmission system with a wearable squeezing display that works under an existing online communication system. Half of 32 participants used the wearable tactile display and the other did not use it. Psychological state and behavioral measures were conducted at the pre and post communication. The results showed the state change with tactile stimulus, and the possibility of behavior change was discussed.

Keywords: Haptics · Remote communication · Behavioral economics

1 Introduction

Online communication is attracting more attention, partly due to the spread of COVID-19. Audio-visual sense is dominantly utilized for communication, but the tactile sense might be also relevant to emotion and interpersonal relationship [1, 2]. Previous studies often used specific communication systems to establish the interaction involving tactile stimuli. However, vibration signals can be transmitted via one of audio stereo channels in an existing online communication system, implying that many people can physically connect by augmenting interaction.

Considering hugs and handshakes, this study developed a pressure transmission system that can be adopted to an existing online communication system. We quantitatively evaluated emotional and behavioral changes caused by the transmission of pressure stimulus using the developed system.

2 Remote Audio-visual-tactile Communication System

We assembled a wearable squeezing display to present pressure stimulation [3]. It is worn on the wrist and a flat rubber is tightened by winding a servo motor.

This Work Was Supported by JSPS KAKENHI No.21H05071

H. Seifi et al. (Eds.): EuroHaptics 2022, LNCS 13235, pp. 422–425, 2022.
https://doi.org/10.1007/978-3-031-06249-0

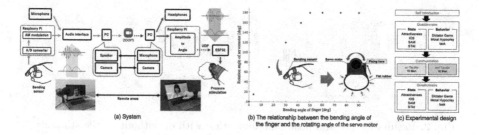

Fig. 1. System and experimental design

The left and right channels of an audio interface are employed for tactile and audio information, respectively, with Zoom (Zoom Video Communication Inc.) (Fig. 1(a) and (b)). The pressure stimulus is presented in response to the action of grasping the hand, which is measured by a bending sensor attached at the partner's finger. The bending information is converted to an amplitude-modulated audio wave using 1000 Hz carrier frequency to send via Zoom, and it is reconverted into the rotation angle of the squeezing display. This enables us to transmit DC components using the existing online communication system.

3   Experiment

32 male participants in their twenties and one female partner aged twenty-two conducted the experiment. Half of the participants involved the tactile stimulus, and the others did not involve, all with the same partner. Figure 1(c) shows the experimental design. First, the participants introduced themselves to the remote partner for about 30 s. 3 min communication was adopted, where the participants answered questions that has little effect on the psychological distance [7] with the partner on a one-on-one basis. The tactile stimulus was presented simultaneously with reactions such as nodding of the partner, while the bending action of the partner's finger was not seen. The participants were asked to answer questionnaires to measure their state and behavioral measures before and after the communication. To measure the state, the Self-Assessment Manikin (SAM:[4]), attractiveness evaluation with a five-point Likert scale, IOS [5] on psychological distance, and Y-1 evaluation of STAI [6] on anxiety were adopted. To measure the behavioral measures, the participants conducted a moral hypocrisy task that the participant selected a long or short task and dictator game in which the participants was asked whether they gave the partner 0,100, 200, 300, 400, or 500 JPY of the reward and take the rest. The order of measures and questionnaires were randomized. For the analysis, a Wilcoxon singed rank test with Bonferroni correction was adopted for each measure.

4   Results and Discussion

Figure 2 shows the results for the pre and post communication in each condition. The moral hypocrisy task yielded no difference. The IOS significantly changed

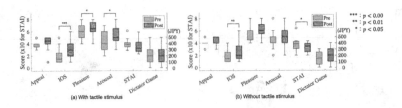

Fig. 2. Results for pre and post communication with or without tactile stimulus.

positive from the pre communication to the post for both conditions with and without tactile stimulus. In the present experiment, one woman was employed as the partner and all the participants were men. Besides, the attractive score was relatively high. Thus, the communication might have affected IOS positively and psychological distance was significantly closer after the communication. There was also a trend that dictator game amount increased with the rise of IOS score. The effect of tactile stimulus on IOS will be investigated by increasing conditions of partners and the number of participants. Scores in pleasure and arousal based on the SAM test were significantly increased after the conversation with tactile stimulus. It may relate to the pressure given by the squeezing display and how attractive the partner was. Regarding the condition without tactile stimulus, there was a significant difference in STAI, indicating that anxiety was reduced by the communication, whereas there was no significant difference for the condition with tactile stimulus. A possible reason is unfamiliarity to tactile stimulus for the participants. A tactile feeling with the proposed system will be investigated.

This study has a lot of limitations. The communication partner was fixed and all participants were men as a first step. Additionally, the reaction of the partner, visibility of partner's finger, and the duration of communication might influence the state and behavior. Although the resulting state and behavior gave consistent interpretations, indicating the possibility of the effect with tactile stimuli, various conditions are required to clarify the effect with deeper analysis.

References

1. Field, T.: Touch for socioemotional and physical well-being: a review. Dev. Rev. **30**(4), 367–383 (2010)
2. Shiomi, M., Sumioka, H., Ishiguro, H.: Survey of social touch interaction between humans and robots. J. Robot. Mechatron. **32**(1), 128–135 (2020)
3. Salvietti, G., Iqbal, M.Z., Prattichizzo, D.: Bilateral haptic collaboration for human-robot cooperative tasks. IEEE Robot. Autom. Lett. **5**(2), 3517–3524 (2020)
4. Lang, P.J.: Behavioral treatment and bio-behavioral assessment. In: Sidowski, J.B., Johnson, J.H., Williams, T.A. (eds.) Technology in Mental Health Care Delivery Systems, pp. 119–167. Norwood, NY (1980)
5. Aron, A., Aron, E.N., Smollan, D.: Inclusion of other in the self scale and the structure of interpersonal closeness. J. Pers. Soc. Psychol. **63**(4), 596–612 (1992)

6. Spielberger, C.D., Gorsuch, R.L.: State-Trait Anxiety Inventory for Adults: Manual and Sample: Manual, Instrument and Scoring Guide. Consulting Psychologists Press, Leeds (1983)

7. Aron, A., Melinat, E., Aron, E.N., Darrin Vallone, R., Bator, R.J.: The experimental generation of interpersonal closeness: a procedure and some preliminary findings. Pers. Soc. Psychol. Bull. **23**(4), 363–377 (1997)

Using Digital Image Correlation to Quantify the Effects of Thin Films on Skin Deformation

Anika Kao$^{(\boxtimes)}$ and Gregory Gerling

University of Virginia, Charlottesville, VA 22903, USA
ak4hz@virginia.edu

Abstract. As new haptic technologies emerge, such as wearables and nano tattoos, that apply thin films to the skin, we need to understand their effect on skin mechanics and perception. The work herein describes the measurement of changes in finger pad skin deformation upon the application of three thin films (ink, paint, and nitrile) with thickness from 2.5 to 127 μm. Using the non-contact optical tracking method of 3D digital image correlation, attenuation of the skin's deformation is compared via four biomechanical metrics. The preliminary findings indicate that surface strain and lateral movement of the skin are attenuated by additional thickness. While a larger cohort of participants and trials are yet needed, this approach demonstrates an ability to differentiate minor changes in skin deformation due to thin films, which are likely to impact tactile acuity.

Keywords: Skin Mechanics · Tactile Acuity · Digital Image Correlation

1 Introduction

We readily differentiate various textures and materials with high sensitivity [1]. Our tactile acuity is shaped by a number of factors, including finger pad stiffness, humidity, and size, which influence skin deformation, and thereby, what is encoded by peripheral afferents [2]. With the emergence of new haptic devices, such as augmented reality wearables [3] and nano tattoos [4], it is important to understand how thin films impact skin deformation. While prior studies have investigated the effects of films on pleasantness [5], this work aims to quantify underlying changes in skin deformation that ultimately enable tactile acuity.

2 Methods

This work adopts a non-contact optical tracking method called 3D digital image correlation (DIC) to measure the deformation of the finger pad surface upon the application of three different thin films. Using a three-camera setup (Fig. 1A), the skin deformation upon indentation of a rigid tip (diameter = 0.83 mm, 4 mm/s, 2 s hold at 2.5 mm depth) is measured and characterized using four biomechanical metrics. The experimental setup, including camera alignment and finger placement, follows prior work [6]. Three thin films of varying thickness were applied to one participant's finger

© The Author(s) 2022
H. Seifi et al. (Eds.): EuroHaptics 2022, LNCS 13235, pp. 426–428, 2022.
https://doi.org/10.1007/978-3-031-06249-0

pad in order of increasing thickness. Namely, a white acrylic ink (Liquitex, Cincinnati, OH, USA) (~2.5 μm thick), a white, washable (non-toxic) acrylic paint (Craft Smart, Michaels Stores, Irving, TX, USA) (~25 μm thick), and the smallest digit tip of a size small, black nitrile glove (GlovePlus, AMMEX, Kent, WA, USA) (127 μm thick). Indentations of the rigid tip were run three times per film (n = 3) and soap and water were used to remove the prior film. To attain sufficient speckling patterns for DIC, an airbrush with a portable mini air compressor (Master Airbrush, TCP Global, San Diego, CA, USA) was used to apply a random speckle pattern to the finger pad, after application of each film, Fig. 1B. Approximately 3 min elapsed between film application and speckling to ensure sufficient dry time. Either white or black films were chosen to create a high contrast pattern of black or white speckles, respectively.

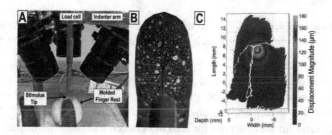

Fig. 1. Experimental apparatus. (A) Indenter with load cell and 0.83 mm stimulus tip. Three cameras aligned around finger pad capture images synchronously for optical tracking using 3D DIC. (B) Index finger in black nitrile glove with white paint speckles applied via airbrush. (C) 3D point cloud showing DIC results of one trial with the glove at t = 1.5 s.

3   Experiments and Results

While the force data (Fig. 2) indicates each film influences the skin mechanics, the DIC tracking method characterizes skin deformation in greater detail.

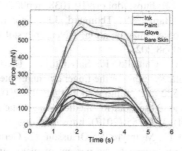

Fig. 2. Measured force on finger pad over time for bare skin and applied films (n = 3). Force ordering aligns with increasing thickness of film (i.e., bare skin < ink < paint < glove).

Figure 3A illustrates the change in 1st principal Lagrangian strain, for one trial in each case of ink, paint, and glove. Measured values during the stimulus hold (t = 1–3 s) were averaged per trial, film, and metric, to generate the plots in Fig. 3B-E.

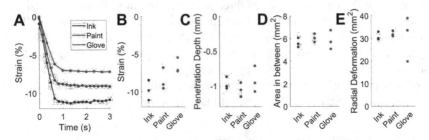

Fig. 3. Four biomechanical metrics across the three thin film cases. (A) Strain over time for one trial per film. (B-E) Average value of each metric during hold (t = 1–3 s) per film and trial.

A systematic change in compressive strain is observed as film thickness increases, Fig. 3A-B. The cases of paint and glove likely constrain the skin more than ink, resulting in less skin stretch and thereby lower compressive strain. In Figs. 3D-E, slight increases between film are observed for area in between and radial deformation metrics. This work is limited to one participant and three trials per film. As such, further studies will require an increased number of participants, trials, and tested films.

Acknowledgements. This work was supported in part by National Science Foundation (Grants IIS-1908115 and NRT-1829004), and in part by the National Institutes of Health (Grant NINDS R01NS105241).

References

1. Skedung, L., Arvidsson, M., Chung, J., et al.: Feeling small: exploring the tactile perception limits. Sci Rep **3**, 2617 (2013). https://doi.org/10.1038/srep02617
2. Delhaye, B.P., Schiltz, F., Barrea, A., Thonnard, J.L., Lefèvre, P.: Measuring fingerpad deformation during active object manipulation. J Neurophysiol **126**(4), 1455–1464 (2021). https://doi.org/10.1152/jn.00358.2021
3. Maisto, M., Pacchierotti, C., Chinello, F., Salvietti, G., De Luca, A., Prattichizzo, D.: Evaluation of wearable haptic systems for the fingers in augmented reality applications. IEEE Trans. Haptics **10**(4), 511–522 (2017). https://doi.org/10.1109/TOH.2017.2691328
4. Srinivas, K.: Nano tattoos as biosensors for medical diagnostic applications. Int. J. Emerg. Res. Manage. Technol. **6**(4), 151–155 (2017). https://doi.org/10.23956/ijermt/sv6n4/116
5. Rezaei, M., Nagi, S.S., Xu, C., McIntyre, S., Olausson, H., Gerling, G.J.: Thin films on the skin, but not frictional agents, attenuate the percept of pleasantness to brushed stimuli. In: IEEE World Haptics Conference (WHC), pp. 49–54 (2021). https://doi.org/10.1109/WHC49131.2021.9517259
6. Kao, A.R., Xu, C., Gerling, G.J.: Using digital image correlation to quantify skin deformation with von frey monofilaments. IEEE Trans. Haptics (2021).https://doi.org/10.1109/TOH.2021.3138350

Engineering Haptic Devices 3rd Edition in 2022 in Open-Access

Thorsten A. Kern$^{(\boxtimes)}$ ⓘ and Alireza Abbasimoshaei ⓘ

Hamburg University of Technology, Hamburg, Germany
t.a.kern@tuhh.de
https://www.hapticdevices.eu

Abstract. *Engineering Haptic Devices* was first published in 2008. Its structure and content were intended as a basic textbook for engineers who wanted to learn about haptics and the design of haptic devices. Its emphasis was on applicability, providing a broad overview of each topic and some in-depth coverage of all areas relevant to the design of mechatronic systems in the field of haptics. Its editions have since been downloaded more than 50,000 times and cited and referenced several hundred times. As part of the *Springer Series on Touch and Haptic Systems* edited by Ferre et al. it is the most cited book in this series and has established itself as one of the must-haves for any hardware designer focusing on haptic devices. The 3rd edition of this book is now being published this year, with some chapters significantly revised and all sections updated with relevant findings of the last few years. The biggest change, however, has been the opportunity to move this book into an open-access format that will allow for much wider distribution than before. This short paper describes the history of the book, the changes made from edition 2 to 3, and outlines a future for further editions.

Keywords: Textbook · Education · Engineering

1 History

The idea for this book was born in 2003. It was intended to fill a gap: The regrettably small number of comprehensive, summary publications on haptics available to, for example, a technically interested person who is confronted for the first time with the task of designing a haptic device. In 2004, apart from a considerable number of conference proceedings, journals and dissertations, there was no document summarising the most important findings of this challenging topic. The first edition was edited by Thorsten A. Kern. It was funded by the German Research Foundation (DFG, grant KE1456/1-1) with a special focus on consolidating the design methodology for haptic devices. In 2008 the German version *Entwicklung Haptischer Geräte* [3] and in 2009 the English version *Engineering Haptic Devices* [2] were published by Springer.

In 2010, the idea of a second edition of the book was born. With Kern's move from university to an industrial employer, attention also shifted from mainly

© The Author(s) 2022
H. Seifi et al. (Eds.): EuroHaptics 2022, LNCS 13235, pp. 429–432, 2022.
https://doi.org/10.1007/978-3-031-06249-0

kinaesthetic to tactile devices. This made severe gaps in the first edition eminent as it was focussing on kinaesthetic devices only. In parallel, science made great strides in understanding the individual tactile modalities and blurring the boundaries between different conceptual approaches to the same perception. This now provided an opportunity to take an engineering approach to a more complete picture. However, it took until 2013 for work to begin on the second edition. In that year, Christian Hatzfeld completed his doctoral thesis on the perception of vibrotactile forces and took the lead in editing this second edition [1]. Like the first edition, this work was also funded by the DFG (grant HA7164/1-1). In a fruitful collaboration between Springer and the series editors, the book was integrated into the *Springer Series on Touch and Haptic Systems* as it was felt that the design of task-specific haptic interfaces would be well complemented by the other works in this series.

In 2020, a new opportunity for this book arose when Kern returned to academia as a full professor at the Hamburg University of Technology. Despite a detour into the automotive world of visible displays, he returned to his scientific roots and resumed his work on the design of haptic devices and actuators. This has also led him to revisit some of the content of this book with some distance, as he now sees more clearly how the global community has evolved and professionalized, but also realizes what questions remain. Dr. Alireza Abbasimoshaei, an experienced researcher who has made a name for himself in the field of rehabilitation robotics, has agreed to assist with the editorial portion of the work. Fortunately, we have also found a strong supporter of haptic research in Grewus GmbH focussing on the development of tactile system solutions, and with their help we have succeeded in making this edition of the book an open access publication.

2   Changes Between Edition 2 and 3

The general structure of the book remained identical between edition 2 and 3 (Table 1) with all content updated and some areas rewritten from scratch:

Control. The chapter on control of haptic devices has been expanded to provide not only a general introduction to control theory, but also to add some relevant concepts related to the design of control laws and the approaches specific to haptic systems. A large part is devoted to the control of telemanipulation systems. Some basics on control design for rehabilitation systems complete the picture.

Kinematic Design. While the first edition focused on serial kinematic systems and the second edition focused on parallel kinematic systems, this new version of the chapter covers both types of construction and their respective characteristics equally in the context of kinematic synthesis for haptic systems. Many details and examples cover topics of workspace design and dynamics, and some examples are given of the use of simulations for rapid concept verification.

Haptic Software Design. Previous editions focused mainly on virtual reality applications, Arsen Abdulali used his insights in the third edition to explain the state of the art of haptic rendering and synthesis of kinesthetic and tactile feedback based on haptic models. After a basic orientation in the software domain and with many examples of how to actually realize haptic rendering, this chapter also includes a relevant outlook on current research with data-driven models.

Industrial Haptics. Not a chapter by itself but influencing the scope in many areas of the book industry standards on design-methods, concept evaluations and product verifications were added. By the strong involvement of Jörg Reisinger the Requirements-Identification- and the Evaluation-Chapter was significantly extended to present industrial methods otherwise hidden to the general public.

Table 1. Content-structure of *engineering haptic devices, edition 3*

Part I	Basics
1	Motivation and application of haptic systems
2	Haptics as an interaction modality
3	The user's role in haptic system design
4	Development of haptic systems
Part II	Designing haptic systems
5	Identification of requirements
6	General system structures
7	Control of haptic systems
8	Kinematic design
9	Actuator design
10	Sensor design
11	Interface design
12	Haptic software design
13	Evaluation of haptic systems
14	Examples of haptic system development
15	Conclusion

3 Future Editions

Considering the history of the book and the employment-situation of the editors we are planning to release a new edition every five to seven years. Whereas the structure will probably not change much, an update of each chapter and a systematic strategy for major revisions of 20 to 30% of the chapters with every new edition is planned to secure the future of this series. As the book benefited a lot from experts taking responsibility to rewrite and update areas, the editors welcome to record interest of senior professionals to participate in future editions.

References

1. Hatzfeld, C., Kern, T.A. (eds.) Engineering Haptic Devices: A Beginner's Guide. Springer, London, London (2014). https://doi.org/10.1007/978-1-4471-6518-7
2. Kern, T.A. (ed.) Engineering Haptic Devices: A Beginner's Guide for Engineers. Springer, Berlin, Heidelberg (2009). https://doi.org/10.1007/978-3-540-88248-0
3. Kern, T.A. (ed.) Entwicklung Haptischer Geräte: Ein Einstieg für Ingenieure. Springer, Berlin, Heidelberg (2009). https://doi.org/10.1007/978-3-540-87644-1

Wearable Haptic Device for Individuals with Congenital Absence of Proprioception

Sreela Kodali[1(✉)], Allison M. Okamura[1], Thomas C. Bulea[2],
Alexander T. Chesler[2], and Carsten G. Bönnemann[2]

[1] Stanford University, Stanford, CA 94305, USA
{kodali,aokamura}@stanford.edu
[2] National Institutes of Health, Bethesda, MD 20892, USA
{thomas.bulea,alexander.chesler,carsten.bonnemann}@nih.gov

Abstract. A rare genetic condition, PIEZO2 loss of function (LOF) is characterized by absence of proprioception and light touch, which makes functional tasks (e.g., walking, manipulation) difficult. There are no pharmacological treatments or assistive technologies available for individuals with PIEZO2-LOF. We propose a sensory substitution device that communicates proprioceptive feedback via detectable haptic stimuli. We created a wearable prototype that maps measurements of elbow movement to deep pressure applied to the forearm. The prototype applies up to 18 N, includes an embedded force sensor, and is programmable to allow for various angle-to-pressure mappings. Future work includes comparing proprioceptive acuity and movement ability with and without the device in healthy and PIEZO2-LOF individuals, developing low-profile devices using soft robotics, providing sensory substitution for multiple joints simultaneously, and encoding additional aspects of joint dynamics.

Keywords: Sensory substitution · Wearable devices · Proprioception

1 Introduction

Proprioception can be considered our "sixth sense." It provides continuous information on body position and movement vital to motor control and coordination, balance, muscle tone, postural reflexes, and skeletal alignment. Many neuromuscular disorders arise from dysfunction of motor efferents, but a deficiency of afferent proprioceptive sensory input is another, often overlooked, cause of impairment that can severely impact motor function, even when strength is preserved. Proprioception in humans is entirely dependent on the non-redundant mechanosensor PIEZO2; individuals with recessive PIEZO2-LOF show complete congenital absence of proprioception leading to motor and functional impairment [3]. Individuals with PIEZO2-LOF also lack vibratory sense and discriminatory touch perception specifically on glabrous skin, although deep pressure, temperature, and some pain sensation is preserved [1–4]. No therapeutic or assistive technology options currently exist for individuals with PIEZO2-LOF. Our goal

H. Seifi et al. (Eds.): EuroHaptics 2022, LNCS 13235, pp. 433–435, 2022.
https://doi.org/10.1007/978-3-031-06249-0

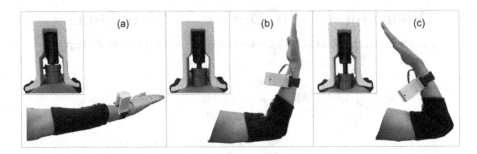

Fig. 1. Prototype device with elbow angle measurement and tactor mounted on the forearm. As elbow angle becomes more acute, the tactor provides increased pressure.

is to design and test a wearable haptic device that enables proprioceptive feedback using preserved sensory input modalities and evaluate its efficacy to enable intuitive control of limb movement in individuals with PIEZO2-LOF.

2   Prototype and Preliminary Results

Because individuals with PIEZO2-LOF have intact deep pressure sensation, we designed a prototype device that transduces elbow angle information to deep pressure stimuli applied to the forearm. The haptic feedback component consists of a linear actuator and cylindrical tactor housed in a plastic enclosure worn flush against the surface of the arm. The device is fastened in place with wide hook-and-loop straps. As the actuator extends its position, the tactor presses directly on the skin and applies a deep pressure stimulus. We used an Actuonix PQ12-P micro linear actuator because of its built-in position sensor, high output force (18 N), and reliable control. We embedded a low-profile SingleTact capacitive force sensor (diameter 15 mm, force range 45 N) in the tactor to measure the applied pressure in real time.

We measured elbow angle with a resistive flex sensor (20–40 kΩ range in series with 42 kΩ resistor as a voltage divider) sewn into a fabric sleeve. A microcontroller maps the elbow angle to a pressure stimulus, sends a position to the linear actuator, and records the force and actuator position. The real-time feedback and precise control of the actuator will allow us to explore different stimulus patterns. Figure 1 depicts the device behavior when the angle-to-pressure mapping is a constant 0.123 mm/deg. When the elbow is fully extended and arm is outstretched, the actuator is retracted and applies no pressure (Fig. 1a). When the elbow and arm are bent, the actuator is proportionately extended and applies pressure (Fig. 1b). When the elbow is fully flexed, the actuator is maximally extended and applies the most pressure (Fig. 1c).

Figure 2 shows measurements of elbow angle, actuator position, and force during repetitions of the movement shown in Fig. 1. Our results confirmed the device functioned as expected. The actuator position and force are affected by noise in the flex sensor and friction in the actuator.

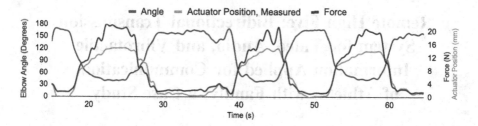

Fig. 2. Filtered measurements of elbow angle, actuator position, and force recorded during use of the prototype. Acute angles are flexion and 180° is full extension.

In addition, to verify force sensor functionality, we measured the force-displacement relationship of the device when mounted on two locations on the body: the ventral side of the forearm and the dorsal side of the hand. Because the hand is locally stiffer than the forearm, we expected that the forces would be larger for a given displacement at the hand compared to the forearm. The measurements corroborated this; we identified the arm and hand surfaces with stiffness values of 465.6 and 8115.4 N/m, respectively.

3 Conclusions and Future Work

We created a wearable sensory substitution device for eventual use by individuals with PIEZO2-LOF to convey proprioceptive feedback via haptic stimuli. Characterizing the sensory input of PIEZO2-LOF patients is ongoing, so we designed our prototype device to be programmable and cover a wide force range. The device successfully mapped elbow joint angle to single site deep pressure stimuli on the forearm. For our next steps, we will replace the resistive flex sensor with a robust capacitive sensor for angle measurement and use the programmable device to evaluate different designs (e.g. stimulus patterns, tactor sizes, mappings) and quantify their impact on PIEZO2-LOF patients' proprioceptive acuity. We will also explore the use of multiple deep pressure units, spatiotemporal stimuli patterns, and new designs based on soft robotics techniques. If the elbow device is effective, we will proceed to encode multiple degrees of freedom of joint movement for both the upper and lower limbs.

References

1. Case, L.K., et al.: Innocuous pressure sensation requires a-type afferents but not functional PIEZO2 channels in humans. Nat. Commun. **12**(1), 1–10 (2021)
2. Chesler, A.T., Szczot, M.: Portraits of a pressure sensor. eLife **7**, e34396 (2018)
3. Chesler, A.T., Szczot, M., Bharucha-Goebel, D., et al.: The role of PIEZO2 in human mechanosensation. N. Engl. J. Med. **375**(14), 1355–1364 (2016)
4. Szczot, M., Liljencrantz J, Ghitani N, et al.: PIEZO2 mediates injury-induced tactile pain in mice and humans. Sci. Transl. Med. **10**(462), eaat9892 (2018)

Remote High Five: Bidirectional Transmission System for Video, Audio, and Vibrotactile Information Applied for Communication of Athletes with Family; a Case Study

Kakagu Komazaki$^{(\boxtimes)}$ and Junji Watanabe

NTT Communication Science Laboratories, Nippon Telegraph and Telephone Corporation, 3-1 Morinosato-Wakamiya, Atsugi 243-0198, Kanagawa, Japan
`kakagu.komazaki.sn@hco.ntt.co.jp`

Abstract. The high five is a widely used communication gesture that promotes cooperation and trust between people. This paper presents a remote communication system that enables people to experience the feeling of high fiving remotely. Each remote location is equipped with a video monitor along with a transparent surface placed in front. When one user high fives the tactile surface, the corresponding surface of the other user vibrates to convey the sensation. We verified the feasibility of the proposed system for delivering an emotional experience among the athletes and their family members by installing this system at a sports event.

Keywords: Remote communication · Vibration · Multi-sensory experience

1 Introduction

The high five is a mode of physical communication in which two people tap their palms together at face level. It is one of the major celebratory touches and is observed in many scenes. For instance, when a big play is made in a game, players on the court as well as the spectators high five each other. It is generally believed that the high five promotes cooperation and mutual trust among teammates and improves team performance [1]. However, the recent COVID-19 pandemic has severely limited physical contact and made it difficult to communicate emotions through physical touch.

To compensate for the lack of physical communication, remote communication technologies have been applied to share emotions. For example, Olympic athletes communicate with their friends and families through video conferencing after their performance [2]. However, such communication fails to create a sense of togetherness because the people involved cannot touch each other. Alternatively, it has been suggested that a better sense of the presence of a person can be achieved by transmitting vibrations through video calls to promote the feeling of kinship [3]. If the vibrotactile information is bidirectionally transmitted during remote communication under highly stimulating situations such as high fiving, people can experience a wide range of emotions remotely, including the elation and surprise elements associated with sports.

© The Author(s) 2022
H. Seifi et al. (Eds.): EuroHaptics 2022, LNCS 13235, pp. 436–438, 2022.
https://doi.org/10.1007/978-3-031-06249-0

In this study, we propose a bidirectional transmission system for video, audio, and vibrotactile information that can virtually transmit high fives. We tested this system in a real-life scenario, wherein an athlete was being cheered by a family member, to examine the feasibility of the system for promoting cooperation and mutual trust between these individuals.

2 System Overview

Figure 1 shows an overview of the system. A pair of visual devices along with a transparent surface in front (Fig. 1; left) is connected to a network. The camera placed between the transparent board and monitor captured one user and transmitted the information to the monitor on the other side. The vibration of the transparent surface was recorded using a vibration sensor and transmitted to the vibration speaker on the other side in real-time. When users stood in front of these devices and touched the tactile surfaces, they could see each other on the monitor and experience the sense of touch through the vibrating board. If one user high fived the transparent board, the other user could feel the vibration and respond to it and if users tap each other at the same time, they can feel the vibrations together. This communication through vibro-tactile transmission is what makes virtual high fives possible. In addition, the strength of the impact on one board is translated into vibration intensity of the board on the other side—if users hit the board harder, the vibration is stronger and vice versa.

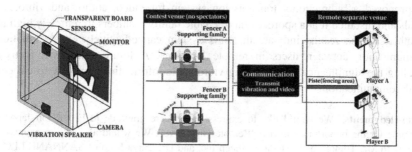

Fig. 1. (Left) Configuration of the device. (Right) Installed system at an actual sports event.

3 Application Example

We installed this system at the final bout of the 73rd All Japan Fencing Championships, 2020. During this bout, the family members of two fencers were in remote locations. The "Remote High Five" systems were installed at the fencing match venue (on the side of the piste) and in the hotel room where family members were watching the live broadcast of the bout (Fig. 1; right).

To focus on the momentous feelings that the fencers and family members might want to convey to each other, we targeted two specific times for emotions: immediately before and after the bout. The fencers and family members high fived through the system at these times. One fencer communicated with her son (Fig. 2), whereas the other fencer communicated with her father.

After the bout, the winner was interviewed. The winner said 'Actually, I'm most happy to have people cheer for me at the venue, but at a time like Covid-19 pandemic, by doing something like this (system), I can feel closer to my family even when they are far away. I felt as if my children were close to me.' She added 'I felt reassured, and I was able to fight as if my family was behind me.' These comments highlight the feasibility of this system for realizing cooperation and mutual trust in certain situations.

Fig. 2. Application scenario. Player (right) and her son (left) high fived using the system.

4   Conclusion and Future Work

We proposed a bidirectional transmission system for video, audio, and vibrotactile signals and installed it at a sports event to enable family members to remotely high five the athletes. The results indicate that this system can effectively enable emotional communication between users in remote locations. A future study will focus on revealing the significance of conveying vibrotactile information remotely to aid emotional communication.

Acknowledgments. We would like to express our sincere gratitude to Federation Japonaise d'Escrime for the opportunity to use "Remote High Five". We would also like to express our gratitude to Mr. Tomofumi Yoshida of Sheep Inc. and Dr. Daiya Kato of SANNANE LLC for their assistance with system construction.

References

1. Linden, D.J.: Touch: the science of the hand, heart, and mind (2015)
2. Olympics. https://olympics.com/athlete365/athlete-moment/. Accessed 27 Feb 2022
3. Hayakawa, H., et al.: Design of interactive visuo-audio-tactile media "public booth for vibrotactile communication" for telecommunication with high presence. Trans. Virtual Reality Soc. Jpn **25**(4), 412–421 (2020). (in Japanese)

Rolling Handle for Hand Motion Guidance and Teleoperation

Lisheng Kuang[1]($\boxtimes$), Maud Marchal[2,3], Paolo Robuffo Giordano[1],
and Claudio Pacchierotti[1]

[1] CNRS, Inria, IRISA, University of Rennes, Rennes, France
{Lisheng.Kuang,PaoloRobuffo.Giordano,Claudio.Pacchierotti}@irisa.fr
[2] INSA Rennes, CNRS, Inria, IRISA, University of Rennes, Rennes, France
Maud.Marchal@irisa.fr
[3] IUF France, Paris, France

Abstract. This paper presents a grounded haptic device able to provide force feedback. The device is composed of a biaxial rocker module and a grounded base which houses two servomotors actuating a mobile platform through three constrained coupling structures. The mobile platform can apply kinesthetic haptic feedback to the user hand, while the biaxial rocker module has two analog channels which can be used to provide inputs to external systems.

1 Introduction

Grounded haptic feedback systems have been used in a plethora of applications, including robotic teleoperation and guidance. Representative examples of such interfaces include the Virtuose (Haption, FR), Omega.x (Force Dimension, CH), Falcon (Novint Tech., USA) and Phantom (Geomagic, 3D Systems, USA) series. Most common haptic devices are designed as robots of the impedance type, having small amounts of inertia and friction, enabling to freely and naturally move in space when no force feedback is provided. In such systems, the user controls the position of the device that, in turn, provide a force feedback.

In this paper, we present the preliminary design of a force feedback haptic handle for virtual rendering and robotic teleoperation. It improves upon the handle presented in [2], that did not enable the user to provide any input. With respect to [2], the proposed device includes a biaxial rocker module and it is able to provide significantly larger forces.

2 Rolling Handle with Force Feedback

2.1 The Parallel Mechanism

The design of the bilateral handle is inspired from [2], for which the mechanism was in turn originally inspired by [3]. Three identical supporting legs, as shown

---

This work has received funding from the Inria Défi project "DORNELL" and the China Scholarship Council No. 201908440309.

H. Seifi et al. (Eds.): EuroHaptics 2022, LNCS 13235, pp. 439–442, 2022.
https://doi.org/10.1007/978-3-031-06249-0

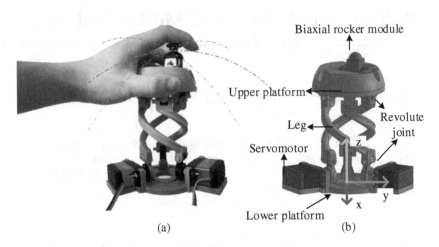

Fig. 1. (a) The device in its resting position, with the lower platform secured to an external support and the user hand posed on the upper one. (b) CAD of the device. The grey dotted lines show the surface of the sphere on which the upper platform moves. The user can provide directional input information through the biaxial rocker module.

in Fig. 1, are evenly placed around the z axis of the two platforms, forming an interlaced structure with no interference between them. Two revolute joints on the two sides of the leg generate reciprocal force and constrain the mobility between each linkage.

Gruebler-Kutzbach criterion (G-K criterion) or modified G-K can be used to analyze the mobility of the proposed handle, achieving a motion along 2°C of freedom (DoF). The motion of the upper platform is confined on the surface of a sphere centered in the center of the lower platform.

Two servomotors are installed on the lower platform, with the motor shaft connected to the distal revolute joint on the leg, providing 2-DoF actuation. The two legs equipped with the servomotors have an active rotation on the lower joint and a passive rotation on the other three joints. By varying the actuation of the servomotors, the upper platform can reach any position within its workspace (the surface of a sphere, as mentioned above and shown in Fig. 1).

2.2   User Input for Bilateral Control

On the top of the upper platform, we installed a biaxial rocker module as to enable the user to provide directional 2-D input. This module has one button digital output and (X, Y) 2-axis analog output through rocker potentiometer. The two analog channels can also be used to represent the rotation angle about two mutually perpendicular axes, for instance, two of the Euler angles ($Roll - Pitch - Yaw$) in 3D space. This input can also be used to control the free-space motion of the handle, similarly to a standard impedance-type haptic interface.

3   Use Cases

We develop a goal-following game to test the effectiveness of the haptic force feedback generated by the handle as well as the intuitiveness of using the biaxial rocker module. A 2D GUI interface was developed with Matlab. The servomotors of the handle were controlled through an Arduino board.

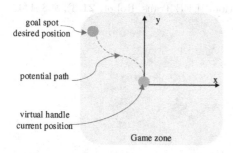

Fig. 2. The handle moves towards the goal point (in green, invisible to the user) to guide the user towards it. The closer the position of the user in the virtual environment (yellow point) to the goal spot, the smaller tilt of the handle with respect to its vertical resting position.

Figure 2 shows the considered 2D environment. The current position of the user in the virtual environment is shown in yellow, controlled by the user through the biaxial rocker module. A random goal point, not visible by the user, is generated. The handle tilts toward the (invisible) goal point, guiding the user towards the objective position. The closer the position of the user in the virtual environment to the goal spot, the smaller tilt of the handle with respect to its vertical resting position. The task ends when the user reaches the goal point.

Preliminary tests indicated that the device is easy to use and intuitive to follow. An extensive human subjects will be subject of future work.

4   Conclusion and Future Work

The proposed device is inspired from our previous work presented in [2]. We improved upon its design by including more powerful motors, able to provide force feedback to the user, a biaxial rocker module, to enable the user to provide 2D navigational input to the system, and making the design more ergonomic and comfortable to use.

In future work, we will integrate the proposed device in a power wheelchair, so as to provide the driver with information about the presence of surrounding obstacles [1]. This scenario is inspired by the collaborative Inria project DOR-NELL, where we aim at developing a multisensory haptic handle for helping disabled people using mobility aids such as power wheelchairs, white canes, and walkers.

References

1. Devigne, L., et al.: Power wheelchair navigation assistance using wearable vibrotactile haptics. IEEE Trans. Haptics **13**(1), 52–58 (2020)
2. Kuang, L., Marchal, M., Aggravi, M., Robuffo Giordano, P.: Design of a 2-DoF haptic device for motion guidance. In: Proc. Eurohaptics (2022)
3. Okada, M., Nakamura, Y.: Development of a cybernetic shoulder-a 3-DoF mechanism that imitates biological shoulder motion. IEEE Trans. Robot. **21**(3), 438–444 (2005)

Is Tactile Shape Coding Invariant Against Small Location Shifts on the Skin?

Scinob Kuroki[1(✉)] and Shin'ya Nishida[2]

[1] NTT Communication Science Laboratories, 3-1 Morinosato-Wakamiya,
Atsugi, Kanagawa, Japan
scinob@gmail.com
[2] Kyoto University, Yoshida Hon-machi, Sakyo, Kyoto, Japan

Abstract. A preliminary experiment was conducted to investigate the nature of location invariance of shape perception in touch. The small bar shape was presented on the fingertip by means of a dense pin-array stimulator. Participants were asked to report which of the two sequential pattern presentation groups contained the oddball pattern. In one group, the same pattern (e.g., one vertical bar) was presented sequentially six times at short intervals at a fixed location on the fingertip, and in the other group, the same five patterns and one different pattern (e.g., one oblique bar) were presented. The obtained performance of discriminating shapes was compared to that when the patterns were presented at random locations on the fingertip. We found that the discrimination performance dropped when the pattern had to be recognized at new locations. This result suggests that the tactile system cannot easily find the shape identity over different locations, even when the shape is very simple and position change is within a single finger.

Keywords: Orientation perception · Shape perception · Constancy · Somatotopy

1 Introduction

We can perceive complex surface patterns, such as Braille, by moving our hands to touch the surface. While surfaces are usually fixed in the world coordinate, our mechanoreceptors beneath the skin are fixed in the somatotopic coordinate. How we touch dramatically changes where on the skin the deformation occurs and which mechanoreceptors to be fire. Meanwhile, haptic pattern discrimination performance should be at least partly skin location invariant, since changes only in where on the skin the pattern is transcribed should not affect the estimated pattern in principle. This issue is not unique in tactile modality, as previous studies of visual perception have revealed location invariant coding for simple features such as lines and angles [1]. Though it sounds reasonable to assume such mechanism also in touch, to our best knowledge, direct behavioural evidence was lacking mainly due to the difficulty in stimulus presentation.

In this study, we try to directly address whether we have location invariant (in other words, somatotopy invariant) shape coding in touch by examining the shape

H. Seifi et al. (Eds.): EuroHaptics 2022, LNCS 13235, pp. 443–446, 2022.
https://doi.org/10.1007/978-3-031-06249-0

discrimination ability of the stimuli presented on different local skin sites within a single finger. The small (~ 10 mm) bar shape was presented on the finger pad with spatial jitter by means of a dense pin-array stimulator. If we have robust location invariant shape coding in touch, the performance would be stable regardless of the jitter size.

2   Methods

The stimuli were generated by the 'Latero' (Tactile Lab, Montreal, Canada) [2]. An array of 64 pins constructs 10 x 10 mm contact surfaces with a spatial period of $\sim$ 1.2 mm. Each pin can independently move the contacted surface of the skin laterally.

The presented patterns were either vertical (V), left oblique (L), or right oblique (R) bar shapes, presented by four pins (6–8 mm long) vibrating laterally at 40 Hz. The stimulus duration for each pattern and the interval between patterns was 0.1 s.

One trial consisted of two kinds of patterns: a base pattern and a target pattern. There were three different combinations of base and target (B-T) pair: a vertical target in oblique base patterns (O-V), an oblique target in vertical base patterns (V-O), and an oblique target in inverted oblique base patterns (L-R). In each trial, two groups of stimuli, each consisting of six patterns, were presented with an inter-group interval of 2 s. One group consisted of six base patterns, and the other group consisted of five base patterns and one target pattern (Fig. 1A). In which group (1st or 2nd) and wherein the group (2nd to 5th) the target pattern was presented was randomized. Note that the target pattern was presented at neither first nor last of the group.

The variability of where the stimuli were presented was defined in three levels by the jitter parameter. In the condition of zero jitter (J0), all patterns belonging to one group were presented to the same location of the skin (although the target and base patterns deformed the skin in slightly different locations). Their locations were different in the first and second groups. In the condition of x/y jitter (Jx/Jy), patterns belonging to a group were presented on a location fixed in the long/short axis while randomized in the short/long axis. The shift amount was within 6 mm. In the condition of xy jitter (Jxy), patterns belonging to a group were presented on a pseudo-random location. No successive patterns were presented on the same location except J0 condition.

A participant sat at a table with the index or middle finger of their left hand placed on the stimulator. The task was two-interval forced choice (2-IFC) to report which group contained oddball (target) pattern with different angle by pressing "1" or "2" on the keyboard. After each keypress, the feedback signal was provided as a beep sound when the response was incorrect. After a 2-s blank period, the next trial started.

Seven participants (ages 20–38, three females, two left-handed), including the first author, participated in the experiment. Each participant completed at least 16 repetitions for each combination of the base-target pair and the spatial jitter.

3 Results and Discussion

Although the number of participants was not yet sufficient, our tentative results suggest two main things. First, discrimination of simple shapes on the skin was equal or easier when the shapes are presented in the same location (J0) than when they were presented in different locations. A two-way repeated ANOVA shows that there was a significant effect for the jitter size ($p < 0.001$), while not for the pattern pair ($p = 0.055$) nor the interaction effect ($p = 0.064$). As a post-hoc analysis, multiple comparisons for the jitter size showed that the performance in J0 condition was better than those in Jy and Jxy conditions. Note, however, that the number of subjects reported here is not sufficient at this point and the results of the statistical tests are only approximate since a power analysis showed that the required number of participants is at least 16.

Secondly, there were large individual differences in the ability to discriminate the shapes. One participant showed high performance overall, with a negligible drop according to the increase in spatial jitter for V-O and L-R pairs (although there was some drop for O-V pair). Meanwhile, another participant was unable to discriminate the shape difference at all even in the no spatial jitter condition for V-O and L-R pairs. Such large individual differences were unexpected for the basic task of discriminating the angles of a simple shape, a task that seems to be essential in daily activities. This point awaits further investigation in the future.

In summary, current results are consistent with the possibilities that the tactile system lacks or is incomplete in location invariant shape coding.

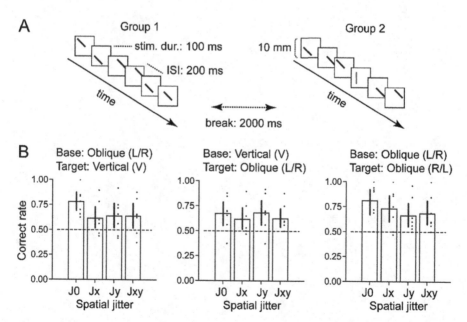

Fig. 1. A: Trial sequence of O-V and Jxy condition. The task was 2-IFC to report which group contained an oddball pattern. B: Results of the experiment. Error bars denote 95% confidence interval, and dots denote individual data averaged across repetitions for each participant.

References

1. Cavanagh, P.: Size and position invariance in the visual system. Perception **7**(2), 167–177 (1978)
2. Wang, Q., Hayward, V.: Biomechanically optimized distributed tactile transducer based on lateral skin deformation. Int. J. Robot. Res. **29**(4), 323–335 (2010)

Precision Grip Discrimination of Object Thickness

Alastair J. Loutit[1,2(⊠)], Siddhartha Pande[2], Naqash Afzal[2],
Pouya Abdollahzadeh[3], Richard M. Vickery[1,2],
and Ingvars Birznieks[1,2]

[1] Neuroscience Research Australia, Sydney, Australia
a.loutit@neura.edu.au, {r.vickery,i.birznieks}@unsw.
edu.au
[2] School of Medical Sciences, UNSW Sydney, Sydney, Australia
{s.pande,h.afzal}@student.unsw.edu.au
[3] Urmia University of Technology, Urmia, Iran

Abstract. Humans can discriminate the thickness of very thin objects (~ 0.1 mm), like our awareness when gripping one or more sheets of paper when turning book pages, or counting banknotes. Fingertip separation generates proprioceptive signals from the joint and muscle receptors of the fingers, and cutaneous receptors overlying the skin covering these joints. These inputs are integrated in the central nervous system to enable humans to perceive object thickness. However, in the case of very thin objects it is unclear whether thickness discrimination is based on inherent physical properties of such objects instead of explicitly evaluating its thickness. Thinner objects may provide additional discriminable sensory cues like their pliability, vibration transmission, or heat transfer. To evaluate subject performance when no such cues are available, we investigated the accuracy of human thickness discrimination of two separated parallel-aligned 0.7 mm steel plates using a pinch grip formed with the thumb and index finger. The separation of the plates was precisely controlled using a robotic positioning device. We introduce our psychophysics experiments in which participants (n = 6) were presented with pairs of separation distances and indicated which of two separations they perceive as wider in a two-alternative forced choice paradigm. Each stimulus pair comprises a reference and comparison width, with seven reference widths ranging from 2.5–80 mm. The smallest median just noticeable difference was 0.75 mm (range 0.42–3.7) with 2.5 mm aperture. The pinch grip aperture discrimination did not follow Weber's Law, as just noticeable differences for reference widths narrower than 10 mm did not decrease in proportion to decreasing reference widths. Thus, the influence of inherent cues and fingertip aperture in discrimination of object thickness remains to be tested.

1 Introduction

We can perceive minute differences in our finger position, which enables fine object manipulation, discriminate the thickness of very thin objects (~ 0.1 mm), like being aware gripping one or more sheets of paper when turning book pages, counting

H. Seifi et al. (Eds.): EuroHaptics 2022, LNCS 13235, pp. 447–451, 2022.
https://doi.org/10.1007/978-3-031-06249-0

banknotes or manipulating a deck of cards. Fingertip separation generates proprio-ceptive signals from the joint and muscle receptors of the fingers, and cutaneous receptors overlying the skin covering these joints. John et al. [1] found that participants using a pinch grip on pairs of flat metal plates (a reference and comparison plate) could discriminate the difference in their thickness with a just noticeable difference (JND) of 0.075 mm with a 0.2 mm thick reference object. In another study, one subject was able to discriminate thickness differences as small as 0.03 mm [2], however, it was acknowledged that for plates narrower than 0.5 mm subjects could have base their decision on plate deformation. Similarly other confounding factors, such as heat transfer from the fingertips through the object, or vibrations spreading through the object determined by the rigidity may often aid discrimination performance. In the John, et al. [1] study, the authors sought to control for confounding variables by minimizing visual and auditory cues with a curtain blind and earphones, minimized thermal conduction cues equilibrating the temperature of the plates and fingertip skin. Nevertheless, it is possible that the thin plates underwent mechanical deformation during the pinching task. Here, we aim to address limitations in previous studies of object thickness discrimination by using a novel method incorporating two separate, parallel-facing steel plates (0.7 mm thick). This allows investigation of thickness dis-crimination based exclusively on sensory signals about pinch grip separation distance. The limitation of this technique is that very small grip apertures cannot be tested, but insight could be obtained based on extrapolation based on smallest discrimination limits and observation whether data would follow Weber's Law.

2 Methods

2.1 Participants

Six healthy right-hand dominant subjects (aged 21–28, 4 female) participated in psy-chophysics experiments. The experimental protocol was approved by the Human Research Ethics Committee (HC180109) of UNSW Sydney and written consent was obtained from subjects before experimentation began.

2.2 Apparatus

Two flat steel plates (dimensions) were mounted parallel to each other. Each plate was fixed to clear acrylic Perspex block with space for a finger to grip the flat surface. One plate was mounted to a rigid stationary frame, while the other was mounted on a Hexapod H-850 (Physikinstrumente, Karlsruhe, Germany) robotic manipulator capable of controlling the separation between the plates with 1 μm precision.

2.3 Psychophysics

Participants wore a blindfold and noise-canceling headphones, which were used to give automated voice commands and mask any auditory or visual cues from the robotic manipulator's movement. On voice command, participants pinched the flat surface of the steel plates with index finger and thumb, and then retracted their hand. A two-alternative forced-choice protocol was used in which the stimulus widths were presented in pairs, one reference width and one comparison width, and participants indicated which grip aperture they perceived to be wider. We tested seven reference stimulus widths ranging from 2.5–80 mm, each with six comparison apertures, shown in Table 1. Reference-comparison pairs were presented 20 times each in pseudorandom order, half with the reference presented first and half with the reference second (120 total). Participants pressed a button to indicate whether the first or second grip aperture of each stimulus pair was wider.

Table 1. Reference and comparison widths

Reference width (mm)	Comparison widths (mm)
2.5	2.0, 2.25, 2.75, 3.0, 3.25, 3.5
3.5	2.0, 2.5, 3.0, 4.0, 4.5, 5.0
6.0	3.75, 4.75, 5.5, 6.5, 7.25, 8.25
10.0	7.5, 8.5, 9.25, 10.75, 11.5, 12.5
20.0	16.5, 18.0, 19.0, 21.0, 22.0, 23.5
40.0	36.0, 38.0, 39.0, 41.0, 42.0, 44.0
80.0	75.0, 77.0, 78.5, 81.5, 83.0, 85.0

2.4   Analysis

We used the *findchangepts* MATLAB function (Mathworks, Natick, MA, USA) to identify an obvious change in the Weber's Fraction data. After statistically identifying the inflection point, we used piecewise linear regression to compare the two regression lines.

3   Results

The subject's performance was quantified as the proportion of trials in which they reported the comparison stimulus to be wider than the reference stimulus. Thus, correct judgment of the stimulus width as always greater than the reference would render a value of 100%, but narrower 0%. Figure 1A shows example responses from the 10 mm reference stimulus width comparisons. With just 0.75 mm difference from the reference width, participants typically performed at or above 75% accuracy and followed an approximately sigmoidal curve. Logit scaled regression confirmed a significant positive correlation between stimulus width and the logit scaled responses (Fig. 1B) and demonstrates that variability among participants was small for the 10 mm reference stimulus width. We investigated the just noticeable differences (JNDs) for each comparison stimulus width, and whether Weber's Law applies to stimulus width discrimination (Fig. 2A and B). If data would follow Weber's Law the JND would be a constant proportion of the reference stimulus width, for example, we would expect the

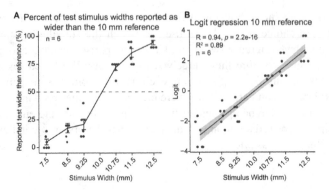

Fig. 1. Example responses from the 10 mm reference stimulus width. (A) shows the response proportions from six participants and (B) is the logit scale transormation of (A).

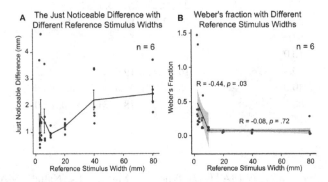

Fig. 2. The just noticeable difference (JND) and Weber's Fraction for the range of reference apertures tested. A) The JND for apertures below 10 mm is highly variable, but increases linearly for apertures greater than 10 mm. B) Weber's constant is violated for apertures below 10 mm, but is consistent above 10 mm.

JND with the 80 mm reference to be 32 times larger than with the 2.5 mm reference. Obviously, this wasn't the case (see Fig. 2A)—the median JND with the 2.5 mm reference was 0.75 mm (range 0.42–3.7 mm) vs 2.3 mm (range 1.7–3.7 mm) with the 80 mm reference. It was determined that changes in Weber's fraction expressed as JND/stimulus width could be described as a piece-wise function with a breakpoint at 10 mm (Fig. 2B). Linear regression analyses show that at 10 mm and above the slope is not different from 0 ($R = -0.08$, $p = .72$) and thus JND is determined by Weber's Law, while slope is negative ($R = -0.44$, $p = .03$) for stimulus widths < 10 mm.

4 Discussion

We aimed to investigate human ability to discriminate object width by means of evaluating grip aperture in the absence of any other associated cues which might signal differences between object thickness, for example pliability or the ability to spread vibrations. In this context we wanted to examine whether previously reported JND values of 0.075 mm reported at 0.2 mm stimulus width by John et al. (1989) can be reasonably attributed to the finger joint angle changes. We were able to eliminate associated cues by using two parallel, separated surfaces. The smallest just noticeable difference in our study was 0.75 mm (median, range 0.42–3.7) obtained with the 2.5 mm stimulus width. To be able to extrapolate the JND values for 0.2 mm stimulus width, which is too small to test with our apparatus, we would have to observe the relationship between the JND and stimulus width. If JND would follow Weber's Law then by extrapolation this would match the findings of John et al. (1989). However this wasn't the case. The Weber's fraction was constant only with apertures wider than 10 mm. Below 10 mm we observed considerable variability between subjects and while we did observe the smallest JND at the smallest aperture it seemed that the maximum sensory resolution was achieved and there was no evidence that JND could further decrease in proportion to the stimulus width which, by extrapolation, would suggest that subjects could detect 0.075 mm differences in thickness of 0.2 mm thick objects.

This indicates that extrapolation in the submillimeter range cannot be applied and current investigations should be extended into submillimeter range stimulus widths using different experimental procedures eliminating subject's ability to rely on other cues than grip aperture.

References

1. John, K.T., Goodwin, A.W., Darian-Smith, I.: Tactile discrimination of thickness. Exp. Brain Res. **78**(1), 62–68 (1989). https://doi.org/10.1007/BF00230687
2. Ho, C.-H., Srinivasan, M.A.: Human haptic discrimination of thickness. Master of Science in Mechanical Engineering, The Research Laboratory of Electronics, Massachusetts Institute of Technology, Cambridge, Massachusetts (1996)

Use of Multiple Frequencies of Ultrasound in Midair Haptic Stimulation

Saya Mizutani[1]([✉]) [ID], Shun Suzuki[2] [ID], Atsushi Matsubayashi[2],
Tao Morisaki[2] [ID], Yutaro Toide[2] [ID], Masahiro Fujiwara[1,2] [ID],
Yasutoshi Makino[1,2] [ID], and Hiroyuki Shinoda[1,2] [ID]

[1] The University of Tokyo, 7-3-1 Hongo, Bunkyo-ku, Tokyo 113-0033, Japan
mizutani@hapis.k.u-tokyo.ac.jp
[2] The University of Tokyo, 5-1-5 Kashiwanoha, Kashiwa-shi,
Chiba 277-8561, Japan

Abstract. We proposed a tactile stimulation method using multiple frequency ultrasound to shift the ultrasound focal point without changing the phase. This method can be presenting stable stronger tactile stimuli than any conventional stimulation method using phase switching. In the user study, it was found that the stimulus of the proposed method was perceived significantly stronger at the palm than that of conventional methods in 200 Hz.

Keywords: Multiple frequencies · Stronger perceptual intensity · No phase update

1 Introduction

Conventional tactile presentation using ultrasound is realized by delivering single frequency ultrasound waves from each transducer of airborne ultrasound phased array (AUPA) with appropriate phase to present acoustic radiation pressure on the skin surface. The maximum force presented by one current AUPA is not strong. Therefore, it is a practically important issue to find a stimulus method that can be perceived stronger with limited power.

Recently, Lateral Modulation (LM) [1], and Spatiotemporal Modulation (STM) are mainly used. These methods can present tactile sensation by moving stimulus points discretely along the skin with constant acoustic power. To move the stimulus point, the transducers transmit the ultrasound of single frequency ω with changing the driving phase φ as shown in Eq. (1),

$$v_i(t) = A_i e^{j\varphi_i(t)} e^{j\omega t} \qquad (1)$$

where i means the i th transducer in the phased array, and A_i means the ultrasound amplitude to be transmitted from i th transducer. In this paper, the phase-switching LM stimulus described above is called Single Frequency Lateral Modulation (SFLM).

© The Author(s) 2022
H. Seifi et al. (Eds.): EuroHaptics 2022, LNCS 13235, pp. 452–455, 2022.
https://doi.org/10.1007/978-3-031-06249-0

2   Proposed Method

In this study, we propose a method of allocating transducers that are constantly driven at different frequencies in AUPA. This method makes it possible to generate a continuously moving distribution even in a steady state driving where the device phase does not change. Since it is possible to create a pattern that moves repeatedly without performing rapid phase switching, it is possible to avoid the problems of radiation pressure decline so that this method can be presenting stable stronger tactile stimulation than any SFLM stimuli.

As an easy method to implement, we consider a method of allocating frequencies to each phased AUPA unit (hereinafter referred to as "unit") of the multi-phased array.

Consider the case where the ultrasound transmitted from each unit form a focal point at a common point $F(x_f, y_f, z_f)$ with the configuration shown in Fig. 1a. The minimum configuration to realize LM that slightly vibrates the focal point from side to side is to vibrate units 1 and 3 with the frequencies $f + \Delta f$ and $f - \Delta f$, respectively. In this case, the standing wave moves in one direction at the frequency Δf. Here, the inclination of the unit θ is neither $-\frac{\pi}{2}$ nor $\frac{\pi}{2}$. When the central unit is added and driven by f, the pressure antinodes of a part of the standing wave is emphasized. In this case, the focal position at each time is as shown in Fig. 1b. This corresponds to LM, where the focal point moves spatially continuously. In this paper, this proposed method is called Multiple Frequency Lateral Modulation (MFLM).

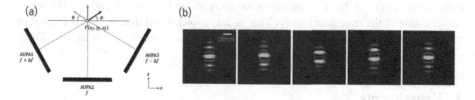

Fig. 1. (a) System configuration to consider. (b) The simulation results of acoustic pressure amplitude distribution on the display plane with 1 cm square dotted frame.

2.1   Focal Point Movement Speed

In case the focal point is far enough from each AUPA, the ultrasound arriving from each transducer can be assumed to have only a wave vector in the direction perpendicular to each device surface. Therefore, let the intersection angle between these plane waves and the $z = z_f$ plane is θ, the sound pressure p vicinity the focal point is expressed as follows,

$$p = a\cos(\omega_1 t - k_1 x) + a\cos(\omega_2 t + k_2 x),$$
$$= 2a\cos\left(\frac{\omega_1 - \omega_2}{2}t - \frac{k_1 + k_2}{2}x\right)\cos\left(\frac{\omega_1 + \omega_2}{2}t - \frac{k_1 - k_2}{2}x\right), \tag{2}$$

where a is the amplitude, ω_1, ω_2 are angular frequencies and $k_1, k_2(> 0)$ are wave numbers in x direction. In Eq. (2), $\cos\left(\frac{\omega_1 - \omega_2}{2}t - \frac{k_1 + k_2}{2}x\right)$ and $\cos\left(\frac{\omega_1 + \omega_2}{2}t - \frac{k_1 - k_2}{2}x\right)$ represents the amplitude of the gentle carrier component and the envelope component near the frequency f, respectively. When the amplitude of the carrier wave component is maximized, the sound pressure is maximized. Therefore, focusing on the maximum sound pressure point,

$$\left|\cos\left(\frac{\omega_1 - \omega_2}{2}t - \frac{k_1 + k_2}{2}x\right)\right| = 1, \quad \frac{\omega_1 - \omega_2}{2}t - \frac{k_1 + k_2}{2}x = 2\pi\sigma(\sigma \in \mathbb{Z}) \quad (3)$$

migration velocity of the focal point in the x direction on the $z = z_f$ plane can be expressed as follows,

$$\frac{dx}{dt} = \frac{\omega_1 - \omega_2}{k_1 + k_2} = \frac{\Delta f}{f} \cdot \frac{c}{\cos\theta}. \quad \left(\theta \neq -\frac{\pi}{2}, \frac{\pi}{2}\right) \quad (4)$$

2.2 Spatial Pattern and Valid Range

Focusing on the sound pressure distribution that appears on the $y = 0$ plane at time $t = 0$, ultrasound with frequencies $f + \Delta f$ and $f - \Delta f$ create a distribution like a standing wave. Therefore, the interval b of antinodes can be calculated as $\frac{c}{2f\cos\theta}$. The focal point diameter w generated by ultrasound of frequency f is given by $\frac{2\lambda r}{d}$, where λ is the wavelength, r is the distance from the array to the focal point, and d is the width of one side of the transducer array. That is, a spatiotemporal distribution is formed in which a pattern of period b flows in one direction in a window of diameter w. Then, some of the belly of the standing wave is emphasized by f.

3 Experiments

In this experiment, MFLM and SFLM where the focal movement in the same way were compared in terms of physical sound pressure and perceptual threshold using 3 AUPAs.

The sound pressure was measured with standard microphone (4138-A-015, Brüel & Kjær) at the center of the focal trajectory of LM stimuli as shown in Fig. 2a. The frequency of the focal shift was 200 Hz, so that the phase update rate of the SFLM was 4400 fps. The results are shown in Fig. 2b. The maximum sound pressure amplitudes of MFLM and SFLM were about 10.76 kPa and about 9.74 kPa, respectively.

We measured relative thresholds of LM stimuli at the palm by method of limits at focal movement frequencies of $8, 48, 96, 200$ Hz. Nine people joined in this experiment. They opened their hand and fixed it in the air with putting their wrist on the wrist rest. The results are shown in Fig. 2c. Note that "the maximum output power of the device" is set to 0 dB here. As the result of one-sided t-test with a significance level of 1%, the threshold of MFLM was significantly lower than the SFLM at 200 Hz ($p = 2 \times 10^{-6}$).

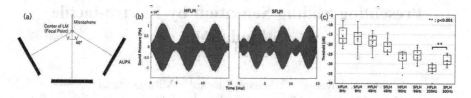

Fig. 2. (a) Measurement system. (b) Measured sound pressure of SFLM and MFLM at 200Hz. (c) Relative thresholds of MFLM and four types of SFLM in 200 Hz at the palm.

4 Conclusion

In this study, we proposed a tactile stimulation method using multiple frequencies. This proposed method can create a distribution that moves continuously without changing the driving phase, which can prevent pressure amplitude drop by quick drive phase switching. In the experiments, it was found that the maximum sound pressure amplitude of MFLM is larger than that of SFLM and MFLM stimulus perceived stronger than SFLM at the palm in 200 Hz. That is, at least more than 200Hz, it is considered that MFLM is perceived stronger than SFLM because of no phase switching.

Reference

1. Takahashi, R., Hasegawa, K., Shinoda, H.: Tactile stimulation by repetitive lateral movement of midair ultrasound focus. IEEE Trans. Haptics **13**(2), 334–342 (2019)

Presenting Sliding Sensation by Electro-tactile Stimulation

Shota Nakayama[1]([⊠]), Seitaro Kaneko[1,2], Mitsuki Manabe[1],
Keigo Ushiyama[1], Masahiro Miyakami[1], Akifumi Takahashi[1,2],
and Hiroyuki Kajimoto[1]

[1] The University of Electro-Communications, 1-5-1 Chofugaoka, Chofu, Tokyo,
Japan
{nakayama,kaneko,manabe,ushiyama,miyakami,
a.takahashi,kajimoto}@kaji-lab.jp
[2] JSPS Research Fellow, Tokyo, Japan

Abstract. This paper reports a preliminary trial on generating an illusory sliding-force sensation, a force sensation that occurs without the use of physical force, using electrical stimulation. Considering that the asymmetric vibration generates an illusory force, we have designed an electrical stimulation pattern by simulating the activity of skin receptors.

Keywords: Electrical stimulation · Force sensation · Sliding illusion

1 Introduction

Wearable-type haptic displays are attractive because they are lightweight and small sized; however, they have difficulty presenting external force as compared to desktop-type tactile displays. Therefore, many methods have been proposed to generate force sensation using illusory phenomena by skin sensation. A typical technique involves using asymmetric vibration [1–3]. When a weight is vibrated to be driven quickly forward and slowly in the reverse direction, the illusion of being pulled is generated on the hand grasping the transducer. However, this technique involves a strong vibration sensation that propagates over the entire hand.

We have proposed a method to solve this problem by using a device that generates an illusory-force sensation through electrical stimulation [4]. The proposed method successfully generates an illusory-force in a direction normal to the skin surface. This paper is our next attempt to generate illusory-force sensation in the tangential direction.

2 Method

2.1 Electrical Stimulation Device

Electrical stimulation was performed using the electrical stimulator developed by Kajimoto [5]. This stimulator is divided into a control unit that determines the current and stimulation pattern, and an electrode unit comprising electrodes and switching circuits. The control unit is connected to a PC through a USB connection.

In the electrode unit (Fig. 1(a)), electrodes are attached to the top and bottom of a small box (4 cm × 3 cm × 1 cm, Fig. 1(b)). Sixty-three (7 × 9) circular electrodes (1.4 mm in diameter) are placed on one electrode board at 2 mm center-to-center intervals. The weight of the complete grasping part is 17 g.

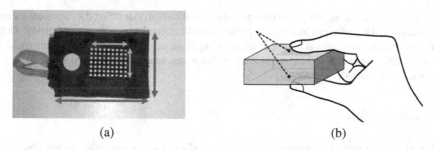

(a)   (b)

Fig. 1. (a) Electrode unit. (b) Structure of the grasping part.

2.2 Stimulus Pattern

Figure 2 shows the stimulus pattern, designed based on findings of Kaneko et al. [6]. They reported that asymmetrical vibration of the weights causes temporal bias in the skin deformation of the finger pad. Specifically, when an illusory-force sensation is generated from the base to the tip of the finger, a large skin tension is generated at the base of the finger with a large compression generated at the fingertip.

We focused on the activity of Meissner corpuscles and Merkel cells to apply this phenomenon to electrical stimulation. Meissner corpuscles are more sensitive to skin compression than tension during the beginning of slip [7], whereas Merkel cells are active in both tension and compression. As for electrical stimulation, anodic stimulation tends to generate Meissner-related sensation (vibration), and cathodic stimulation tends to generate Merkel-related sensation (pressure) [8].

Therefore, we hypothesized that the illusory sliding force is generated by cathodic stimulation to the area with strong tension and combined anodic and cathodic stimulation to the area with strong compression.

In our preliminary trial conducted within authors, five out of six participants felt the illusory force. Out of these five, three felt the illusory sliding-force sensation while the remaining two felt the illusory sensation normal or oblique to the skin surface, possibly due to the difference in sensitivity between the thumb and index finger. The activity of deep mechanoreceptors such as Ruffini endings and Pacini corpuscles may be necessary for the generation of the illusory-sliding-force sensation.

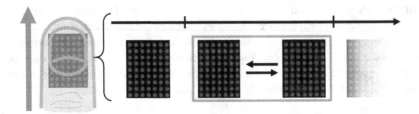

Fig. 2. Stimulus pattern. The same pattern is stimulated on both sides. Red dots represent anodic stimulation points, whereas blue dots represent cathodic stimulation points, respectively. Only one electrode is stimulated at a time; it takes 200 us to stimulate one electrode. As this speed is imperceptible, we feel as if the patterns are stimulated simultaneously. The maximum current for electrical stimulation is 6 mA.

3   Conclusion

This study attempted to generate the illusory sliding-force sensation using electrical stimulation. The stimulus pattern was designed based on the previous measurement of skin under asymmetric vibration—an existing method for generating illusory-force sensation. Our preliminary trial conducted within authors showed that three out of six felt illusory-sliding-force sensation.

However, our proposed stimulation pattern generates limited sensation compared to the illusory-force sensation by asymmetric vibration. Therefore, we will continue studying the stimulus patterns generating more distinct illusory-sliding-force sensations.

Acknowledgements. This work was supported by JSPS KAKENHI Grant Number JP18H04110.

References

1. Amemiya, T., Ando, H., Maeda, T.: Virtual force display: direction guidance using asymmetric acceleration via periodic translational motion. In: First Joint Eurohaptics Conference and Symposium on Haptic Interfaces for Virtual Environment and Teleoperator Systems. World Haptics Conference, pp.619–622 (2005)
2. Rekimoto, J.: Traxion: a tactile interaction device with virtual force sensation. In: ACM SIGGRAPH 2014 Emerging Technologies, New York, p. 1 (2014)
3. Culbertson, H., Walker, J.M., Raitor, M., Okamura, A.M.: WAVES: a wearable asymmetric vibration excitation system for presenting three-dimensional translation and rotation cues. In: Proceedings of the 2017 CHI Conference, New York, pp. 4972–4982 (2017)
4. Nakayama, S., Manabe, M., Ushiyama, K., Miyakami, M., Takahashi, A., Kajimoto, H.: Pilot study on presenting pulling sensation by tactile electrical stimulation. In: EuroHaptics 2022 Conference, Hamburg, Germany (2022). (accepted)
5. Kajimoto, H.: Electro-tactile display kit for fingertip. In: 2021 IEEE World Haptics Conference (WHC), p. 587 (2021)

6. Kaneko, S., Nakamura, T., Kajimoto, H.: Measurement of skin under virtual force illusion caused by asymmetric acceleration vibration. Submitted to IEEE Transactions on Haptics
7. Delhaye, B.P., Schiltz, F., Barrea, A., Thonnard, J.-L., Lefèvre, P.: Measuring fingerpad deformation during active object manipulation. bioRxiv (2021)
8. Yem, V., Kajimoto, H.: Comparative evaluation of tactile sensation by electrical and mechanical stimulation. IEEE Trans. Haptics 10(1), 130–134 (2017)

Uncoupled Stability Dynamic Range for Kelvin- Voigt and Hunt-Crossley Virtual Environments

Seanna Oliver$^{(\boxtimes)}$, Leonam Pecly, and Keyvan Hashtrudi-Zaad

Department of Electrical and Computer Engineering, Queen's University,
Kingston, Canada
17smo5@queensu.ca

Abstract. Kelvin-Voigt (KV) and Hunt-Crossley (HC) are models commonly used to simulate viscoelastic environments in haptic simulation systems. Due to the sample-and-hold process, the range of dynamics - viscosity and elasticity, that can be rendered in a stable way is limited. In this paper, we experimentally evaluate the range of viscosity and elasticity that result in uncoupled stability. We also propose a method to map the KV parameters to the HC parameters space, and vice-versa.

1 Introduction

Haptic simulation systems allow users to interact with virtual environments (VEs), interfacing with robotic mechanisms, known as haptic devices. Due to the sample-and-hold process, the range of the VE dynamics with which the users can interface in a stable manner is limited. This limitation degrades the sense of realism and limits the applications of these systems. Viscoelastic models are commonly used to implement VEs.

Kelvin-Voigt (KV) is a linear damper-spring model commonly utilized to emulate viscoelastic environments [1]. The stability of haptic simulation systems with the KV VE model, when user is interfaced (coupled) or not interfaced (uncoupled) has been significantly studied. Since users tend to stabilize their interaction with objects, uncoupled stability is considered as a more stringent stability condition [1–3]. HC is a nonlinear model that does not have the inconsistencies of the KV model and may better represent environments with substantial viscous effects [4].

Although the uncoupled stability range of the viscosity and elasticity components of the KV VEs, i.e. stiffness and damping, have been experimentally studied through analytical and experimental determination of the damping and stiffness coefficients, such range has not been determined for the dynamic parameters of the full HC model. Furthermore, since the interpretation of the dynamic parameters of the KV and HC models are different, given the parameters of one model, a means to determine the equivalent dynamic parameter in the other model helps with comparison and interpretation of the results.

Supported by Natural Sciences & Engineering Research Council of Canada.

H. Seifi et al. (Eds.): EuroHaptics 2022, LNCS 13235, pp. 460–463, 2022.
https://doi.org/10.1007/978-3-031-06249-0

In this work, the uncoupled stability range of elasticity K and viscosity B will be determined through experiment for both the KV and HC models, and simulations will be used to produce the necessary data for identification, needed to map the dynamic parameters from KV to HC, and vise-versa.

2   Methodology

The environment forces are generated by the KV and HC models according to

$$F_{KV} = \begin{cases} K_{KV}x + B_{KV}\dot{x} & x \leq 0 \\ 0 & x < 0 \end{cases} ; \quad F_{HC} = \begin{cases} K_{HC}x^n + B_{HC}\dot{x}x^n & x \leq 0 \\ 0 & x < 0 \end{cases}, \quad (1)$$

where K_{KV} and B_{KV} are the KV stiffness and damping parameters, respectively, x is the position, K_{HC} and B_{HC} are the HC elastic and viscous parameters, respectively, and n is the HC exponent.

The range of K and B values for uncoupled stability for each model were identified experimentally using a QET, a one degree-of-freedom platform, as shown in Fig. 1. The models were implemented on the QET with velocity computed from backward difference (BD), and for each B value, the K value was increased from zero until instability was reached. The system was considered unstable if the amplitude of the position was greater than 2% of the initial position after 4 s. For HC, the exponent was kept at one.

All of the stable experimental KV parameter sets (K_{KV}, B_{KV}) were used to find their equivalent HC parameter set $(\tilde{K}_{HC}, \tilde{B}_{HC}, \tilde{n}_{HC})$ that generate a similar force in the least mean square error sense. As shown in Fig. 1b, this was accomplished by simulating the uncoupled haptic interaction between the KV model and the QET model, in order to generate position, velocity and force needed to identify the HC model parameters $\tilde{K}_{HC}, \tilde{B}_{HC}$, and $\tilde{n}_{HC}$ [5, 6]. A dual process was repeated for the stable experimental K_{HC}, B_{HC} and $n_{HC}=1$, which were mapped to KV to find their equivalent KV values.

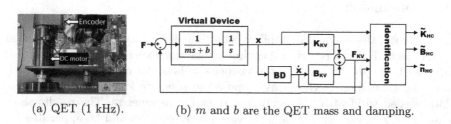

(a) QET (1 kHz).   (b) m and b are the QET mass and damping.

Fig. 1. QET and the process for mapping KV to HC parameters.

3   Results and Analysis

Figure 2.a-b show the experimental uncoupled stability KV range and their equivalent HC K and B values mapped from KV. Figure 2.c-d show the experimental uncoupled stability HC range and their equivalent KV K and B values.

The accuracy of the mapping from KV to HC and vise-versa are shown by the force %RMSE heat maps in Fig. 2.c,f, where $\%RMSE = 100\sqrt{\sum(F - \hat{F})^2}$ $/\sqrt{\sum F^2}$. The %RMSE ranges from 0 to 4. Comparing Fig. 2.a and 2.c, the KV model simulated a larger elastic range, whereas the HC model implements a larger viscous range. This is likely because for the HC model, as penetration reduces so does the coefficient of velocity, that is $B_{HC}x^n$. This results in the HC model being unable to maintain stability for highly elastic VEs.

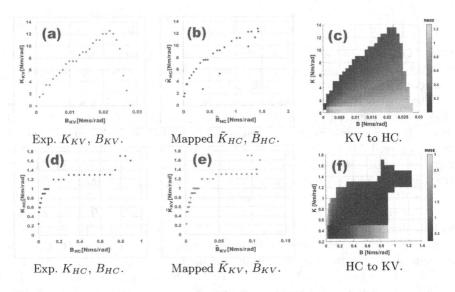

Fig. 2. Experimental and mapped KV, HC parameters and %RMSE heat maps.

4   Conclusion

In this work, the uncoupled stability conditions for a 1-DOF haptic simulation system using KV and HC virtual environments were experimentally determined. In order to be able to compare the viscoelastic behavior offered by each model, a virtual simulation method was proposed to generate kinetic and kinesthetic data needed to identify the equivalent dynamic parameters. The results show that the KV model offers a larger elastic range, which would be applicable for interactions with stiffer environments, and the HC model offers a larger viscous range, which would be applicable for interactions with compliant environments, such as soft tissue.

References

1. Colgate, J.E., Schenkel, G.G.: Passivity of a class of sampled-data systems: application to haptic interfaces. J. Rob. Sys. **14**(1), 37–47 (1997)
2. Gil, J., et al.: Stability analysis of a 1 DoF haptic interface using the Routh-Hurwitz criterion. IEEE Trans. Cont. Sys. Tech. **12**(4), 583–88 (2004)
3. Luna, V., Ozdil, P.G., Hashtrudi-Zaad, K.: Effect of direct velocity measurement on the stability of haptic simulation systems. Proc. IEEE Haptics Symp (2020)
4. Huang, Y., et al.: Contact dynamics of massage compliant robotic arm and its coupled stability. In 2014 IEEE International Conference on Robotics and Automation (ICRA), pp. 1499–1504 (2014)
5. Pecly, L., Souza, M.L.O., Hashtrudi-Zaad, K.: Offline and online synchronization of position derived signals for identification of dynamic systems. Proc. Latin Am. Rob. Symp. 1–6 (2017)
6. Haddadi, A., Hashtrudi-Zaad, K.: Real-time identification of hunt-crossley dynamic models of contact environments. IEEE Trans. Rob. **28**(3), 555–66 (2012)

Fashion Touch. Surface Haptics in Fashion E-commerce

Michela Ornati[1,2]([⊠]) [iD]

[1] Università Della Svizzera Italiana, 6900 Lugano, Switzerland
michela.ornati@usi.ch
[2] University of Applied Sciences and Arts of Southern Switzerland,
6928 Manno, Switzerland

Abstract. The sense of touch is an essential part of the fashion experience. In the digital domain, however, consumers cannot perceive tactile garment qualities as they would in a retail store. The acceleration of online commerce during Covid-19 has spotlighted this issue. Thus, surface haptic technologies which enrich visual and textual digital content with touch feedback might be of interest to the fashion sector. The author's research explores this possibility by engaging consumers and industry executives by using two surface haptic devices, TanvasTouch® and WeArt. This work-in-progress paper illustrates how the devices were leveraged for research and describes some preliminary findings.

Keywords: Touch · Haptic technologies · Fashion digital communication · e-commerce

1 Introduction

The sense of touch is central to the fashion and luxury experience: dress is experienced with and on the body. Actively touching and physically trying on a garment is an essential part of the customer journey. In the digital domain, however, fashion is hands-off – consumers cannot actively sense, perceive and understand tactile garment qualities in an e-commerce website as they would in a retail store. The lack of sensory inputs in the online domain has long held back luxury and fashion brands from pursuing e-commerce strategies. During the Covid-19 lockdowns, however, firms had to shift customer operations almost entirely to digital channels. Digital acceleration during the pandemic spotlighted issues connected to lack of embodiment in e-commerce, such as the unsustainable and costly phenomena of product returns.

In this scenario, technologies which enrich the audio-visual digital experience by adding touch feedback might be of interest in the fashion and luxury sector. The author's doctoral research [1, 2] explores such a possibility, focusing on surface haptics. The objective is threefold. First, to assess in what way the lack of physical interaction is currently addressed by fashion and luxury brands in the online domain [3]. Second, to investigate consumer reactions to the introduction of haptic feedback in a fashion e-commerce context [4]; and third, to explore executives' opinion on the role

H. Seifi et al. (Eds.): EuroHaptics 2022, LNCS 13235, pp. 464–467, 2022.
https://doi.org/10.1007/978-3-031-06249-0

these technologies may play in addressing the sensory limits of the online experience. To pursue these last two objectives the author adopted a pragmatic and qualitative research design which includes the use of two surface haptic devices: TanvasTouch® (www.tanvas.co) and the WeArt (www.weart.it) Touchkey. The haptic feedback experience enabled by the devices is partially customized and the devices are used as prompts within focus groups and in-depth interviews, as illustrated in the next sections.

2 FashionTouch E-commerce Haptic Simulation

As described above, the research design required simulating a fashion e-commerce website on the TanvasTouch and WeArt surface haptic devices. To create the mock website's content from scratch, five women's and three men's garments (including a knitted cotton top, corduroy pants, a woolen houndstooth skirt and a jeans jacket) were purposefully selected based on diverse material and surface characteristics, then the garments were photographed on real models. Subsequently, a graphic designer created a home page showcasing the eight garments, plus a product page for each garment. The latter featured five different views of each garment and a detailed image of the corresponding material. To enhance these material images with haptic effects each device required a different approach, as detailed in the following paragraphs.

TanvasTouch enables precise fingertip tracking and simultaneous surface haptic rendering. It can be programmed to accurately deliver real-time variable-intensity friction and electrostatic haptic feedback within a specific area of the touchscreen. In collaboration with colleagues working in the USI eLab[1], the zoom-in image of each garment material was rendered in black and white, enhanced and uploaded to the software environment of the device, where surface texture characteristics were matched with modulated, ad-hoc haptic feedback effects. The finished, full-color mock website – entitled *FashionTouch* – was uploaded locally on a personal computer and simultaneously displayed on the paired TanvasTouch display. Users navigating the website and accessing a product page can click onto any garment image to access the zoom-in of its material. Then, by stroking the display screen, they can experience a simulation of the material's surface effects[2].

WeArt currently features a wearable haptic system which reproduces tactile cues (forces, textures, and temperature changes) on the wearer's skin in virtual reality environments [5]. However, in this study an earlier, non-wearable Touchkey was used which similarly features incorporated force feedback, texture-based vibrations, and thermal cues. The haptic feedback effects for the WeArt interaction were recorded in the supplier's laboratories directly from the original garments and synched with a graphical video rendition of the stroking gesture. The mock website and the interaction simulation for each garment were made accessible on an Apple iPad (using a TestFlight

---

[1] https://www.usi.ch/en/university/info/elab.

[2] See: https://youtu.be/NEtf1d53eZ8.

application) and paired to the WeArt Touchkey via bluetooth. Thus, when users place a finger on the Touchkey with one hand and choose a garment on the iPad screen with the other, they simultaneously see a pointer moving across the material and feel the corresponding force-feedback effects under their fingertips[3].

3　Early Research Results and Discussion

The author has leveraged the *FashionTouch* experience on TanvasTouch and on WeArt as a prompt in focus groups and in-depth expert interviews, in pursuit of the second and third research objectives described in Sect. 2 above. In other words, the devices are not used to evaluate the technologies per se, nor their maturity for the luxury and fashion sector, but to stimulate research participants' thoughts and opinions on the possibility and potential value of enhanced surface interaction in an e-commerce brand setting. Participants in the first focus group – conducted in late 2019 using TanvasTouch – expressed interest in the future of sensory enrichment via haptic technologies, but held reservations regarding their haptic experience. These included feeling constrained by the flat, two-dimensional surface – which limits the gestures one would habitually adopt to handle textiles (e.g., stroking vs. grasping) – as well as being unable to adequately distinguish different garment materials [6]. In the course of 2021, the author conducted fourteen expert interviews with luxury and fashion digital marketing executives in Italy and Switzerland, using both TanvasTouch and the WeArt Touchkey. While interacting with the technologies, experts expressed reservations similar to those of focus group participants. However, early research insights suggest decision-makers are very attentive to any haptic technology development which might enhance the customer experience – either in digital or phygital (retail) contexts – underscoring the relevance of haptic research for the fashion and luxury industry.

References

1. Ornati, M.: Touching the cloth: haptics in fashion digital communication. In: Kalbaska, N., Sádaba, T., Cominelli, F., Cantoni, L. (eds.) FACTUM 2019, pp. 254–258. Springer, Cham (2019). https://doi.org/10.1007/978-3-030-15436-3_23
2. Ornati, M.: A true feel: re-embodying the touch sense in the digital fashion experience. In: Cinque, T., Vincent, J.B. (eds.) Materializing Digital Futures: Touch, Movement, Sound and Vision, pp. 205–222. Bloomsbury Academic, New York (2022)
3. Ornati, M.: Touch in text. The communication of tactility in fashion e-commerce garment descriptions. In: Sádaba, T., Kalbaska, N., Cominelli, F., Cantoni, L., Torregrosa Puig, M. (eds.) Fashion Communication, pp. 29–40. Springer, Cham (2021). https://doi.org/10.1007/978-3-030-81321-5_3

---

[3] See: https://youtu.be/_wt_6IG-NU8

4. Ornati, M., Cantoni, L.: fashiontouch in e-commerce: an exploratory study of surface haptic interaction experiences. In: Nah, F.-H., Siau, K. (eds.) HCII 2020. LNCS, vol. 12204, pp. 493–503. Springer, Cham (2020). https://doi.org/10.1007/978-3-030-50341-3_37
5. Gioioso, G., Pozzi, M., Aurilio, M., Peccerillo, B., Spagnoletti, G., Prattichizzo, D.: Using wearable haptics for thermal discrimination in virtual reality scenarios. In: Kajimoto, H., Lee, D., Kim, S.-Y., Konyo, M., Kyung, K.-U. (eds.) AsiaHaptics 2018. LNEE, vol. 535, pp. 144–148. Springer, Singapore (2019). https://doi.org/10.1007/978-981-13-3194-7_32

The Influence of Aging on Perceptual Grouping in Haptic Search

K. E. Overvliet[✉]

Utrecht University, Experimental Psychology and Helmholtz Institute,
Utrecht, The Netherlands

Abstract. Perceptual grouping speeds up haptic search. This has particularly been shown for grouping of distractors by similarity and good continuation [1]. Here, we investigated the effect of aging on grouping in haptic search. We reasoned that because older adults have less cognitive resources available for processing perceptual information, they would benefit more from grouping as compared to younger adults. We tested this hypothesis in a haptic search task in which proximity, similarity and good continuation of the distractors were manipulated. We found that older adults indeed show a larger effect of distractor similarity on search times as compared to younger adults, where similar distractors were processed faster than dissimilar distractors. However, older adults showed an opposite effect of grouping by proximity, where items that were further apart were processed faster. This may be caused by a strong bowed spatial position effect in older adults: stimuli that are closer to each other are more difficult to discriminate. We conclude that haptic perceptual grouping by similarity has larger benefits in elderly as compared to younger adults.

Keywords: Haptic perception · Aging · Perceptual grouping

1 Introduction

A well-known phenomenon in the process of aging is the decrease in cognitive resources that are available for perception, action, and other processes [2]. One strategy limiting the amount of cognitive resources that are needed is just doing one task simultaneously, another possibility may be to more efficiently use those resources. The aim of the current study is to investigate whether perceptual grouping reduces the amount of cognitive resources needed for haptic perception in older adults.

When only limited cognitive resources are needed to perform a certain task, age differences tend to be small, while when a task is complex, age differences arise [3]. This also holds in haptic 2D shape recognition: age differences in recognizing tangible line drawings of simple shapes were much smaller as compared to more complex representations of everyday objects, not only in terms of reaction times but also in accuracy [4].

In the current study we adopted the paradigm as used in the Overvliet et al. [1] study, where we show that perceptual grouping speeds up haptic search. If grouping reduces the amount of cognitive resources needed to perform such perceptual tasks, we

H. Seifi et al. (Eds.): EuroHaptics 2022, LNCS 13235, pp. 468–471, 2022.
https://doi.org/10.1007/978-3-031-06249-0

expect older adults to benefit more from perceptual grouping as compared to younger adults.

2   Method

2.1   Participants

Twelve younger volunteers (mean age 24.67 ± 3.82, 1 left-handed, 4 males, 8 females) and 12 older volunteers (mean age 74.75 ± 3.67, 1 left-handed, 5 males, 7 females) from the university community were paid 8 euros per hour for their participation. The study was approved by the Medical Ethical Committee of the University Hospital Gasthuisberg (Leuven), and the participants gave written informed consent before starting the experiment. All participants had normal or corrected to normal vision but were blindfolded during the experiment.

2.2   Setup and Procedure

The task for the participant was to search for a target by moving two of their fingers down two columns of ten horizontal and vertical tangible line segments (length:1.5 cm, width: 1.4 mm, height: about 1 mm). A target was defined as a pair (one item from each column) that was different from the distractor pairs. The target could be located between position two and ten, counted from the top. We varied similarity, good continuity, and proximity of the distractor pairs (see legend of Fig. 1). We varied similarity by having similar distractor pairs and a dissimilar target pair (similar) or vice versa (dissimilar). For good continuity, we varied whether the distractors were aligned (vertical) or not-aligned (horizontal; in the similar conditions only). Lastly, we varied proximity in two ways: For spatial proximity (Experiment 1A) we placed the items close or far apart and for somatotopic proximity (Experiment 1B) the pairs of fingers used was varied: fingers from the same hand (near) or from opposite hands (far).

3   Results

Exploration time for different target positions was standardized by fitting a regression line through the data and the slopes of the different conditions were compared (Fig. 1). We ran a mixed-model ANOVA with factors age group (young vs. old), proximity (near vs. far) and similarity (different, same vertical, same horizontal). We specified contrasts to test the effect of similarity (different vs. same horizontal and same vertical) and good continuity (same horizontal vs. same vertical).

For Experiment 1A, we found significant main effects for age group F $(1,22) = 24.72$, $p < .0001$, $\eta_p^2 = .53$, spatial proximity, $F(1,22) = 6.26$, $p < .05$, $\eta_p^2 = .22$, and similarity $(F(2,44) = 90.27, p < .0001, \eta_p{}^2 = .80)$. Moreover, we found significant effects for the same-different and the good continuity contrasts $f(F(1,22) = 135.85$, $p < .0001$, $\eta_p^2 = .86$ and $F(1,22) = 8.70$, $p < .01$, $\eta_p^2 = .28$ respectively). For Experiment 1B, we found main effects for age group F

$(1,22) = 53.21$, $p < .0001$, $\eta_p^2 = .71$,   somatotopic   proximity $F(1,22) = 14.88$, $p < .001$, $\eta_p^2 = .40$ and similarity $(F(2,44) = 75.51, p < .0001, \eta_p^2 = .77)$. We again found an effect for the same-different and the good continuity contrasts $(F(1,22) = 7.53$, $p < .05$, $\eta_p^2 = .26$ and $F(1,22) = 111.93$, $p < .0001$, $\eta_p^2 = .84$ respectively).

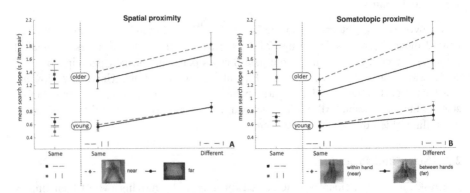

Fig. 1. The mean slopes of the regression lines for spatial proximity condition (A) and somatotopic proximity condition (B). Error bars indicate the standard errors of the mean over participants. The left panels in both figures show the good continuation manipulation. The pictures in the legend show the stimulus materials and proximity manipulations.

4   Discussion

We found that older adults indeed show a larger effect of distractor similarity on search times as compared to younger adults, where similar distractors were processed faster than dissimilar distractors. However, older adults showed an opposite effect of grouping by spatial proximity, where items that were further apart were processed faster. This may be caused by a strong bowed spatial position effect in older adults: stimuli that are closer to each other are more difficult to discriminate [5]. Age-related decline in tactile sensitivity and processing speed may explain general slower exploration by older adults [2], but it does not explain larger grouping effects. We conclude that haptic perceptual grouping by similarity has larger benefits in elderly as compared to younger adults.

References

1. Overvliet, K.E., Krampe, R.T., Wagemans, J.: Perceptual grouping in haptic search: the influence of proximity, similarity, and good continuation. J. Exp. Psychol. Hum. Percept. Perform. **38**(4), 817–821 (2012)
2. Salthouse, T.A.: The processing-speed theory of adult age differences in cognition. Psychol. Rev. **103**(3), 403–428 (1996)

3. Gick, M.L., Craik, F.I.M., Morris, R.G.: Task complexity and age-differences in working memory. Mem. Cognit. **16**(4), 353–361 (1988). https://doi.org/10.3758/BF03197046
4. Overvliet, K.E., Wagemans, J., Krampe, R.T.: The effects of aging on haptic 2D shape recognition. Psychol. Aging **28**(4), 1057–1069 (2013)
5. Adam, J.J., van Boxtel, M.P.J., Houx, P.J., van Gerven, P.W.M., Jolles, J.: Perceptual and motor factors mediate the bowed spatial position effect in ageing. Eur. J. Cogn. Psychol. **18** (5), 673–685 (2006)

Investigating Haptic Feedback and Arousal for Motor Skill Training in Virtual Reality

Unnikrishnan Radhakrishnan[✉](ID), Konstantinos Koumaditis(ID), and Francesco Chinello(ID)

Department of Business Development and Technology (xR2 Lab), Aarhus University, Aarhus, Denmark
{unnik,kkoumaditis,chinello}@btech.au.dk

Abstract. In this paper, we discuss the results of an experiment comparing physiological arousal levels between participants performing a motor skill task in immersive VR and physical conditions and future directions for investigating the relationship between haptic feedback modality, arousal, and training effectiveness.

Keywords: Haptic feedback · Physiological arousal · Motor skill training · Virtual reality

1 Introduction

Immersive Virtual Reality (IVR) based training is gaining acceptance across industries. Haptic feedback could play an important role in making IVR training more effective, especially when it comes to motor skills [4]. Physiological arousal levels might play a role in affecting training performance, for example, the hypothesized inverted U-shaped relation between performance and arousal [1] which has been validated in literature [2]. A systematic literature review on the use of haptics in conjunction with IVR for industrial skills [3] found that 35% of publications used at least one of the following haptic modalities - portable, grounded, pseudo-haptics, and wearable haptics. However, though 14% of the publications used biosensors for measuring physiological arousal, there were none that investigated the relationship between arousal and haptics technologies in motor skill training use cases. Though the intersection of haptics and physiological arousal has been investigated in some studies [5, 6], there has been no major exploration of its application in the motor skill training in immersive VR.

2 Methods

The popular Buzz Wire (also known as 'Wire Loop') game was adapted as a motor skill training task. The objective of the task is for the participant to move a loop across a wire from beginning to end without making contact (see Fig. 1). A previous pilot study discusses the design of the physical and IVR setups used in this experiment [7]. Two physiological measures of arousal are

© The Author(s) 2022
H. Seifi et al. (Eds.): EuroHaptics 2022, LNCS 13235, pp. 472–474, 2022.
https://doi.org/10.1007/978-3-031-06249-0

Fig. 1. Physical (left) and Immersive VR (right) conditions

recorded while the task is performed - Electrodermal Activity (EDA) measured from the fingers using the Shimmer GSR+, and Heart Rate Variability (HRV) data measured using the Polar H10 ECG strap worn on the chest. In EDA, an increase in arousal leads to an increase in electrical conductivity of the skin as measured by Skin Conductance (SC) level and the peak amplitude of the Skin Conductance Response (SCR) [8]. In HRV, changes in arousal levels are indicated by variation in heartbeats [8]. Popular measures of HRV include time-domain metrics like Mean IBI (Inter-beat Interval), RMSSD (Root Mean Square of Successive Difference), SDNN (Standard Deviation of NN Intervals), and frequency domain measures like HFN/LFN (Normalized High/Low Frequency Component) and LF/HF Ratio [8].

Participants were randomly divided into two conditions - IVR (N = 33) where the task was performed using the Oculus Quest (Rift mode) and physical condition (N = 39) where the task was performed using a physical setup. In both conditions, the participants received feedback when contact is made between the loop and the wire, in the form of an 'augmented' haptic vibration from the quest controller, an LED light turning on behind the wire, and a beep sound played over headphones.

Table 1. Arousal Metrics (Means) in IVR and Physical Conditions. ‡ - baseline corrected values, + - Significance at (α=0.10), * - Significance at (α=0.05)

Arousal metric	Source	VR	Physical	p-value
SC Level ‡	EDA	2.28	2.68	0.1108
SCR Amplitude	EDA	0.16	0.23	0.2344
Heart Rate ‡	HRV	−1.45	0.58	**0.056+**
IBI ‡	HRV	16.36	4.14	0.2864
RMSSD ‡	HRV	−31.64	−32.89	0.1598
SDNN ‡	HRV	−27.22	−28.25	0.3456
HFN ‡	HRV	9.94	5.59	0.2149
LFN ‡	HRV	45.21	54.58	**0.0135***
LF/HF Ratio ‡	HRV	−1.19	−0.85	0.2765

3 Results and Future Work

A 10-second post-contact window for EDA and HRV signals was extracted for each contact recorded in the system. Arousal metrics were then averaged across all contacts per participant for statistical analysis. Table 1 shows the results of the non-parametric independent sample (Mann-Whitney U) tests performed to compare arousal levels between participants in the two conditions. Participants in the physical condition showed significantly more arousal in terms of Heart Rate ($\alpha=0.10$) and Low Frequency (HRV) signals ($\alpha=0.05$) in addition to the near significance trend in the SC level between conditions.

The differences in arousal observed in the experiment might be explained by the differences inherent in the medium of training, i.e. a difference in the medium (IVR vs physical), or perhaps due to the presence of a real wire in the physical condition providing an additional 'natural' haptic feedback. These limitations may be addressed by designing the haptic feedback in the physical condition to better mimic the vibratactile feel provided in the IVR condition, or by limiting the medium to IVR alone. Future experiments can then vary the haptic modality (portable, grounded, etc.) and then study its relation to arousal and the effectiveness of training.

References

1. Yerkes, R.M., Dodson, J.D.: The relation of strength of stimulus to rapidity of habit-formation, 459–482 (1908)
2. Quick, J.A., Bukoski, A.D., Doty, J., Bennett, B.J., Crane, M., Barnes, S.L.: Objective measurement of clinical competency in surgical education using electrodermal activity. J. Surg. Edu. **74**(4), 674–680 (2017)
3. Radhakrishnan, U., Koumaditis, K., Chinello, F.: A systematic review of immersive virtual reality for industrial skills training. Behav. Inf. Technol. **40**(12), 1310–1339 (2021)
4. Zhou, M., Tse, S., Derevianko, A., Jones, D.B., Schwaitzberg, S.D., Cao, C.G.L.: Effect of haptic feedback in laparoscopic surgery skill acquisition. Surg. Endosc. **26**(4), 1128–1134 (2012)
5. Krogmeier, C., Mousas, C., Whittinghill, D.: Human-virtual character interaction: toward understanding the influence of haptic feedback. Comput. Anim. Virtual Worlds **30**(3–4), e1883 (2019)
6. Giroux, F., et al.: Haptic stimulation with high fidelity vibro-kinetic technology psychophysiologically enhances seated active music listening experience. In: 2019 IEEE World Haptics Conference (WHC), pp. 151–156. IEEE (2019)
7. Radhakrishnan, U., Blindu, A., Chinello, F., Koumaditis, K.: Investigating motor skill training and user arousal levels in VR: pilot study and observations. In: 2021 IEEE Conference on Virtual Reality and 3D User Interfaces Abstracts and Workshops (VRW), pp. 625–626. IEEE (2021)
8. Cacioppo, J.T., Tassinary, L.G., Berntson, G., et al.: Handbook of psychophysiology. Cambridge University Press (2007)

Safe Movement for the Blind Using Localised Haptic Stimulation

Jaafar Rammal$^{(\boxtimes)}$ and Robert Spence$^{(\boxtimes)}$

Imperial College London, Exhibition Road, London SW7 2AZ, England
jr4918@ic.ac.uk, r.spence@imperial.ac.uk
https://jaafarrammal.com, https://www.imperial.ac.uk/people/r.spence

Abstract. Blind and visually impaired people have understandable difficulty in establishing, with an acceptable level of confidence, a safe direction in which to move in an environment populated with obstacles such as people, furniture, and buildings [2].

The solution being explored is a system that automatically computes a safe movement direction and seamlessly communicates that movement to the user using a wearable device. This paper focuses on how the wearable device - a head band - communicates safe movements to the user. The device is being tested with volunteers to evaluate its reliability, effectiveness, and other parameters.

Keywords: Haptics · Blind · Communication

1 Existing Solutions

A blind or visually impaired person can establish a safe movement following different common approaches. In one approach, the user has the demanding tasks of forming a mental model of their environment by sampling it in some way, and then of making a personal estimate of a safe direction in which to move using that model. The cognitive demands placed upon a user by this approach should not be underestimated. This approach places huge cognitive demands upon a user and can lead to less than 100% confidence in the decision made [5]. A simple example of this approach is provided by the familiar white cane.

By contrast, simple examples of another approach, which we are investigating, include the familiar guide dog, or the arm of a companion. Both of these examples involve very little cognitive effort for the blind user as well as enhanced confidence in movements compared with the first approach.

2 Our Proposal

The approach we are investigating requires the safe movement to be communicated to a user by means of a wearable device. Components that have earlier been explored with this purpose in mind include vibrating [4, 1] and electrotactile [3] interfaces. In our case the device is a vibrating head band using haptic stimulation to communicate the direction of a safe movement to the user.

© The Author(s) 2022
H. Seifi et al. (Eds.): EuroHaptics 2022, LNCS 13235, pp. 475–477, 2022.
https://doi.org/10.1007/978-3-031-06249-0

2.1 The Device - Head Band

Such a device is expected to be attractive for many reasons in addition to its important replacement of cognitive effort by haptic perception: it is unobtrusive, non-invasive (e.g., does not interfere with user's hearing), and is expected to be economical to manufacture (using e.g., 3D printing) as seen in Fig. 1.

Fig. 1. Headband with vibrators (left) and 3D printed prototype worn by a user.

We would not be surprised if the user chose to additionally carry the conventional white cane, partly to increase confidence in the safety of the suggested movement, and for passers-by to realise the user is blind or visually impaired.

2.2 User Experience - Localised Haptic Stimulation

Irrespective of how a safe direction of movement is automatically estimated, a crucial component of our system is one which signals, to the user, a recommended change in their direction of movement. Our proposed 'head band' contains vibrating motors, one of which will, at any one time, indicate a head movement associated with a safe direction in which the user can move. The user's instantaneous perception of a vibration is expected to be followed by turning the head until it is pointing in the suggested direction, as seen in Fig. 2. Freedom of the user to rotate their head while walking is not compromised, as the head band will still indicate the correct safe direction relative to the head's rotation.

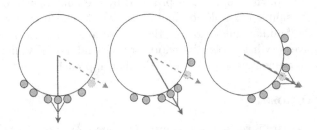

Fig. 2. Top view of a user (big circle) with the head-band's vibrating motors (small circles). The motors' state (light color = vibrating), the user's head direction (full arrow), and the safe movement direction (dotted arrow) are also shown

The vibration protocol of the motors is meant to encode minimal yet sufficient information. The protocol is characterized with an activation of the vibration only when required with (1) an alert signal (e.g., two quick vibrations), (2) a pulsed, continuous vibration indicating the safe direction, and (3) a completion signal. There would be no vibrations otherwise unless requested by the user to check if the system is operational. Finally, the vibrations' frequency is proportional to the magnitude of the movement change and the proximity of obstacles.

3   Evaluation

The evaluation, initially exploratory with ten users, will focus on the wearable's effectivness in communicating safe movements, hence other navigation elements (locations of user, destination, and obstacles) will be assumed. The user will navigate around the obstacles to reach the destination, and the device will update the safe direction only when an obstacle is within the radius of change (Fig. 3). This allows us to measure the device's effectiveness when both simple and abrupt direction changes are required. The evaluation will first involve sighted users to validate the device's reliability and performance, before validating the experience with visually impaired people. University ethical guidelines will be followed.

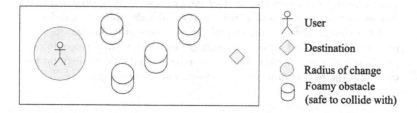

Fig. 3. Testing framework to evaluate the device's effectiveness.

References

1. Booth, J.: The Sixth Sense - Electronically Augmented Perception. Master's thesis, Imperial College London (2017)
2. Golledge, R.G.: Geography and the disabled: a survey with special reference to vision impaired and blind populations. Trans. Inst. Brit. Geogr. **18**(1), 63–85 (1993). http://www.jstor.org/stable/623069
3. Kajimoto, H., Kanno, Y., Tachi, S.: Forehead electro-tactile display for vision substitution. In: Proc. EuroHaptics, p. 11. Citeseer (2006)
4. Katzschmann, R.K., Araki, B., Rus, D.: Safe local navigation for visually impaired users with a time-of-flight and haptic feedback device. IEEE Trans. Neural Syst. Rehabil. Eng. **26**(3), 583–593 (2018). https://doi.org/10.1109/TNSRE.2018.2800665
5. Tversky, B.: Cognitive maps, cognitive collages, and spatial mental models. In: Frank, A.U., Campari, I. (eds.) COSIT 1993. LNCS, vol. 716, pp. 14–24. Springer, Heidelberg (1993). https://doi.org/10.1007/3-540-57207-4_2

An Immersive Virtual Paleontology Application

Lucas Rodrigues[(✉)], John Nyakatura, Stefan Zachow,
and Johann Habakuk Israel

Cluster of Excellence "Matters of Activity. Image Space Material", Berlin, Germany
lucas.siqueira.rodrigues@hu-berlin.de

Abstract. Virtual paleontology studies digital fossils through data analysis and visualization systems. The discipline is growing in relevance for the evident advantages of non-destructive imaging techniques over traditional paleontological methods, and it has made significant advancements during the last few decades. However, virtual paleontology still faces a number of technological challenges, amongst which are interaction shortcomings of image segmentation applications. Whereas automated segmentation methods are seldom applicable to fossil datasets, manual exploration of these specimens is extremely time-consuming as it impractically delves into three-dimensional data through two-dimensional visualization and interaction means. This paper presents an application that employs virtual reality and haptics to virtual paleontology in order to evolve its interaction paradigms and address some of its limitations. We provide a brief overview of the challenges faced by virtual paleontology practitioners, a description of our immersive virtual paleontology prototype, and the results of a heuristic evaluation of our design.

Keywords: Haptics · Virtual reality · Virtual paleontology

1 Introduction

Virtual paleontology is the study of digital fossils through the use of three-dimensional visualization systems, comprising a powerful tool set for the analysis of paleontological data [2]. The field is growing in relevance due to the clear advantages of non-destructive exploration of digital representations of fossils [6]. Despite its advancements, virtual paleontology still faces technological challenges that constrain it as "an expensive and time-consuming undertaking" [8]. Researchers strive to partition fossil data into meaningful regions, as automated segmentation methods are seldom applicable to paleontological datasets [2]. Virtual Paleontology currently relies on semi-automated image segmentation, which obliges researchers to perform tedious and laborious manual processes [8]. Paleontological image datasets are difficult to explore manually due to their low attenuation contrast [2]. Additionally, virtual paleontologists must segment volumes through two-dimensional slices, which is inappropriate for their tasks [3].

© The Author(s) 2022
H. Seifi et al. (Eds.): EuroHaptics 2022, LNCS 13235, pp. 478–481, 2022.
https://doi.org/10.1007/978-3-031-06249-0

2 Methodology

Application Design: We propose a novel design that aspires to evolve the virtual paleontology paradigm and overcome its limitations. Differently from the common additive image segmentation approach, our prototype enables users to subtractively carve away undesirable artifacts and formations surrounding objects of interest within volumes. We emulate some processes commonly employed by fossil preparators working with physical specimens, as we aspire to make the segmentation process more intuitive for paleontology researchers. We designed a real-time interactive visualization of a paleontological image dataset, which is presented within a virtual preparation lab scene rendered by a head-mounted display. Our prototype leverages haptics to enable researchers to establish better cognitive models of image datasets' physical structures [4]. Our simulation uses voxel intensity data to calculate the spring forces yielded by collisions between a virtual probe and the image dataset's surface. Spring forces repel the probe from voxel locations proportionally to the product of gain and distance between the positions of affected voxels and the probe using the following equation [1]:

$$F = k(P - X) \tag{1}$$

Spring force (F) is a function of gain (k) and voxel position (P) minus probe position (X). Gain assumes voxel intensity values and invokes OpenHaptics to translate these forces to a Phantom device and instruct it to produce mechanical movements opposing the forces applied by users onto the image dataset volume.

Evaluation: We assessed our prototype through Nielsen's Heuristic Evaluation [5]. This method was chosen because virtual paleontology image segmentation is a highly-specific use case, and proposing a design that aims to change its paradigms required timely feedback from subject-matter and usability experts. Following Nielsen's recommendations, three experts evaluated our prototype: E1 is a tenured professor specialized in functional morphology and biomechanics from a visual studies perspective. E2 leads a research group focused on visual and data-centric computing. E3 is a doctoral researcher specialized in virtual reality and human-computer interaction. We have chosen usability principles described in *Heuristic evaluation of virtual reality applications* [7]. The evaluation sessions involved the segmentation of a paleontological CT scan using a 3DSystems Phantom Touch force-feedback device. Sessions lasted around 60 min.

3 Results

Compatibility with the user's task and domain:
Our haptic feedback rendering was deemed inappropriate for the segmentation task. E1 suggested generating forces based on edge detection to prevent accidental segmentation of neighboring areas. E2 considered our haptic rendering too abrupt for a precision task. E3 suggested basing force calculations on gradients instead of intensity values. E1 and E2 found the haptic feedback too coarse for precise contouring.

Natural expression of movement:
This heuristic has been breached, as hand movement is restricted by our device's movement range. E1 believes that allowing users to move the dataset volume could overcome movement limitations.
Close coordination of action and representation.
E1 and E3 stated that this heuristic was breached because they experienced delays in haptic feedback rendering, which was not present in the visual representation.
Realistic feedback:
E1 and E2 considered that haptic feedback did not match real-world expectations. E2 added that our simulation is generating discrete forces, whereas continuous forces would create a stable tactile experience.
Support for learning:
E1 found it unclear that the probe's tooltip could constantly modify the volume, and he recommended linking activation to the haptic pen's button as a metaphor to the power button on an air scribe.

4   Conclusions

We developed an immersive virtual paleontology prototype that translates voxel intensity values to haptic forces. A heuristic evaluation has allowed us to detect important issues during the early stages of our user-centered design process. Most reported issues were related to incompatibility with users' tasks and domains, as our prototype aims to change an image segmentation paradigm for the virtual paleontology discipline. Evaluations have taught us valuable lessons on the discipline's workflows and how our prototype needs to be adapted to become useful to virtual paleontologists. Other heuristics covered general user interface issues in our immersive approach to image segmentation. Our next steps will be to address the issues found in our prototype and to conduct other rounds of heuristic evaluations, then we will proceed to conduct usability testing.[1]

References

1. Chan, L.S.H., Choi, K.S.: Integrating physX and OpenHaptics: efficient force feedback generation using physics engine and haptic devices. In: 2009 Joint Conferences on Pervasive Computing (JCPC), pp. 853–858. IEEE (2009)
2. Cunningham, J.A., Rahman, I.A., Lautenschlager, S., Rayfield, E.J., Donoghue, P.C.: A virtual world of paleontology. Trends. Ecol. Evol. **29**(6), 347–357 (2014)
3. van Dam, A., Herndon, K., Gleicher, M.: The challenges of 3D interaction, p. 469 (10 1994). https://doi.org/10.1145/259963.260500

---

[1] The author acknowledges the support of the Cluster of Excellence Matters of Activity. Image Space Material funded by the Deutsche Forschungsgemeinschaft (DFG, German Research Foundation) under Germany's Excellence Strategy – EXC 2025 – 390648296.

4. Massie, T.H., Salisbury, J.K., et al.: The phantom haptic interface: a device for probing virtual objects. In: Proceedings of the ASME Winter Annual Meeting. vol. 55, pp. 295–300. Chicago, IL (1994)
5. Nielsen, J., Molich, R.: Heuristic evaluation of user interfaces. In: Proceedings of the SIGCHI Conference on Human Factors in Computing Systems, pp. 249–256. CHI 1990, Association for Computing Machinery, New York, NY, USA (1990)
6. Racicot, R.: Fossil secrets revealed: X-ray CT scanning and applications in paleontology. Paleontol. Soc. Pap. **22**, 21–38 (2016)
7. Sutcliffe, A., Gault, B.: Heuristic evaluation of virtual reality applications. Interact. Comput. **16**(4), 831–849 (2004)
8. Sutton, M., Rahman, I., Garwood, R.: Virtual paleontology–an overview. Paleontol. Soc. Pap. **22**, 1–20 (2016)

Mapping of Vibrotactile Stimuli into a Perceptual Space

Rubén Rodriguez[1], Santiago Cuevas[1], Jose Lobera[2],
Damián Hernández Lahme[1,2], Juan José Zárate[3],
and Inés Samengo[1,2(✉)]

[1] Centro Atómico Bariloche, Instituto Balseiro, San Carlos de Bariloche,
Argentina
ines.samengo@gmail.com
[2] Department of Medical Physics of Centro Atómico Bariloche, and CONICET,
San Carlos de Bariloche, Argentina
[3] Department of Computer Science of ETH Zurich, Zürich, Switzerland

Abstract. To transmit tactile information with a discrete set of symbols, the alphabet of the code has to pack as many stimuli as can be reliably discriminated. Here we introduce a notion of perceptual space to construct such a code, defined in terms of the probabilities that pairs of neighboring stimuli are perceived as equal. We perform discrimination experiments exhibiting the well-known effects described by Weber's law, and identify the physical properties of vibrotactile stimuli that most efficiently modulate perceptual discriminability. We conclude that widely used libraries of vibration profiles contain stimuli that are far from homogeneously discriminable. We present our preliminary results as a starting point for the construction of optimal codes.

Keywords: Code · Discriminability · Haptic library

1 Introduction

The degree of similarity with which two stimuli are perceived is a complex function of the physical parameters of the stimuli [1, 2]. Yet, an optimal vibrotactile code should make use of an alphabet that contains as many elementary stimuli as can be reliably discriminated, using all the available stimulus space. We are interested in developing systematic methods to achieve optimal alphabets. Here we present preliminary work aimed to construct a *perceptual space*. We derive a transformation that maps a representation of the stimuli in terms of their physical parameters into a perceptual space, where the Euclidean distance between pairs of stimuli is proportional to their discriminability. To this end, we define a notion of perceptual distance between stimuli that can be obtained from discrimination tests performed with a subset of the effects produced by a widely used haptic controller. We show that the discriminability varies widely across the space, underscoring the importance of the perceptual perspective. Our preliminary work opens the door for a more general approach to designing efficient vibrotactile codes.

H. Seifi et al. (Eds.): EuroHaptics 2022, LNCS 13235, pp. 482–485, 2022.
https://doi.org/10.1007/978-3-031-06249-0

2   Methods

The widely-used haptic controller DRV2605L from Texas Instrument provides 123 predefined stimuli parsed into categories such as "click", "buzz", "ramps", etc. (two examples in Fig. 1B). We used this library to characterize the perceptual space.

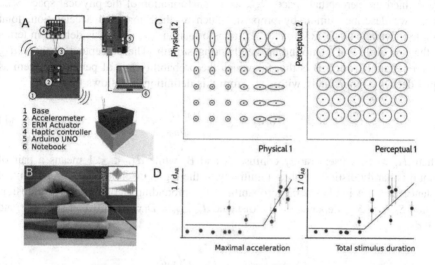

Fig. 1. Experimental paradigm of the (A) mechanical and (B) perceptual measurements. (C) Each ellipse represents the region of the space of stimuli that is confounded with the central stimulus. Ellipses vary from point to point in S_{phs}, and become homogeneous and isotropic in S_{per}. (D) Dependence of the population averaged JND (proportional to the inverse of the perceptual distance) with two physical properties of stimuli. Lines: thresholded Weber's law.

Physical Tests: we measured the acceleration produced by an ERM for each stimulus of the library. Categories could be differentiated by the shape of the vibration's envelope. For example, while the envelope of *Clicks* rises abruptly and then decays almost immediately, that of *Buzzes* remains fairly constant throughout the duration of the effect. Different stimuli in one category varied in intensity or duration. Acceleration data were collected as in [3]. We used a 3D printed base supported by a sponge, to which the vibrotactile actuator was attached (Pololu-1638, shaftless vibration motor 10×2.0 mm) together with an accelerometer (MPU-6050) as in Fig. 1A top.

Perceptual Tests: Here we report the results of work-in-progress obtained over a subset of 20 stimuli of the library that were evaluated by 20 participants. Each trial consisted of a pair of stimuli, A and B, presented in rapid succession to participants, who were instructed to decide whether they perceived them as equal or not (Fig. 1B). A randomized half of the trials contained two equal stimuli, whereas the other half, two different ones. Each pair of different stimuli was tested 10 times per subject, and the stimuli conforming a pair were first neighbors in physical space.

Perceptual distances are estimated from the probability p_{AB} of perceiving stimuli A and B as equal, estimated with the Laplace estimator from the sampled responses.

3 Results, Discussion and Future Work

We defined the perceptual space, S_{per}, as a transformation of the physical space S_{phy}, where we describe stimuli by properties such as their maximal acceleration, total duration, average slope, RMS, etc. The transformation $S_{phy} \rightarrow S_{per}$ is defined in terms of the probability of confounding neighboring stimuli. The perceptual distance d_{AB} between stimuli A and B is that for which the probability p_{AB} of perceiving them as equal decays exponentially with d_{AB}. From this definition, it follows that

$$dAB = \frac{1}{2} \log \frac{p_{AA}p_{BB}}{p_{AB}p_{BA}} \tag{1}$$

When $d_{AB} >> 1$, users rarely confuse A and B, while $d_{AB} << 1$ means a pair of stimuli is hardly distinguishable. Intuitively, if the log is taken in base 2, a perceptual distance of 1 *bit* implies a 0.5 probability of confounding the stimuli. The transformation $S_{phy} \rightarrow S_{per}$ depends on the distance $d_{x, \; x+dx} = D(x) \, dx$ between stimuli x and $x + dx$,

$$S_{per}(S_{phy}) = \int_{s_0}^{s_{phy}} D(x)dx \tag{2}$$

Figure 1C shows the transformation of inhomogeneous discrimination ellipses in S_{phy} into homogeneous perceptual regions. We set the discrimination threshold around each stimulus as a circle of unit radius in S_{per}. The corresponding ellipse in S_{phy} defines the path in each direction from x to a just-noticeable difference (JND). Since the distance between x and $x + dx$ is proportional to $D(x)$, the length dx that has to be traveled to arrive at a unit distance (that is, the JND) is proportional to $1/D(x)$.

We used d_{AB} as a local estimator of $D(x)$ in the region where stimuli A and B live, so that the JND becomes proportional to $1/d_{AB}$. We evaluated the degree up to which different physical parameters modulated $1/d_{AB}$ for the tested pairs. The two most relevant properties were the maximal acceleration and the total duration of the stimuli. The relation between these physical properties and $1/d_{AB}$ follows Weber's law (Fig. 1D), with the addition of a minimal threshold for small values. Indeed, the circles in the right panel of Fig. 1C are obtained from the ellipses on the left by the transformation of Eq. (2), when $D(X)$ is defined by the inverse of the experimentally obtained JNDs of Fig. 1D. Perceptual spaces are thus derived from behavioral tests.

Acknowledgments. We thank the support of the Swiss Leading House for the Latin American Region (No. SMG1911), ERC (OPTINT StG-2016–717054) and ANPCyT Grant n. 2113.

References

1. Merchel, S., Altinsoy, M.E.: Psychophysical comparison of the auditory and tactile perception: a survey. J. Multimodal User Interfaces **14**(3), 271–283 (2020). https://doi.org/10.1007/s12193-020-00333-z
2. Choi, S., Kuchenbecker, K.J.: Vibrotactile display: Perception, technology, and applications. Proc. IEEE **101**(9), 2093–2104 (2012)
3. Application report SLOA194, Texas Inst. (2014). https://www.ti.com/lit/pdf/sloa194

Perception of Guitar Strings on a Flat Visuo-Tactile Display

Baptiste Rohou–Claquin[1]([✉]), Malika Auvray[1], Jean-Loïc Le Carrou[2], and David Gueorguiev[1]

[1] CNRS, Institut des Systèmes Intelligents et de Robotique, Sorbonne Université, 4 Place Jussieu, 75005 Paris, France
rohouclaquin@isir.upmc.fr
[2] CNRS, Institut Jean Le Rond d'Alembert, équipe LAM, Sorbonne Université, 4 Place Jussieu, 75005 Paris, France

Abstract. There is a rapid growth of interest in virtual musical instruments, partly due to the advantages they present over their physical counterparts, such as their portability. However, these instruments still fail to provide a genuine haptic feedback, which might weaken user's experience. This study investigates how vibrations added to a visual feedback provide a realistic sensation of plucking a guitar string. Participants' abilities to recognize different guitar strings and their feeling of realism were collected as a function of different rendering types. In particular, five different ways to record and translate the vibrations induced by the plucking of guitar strings were tested. Three were connected to the moving right hand, one was recorded on the left hand kept still on the guitar's neck, and the last one was directly recorded on the neck of the guitar. The results revealed that discrimination of the rendered string occurred only for recordings on the motionless left hand and on the guitar neck. These two conditions were also rated as being the most realistic, regardless of whether participants were acquainted with musical instruments. Overall, this study shows that distinct guitar strings can be provided by vibrotactile feedback and that the recording techniques impact their perceived realism.

Keywords: Vibrotactile rendering · Guitar string discrimination · Visuo-tactile perception

1 Introduction

The field of digital instruments is fastly growing, bringing many possibilities that were previously impossible with physical instruments such as the possibility to be adapted for people with disabilities [1]. Until recently, one of the main disadvantages of this technology lied in the lack of haptic sensations, compared to physical instruments. This resulted in decreased realism of user's experience,

Supported by UFR d'ingénierie (Sorbonne Université) and ANR Maptics.

H. Seifi et al. (Eds.): EuroHaptics 2022, LNCS 13235, pp. 486–489, 2022.
https://doi.org/10.1007/978-3-031-06249-0

diminished quality of play, and slowing down of their adoption [2]. To overcome these drawbacks and to improve users' sensations, haptic feedback is starting to be implemented into virtual instruments, which is an important progress in this field [3]. To push the possibilities opened by haptic feedback further, our study aims at understanding how the vibrations of a guitar string can be captured and reproduced in a natural way on a virtual digital instrument.

2 Materials and Methods

For the creation of the stimuli, guitar string vibrations were recorded on the index finger nail of both hands and directly on the guitar neck with an accelerometer (PCB piezotronics, model 352A21). The data were collected via a data acquisition system with a sampling frequency 6250 Hz. The sensor was always placed on the index finger and the guitar was held in a natural way (i.e., the right hand plays a string and the left hand holds the neck and pushes the string behind the fret to define the vibrating length of the string). For each sensor position, three different notes were recorded: High (note E_4) 329 Hz, Medium (note G_3) 196 Hz and Low (note A_2) 110 Hz.

In addition to these three frequencies, two additional conditions were used, corresponding to a sample on the right hand, to which 70 Hz filtering were applied: a low-pass one to isolate kinesthetic components and a high-pass one to isolate purely vibrotactile components. This was done to verify whether the frequency pattern of the hand's movements is perceived in a kinesthetic or a vibratory way. Furthermore, the visual vibration corresponding to the strings' theoretical frequencies was implemented following the equation of a vibrating string fixed at both guitar's ends. For the haptic feedback, a tactuator MM3C (Tactile Labs, Canada) was used with a 7-inch touch screen that displayed visual simulations of the vibrations.

21 participants (7 women and 14 men, mean age of 31 years) completed the study. The experiment consisted in five experimental blocks (one for each type of feedback) made of 10 repetitions of each of the three notes, with the participants completing 150 trials in total. The blocks and the notes within the blocks were pseudo-randomized. Six training trials were presented at the start of each block. For each trial, the participants were asked to recognize which of the three notes (E_4, G_3, A_2) was played. Then, at the end of each block, the participants were asked to rate the realism of the feedback they had just experienced in comparison with the sensation of a physical guitar. They provided their responses using a scroll bar between 0 and 100%.

3 Results

Two ANOVAs were conducted on the scores related to frequencies recognition (Fig. 1A) and to ratings about the feedback realism, respectively (see Fig. 1B). The results from the analyses show, for both measurements, a significant effect of the feedback type ($p < 0.0001$ and $p = 0.0002$). The two types of feedback related to the left hand were better perceived and rated as being more realistic than the other ones (Fig. 1C). The majority of users preferred the guitar's feedback and rated the left hand's feedback as second. It is worth mentioning that, despite the pseudo-randomized conditions, a learning phenomenon was observed for the discrimination task (Fig. 1D). Future studies should dig further into the mechanisms underlying this phenomenon, such as generalized versus specific learning processes.

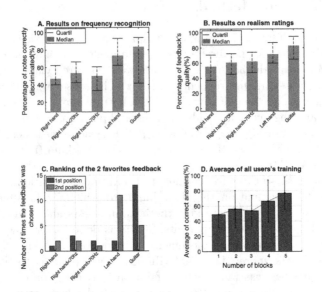

Fig. 1. A) Discrimination results for the 5 types of feedback. (mean ± sd) B) same graph for the quality of the feedback given by the users. C) Barplot of the two preferred feedbacks. D) Discrimination score related to the order of the block in the experiment.

4   Discussion

Our study shows that people are able to recognize among three vibrotactile renderings of guitar strings (high, medium, and low), in particular when they are recorded on the left hand or on the guitar neck. These two types of feedback correspond to the renderings that are considered as being the most realistic. These results provide a first step toward a more realistic design of virtual guitar instruments.

References

1. Frid, E.: Accessible Digital Musical Instruments-A Review of Musical Interfaces in Inclusive Music Practice. Multimodal Technologies and Interaction 3.3 (July 2019), p. 57 (2019). https://doi.org/10.3390/mti3030057

2. Papetti, S., Saitis, C. (eds.): Musical Haptics. SSTHS, Springer, Cham (2018). https://doi.org/10.1007/978-3-319-58316-7
3. Fontana, F., Passalenti, A., Serafin, S., Paisa, R.: Keytar: melodic control of multi-sensory feedback from virtual strings. In: Proceedings of the International Conference on Digital Audio Effects, (DAFx-19), Birmingham, UK, pp. 2–6 (2019)

Pseudo Forces from Asymmetric Vibrations Can Modulate Movement Velocity

Nihar Sabnis[✉], Eline van der Kruk, David Abbink, and Michaël Wiertlewski

Delft University of Technology (TU Delft), Delft, The Netherlands
{e.vanderkruk,d.a.abbink,m.wiertlewski}@tudelft.nl

Abstract. We demonstrate vibrotactile haptic guidance that uses the illusion of pseudo-forces created by asymmetrical vibrations. This particular vibration pattern is zero-mean, but asymmetrical, such that half of the cycle's peak acceleration is significantly higher than the other half, with the amount of asymmetry being tunable. During our experiment, 19 participants correctly perceived the direction of the pseudo force. Moreover, we observed that the elbow flexion velocity was naturally modulated by the amount of asymmetry in the waveform (Pearson's $r = 0.84, p < 0.01$), despite participants not having been provided specific instructions. These results pave the way for an embedded, unobtrusive and effective method to provide movement guidance using pseudo-forces.

Keywords: Pseudo forces · Asymmetric vibrations · Vibrotactile guidance

1 Introduction

Haptic guidance can help users to execute the right movement at the right pace. This guidance typically uses desktop haptic devices or exoskeletons to apply directional forces as a function of the user's movement. However, these devices are complex and costly, limiting their wide-scale adoption. In contrast, some researchers have turned to vibrotactile feedback using lightweight and inexpensive actuators. But most attempts to guide movement with vibrotactile feedback are cognitively demanding and less intuitive. We postulate that these vibrotactile sensations are not integrated as guiding signals because they are incongruent to the sensation of slow-varying forces present in natural physical interactions.

Amemiya et al. showed that asymmetric vibrotactile signals, where one half-cycle had a significantly larger amplitude than the other half, could generate the illusion of a pseudo-force [2]. Subsequent studies have demonstrated that these pseudo forces can provide directional cues [1, 3]. In light of these results, we ask: could we use asymmetric vibrations to elicit a sensation of pseudo-force to guide the speed and direction of the user's movement?

In this preliminary study, we explore how the amount of asymmetry of the waveform can modulate the velocity at which a participant performs elbow flexion. Affecting the dynamics of movements can enable vibrotactile haptic guidance

© The Author(s) 2022
H. Seifi et al. (Eds.): EuroHaptics 2022, LNCS 13235, pp. 490–493, 2022.
https://doi.org/10.1007/978-3-031-06249-0

in situations where conventional haptic guidance is difficult to implement, such
as in sports training and remote physiotherapy.

2   Methods

Asymmetric waveform generation: The waveform played by the vibrotactile actuators is the second time derivative of the motion equation proposed in
Amemiya et al. [2] that models a slider-crank mechanism, see Fig. 1a.

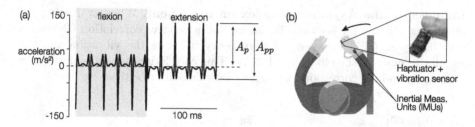

Fig. 1. (a) Example of a waveform with an asymmetry index of 0.61. (b) Experimental setup including the vibrotactile actuator, accelerometer and inertial measurement
units.

Asymmetry Index. To quantify the amount of asymmetry of the vibration,
we define the asymmetry index, $\Gamma = A_p/A_{pp}$ where, A_p is the max amplitude
and (A_{pp}) is the peak to peak amplitude. Since the shape of the waveform is a
function of frequency, its asymmetry index also depends on the frequency. This
dependency allows us to tune the asymmetry index by changing the frequency.

Experimental protocol. Participants faced a table marked with resting position, Fig. 1-b. A wide bandwidth vibrotactile actuator (Haptuator) was attached
to their thumb. The actual vibration was measured by an accelerometer, from
which the asymmetry index was computed. Two inertial measurement units
(Xsens Dots) placed on the bicep and forearm, measured the elbow flexion velocity.

Conditions. We generated waveforms with frequencies ranging 10 Hz 100 Hz,
with an interval 10 Hz. These waveforms had asymmetry indices ranging from
0 to 0.62. For each frequency, vibrations with a duration of 3 s were randomly
provided as input to the participants.

Participants. Nineteen (17M, 2F) adults, from 22 and 30-year-old, participated
in the experiment. They wore ear protection and were blindfolded to avoid audio
or visual cues. For every trial, participants were instructed to "follow the force",
without any instructions about the direction or speed.

3   Results

Perception of the direction. During pilot studies, we validated the perceived direction of pseudo forces. A 2-AFC protocol showed that participants correctly perceive directional cues in the elbow flexion or extension direction for asymmetric vibration ($96 \pm 4.66\%$). In contrast, with symmetric vibration, the rate of success was at chance level ($49 \pm 5.61\%$) showing significant differences of performance between both vibrations ($t_{36} = 27.9$, $p < 0.01$), see Fig. 2a. These results are in line with previous reports using pseudo forces for directional guidance [2, 3].

Influence of the Asymmetry index on movement. When evaluating the movement velocity for each condition, we find a positive correlation (Pearson's $r = 0.84$, $p < 0.01$) between the asymmetry index of the vibration and the angular velocity of the elbow, see Fig. 2b. This correlation indicates that larger asymmetry indices correspond to larger elbow flexion velocities.

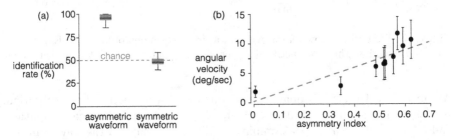

Fig. 2. (a) Influence of symmetry on the identification of direction. (b) The angular velocity of elbow flexion is correlated to the asymmetry index.

4   Discussion and Conclusion

We demonstrate an intuitive method to modulate users' elbow flexion velocity by varying the amount of asymmetry. This can be due to greater skin deformation caused by the attached haptuator when vibrated with a waveform having higher asymmetry index. These findings point toward ways to modulate the intensity and direction of pseudo forces by varying the shape of the waveform, thus providing an effective way to guide movements of users. Nonetheless, the velocities are variable between participants and their precise control would require clear instructions.

In future research, we will investigate how to tune the asymmetry index independently of the frequency and study the effect of varying the pseudo-forces as a function of the user's movement. Our findings show that pseudo-forces can be used to provide remote movement guidance with inexpensive and wearable devices, thus allowing haptic feedback in daily activities.

References

1. Amemiya, T., Gomi, H.: Distinct pseudo-attraction force sensation by a thumb-sized vibrator that oscillates asymmetrically. In: Auvray, M., Duriez, C. (eds.) EURO-HAPTICS 2014. LNCS, vol. 8619, pp. 88–95. Springer, Heidelberg (2014). https://doi.org/10.1007/978-3-662-44196-1_12
2. Amemiya, T., Maeda, T.: Directional force sensation by asymmetric oscillation from a double-layer slider-crank mechanism. J. Comput. Inf. Sci. Eng. **9**(1) (2009)
3. Culbertson, H., Walker, J.M., Raitor, M., Okamura, A.M.: Waves: a wearable asymmetric vibration excitation system for presenting three-dimensional translation and rotation cues. In: Proceedings of the 2017 CHI Conference on Human Factors in Computing Systems, pp. 4972–4982 (2017)

Tactile Friction of UV-Curable Coatings and Its Relation to Surface Hardness

Thomas Ules[✉], Michael Grießer, Sandra Schlögl,
and Dieter P. Gruber

Polymer Competence Center Leoben GmbH, Roseggerstraße 12,
A-8700 Leoben, Austria
thomas.ules@pccl.at

Abstract. This contribution examines the relationships between surface hardness, perceived friction and tactile friction of selected UV-curable coatings. For this purpose, the chemical composition of the investigated coatings was varied on the basis of curing agent and binder resin, as well as selected fillers were added. This allowed the adjustment of various surface parameters such as hardness, surface free energy and roughness over a wider range. Tribological tests were carried out in which the coefficient of friction between the finger pad and the respective coating was measured using a specially developed measurement setup that enabled reproducible results. The surface hardness of the coatings was investigated via nanoindentation measurements. In addition, the coatings were evaluated with regard to perceived friction. The results showed a good correlation between the perceived friction, the coefficient of friction and the surface hardness in terms of the indentation modulus. Thereof, particularly important is the surface hardness in terms of indentation modulus.

Keywords: Haptics · Tactile friction · Physical parameters

1 Introduction

An important aspect of haptics research is the development of new materials or coatings with optimized tactile property profiles for the respective applications. In order to facilitate the production of these materials, it is necessary to bring tactile sensation to a quantifiable parameter level. One of the five basic psychophysical dimensions of tactile perception is the perceived friction [1]. The correlation of tactile friction sensation with the coefficient of friction is often unsatisfying due to the strong dependence of the mechanical properties of human skin on its hydration level and its impact on contact area formation [2].

For this reason, this work presents a methodology that allows comparable friction measurements that correlate well with perceived friction. Furthermore, the indentation moduli of the sample surfaces are presented and correlated with the perceived friction and the measured friction coefficient.

© The Author(s) 2022
H. Seifi et al. (Eds.): EuroHaptics 2022, LNCS 13235, pp. 494–497, 2022.
https://doi.org/10.1007/978-3-031-06249-0

2 Materials and Methods

For the production of the sample coatings, different curing agents and binder resins were used. The varying molecular weight and functionality of the used (meth)acrylate resins and the thiol crosslinkers allowed the generation of 10 samples with different chemical network densities. This enabled a greater variation of the physical properties, in particular the surface hardness of the coatings. The measurement of surface hardness was performed by nanoindentation (Anton Paar UNHT[3]) and the values were calculated with the Oliver-Pharr method.

The friction measurements were conducted on a test rig described in [3]. A normal force of 0.5 N during dynamic movement was applied and the moisture of the finger skin was monitored with a corneometer (Courage + Khazaka CM 825) before and after each measurement. To reduce the influence of varying skin hydration and interfacial moisture levels on the outcome of the friction measurements, the coatings were investigated pairwise. For this purpose, respectively two samples were mounted one after the other on the sample holder, which moves linearly back and forth under the human finger. The developed method allowed to measure two samples with comparable hydration levels of the human finger. After measuring all sample combinations, the coatings were ranked regarding their coefficient of friction. The ranking was obtained by pairwise comparison of the 10 surfaces. This led to information about how often each surface revealed higher friction to another and these frequencies were then assigned rank numbers according to their magnitude.

To evaluate the tactile friction perception a pairwise comparison test was conducted. All possible sample pairs were presented to a blindfolded subject and rated with regard to the perceived friction. This results in a ranking of the 10 samples regarding the tactile friction perception.

3 Results and Discussion

A summary of the experimental results is displayed in Fig. 1. The data shown in the diagram were measured by one subject and each surface was measured nine times. Instead of the ranking values of the samples with regard to the coefficient of friction, the average values are shown, as in this case the average values resulted in the same ranking but give a more detailed insight into the differences in the friction properties. A clear trend towards rising friction with decreasing surface hardness is observed.

A linear fit between indentation modulus and the measured friction coefficient showed a good correlation, yielding a correlation coefficient of 0.89. The samples S2, S3 and S6 were not included in the linear fit as they reveal a higher surface roughness compared to the other coatings and are therefore shifted towards lower friction values due to reduced contact area.

It should be noted that from the adhesion theory of friction [4] the coefficient of friction is expected to be proportional to $E^{*-2/3}$. It depends non-linearly on the combined Elastic Modulus E^* of the counter surface and finger skin. This can be understood in terms of adhesion dominated friction. In such cases the coefficient of friction is proportional to the contact area which is proportional to surface hardness expressed via the indentation modulus.

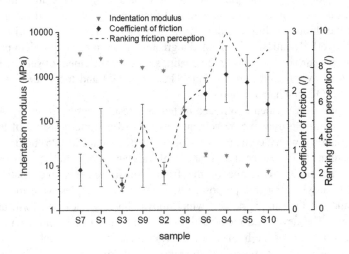

Fig. 1. Indentation modulus, friction coefficient and the resulting ranking order regarding the perceived friction of all 10 sample coatings.

The results of the perception study showed good agreement with the measured friction values. Some mismatch is observed for the coatings ranked highest in the perception study. It is assumed that at the extreme positions it was possibly difficult to rank the samples regarding friction perception as the surfaces felt very similarly due to the high friction.

4 Conclusion

In this contribution the relationship between coefficient of friction of human finger skin and coatings of different chemical composition was presented. It was observed that with increasing indentation modulus of the coatings, the coefficient of friction declines.

It was found that for these types of coatings the perceived friction is indeed closely related to the measured friction coefficient and the surface hardness.

References

1. Okamoto, S., Nagano, H., Yamada, Y.: Psychophysical dimensions of tactile perception of textures. IEEE Trans. Haptics **6**(1), 81–93 (2013)
2. Derler, S., Gerhardt, L.-C.: Tribology of skin: review and analysis of experimental results for the friction coefficient of human skin. Tribol Lett **45**(1), 1–27 (2012). https://doi.org/10.1007/s11249-011-9854-y

3. Ules, T., Hausberger, A., Grießer, M., Schlögl, S., Gruber, D.P.: Introduction of a new in-situ measurement system for the study of touch-feel relevant surface properties. Polymers **12**(6), 1380 (2020)
4. Adams, M.J., Briscoe, B.J., Johnson, S.A.: Friction and lubrication of human skin. Tribol Lett **26**(3), 239–253 (2007). https://doi.org/10.1007/s11249-007-9206-0

Electrotactile Patterns for Single Finger Interactions in VR

Sebastian Vizcay[1], Panagiotis Kourtesis[1], Ferran Argelaguet[1],
Claudio Pacchierotti[2(✉)], and Maud Marchal[3]

[1] Inria, Univ Rennes, IRISA, CNRS, 35042 Rennes, France
{Sebastian.V,Panagiotis.K,Ferran.A}@inria.fr
[2] CNRS, Univ Rennes, IRISA, Inria, 35042 Rennes, France
Claudio.P@inria.fr
[3] Univ Rennes, INSA Rennes, CNRS, Inria, IRISA - France and IUF, Rennes, France
Maud.M@irisa.fr

Abstract. Electrotactile feedback has been proven effective in improving mid-air interactions in Virtual Reality (VR) scenarios. However, the elicited sensation is often described as unnatural. We explore standard stimulation patterns or effects (FXs) found in the literature and how they can get coupled with common VR interactions. We propose 6 implementations of these patterns and, based on our expectation that some couplings work better than others, we evaluated their coherence in an experiment (N = 8).

1 Introduction

Matching user's expectations while interacting in VR is an important factor of presence [2] and agency [1]. Tactile feedback needs to ensure a coherence between the user's interaction and the elicited sensations. While electrotactile feedback has been proven effective in rendering contact information when interacting with virtual objects [3], one of its disadvantages is that it elicits sensations that are often described as unnatural, due to the fact that it directly stimulates the skin nerves endings. Electrotactile feedback is still capable of rendering rich sensations thanks to the high density of actuators, their high wearability, and the wide number of parameters that can be customized. This paper investigates and present preliminary results of how different set of electrotactile actuation parameters can be used to render different tactile sensations and how these are perceived while performing single finger interactions in VR.

2 Methodology

We propose the study of three common single finger interactions: tapping an object, sliding the finger along a surface, and pressing down objects (see Fig. 1-left). From the literature, we collected six common tactile effects rendered using electrotactile feedback.

This work was supported by the European Union's Horizon 2020 research and innovation program under grant agreement No. 856718 (TACTILITY).

H. Seifi et al. (Eds.): EuroHaptics 2022, LNCS 13235, pp. 498–500, 2022.
https://doi.org/10.1007/978-3-031-06249-0

Fig. 1. VR scenario with 3 interactable objects (left) and equipment used (right), with electrode layout in the inset.

We use a custom electrical stimulator (see Fig 1 right) that has up to 32 channels which can be configured as cathode or anode. The stimulator produces biphasic cathodic square pulses with a pulse frequency in the range [1–200] Hz and with pulse widths between [30–500] μs. Amplitude of the pulses can be set between the range of [0.1–9.0] mA. For this particular experiment, we connected a 7-channels electrode (6 cathodes and 1 anode) with the cathodes laid out in a 2×3 matrix.

We designed 6 electrotactile patterns (FXs) which can be easily distinguished. The feedback design was done for a 2×3 pads electrode (see Fig 1 right). The six patterns (or FX) are described in detail in this accompanying technical report [4].

Eight subjects are asked to use their index finger, equipped as shown in Fig. 1, to interact with a virtual environment in three different ways – tapping an object, sliding the finger along a surface, and pressing down objects. We tested the performance of each of these interactions when rendered in the six different ways mentioned above, in order to understand the best way to render such interactions through electrotactile feedback. Each interaction is repeated 6 times per each pattern, leading to 108 interactions in total.

At the end of the task, participants answer a post-experience questionnaire providing us additional feedback regarding the perception of the tactile patterns during the considered interactions.

3 Results and Discussion

The ranking and the distribution of the scores for all interactions are reported in Fig. 2.

For the tapping interaction, the best patterns are the direction, the intensity, and the binary ones. This is also corroborated with the ranking data. We emphasize here that the directional pattern behaves similarly to the intensity one, having as only difference the activation of 2 central active pads rather than only one [4].

For the press interaction, we found similar results, but these are more pronounced given that the interaction span is longer. The directional and intensity patterns are still the best but this time the intensity pattern takes the first place most often in the ranking data.

Finally, for the slide interaction, we see clearly the preference for the directional pattern which is a richer pattern compared to the others, thanks to the additional interaction input data. It is interesting to notice in the ranking data that 2nd and 3rd best are the clockwise and random patterns, indicating that one of the most important factors for rendering a coherent sensation for this particular interaction is having a stimulus that changes location over time.

As next step, we will recruit more participants in order to perform a statistical analysis and verify that the differences are significant. We will also analyze the interaction data that we have collected such as time spent per interaction, object interpenetration, button compression and sliding speed to see if the tactile FXs have an incidence in the way participants interact with the virtual objects.

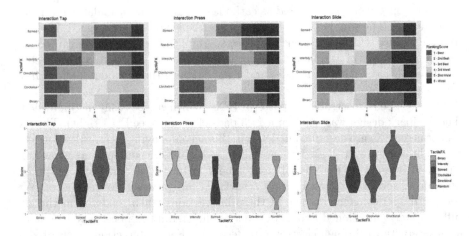

Fig. 2. Ranking order (top) and distribution of the scores (bottom) of the tactile patterns (FXs) per each interaction: tapping an object, sliding the finger along a surface, and pressing down objects. The six considered patterns are detailed in [4]

References

1. Jeunet, C., Albert, L., Argelaguet, F., Lécuyer, A.: "do you feel in control?": towards novel approaches to characterise, manipulate and measure the sense of agency in virtual environments. IEEE Trans. Vis. Comp. Graph. **24**(4), 1486–1495 (2018)
2. Skarbez, R., Brooks, F.P., Jr., Whitton, M.C.: A survey of presence and related concepts. ACM Comput. Surv. (CSUR) **50**(6), 1–39 (2017)
3. Vizcay, S., Kourtesis, P., Argelaguet, F., Pacchierotti, C., Marchal, M.: Electrotactile feedback for enhancing contact information in virtual reality. In: Orlosky, J., Reiners, D., Weyers, B. (eds.) ICAT-EGVE (2021)
4. Vizcay, S., Kourtesis, P., Argelaguet, F., Pacchierotti, C., Marchal, M.: Tactile patterns for electrotactile simulation. Technical report (2022). https://hal.inria.fr/hal-03621989/file/Vizcay_WIP_Eurohaptics21_TactilePatterns.pdf

Sit-By-Me: A Multi-Sensory Feedback Bench for Social Impromptu Interactions

Keyu Wang[✉], Yun Suen Pai, and Kouta Minamizawa

Keio University Graduate School of Media Design, Yokohama, Japan
{doudou,pai,kouta}@kmd.keio.ac.jp

Abstract. We propose Sit-By-Me, a multi-sensory feedback system in the form of a bench where interactions are facilitated merely by passively sitting on the bench with another. It integrates visual, tactile, and audio feedback depending on the sitting pattern, allowing for numerous interaction paradigms between a pair of users. We found that Sit-By-Me could stimulate people's motivation to explore it and thus trigger some simple conversations and discussions. The interactive behaviors of the participants together create musical feedback that enhances the sense of connection between them.

Keywords: Public space · Impromptu interaction · Multi-sensory

1 Introduction

Face-to-face interaction is produced by the mutual influence of the individual's physical presence and his or her body language [5]. It is one of the basic elements of the social system and an important part of individual socialization. It is also essential for the development of communities, groups, and organizations composed of individuals [1]. With interactive systems, people's behaviors in public spaces can be transformed and expanded.

However, it is challenging to initiate a friendly interaction through a interactive system in these public space. Monastero et al. [4] studied how to provide and increase opportunities for individuals in the same space to communicate with each other through personal devices. Yet, people's participation in interactive activities based on daily objects in public spaces, and whether interaction methods in different contexts will affect people's social interaction behavior and experience has not been evaluated.

Kinc et al. [2] and Monastero et al. [3] both designed an interactive bench to promote interaction. However, the interactions were short-lived and were not tested in public spaces. Thus, we propose Sit-By-Me, a public interactive system that employs a bench as the interface. Encouraging connection and consequent interaction through the passive behavior of users sitting on a bench is our core concept.

© The Author(s) 2022
H. Seifi et al. (Eds.): EuroHaptics 2022, LNCS 13235, pp. 501–503, 2022.
https://doi.org/10.1007/978-3-031-06249-0

2   System Design

We used pressure sensors to detect participants' behavior of sitting on the bench or touching the bench surface with their hands. The pressure sensor used in this study adopts polymer thick film (PTF) and the FSR406[1] sensor. Each of them are connected to an Arduino Mega 2560[2]. Applying pressure to the sensing part of the sensor causes the material to deform and thus increase its resistance. We designed a wooden cushion to load the pressure sensor as shown in Fig. 1. This disperses the pressure evenly across the panel.

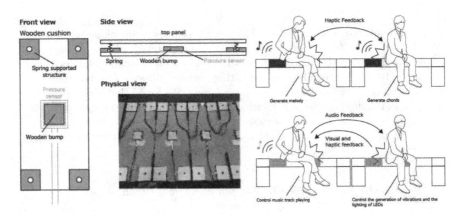

Fig. 1. (left) Wooden cushion structure in the bench to support the (top right) cooperative mode and (bottom right) empathic mode

The audio output of the Sit-By-Me is realized through an external speaker. Because the Arduino microcontroller has limited audio drive, we made a additional Processing[3] software to play the audio feedback. The person sitting on the left side of the bench will send waves to the right, and vice versa. Lastly, we implemented two interaction modes: cooperative mode and empathic mode. For the cooperative mode, the program will detect all cushion units during each detection period. No matter which position the participant made an input, the bench will generate the right music note. For the empathic mode, one of the benches will continue to play piano music while the participant sits down. There are two different piano pieces, a passionate one and a calm one, which can be switched according to where the participant sitting.

3   Preliminary Workshop and Results

We selected a co-working space in a university to test Sit-By-Me and let the participants experience both Cooperative Mode and Empathic Mode for 5 min.

[1] https://akizukidenshi.com/catalog/g/gP-04158/.
[2] https://store.arduino.cc/products/arduino-mega-2560-rev3.
[3] https://processing.org/.

We invited 8 groups (two to a group) of 16 participants (12 females and 4 males) to take part in the test.

In the Empathic Mode test, participants were relatively quick to discover the Sit-By-Me interaction. They spontaneously adjusted their seats and kept touching the surface of Sit-By-Me with their hands. The participants also tended to explore the correlation between the two benches through simple discussion and interactive behaviors at the same time. On the other hand, in the Cooperative Mode test, most of the participants in the test group did not fully understand the principles of music generation within 5 min, which seemed to frustrate them. They found that the interactive behavior of Sit-By-Me could trigger the generation of notes but were confused by the difference in the timing of each note generation. Most of the participants reported that they felt that the generated notes were continuous but could not be identified as a song.

4   Conclusion and Future Works

Through Sit-By-Me, users can interact with the people around them in a collaborative manner by simply sitting or touching. For our future works, we plan to more systematically test and understand how haptic feedback behaves when people start and engage in social interaction.

References

1. Kendon, A., Harris, R.M., Key, M.R. (eds.) Organization of Behavior in Face-to-Face Interaction. De Gruyter Mouton (2011). https://doi.org/10.1515/9783110907643
2. Kinch, S., Grönvall, E., Petersen, M.G., Rasmussen, M.K.: Encounters on a shape-changing bench: exploring atmospheres and social behaviour in situ. In: Proceedings of the 8th International Conference on Tangible, Embedded and Embodied Interaction, pp. 233–240 (2014)
3. Monastero, B., McGookin, D., Takala, T.: "i just leaned on it!" exploring opportunistic social discovery of a technologically augmented cushion. In: Proceedings of the 2020 CHI Conference on Human Factors in Computing Systems, pp. 1–13 (2020)
4. Monastero, B., McGookin, D., Takala, T.: "i just leaned on it!" exploring opportunistic social discovery of a technologically augmented cushion. In: Proceedings of the 2020 CHI Conference on Human Factors in Computing Systems, pp. 1–13. CHI 2020. Association for Computing Machinery, New York (2020). https://doi.org/10.1145/3313831.3376802
5. Sternberg, J.: Misbehavior in Cyber Places: The Regulation of Online Conduct in Virtual Communities on The Internet. Rowman & Littlefield (2012)

ExoSpine: Artificial Muscle-Driven Spine for Posture Correction

Yutong Xie[✉], Yun Suen Pai, and Kouta Minamizawa

Keio University Graduate School of Media Design, Yokohama, Japan
xieyutong741111@gmail.com, {pai,kouta}@kmd.keio.ac.jp

Abstract. We present ExoSpine, a wearable external spine driven by artificial muscle that is soft, flexible, and can provide the appropriate amount of kinaesthetic feedback for posture correction. ExoSpine first detects the user's current posture via a bend sensor installed at the back of the neck, followed by the artificial muscle activation that self-corrects the user's posture when sitting. From our preliminary study, we found ExoSpine to overall perform better than vibrotactile feedback that is similar to conventional posture correctors.

Keywords: Artificial muscle · Posture correction · Kinaesthetic feedback · Artificial spine

1 Introduction

The need to continuously face a display of various shapes, sizes and positions forces the cervical spine to be in undesirable positions for a long period of time. In the past 25 years or more, low back and neck pain prevalency and disability have increased and has become one of the leading causes of disability [5]. Even worse, the rise of COVID-19 has reduced the chance for proper physical activity [2], and the ever expansion of our telecommunication services such as 5G makes the watching online content too convenient to the point of being debilitating to one's health [1] The global incidence rate of neck pain was 16.2% and will undoubtedly bring a huge negative impact to our current social and economic status [3].

Proper physical activity was identified as the only effective method against this [4]. However, researchers and health related companies have also been looking into methods to ensure proper posture throughout the day. Posture correctors are wearable devices that provides this usually in the form of a shoulder brace that can be quite rigid. Thus, we propose ExoSpine, utilizing programmable artificial muscles to drive a correct posture via kinaesthetic feedback. The system uses a trio of artificial muscles at the spine and the lower arms provide enough force to kinaesthetically actuate the user's body to the correct posture. Our system is also safe as artificial muscles are a form of exoskeleton that is soft, flexible and lightweight.

© The Author(s) 2022
H. Seifi et al. (Eds.): EuroHaptics 2022, LNCS 13235, pp. 504–506, 2022.
https://doi.org/10.1007/978-3-031-06249-0

2   Implementation

To kinaesthetically actuate the back comfortably, we surrounds the upper body, shoulder and back as shown in Fig. 1(right) which mimics a conventional posture corrector. We use 7 mm inner diameter air pipes and 8 mm inner diameter fiber pipes for the artificial muscles itself. We also designed and 3D printed several pipe connectors shown in Fig. 1 (left) so that the muscles can be connected in series and transmit compressed air simultaneously across each pipe. To detect cervical curvature, a 112 mm bending sensor[1] is attached to the back of the neck. When the sensor bends, its resistance value changes and this signal is obtained by an Arduino Uno, which sends this information to the pneumatic controller.

Fig. 1. modeling and prototyping process

The controller consists of two solenoid valves. When the resistance of the bend sensor changes, the Arduino Uno send a signal to the L298 N drive chip. The program then reads and records the bending sensor signal transmitted from the serial port 9600 frequency signal through the python program on the computer. The overall delay is about 0.3 to 0.5 s, which is negligible. An air compressor provides an air input with a pressure of about 0.5 KPa. Through the Arduino Uno program, it controls the switch of a 12 V solenoid valve, and then controls the contraction and relaxation of the artificial muscle of the whole structure. We found that our system was able fix the posture of the participant from a 20° bend to 15° in 1 s.

3   Preliminary Study and Results

The goal of the preliminary study is to determine if ExoSpine is able to provide posture correction. To achieve this, we designed a within-subject study to

---

[1] https://www.switch-science.com/catalog/126/.

compare between three conditions: baseline with no feedback, vibrotactile feedback, and Exospine. The vibrotactile posture corrector consists of a 5 V vibration motor and is placed at the back of the neck, similar to the Upright Go S[2]. The participants (3 males, Mean: 26.33, SD: 1.25) were required to sit and perform activities like reading and checking their smartphone while experiencing each condition which lasts about 5 min and a rest period of about 2 min in between. The conditions are also counter balanced.

The Anova test results of each group show that the difference among three groups is statistically significant ($F = 47.556, p < 0.01$). Post hoc Bonferroni adjustment found a statistical significance between the baseline and vibrotactile condition ($p < 0.05$), baseline and ExoSpine ($p < 0.0001$), and vibrotactile with exospine ($p < 0.0001$). This shows that compared with to the baseline condition with no posture corrector, wearing a reminder device or an artificial muscle device can play an effective role in people's sitting posture. Although vibrotactile was effective, the effect is not as good as ExoSpine that actually moves the user directly into the correct posture.

4   Conclusion and Future Works

We present ExoSpine, a wearable spine driven by pneumatic artificial muscles for posture correction. Our initial findings indicate that ExoSpine is potentially more effective than conventional methods of posture correction, and we will further improve the prototype based on the gathered results and feedbacks.

References

1. Al-Hadidi, F., et al.: Association between mobile phone use and neck pain in university students: a cross-sectional study using numeric rating scale for evaluation of neck pain. PloS one. **14**(5), e0217231 (2019)
2. Bouziri, H., Smith, D.R.M., Descatha, A., Dab, W., Jean, K.: Working from home in the time of COVID-19: how to best preserve occupational health? Occup. Environ. Med. **77**(7), 509–510 (2020)
3. Luime, J.J., Kuiper, J.I., Koes, B.W., Verhaar, J.A.N., Miedema, H.S., Burdorf, A.: Work-related risk factors for the incidence and recurrence of shoulder and neck complaints among nursing-home and elderly-care workers. Scand. J. Work, Environ. Health **30**(4), 279–286 (2004)
4. Palmlöf, L., Holm, L.W., Alfredsson, L., Magnusson, C., Vingård, E., Skillgate, E.: The impact of work related physical activity and leisure physical activity on the risk and prognosis of neck pain - a population based cohort study on workers. BMC Musculoskelet. Disord. **17**, 1–11 (2016)
5. Violante, F.S., Mattioli, S., Bonfiglioli, R.: Chapter 21 - low-back pain. In: Lotti, M., Bleecker, M.L. (eds.) Occupational Neurology, Handbook of Clinical Neurology, vol. 131, pp. 397–410. Elsevier (2015). https://doi.org/10. 1016/B978-0-444-62627-1.00020-2, https://www.sciencedirect.com/science/article/ pii/B9780444626271000202

---

[2] https://www.uprightpose.com/.

Design of Robot End-Effector to Be Used in Studying Actuator-Performance for Vibrotactile Perception in a Telemanipulation Setup

Fady Youssef[(✉)] [ID], Maximilian Becker [ID], and Thorsten A. Kern [ID]

Institute for Mechatronics in Mechanics, Hamburg University of Technology,
Hamburg, Germany
f.youssef@tuhh.de
https://www.tuhh.de/imek/

Abstract. This paper presents an overview of a work-in-progress project, where an initial case study is done to study the performance of different vibrotactile actuators. The case study was performed on six participants. Contact forces on different samples were recorded and the participants were asked to judge the change of roughness in the sample using three different actuators. A telemanipulation setup is used to increase the accuracy of recording the contact forces. Two end-effectors were designed based on measured human stiffness in x, y, and z.

Keywords: Actuator performance · Telemanipulation setup · End-effector design · UR10e

1 Introduction

Generating tactile feedback is done using vibrotactile actuators. This class of actuators is widely used in daily life devices, from mobile phones [1] to, even, in sophisticated applications such as robot-assisted surgeries [2]. There are a lot of research areas related to using vibrotactile actuators in perceiving tactile feedback. One specific research area is linked to the use of pens (stylus). The advantage of using pens in haptics, is the simplicity in design and the universal use of pens. Pen-based interface was used to validate the concept of event-based haptics in order to differentiate between real and virtual objects [3], and to create a texture model from recorded acceleration data generated from tool-surface interaction [4]. There is a gap between the performance of vibrotactile and the limit of perception. Using a telemanipulation setup would increase the accuracy of both recording and playing of the acceleration data signals, thus improving the comparison between different actuators.

2 Case Study

An initial case study was conducted on six participants Fig. 1a. The participants were presented with three different actuators, ERM-coin (ERM 1003C A),

ⓒ The Author(s) 2022
H. Seifi et al. (Eds.): EuroHaptics 2022, LNCS 13235, pp. 507–510, 2022.
https://doi.org/10.1007/978-3-031-06249-0

ERM-cylinder (KPD7C-0716) and an Exciter (EXC 221408KF A). Each actuator is mounted on a haptic pen made from aluminum. The participants were asked to hold the haptic pen in a stand still position and judge upon 16 samples per actuator, each sample consists of two different textures Fig. 1b. The judgment is based on the change of roughness at the second half of the sample compared to the first half. The contact forces were recorded using an acceleration sensor. The results of the study showed that the ERM-cylinder performed better than the other two.

One challenge faced in the case study, was maintaining the same speed and contact force while recording the signal. This challenge could be solved by using a telemanipulation setup.

(a)

(b)

Fig. 1. (a): Case study to judge upon the performance of different vibrotactile actuators. (b): Texture sample. The roughness of second half is higher than the first half.

3   Telemanipulation Setup Used

The telemanipulation setup used is shown in Fig. 2. The end-effectors of both robots should be designed in a way that the ideal transparency of the recorded signal is achieved. In this case the transparency is given as follows [5]: $Z_{ss} = Z_{us}$, where Z_{ss} and Z_{us} are the total mechanical impedance on both the sensor side and the user side, respectively. Z_{ss} could be expressed as follows:

$$Z_{ss} = Z_{robot} + Z_{ees} + Z_{hp} \quad (1)$$

where Z_{robot} is the impedance of the robotic arm, Z_{ees} is the impedance of

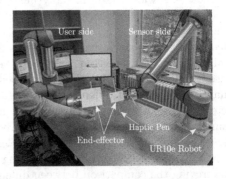

Fig. 2. Telemanipulation setup used: two UR10e robots are used, one for each side of the setup. Haptic Pen is made out of aluminum. The signal on sensor side is recorded using an acceleration sensor.

the end-effector, and Z_{hp} is the impedance of haptic pen. On the user side, in addition to the components in Eq. 1, the impedance of the user is added Eq. 2

$$Z_{us} = Z_{robot} + Z_{eeu} + Z_{hp} + Z_{user} \tag{2}$$

3.1 Measuring Human Hand Impedance

In the frequency range up to 200 Hz, the mechanical impedance of human hand is dominated by elasticity effect, especially in performing precision grasps [5]. This means that:

$$Z_{user} = c_{user} \tag{3}$$

To measure the stiffness in $x, y,$ and z, an experiment was performed. Figure 3 shows the four contact points of a precision grasp. A linear motor moved in a sinusoidal pattern from 0 to 5 mm, and the forces acting on the human hand in each elasticity direction ($+ve$ x, $-ve$ x, .. etc.) were measured. The stiffness of the human hand in three axes were then calculated Table (1).

3.2 End-Effector Design

With adding Z_{user} to the user side, the end effector on the sensor side should be designed to compensate for it. Another stiffness, due to the bandwidth of the system, should be compensated. For a required bandwidth of 500 Hz and a mass of 0.04 kg for the haptic pen, $c_{bw} \approx 0.25 \frac{N}{mm}$. Helical springs with the required compensating stiffness were used. Two end-effectors were designed, one for each side of the system Fig. 2.

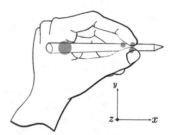

Fig. 3. Sketch of the four contact points in a precision grasp of a human hand.

4 Summary

Initial case study was made to compare between the performance of different vibrotactile actuators using pen-like tool. Telemanipulation setup would increase the accuracy of recording the contact forces. Stiffness of human hand in $x, y,$ and z was calculated. Two end-effectors were designed. The next step would be validation of the designed end-effectors and to perform a case study to compare between the performance of different vibrotactile actuators.

Table 1. Human hand stiffness in $x, y,$ and z

Direction	Stiffness ($\frac{N}{mm}$)
$+ve$ x	1.26
$-ve$ x	1.55
$+ve$ y	0.96
$-ve$ y	0.63
$+ve$ z	1.35
$-ve$ z	2.89

References

1. Yoon, H., Park, S.H.: A non-touchscreen tactile wearable interface as an alternative to touchscreen-based wearable devices. Sensors (Basel) **20**, 1275 (2020). https://doi.org/10.3390/s20051275
2. Saracino, A., et al.: Haptic feedback in the da Vinci Research Kit (dVRK): a user study based on grasping, palpation, and incision tasks. Int. J. Med. Robot. Comput. Assist. Surg. **15**, e1999 (2019). https://doi.org/10.1002/rcs.1999
3. Kuchenbecker, K.J., Fiene, J., Niemeyer, G.: Event-based haptics and acceleration matching: portraying and assessing the realism of contact. In: World Haptics Conference, Pisa, Italy, pp. 381–387 (2005). https://doi.org/10.1109/WHC.2005.52
4. Culbertson, H., Unwin, J., Kuchenbecker, K.J.: Modeling and rendering realistic textures from unconstrained tool-surface interactions. IEEE Trans. Haptics **7**, 381–393 (2014). https://doi.org/10.1109/TOH.2014.2316797
5. Kern, T.A.: Engineering Haptic Devices. Springer, Cham (2009). https://doi.org/10.1007/978-3-540-88248-0

Author Index

Printed in the United States
by Baker & Taylor Publisher Services